Proceedings

16ᵗʰ IEEE Symposium on

Computer Arithmetic

ARITH-16 2003

Proceedings

16th IEEE Symposium on
Computer Arithmetic
ARITH-16 2003

15-18 June 2003
Santiago de Compostela, Spain

Sponsored by
IEEE Computer Society Technical Committee on VLSI

In cooperation with
Universidad de Santiago de Compostela

Ministerio de Ciencia y Tecnología de España

Direccion Xeral de Investigacion e Desenvolvemento de Galicia

The University of California at Los Angeles

AutoPD

Los Alamitos, California

Washington • Brussels • Tokyo

Copyright © 2003 by The Institute of Electrical and Electronics Engineers, Inc.
All rights reserved

Copyright and Reprint Permissions: Abstracting is permitted with credit to the source. Libraries may photocopy beyond the limits of US copyright law, for private use of patrons, those articles in this volume that carry a code at the bottom of the first page, provided that the per-copy fee indicated in the code is paid through the Copyright Clearance Center, 222 Rosewood Drive, Danvers, MA 01923.

Other copying, reprint, or republication requests should be addressed to: IEEE Copyrights Manager, IEEE Service Center, 445 Hoes Lane, P.O. Box 133, Piscataway, NJ 08855-1331.

The papers in this book comprise the proceedings of the meeting mentioned on the cover and title page. They reflect the authors' opinions and, in the interests of timely dissemination, are published as presented and without change. Their inclusion in this publication does not necessarily constitute endorsement by the editors, the IEEE Computer Society, or the Institute of Electrical and Electronics Engineers, Inc.

IEEE Computer Society Order Number PR01894
ISBN 0-7695-1894-X
Library of Congress Number 1063-6889

Additional copies may be ordered from:

IEEE Computer Society	IEEE Service Center	IEEE Computer Society
Customer Service Center	445 Hoes Lane	Asia/Pacific Office
10662 Los Vaqueros Circle	P.O. Box 1331	Watanabe Bldg., 1-4-2
P.O. Box 3014	Piscataway, NJ 08855-1331	Minami-Aoyama
Los Alamitos, CA 90720-1314	Tel: + 1 732 981 0060	Minato-ku, Tokyo 107-0062
Tel: + 1 800 272 6657	Fax: + 1 732 981 9667	JAPAN
Fax: + 1 714 821 4641	http://shop.ieee.org/store/	Tel: + 81 3 3408 3118
http://computer.org/cspress	customer-service@ieee.org	Fax: + 81 3 3408 3553
csbooks@computer.org		tokyo.ofc@computer.org

Individual paper REPRINTS may be ordered at: reprints@computer.org

Editorial production by Frances Titsworth
Cover art production by Joe Daigle/Studio Productions
Printed in the United States of America by Victor Graphics

Table of Contents for Proceedings
ARITH-16

Foreword viii

Dedication ix

Committees x

Additional Reviewers xii

Session 1: Keynote Talk
Chair: Jean-Claude Bajard

Computer Arithmetic — An Algorithm Engineer's Perspective 2
  D. Matula

Session 2: Multiplication
Chair: Luigi Dadda

High-Performance Left-to-Right Array Multiplier Design 4
  Z. Huang and M. Ercegovac

Multiple-Precision Fixed-Point Vector Multiply-Accumulator Using Shared Segmentation ... 12
  D. Tan, A. Danysh, and M. Liebelt

Some Optimizations of Hardware Multiplication by Constant Matrices 20
  N. Boullis and A. Tisserand

A Less Recursive Variant of Karatsuba-Ofman Algorithm for Multiplying Operands
of Size a Power of Two 28
  S. Erdem and Ç. Koç

Session 3: Division
Chair: Paolo Montuschi

Revisiting SRT Quotient Digit Selection 38
  P. Kornerup

SRT Division Algorithms as Dynamical Systems 46
  M. McCann and N. Pippenger

A New Iterative Structure for Hardware Division: the Parallel Paths Algorithm 54
  E. Rice and R. Hughey

Prescaled Integer Division 63
  D. Matula and A. Fit-Florea

Session 4: Floating Point
Chair: Andrew Beaumont-Smith

Hardware Implementations of Denormalized Numbers 70
  E. Schwarz, M. Schmookler, and S. Trong

Representable Correcting Terms for Possibly Underflowing Floating Point Operations 79
    S. Boldo and M. Daumas

High Performance Floating-Point Unit with 116 Bit Wide Divider 87
    G. Gerwig, H. Wetter, E. Schwarz, and J. Haess

The Case for a Redundant Format in Floating Point Arithmetic 95
    H. Fahmy and M. Flynn

Session 5: Decimal Arithmetic and Revisions to the IEEE 754 Standard
Chair: Eric Schwarz

Decimal Floating-Point: Algorism for Computers 104
    M. Cowlishaw

Panel: Revisions to the IEEE 754 Standard for Floating-Point Arithmetic 112
    E. Schwarz

Session 6: Elementary Functions
Chair: Renato Stefanelli, Politecnico de Milano

"Partially Rounded" Small-Order Approximations for Accurate, Hardware-Oriented,
Table-Based Methods 114
    J.-M. Muller

An Overview of Floating-Point Support and Math Library on the Intel®
XScale^{TM} Architecture 122
    C. Iordache and P. Tang

Theorems on Efficient Argument Reductions 129
    R.-C. Li, S. Boldo, and M. Daumas

Accelerating Sine and Cosine Evaluation with Compiler Assistance 137
    P. Markstein

Session 7: Testing and Error Analysis
Chair: Tanya Vladimirova

Worst Cases and Lattice Reduction 142
    D. Stehlé, V. Lefèvre, and P. Zimmermann

Isolating Critical Cases for Reciprocals Using Integer Factorization 148
    J. Harrison

Solving Range Constraints for Binary Floating-Point Instructions 158
    A. Ziv, M. Aharoni, and S. Asaf

A Parametric Error Analysis of Goldschmidt's Division Algorithm 165
    G. Even, P.-M. Seidel, and W. Ferguson

Session 8: Cryptography
Chair: Peter Montgomery

A Unidirectional Bit Serial Systolic Architecture for Double-Basis Division over $GF(2^m)$ 174
    A. Daneshbeh and A. Hasan

Efficient Multiplication in $GF(p^k)$ for Elliptic Curve Cryptography 181
    J.-C. Bajard, L. Imbert, C. Nègre, and T. Plantard

Low Complexity Sequential Normal Basis Multipliers over $GF(2^m)$ 188
    A. Reyhani-Masoleh and A. Hasan

A Low Complexity and a Low Latency Bit Parallel Systolic Multiplier over $GF(2^m)$
Using an Optimal Normal Basis of Type II 196
   S. Kwon

Session 9: Powering, Multiplication, and Counters
Chair: Alexandre Tenca

High-Radix Iterative Algorithm for Powering Computation...204
   J.-A. Piñeiro, M. Ercegovac, and J. Bruguera

On-Line Multiplication in Real and Complex Base 212
   C. Frougny and A. Surarerks

A VLSI Algorithm for Modular Multiplication/Division..220
   M. Kaihara and N. Takagi

Saturating Counters: Application and Design Alternatives 228
   I. Koren, Y. Koren, and B. Oomman

Session 10: Number Systems
Chair: Colin Walter, Comodo Research Labs

Error-Free Arithmetic for Discrete Wavelet Transforms Using Algebraic Integers 238
   K. Wahid, V. Dimitrov, and G. Jullien

On Computing Addition Related Arithmetic Operations via Controlled
Transport of Charge..245
   C. Cotofana, C. Lageweg, and S. Vassiliadis

The Interval Logarithmic Number System 253
   M. Arnold, J. Garcia, and M. Schulte

Scaling an RNS Number Using the Core Function 262
   N. Burgess

Session 11: Modeling and Design of Arithmetic Components
Chair: Peter-Michael Seidel

Energy-Delay Estimation Technique for High-Performance Microprocessor VLSI Adders 272
   V. Oklobdzija, B. Zeydel, H. Dao, S. Mathew, and R. Krishnamurthy

Tutorial: Design of Power Efficient VLSI Arithmetic: Speed and Power Trade-offs..........................280
   V. Oklobdzija and R. Krishnamurthy

Author Index...281

Foreword

Welcome to the 16th IEEE International Symposium on Computer Arithmetic, held in Santiago de Compostela, Spain from June 15 to June 18, 2003. Since 1969, the ARITH symposia have served as the primary forum for presenting scientific work on computer arithmetic, number systems, and the implementation of arithmetic processing components. This year's symposium is dedicated to Michael Flynn, who has made lasting contributions to the fields of computer arithmetic and computer architecture. We are also please to recognize Luigi Dadda as an Honorary Conference Chair to whom the ARITH-9 symposium was dedicated for his pioneering work in computer arithmetic.

This year's technical program includes 34 papers with authors representing 14 countries. The program includes a keynote talk, "Computer Arithmetic: An Algorithm Engineer's Perspective," by David Matula, a tutorial session, "Design of Power Efficient VLSI Arithmetic: Speed and Power Trade-offs," by Vojin Oklobdzija and Ram Krishnamurthy, and a panel session, "Revisions to the IEEE 754 Standard for Floating-Point Arithmetic," chaired by Eric Schwarz. The rest of the technical program is divided into 8 sessions; "Multiplication," "Division," "Floating-Point Arithmetic," "Elementary Functions," "Testing and Error Analysis," "Cryptography," "Powering, Multiplication, and Counters," and "Number Systems."

The technical program contains the latest research in computer arithmetic. Improvements in algorithms and implementations for the basic arithmetic operations are continually being developed to reduce area, delay, and energy consumption. Exciting changes are taking place in floating-point arithmetic, as demonstrated by major revisions to the IEEE 754 floating-point standard. Computing elementary functions in either hardware or software is an area of growing importance, as is the need for new algorithms and implementations for cryptography. The increased complexity of arithmetic algorithms and implementations requires new methods for testing and error analysis. Furthermore, emerging technologies and applications often require specialized number systems to facilitate efficient implementations.

The 34 papers included in these proceedings were selected out of 101 complete papers submitted to ARITH-16. The submission of manuscripts and the review process were done electronically, allowing faster and more efficient processing and a better match between the submitted papers and the reviewers' fields of expertise. For each paper, at least three reviews were solicited. Most reviews were done by the 32 members of the Program Committee, but some were handled by external reviewers. The papers included in these proceedings were selected by 25 members of the Program Committee, who participated in a meeting held in Los Angeles in January, 2003.

The success of a symposium such as this depends on the participation of many individuals. First, we would like to thank all the authors who submitted their research results. We would also like to express our gratitude to the members of the Program Committee and the external reviewers, who contributed so much of their time and provided in-depth reviews of the papers. We sincerely appreciate the direction provided by the members of the Steering Committee. Finally, we are grateful to Tomas Lang for his guidance and support as General Chair, to Milos Ercegovac for graciously hosting the Program Committee Meeting, to Alexandre Tenca for his important work as Publicity Chair, and to Javier Bruguera for his outstanding efforts as Local Arrangements and Finance Chair.

Support for this year's symposium was provided by the IEEE, the IEEE Computer Society, the IEEE Computer Society Technical Committee on VLSI, Universidad de Santiago de Compostela, Ministerio de Ciencia y Tecnologia of Spain, Direccion Xeral de Investigacion e Desenvolvemento de Galicia, the University of California at Los Angeles, and AutoPD.

We hope that you will find this year's program a continuation of the previous ARITH meetings' tradition of excellence.

Jean-Claude Bajard and **Michael Schulte**
Program Chairs

Dedication
Michael J. Flynn

The Proceedings of the 16th Symposium on Computer Arithmetic are dedicated to Michael J. Flynn in recognition of his lasting contributions to the field.

Michael J. Flynn received a BS from Manhattan College, MS from Syracuse University and PhD from Purdue University. He was awarded an honorary DSc from University of Dublin (Trinity Col). He began his engineering career at IBM as a designer of mainframe computers such as the IBM 7090 and later as design manager for the IBM System 360 Model 90 series computers. These processors were IBM's highest performing scientific processors. The Model 90 series incorporated many new (1967) ideas such as data flow out-of-order instruction execution, speculative execution and advanced arithmetic algorithms (fast tree multiplier and binomial series divide). After leaving IBM he joined the EE faculty at Northwestern University in 1966 and moved to Johns Hopkins University in 1970. During this time he introduced the now familiar stream outline of computer organization (SIMD, etc.). In 1970 he co-authored the first published discussion of techniques for issuing multiple instructions in a cycle, now called superscalar design. He became Professor of Electrical Engineering at Stanford in 1975 where he set up the Stanford Emulation Lab. to understand the role and affect of instruction set tradeoffs and the efficient design of microprocessors. He retired from Stanford as Emeritus Professor in 1999.

Early on Mike realized the importance of computer architecture in the development of modern hardware. In the early 1970's he founded both of the specialist organization on Computer Architecture: the IEEE Computer Society's Technical Committee on Computer Architecture and the ACM's SIGARCH. He has had a special interest in computer arithmetic and in 1981 co-authored one of the standard computer arithmetic texts. During the '90s he headed up an interdisciplinary program on sub-nanosecond arithmetic (SNAP). He has authored or co-authored 5 books, over 300 professional papers and holds two patents. He served as Vice President of the IEEE Computer Society and Associate Editor of the Transactions on Computers. He is a fellow of both the IEEE and the ACM and a fellow of the Institution of Engineers of Ireland. He received the ACM/IEEE Eckert-Mauchly Award, the IEEE Computer Society's Harry Goode Memorial Award and the Tesla medal from the International Tesla Society (Belgrade) for outstanding contributions to Computer Architecture.

Mike is an avid gardener with a special affection for camellias and roses.

Steering Committee

Chair

Jean-Michel Muller, *Ecole Normale Superieure de Lyon*

Steering Committee Members

Algirdas Avizienis, *University of California at Los Angeles*

Neil Burgess, *Cardiff University*

Luigi Ciminiera, *Politecnico di Torino*

Milos Ercegovac, *University of California at Los Angeles*

Mary-Jane Irwin, *Penn State University*

Graham Jullien, *University of Calgary*

Simon Knowles, *Eigen Semi*

Israel Koren, *University of Massachusetts*

Peter Kornerup, *University of Southern Denmark*

Tomas Lang, University, *University of California at Irvine*

David Matula, David Matula, *Southern Methodist University*

William McAllister, *Hewlett-Packard*

Vojin Oklobdzija, *University of California at Davis*

Earl Swartzlander, *University of Texas at Austin*

Renato Stefanelli, *Politecnico di Milano*

Naofumi Takagi, *Nagoya University*

Symposium Committee

General Chair
Tomas Lang, *University of California at Irvine*

Program Co-Chairs
Jean-Claude Bajard, *LIRMM CNRS, University of Montpellier II*
Michael Schulte, *University of Wisconsin-Madison*

Local Arrangements and Finance Chair
Javier Bruguera, *University of Santiago de Compostela*

Publicity Chair
Alexandre Tenca, *Oregon State University*

Honorary Chair
Luigi Dadda, *Politecnico di Milano*

Program Committee Members
David Bailey, *Lawrence Berkeley Laboratory*
Andrew Beaumont-Smith, *Intel*
Neil Burgess, *Cardiff University*
Luigi Ciminiera, *Politecnico di Torino*
Milos Ercegovac, *University of California at Los Angeles*
Roger Golliver, *Intel*
David Hough, *Sun Microsystems*
Graham Jullien, *University of Calgary*
Simon Knowles, *Eigen Semi*
Israel Koren, *University of Massachusetts*
Peter Kornerup, *University of Southern Denmark*
David Matula, *Southern Methodist University*
Peter Montgomery, *Microsoft Research*
Paolo Montuschi, *Politecnico di Torino*
Jean-Michel Muller, *Ecole Normale Superieure de Lyon*
Stuart Oberman, *Nvidia*
Vojin Oklobdzija, *University of California at Davis*
Behrooz Parhami, *University of California at Santa Barbara*
Dhananjay S. Phatak, *University of Maryland Baltimore County*
Eric Schwarz, *IBM*
Peter-Michel Seidel, *Southern Methodist University*
Renato Stefanelli, *Politecnico di Milano*
Earl Swartzlander, *University of Texas*
Naofumi Takagi, *Nagoya University*
Colin Walter, *Comodo Research Labs.*
Belle Wei, *San Jose State University*
Paul Zimmermann, *LORIA/INRIA Lorraine*

List of Reviewers

Bill Allombert	Peter Kornerup
Mark Arnold	Fabrizio Lamberti
David Bailey	Tomas Lang
Jean-Claude Bajard	Vincent Lefèvre
Andrew Beaumont-Smith	David Matula
Guido Bertoni	Lee McFearin
Jean-Luc Beuchat	Peter Montgomery
Sylvie Boldo	Bartolomeo Montrucchio
Javier Bruguera	Paolo Montuschi
Luigi Ciminiera	Jean-Michel Muller
Luigi Dadda	Stuart Oberman
Marc Daumas	Holger Orup
Florent de Dinechin	Behrooz Parhami
David Defour	Dhananjay Phatak
Laurent-Stephane Didier	Vincenzo Piuri
Vassil Dimitrov	Nhon Quach
Milos Ercegovac	Fabien Rico
Hossam Fahmy	Patrice Roussel
Alex Fit-Florea	Marty Schmookler
Mike Flynn	Michael Schulte
David Fu	Eric Schwarz
Tom Goff	Peter-Michael Seidel
Roger Golliver	Abhishek Singh
Sameer Halepete	Renato Stefanelli
Guillaume Hanrot	Shane Story
John Harrison	Earl Swartzlander
Anwar Hasan	Naofumi Takagi
David Hough	Lo'ai Tawalbeh
Zhijun Huang	Alexandre Tenca
Laurent Imbert	Mitch Thornton
Cristina Iordache	Arnaud Tisserand
Norbert Juffa	Ramarathnam Venkatesan
Graham Jullien	Colin Walter
Simon Knowles	Belle Wei
Cetin Koc	Gideon Yuval
Israel Koren	Paul Zimmermann

Session 1:
Keynote Talk

Chair: Jean-Claude Bajard

Computer Arithmetic – An Algorithm Engineer's Perspective
Invited Keynote Talk

David W. Matula
Professor of Computer Science and Engineering
Southern Methodist University
Dallas, USA
matula@engr.smu.edu

Abstract

The continuing evolution of hardware speed and expanding storage in cache and memory provides that the choice of fundamental arithmetic algorithms and numeric representations within the arithmetic logic unit must be continually addressed to obtain competitive arithmetic unit implementations. The ongoing expansion of the widely followed IEEE floating-point standard particularly suggests numeric representation and algorithm choices to support a fused multiply-add, and possibly some quad precision capability, must be investigated. Algorithmic novelty is a priority both for competitive advantage and also to avoid being shut out by competitor's intellectual property thrusts.

In this talk we identify some algorithmic goals for improving pipelined arithmetic unit performance including:
- reducing the dependent arithmetic operation penalty,
- reducing the rounding direction computation penalty,
- better integrating divide, square root, and reciprocal instructions into RISC design.

We describe promising research directions on algorithmic tools that can help attain these goals including:
- developing integrated arithmetic algorithms exploiting cost effective redundant representations within the arithmetic unit,
- utilizing concurrent table lookup concepts with supplemental small adders/multipliers to speed up iterative algorithms for divide, square root, and transcendentals.

The flexibilities available for algorithms in hardware provide a rich source for algorithm design of value both in theory and in practice.

David W. Matula received his Ph. D. from U.C. Berkeley in 1966 and has been a Professor of Computer Science and Engineering at SMU in Dallas since 1974. He has published over 100 refereed journal and proceedings papers and has 13 patents on arithmetic unit design and wireless network systems. He has been a consultant with Cyrix/National Semiconductor since Cyrix's founding year, 1988, where he has contributed to the design of several generations of floating-point units. He has had a paper at all 16 of the IEEE Arithmetic Symposia since their inception in 1969.

Session 2:
Multiplication

Chair: Luigi Dadda

High-Performance Left-to-Right Array Multiplier Design

Zhijun Huang and Miloš D. Ercegovac
Computer Science Department
University of California Los Angeles
Los Angeles, CA 90095
{zjhuang, milos}@cs.ucla.edu

Abstract

We propose a split array multiplier organized in a left-to-right leapfrog (LRLF) structure with reduced delay compared to conventional array multipliers. Moreover, the proposed design shows equivalent performance as tree multipliers for $n \leq 32$. An efficient radix-4 recoding logic generates the partial products in a left-to-right order. The partial products are split into upper and lower groups. Each group is reduced using [3:2] adders with optimized signal flows and the carry-save results from two groups are combined using a [4:2] adder. The final product is obtained with a prefix adder optimized to match the non-uniform arrival profile of the inputs. Layout experiments indicate that upper/lower split multipliers have slightly less area and power than optimized tree multipliers while keeping the same delay for $n \leq 32$.

1. Introduction

The three steps of parallel multiplication are denoted as recoding and partial product (PP) generation (PPG), PP reduction (PPR), and final carry-propagate addition (CPA). Based on the approaches to PPR, multipliers are usually classified into: (i) linear array multipliers with logic delay proportional to n, and (ii) tree multipliers with delay proportional to $log(n)$ [12]. The tree reduction treats PP bits either in rows or in columns. Although tree multipliers have the shortest logic delay in the PPR step, they have irregular layout with complicated interconnects. On the other hand, array multipliers have larger delay but offer regular layout and simpler interconnects. As interconnects become important in deep submicron design [22], architectures with regular layout and simple interconnects are desirable. Irregular layouts with complicated interconnects not only demand more physical design effort but also introduce significant interconnect delay and make noise a problem due to several types of wiring capacitance [1, 22].

Modern multiplier designs use [4:2] adders [14] to reduce the PPR logic delay and regularize the layout. To improve regularity and compact layout, regularly structured tree (RST) with recurring blocks [6] and rectangular-styled tree by folding [8] were proposed, at the expense of more complicated interconnects. In [15], three dimensional minimization (TDM) algorithm was developed to design adders of the maximal possible size with optimized signal connections, which further shortened the PPR path by $1 \sim 2$ XOR delays. However, the resulting structure has more complex layout than a [4:2]-adder based tree. In [10], multiplication was divided recursively into smaller multiplications to increase layout regularity and scalability, which essentially resulted in a hierarchical tree structure.

In linear array multiplier design, the even/odd split structure [9] was proposed to reduce both delay and power of conventional right-to-left (R-L) linear array structures. In [13], a *leapfrog* structure was proposed to take advantage of the delay imbalances in adders. in [4], a left-to-right (L-R) carry-free (LRCF) array multiplier was proposed where the final CPA step to produce the MS bits of the product was avoided by using on-the-fly conversion in parallel with the linear reduction. In [2], this LRCF approach was extended to produce $2n$-bit product. It was also discovered that glitches in L-R reduction arrays were smaller than in the conventional R-L arrays, especially for data with large dynamic range [5, 20, 7].

To further reduce the delay of array multipliers while maintaining their regular layout and simple interconnect, this paper proposes split array LRLF (SALRLF) multipliers that combine the advantages of splitting, L-R computation, and leapfrog structure. Two types of splitting are considered: even/odd and upper/lower. Each step of SALRLF is optimized with the primary objective of delay reduction and the secondary objective of power reduction. Logic-level analysis as well as physical layout with guided floorplanning are conducted to compare SALRLF with tree multipliers.

In the following, the multiplicand $X = -x_{n-1}2^{n-1} + \sum_{j=0}^{n-2} x_j 2^j$ and the multiplier $Y = -y_{n-1}2^{n-1} + \sum_{i=0}^{n-2} y_i 2^i$ are integers in the two's-complement form with n being even to simplify description. For logic-level analysis, the delay of a 2-input XOR2 gate, T_{XOR2}, is used as the unit delay. The delay of two-level a complex gate such as AOI22 (AND2-NOR2) is equivalent to T_{XOR2}.

2. Partial Product Generation

Radix-4 recoding is used to reduce the number of PPs to half. After comparing common recoders, we developed a version $neg/two/one\text{-}nf$ ("nf" for $neg\text{-}first$) shown in Fig. 1. The negation operation is done before the selection between $1X$ and $2X$ so that two_i and one_i set PP_i to zero regardless of neg_i for "-0". To generate additional '1' for negative PP_i, a correction bit $c_i = y_{2i+1}(y_{2i}y_{2i-1})'$ is used.

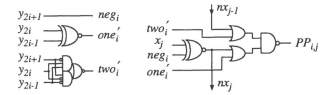

Figure 1: $neg/two/one\text{-}nf$ generator.

Due to shifting, each PP has a 0 between $PP_{i+1,0}$ and c_i. To have a more regular LSB part of each PP, $PP_{i,0}$ is added with c_i bit in advance [18]. The $PP_{i,0}^{(new)}$ and $c_i^{(new)}$ are described as:

$$PP_{i,0}^{(new)} = x_0 \cdot (y_{2i} \oplus y_{2i-1}) \quad (1)$$

$$c_i^{(new)} = y_{2i+1} y_{2i}' y_{2i-1}' + y_{2i+1} x_0' (y_{2i} \oplus y_{2i-1}) \quad (2)$$

Both $c_i^{(new)}$ and $P_{i,0}^{(new)}$ are obtained no later than other PP bits. The generated PP bit-array is arranged in MSB-first or L-R manner as shown in Fig. 2. The grey circles are $PP_{i,0}^{(new)}$ and the white circles are $c_i^{(new)}$.

3. Partial Product Reduction

The delay gap between tree multipliers and array multipliers is mainly due to the linear PPR structure in

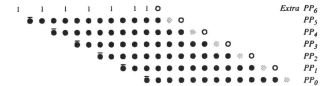

Figure 2: MSB-first radix-4 PP bit array (n=12).

conventional array multipliers. To improve the speed of array multipliers, parallelism is introduced in PPR. In addition, different adder types and the signal flow between them also have impact on delay, area, and power.

3.1. L-R leapfrog (LRLF) structure

To exploit the delay difference between carry and sum signals in adders, the sum signals in the *leapfrog* [13] structure for R-L array multipliers skip over alternate rows. Because all the carry signals propagate through the entire array, the MSBs of final PPR vectors arrive at the same (latest) time. This unfortunately prevents optimization of the final CPA that is possible in tree multipliers. To allow final CPA optimization in linear array multipliers, we combine L-R computation and leapfrog structure resulting in a new L-R leapfrog (LRLF) array multiplier scheme. A LRLF multiplier for PP array of Fig. 2 is shown in Fig. 3. The dashed lines are carries and solid lines are sum signals. Each adder symbol represents either a FA if all three inputs are variables or a HA if one of three inputs is constant.

The power and delay characteristics of LRLF multiplier have been reported in [7]. Here we optimize the [4:2] adder design according to input arrival profiles. The basic [4:2] adder module, M42, is shown in Fig. 4. The delay from any input A, B, C, and D to output Sum or $Cout$ is $3T_{XOR2}$. Each M42 actually has five inputs because there is one intermediate signal Tin. In the $(n-3)$-bit [4:2] adder for LRLF in Fig. 3, more than half 5-input M42s have one or more zero inputs, which can be simplified to have smaller delay. For M42s with one zero input, the simplification is as follows. Assume the four non-zero inputs of a simplified M42 are A, B, C, and D with arrival time (α) relation $\alpha_A \leq \alpha_B \leq \alpha_C \leq \alpha_D$. The order is arbitrary and does not affect the discussion here because all inputs are functionally equivalent. According to input arrival profiles, two designs with different Sum logic are developed: M42L (linear-Sum) in Fig. 5a and M42T (tree-Sum) in 5b. The arrival times of $Tout$ and

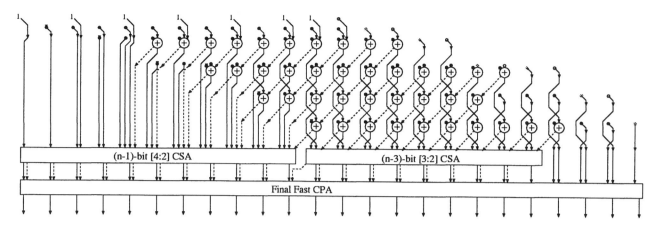

Figure 3: A LRLF array multiplier (n=12).

C_{out} are

$$\alpha_{Tout} = \alpha_B + T_{AND2} \quad (3)$$
$$\alpha_{Cout} = max(\alpha_B + T_{XOR2}, \alpha_D) + T_{AO222} \quad (4)$$

which are smaller than those in M42. In M42L, Sum arrives at

$$\alpha_{linear-Sum} = max(\alpha_{BCX}, \alpha_D) + T_{XOR2} \quad (5)$$

where $\alpha_{BCX} = max(\alpha_B + 2T_{XOR2}, \alpha_C + T_{XOR2})$. In M42T,

$$\alpha_{tree-Sum} = \alpha_D + 2T_{XOR2} \quad (6)$$

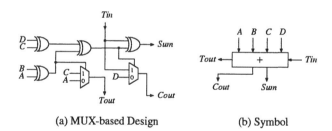

(a) MUX-based Design (b) Symbol

Figure 4: Basic [4:2] adder module M42.

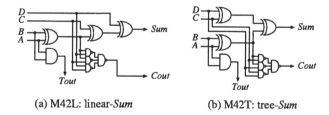

(a) M42L: linear-Sum (b) M42T: tree-Sum

Figure 5: Two simplified M42 designs.

The $(n-1)$-bit [4:2] adder in LRLF is designed from LSB to MSB as follows. For each bit, sort five inputs A, B, C, D, and E (one of them being Tin) according to arrival time and assume that $\alpha_E \leq \alpha_A \leq \alpha_B \leq \alpha_C \leq \alpha_D$. If no input is 0, M42 of Fig. 4 is used. To minimize delay, five inputs E, A, B, C, and D are connected to pin A, B, C, D, and Tin in that order. If two or more inputs are 0s, the adder is reduced to be a FA or HA. The signal flow optimization of FAs will be addressed in Section 3.3. If only one input is 0, it must be E because a constant arrives at the earliest time. Simplified 4-input M42 is used in this case. Four inputs A, B, C, and D are connected to pin A, B, C, and D in that order. M42L is used for faster Sum if $\alpha_{linear-Sum} < \alpha_{tree-Sum}$. Otherwise, M42T is used. The output $Tout$ is fed into the next bit position and the process is repeated. With this optimization, many outputs of the [4:2] adder become available one T_{XOR2} earlier.

3.2. Split array LRLF (SALRLF) structure

The PPR delay of an LRLF array multiplier is about $\lceil \frac{n}{2} \rceil T_{XOR2}$ while that of an $n \times n$-bit radix-4 tree multiplier is $3(\lceil \log_2(\frac{n}{4}) \rceil)T_{XOR2}$. The delay of LRLF is not comparable with that of tree multipliers when $n > 16$. To reduce PPR delay, certain level of parallelism is necessary.

One approach is to split the PP bit array into even PPs and odd PPs, as shown in Fig. 6. In each split part, PPs are shifted four bits each row and reduced into two vectors using a LRLF structure. The final vectors from even and odd parts are merged by a $(2n-3)$-bit [4:2] adder. This algorithm is named even/odd LRLF (EOLRLF).

Another approach is to split the PP bit array into upper PPs and lower PPs, as shown in Fig. 7. In each part, PPs are shifted two bits each row and reduced into two vectors using LRLF. The final vectors from upper and lower parts are merged by a [4:2] adder. To

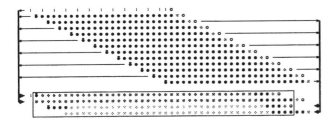

Figure 6: EOLRLF array multiplier (n=24).

reduce the size of this adder, the highest carry bit from the right-side [3:2] CSA in LRLF is fed into the left-side [4:2] CSA as Tin instead of being a bit of the final CPA input. In this way, only a $(n+2)$-bit [4:2] adder is required. This algorithm is named upper/lower LRLF (ULLRLF).

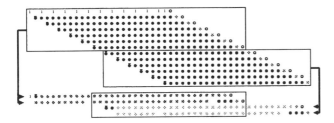

Figure 7: ULLRLF array multiplier (n=24).

The PPR delay of SALRLF is about $(\lceil \frac{n}{4} + 3 \rceil \sim \lceil \frac{n}{4} + 4 \rceil)T_{XOR2}$, depending on the type of adders used. For $n \leq 32$, the delay is $< 11 \sim 12$ while the best result of a tree multiplier is ≤ 9. Further splitting of the PP array reduces the layout regularity and will not be considered. Instead, optimization of FAs and final CPA as well as floorplanning will be used to narrow the remaining gap. In EOLRLF, the arrival profile of PPR final vectors has fewer latest-arriving bits than that in tree multipliers. Fig. 8 shows the PPGR delay profiles in a 32×32-bit TDM multiplier, an EOLRLF, and a ULLRLF. The number of latest-arriving bits in EOLRLF is 5 while this number is 8 in TDM. The bit delay distribution in EOLRLF is also more regular. Most bit groups in EOLRLF have 4-5 bits. But the group size varies a lot in TDM. The final adder design could exploit these better-shaped arrival profiles in EOLRLF to reduce delay. Compared with EOLRLF, ULLRLF has two main advantages. First, the shifting distance between PPs in each upper/lower part is 2 positions instead of 4, which leads to simpler interconnects. Second, the final [4:2] adder in ULLRLF is only $(n+2)$-bit in contrast to $(2n-3)$-bit in EOLRLF. On the other hand, URLRLF has a worse arrival profile than EOLRLF. However, such a profile only leads to just one T_{AO21} delay, which will be explained in Section 4. Our

detailed layout experiments indicate that EOLRLF is worse than URLRLF in all measurements. Therefore, we choose ULLRLF in the following discussion.

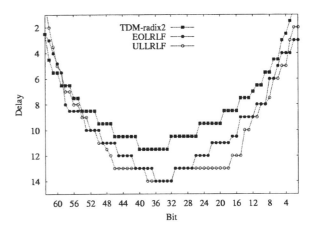

Figure 8: PPGR delay profiles (n=32).

3.3. Optimization of FAs

In array multipliers, the basic components for PPR are full adders (FA). Two common FA structures, FA-MUX and FA-ND3, are shown in Fig. 9. Compared with FA-ND3, FA-MUX typically has smaller area even if pass transistors are not used. Since FA is the most used element in array multipliers, smaller FA would lead to smaller overall area, which is also helpful in the reduction of power consumption and interconnect delay. As to logic delay, however, FA-NAND3 is better than FA-MUX because the delay from all inputs to $Cout$ is T_{AO222} ($T_{AO222} \approx T_{XOR2}$).

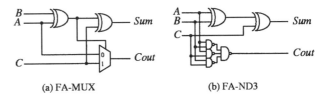

Figure 9: Two FA structures.

Because of the different characteristics of FA inputs, it is possible to optimize signal flow with respect to propagation delay. This technique has been applied in TDM tree multipliers [15]. In addition to delay, signal flow optimization affects power [7]. Assume the three input signals to FA are Ain, Sin, Cin. These input signals are sorted according to their arrival times. We assume that the α relationship is $\alpha_{Ain} \leq \alpha_{Cin} \leq \alpha_{Sin}$. The order is arbitrary since the inputs are functionally equivalent. In FA-ND3, Sin is connected to pin

C. There is no restriction on the connections between $Ain(Bin)$ and pin $A(B)$ unless transistor-level difference between A and B is considered. In FA-MUX, Sin is also connected to pin C. Between Ain and Bin, the signal with less switching activity is connected to pin A for power saving because pin B has less load capacitance and is used for the one with higher switching activity. Since PP bits arrive at the earliest time and never change after PPG, they are connected to A pins. This signal flow optimization technique is named *CSSC* to reflect the interchange of *sum* and *carry* signals. In the experiments section, we show the delay effects of FA selection and CSSC optimization in LRLF and ULLRLF array multipliers.

4. Final Adder

Final adders are optimized to match the non-uniform input arrival profiles. The optimal final adder for tree multipliers is CSMA based design [16]. Efficient design of on-the-fly converter for L-R array multipliers also corresponds to a multi-level carry-select (CSEL) or conditional-sum (CSUM) adders [11]. In [19], generalized earliest-first (GEF) algorithm was proposed to design CSUM for arbitrary input arrival profile. The similarity between CSUM and prefix adder (PFA) is also shown in [19] where PFA is called CLA.

We followed the GEF algorithm and chose PFA for final addition because the PFA operators, AO21 and AND2, are simpler than the basic CSUM operators – a pair of MUX21. Two lists, *Plist* and *Tlist*, are maintained in GEF. All (G, A) signal pairs are initially put into *Plist* and sorted according to arrival times. The earliest pairs are then moved to *Tlist*. Adjacent bit pairs in *Tlist* are retrieved and merged from left to right. The merged pairs are put back into *Plist*. The iteration continues until the generation of the MSB carry bit. Other carry bits are generated using existing (G, A) bits. A PFA example for a hill-shaped arrival profile is shown in Fig. 10. Black nodes in PFA are computation cells and white nodes have no logic or only buffers. In the original GEF, the merging is conducted from from right to left. Because of different input-output delays in operator '•', the left-to-right merging in *Tlist* leads to $0.5 T_{XOR2}$ delay improvement.

Let W_{max} be the largest number of adjacent signals that arrive at the same time. If these W_{max} signals are also the latest arriving signals in a hill-shaped arrival profile, the delay of PFA for such a profile can be estimated as

$$T_{PFA} = (\log_2(W_{max}) + 2) T_{AO21} + T_{XOR2} \quad (7)$$

which is not directly related to the adder width $2n$. A small W_{max} would lead to a small T_{PFA}. However, the difference in T_{PFA} is just one T_{AO21} for most schemes in our study because of the logarithmic relationship. One T_{AO21} delay could be further eliminated from T_{PFA} if carry-select adders are used for the final stages of the left part in hill-shaped arrival profiles [3].

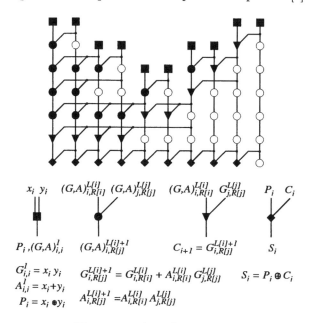

Figure 10: A PFA example.

5. Experiments

To compare the proposed ULLRLF with tree multipliers, logic-level delay analysis is first conducted. Actual VHDL implementation and physical layout are then performed on Synopsys and Cadence design platforms.

5.1. Delay comparison at logic level

VHDL generation programs for both LRLF and ULLRLF algorithms have been written with the flexibility of FA selection and signal flow optimization. The comparison results at logic level without wiring effects are normalized to T_{XOR2} and listed in Table 1. T_{GR} is the delay of PPG and PPR. T_A is the delay of the final adder. For LRLF, the use of FA-ND3 rather than FA-MUX reduces PPR delay and the overall delay by 1 T_{XOR2}. CSSC reduces the delay by 1 T_{XOR2} except for 48-bit LRLF-ND3 where the reduction is 2. For ULLRLF, FA-ND3 reduces one T_{XOR2} in PPR, but not the overall delay. CSSC only reduces PPR delay in ULLRLF-MUX by 0.5 and also has no effect on the overall delay. This is because varying FAs and applying CSSC change the input arrival profiles of the final adder and affect T_A by up to 1 T_{XOR2}. Even if there is little delay advantage, however, it is still useful to apply CSSC for power reduction [7]. We have also noticed that CSSC and FA-ND3 could help EOL-RLF achieve 0.5 ~ 2 less logic delay than ULLRLF. However, ULLRLF outperforms EOLRLF after layout because of smaller area and simpler wiring. Finally, it is worthwhile to note that T_A does have little relation with the adder width as explained in Eq. 7.

Using the results from Table 1, we now compare the delays of LRLF/ULLRLF with tree multipliers. Radix-2 and radix-4 TDM schemes [15][18] are chosen because they are the best tree multipliers to our knowledge. In addition, tree multipliers based on [4:2] and [3:2] CSAs are also used for comparison as they have more regular structures. To avoid the delay due to the extra row $PP[n/2]$ in radix-4 two's-complement multipliers, the reduction of 9 PPs from $PP[n/2-8]$ to $PP[n/2]$ is based on a [9:4] adder with only $3T_{XOR2}$ delay, as illustrated in Fig. 11. The $3T_{XOR2}$ delay is achieved as follows. All FAs except the right most one in the shaded [3:2] CSA are simplified into HAs with half delay as they have constant inputs. Inputs of the second-level [3:2] adders are properly optimized so that each FA has one input arriving at least T_{XOR2} later than the other two inputs. This late input is connected to pin C of FAs to ensure one T_{XOR2} delay. To distinguish from other tree multipliers, the radix-4 tree multiplier using this special [9:4] adder is named *tree9to4*.

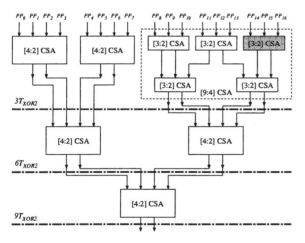

Figure 11: Tree PPR with $9T_{XOR2}$ delay.

The logic delay comparison results are given in Table 2. The blank boxes with '–' are because T_{GR}s or delay profiles from PPR are not available from literature. The original TDM-radix4 data in [18] are normalized to our measurement base. It is shown that the radix-4 tree multipliers based on our [9:4] adder design have almost the same T_{PPGR} as TDM schemes. For $n \leq 32$, ULLRLFs have 0.5 ~ 1.5T_{XOR2} more delay than tree multipliers. For larger precisions, ULLRLF shows 23% more gate delay for 48 × 48-bit multiplication and 28% more for 54 × 54-bit multiplication.

5.2. Simulation with physical layout

For more realistic evaluation, structural VHDL designs are compiled and mapped into Artisan TSMC 0.18μm 1.8-Volt standard-cell library [23] using Synopsys Design Compiler. For a fair comparison of different schemes, [3:2] FA cells in the library are not used because there is no [4:2] adder cells. Buffers are inserted automatically by Design Compiler. Two radix-4 schemes for 24×24-bit and 32×32-bit multiplication are compared: tree9to4 and ULLRLF-MUX-CSSC. tree9to4 has the similar delay as TDM but is more regular. MUX-CSSC based designs are chosen because it has smaller area and CSSC is good for power. CSSC is also applied in tree9to4. Standard-cell based automatic layout is first conducted using Cadence Silicon Ensemble. Interconnect parameters are extracted from layout and back-annotated into Synopsys tools for delay and power calculation. Power consumption is measured at 100MHz with 500 pseudo-random data. The results are shown in Table 3. For 24-bit, URLRLF is better than tree9to4 in area, delay, and power, with up to 7% improvement. For 32-bit, URLRLF has 3% less area and 10% less power than tree9to4 while keeping similar delay.

Table 3: Comparison after automatic layout

Schemes	Area(μm^2)	Delay(ns)	Pwr(mW)
tree9to4-24b	56,126	6.01	26.27
ULLRLF-24b	54,014	5.88	24.59
tree9to4-32b	108,324	7.18	45.12
ULLRLF-32b	105,380	7.25	40.65

We have also experimented layout with guided floorplanning for 32 × 32-bit multipliers. The floorplan of tree9to4 is shown in Fig. 12, which is based on H-tree for symmetry and regularity [17]. The row utilization rate has to be relaxed to 63% from 70% in automatic layout for routability. In addition, all blocks have to be assigned to specific regions for delay reduction. The floorplan of ULLRLF is shown in Fig. 13. The left-side [4:2] CSA in each part is distributed into PPR rows. The row utilization remains at 70%. Regions are assigned for two big upper/lower blocks, final

Table 1: The delay effects of FA type and CSSC in LRLF/ULLRLF

Schemes	n = 24			n = 32			n = 48		
	T_{GR}	T_A	Total	T_{GR}	T_A	Total	T_{GR}	T_A	Total
LRLF-MUX	15	5	20	19	5	24	27	5	32
LRLF-MUX-CSSC	14	5	19	18	5	23	26	5	31
LRLF-ND3	14	5	19	18	5	23	26	5	31
LRLF-ND3-CSSC	13	5	18	17	5	22	24.5	4.5	29
ULLRLF-MUX	12	5	17	14	6	20	18	6	24
ULLRLF-MUX-CSSC	11.5	5.5	17	13.5	6.5	20	17.5	6.5	24
ULLRLF-ND3	11	6	17	13	6	19	17	7	24
ULLRLF-ND3-CSSC	11	6	17	13	6	19	17	7	24

Table 2: Delay comparison of tree multipliers and LRLF/ULLRLF

Schemes	n = 24			n = 32			n = 48			n = 54		
	T_{GR}	T_A	Total	T_{GR}	T_A	Total	T_{GR}	T_A	Total	T_{GR}	T_A	Total
TDM-radix2	10.5	–	–	11.5	6	17.5	13.5	6	19.5	13.5	6	19.5
TDM-radix4	11	5.5	16.5	12	6	18	–	–	–	–	–	–
Tree9to4	10	6	16	11	7	18	13	7	20	14	7	21
LRLF	13	5	18	17	5	22	24.5	4.5	29	26.5	4.5	31
ULLRLF	11	6	17	13	6	19	17	7	24	18	7	25

[4:2] adder, and CPA. The results are shown in Table 4. The delay is improved by 4% from automatic layout. For ULLRLF, there is no cost in area for this delay improvement. For tree9to4, the area increases 10%. After layout with guided floorplanning, ULLRLF and tree9to4 has similar delay while tree9to4 has 15% more area and 9% more power.

Table 4: Comparison after guided layout

Schemes	Area(μm^2)	Delay(ns)	Pwr(mW)
tree9to4-32b	120,958	6.90	45.53
ULLRLF-32b	105,380	6.99	41.72

6. Conclusions

We have studied left-to-right split array multiplier schemes EOLRLF/ULLRLF. An efficient radix-4 recoding logic generates the partial products in a left-to-right order. These partial products are split into upper/lower or even/odd groups. These two groups are reduced in parallel using the L-R leapfrog structure with optimized adder modules and signal flows. Results from the two groups are merged using a [4:2] adder. The final adder is a prefix adder optimized to match non-uniform input arrival profile. We find that upper/lower splitting outperforms even/odd splitting after layout although even/odd splitting is a little bet-

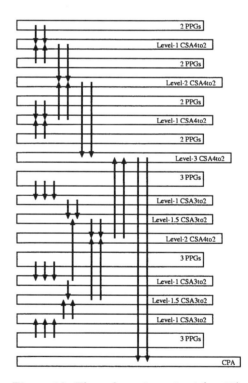

Figure 12: Floorplan of tree9to4 (n=32).

ter in gate delay at logic level. Layout experiments for $n = 24$ and $n = 32$ indicate that ULLRLF multipliers have slightly less area and power than optimized tree multipliers while keeping similar delay. We con-

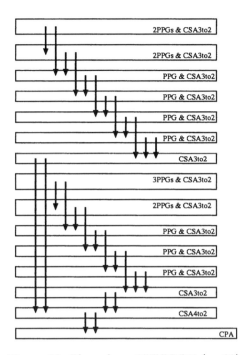

Figure 13: Floorplan of ULLRLF (n=32).

clude that ULLRLF array multipliers and tree multipliers are similar in major performance characteristics for $n \leq 32$ if standard-cell based automatic layout is conducted.

Acknowledgments

The authors would like to thank Dr. Wen-Chang Yeh and Dr. Paul Stelling for their help. This work has been supported in part by Raytheon Company, Fujitsu Laboratories of America, and the State of California MICRO program.

References

[1] K.C. Bickerstaff, E.E. Swartzlander, Jr., and M.J. Schulte, "Analysis of column compression multipliers," in *Proc. 15th IEEE Symp. Computer Arithmetic*, pp.33-29, 2001.

[2] L. Ciminiera and P. Montuschi, "Carry-save Multiplication Schemes Without Final Addition," *IEEE Trans. Comput.*, vol. 45, no. 9, pp. 1050-1055, Sept. 1996.

[3] Y. Choi and Earl E. Swartzlander, Jr., "Design of a hybrid prefix adder for non-uniform input arrival times," in *Proc. SPIE 2002 Advanced Signal Processing Algorithms, Architectures, and Implementations XII*, July 2002.

[4] M.D. Ercegovac and T. Lang, "Fast multiplication without carry-propagate addition," *IEEE Trans. Comput.*, vol.39, no.11, pp.1385-1390, Nov. 1990.

[5] A. Goldovsky, *et. al.*, "Design and implementation of a 16 by 16 low-power two's complement multiplier," in *Proc. 2000 IEEE Int. Symp. Circuits and Systems*, vol.5, pp.345-348, 2000.

[6] G. Goto, *et. al.*, "A 54*54-b regularly structured tree multiplier," *IEEE J. Solid-State Circuits*, vol.27, no.9, pp.1229-1236, Sept. 1992.

[7] Z. Huang and M.D. Ercegovac, "Low power array multiplier design by topology optimization," in *Proc. SPIE 2002 Advanced Signal Processing Algorithms, Architectures, and Implementations XII*, July 2002.

[8] N. Itoh, *et. al.*, "A 600-MHz 54*54-bit multiplier with rectangular-styled Wallace tree," *IEEE J. Solid-State Circuits*, vol.36, no.2, p.249-257, Feb. 2001.

[9] J. Iwamura, *et. al.*, "A high speed and low power CMOS/SOS multiplier-accumulator," *Microelectronics Journal*, vol.14, no.6, pp.49-57, Nov.-Dec. 1983.

[10] J. Kim and E.E. Swartzlander, Jr., "Improving the recursive multiplier," in *Proc. 34th Asilomar Conf. Signals, Systems and Computers*, pp.1320-1324, Nov. 2000.

[11] R.K. Kolagotla, H.R. Srinivas, and G.F. Burns, "VLSI implementation of a 200-MHz 16*16 left-to-right carry-free multiplier in 0.35 mu m CMOS technology for next-generation DSPs," in *Proc. IEEE 1997 Custom Integrated Circuits Conf.*, pp.469-472, May 1997.

[12] I. Koren, *Computer Arithmetic Algorithms*, Prentice Hall, Englewood Cliffs, New Jersey, 1993.

[13] S.S. Mahant-Shetti, P.T. Balsara, and C. Lemonds, "High performance low power array multiplier using temporal tiling," *IEEE Trans. Very Large Scale Integration (VLSI) Systems*, vol.7, no.1, p.121-124, March 1999.

[14] M. Nagamatsu, *et. al.*, "A 15-ns 32*32-b CMOS multiplier with an improved parallel structure," *IEEE J. Solid-State Circuits*, vol.25, pp.494-497, Apr. 1990.

[15] V.G. Oklobdzija, D. Villeger, and S.S. Liu, "A method for speed optimized partial product reduction and generation of fast parallel multipliers using an algorithmic approach," *IEEE Trans. Comput.*, vol.45, no.3, pp.294-306, March 1996.

[16] P.F. Stelling and V.G. Oklobdzija, "Designing optimal hybrid final adders in a parallel multiplier using conditional sum blocks," in *Proc. 15th IMACS World Congress Scientific Computation, Modeling, and Applied Math.*, Aug. 1997.

[17] J.D. Ullman, *Computational Aspects of VLSI*, Computer Science Press, Inc., 1983.

[18] W.-C. Yeh and C.-W. Jen, "High-speed Booth encoded parallel multiplier design," *IEEE Trans. Comput.*, vol.49, no.7, pp.692-701, July 2000.

[19] W.-C. Yeh, *Arithmetic Module Design and its Application to FFT*. Ph.D. dissertation, National Chiao-Tung University, 2001.

[20] Z. Yu, L. Wasserman, and A.N. Willson, Jr. "A painless way to reduce power dissipation by over 18% in Booth-encoded carry-save array multipliers for DSP," in *2000 IEEE Workshop on SiGNAL PROCESSING SYSTEMS*, pp.571-580, Oct. 2000.

[21] R. Zimmermann, *Binary Adder Architectures for Cell-Based VLSI and their Synthesis*. Ph.D. dissertation, Swiss Federal Institute of Technology, Zurich, 1997.

[22] *International Technology Roadmap for Semiconductors – Interconnect*, 2001 Edition.

[23] *TSMC 0.18μm Process 1.8-Volt SAGE-X Standard Cell Library Databook*. Artisan Components, Inc., Oct. 2001.

Multiple-Precision Fixed-Point Vector Multiply-Accumulator using Shared Segmentation

Dimitri Tan[1], Albert Danysh[2], Michael Liebelt[3]

Motorola Inc.[1,2], University of Adelaide, Australia[1,3]
Dimitri.Tan@motorola.com[1], Albert.Danysh@motorola.com[2], Michael.Liebelt@adelaide.edu.au[3]

Abstract

This paper presents a 64-bit fixed-point vector multiply-accumulator (MAC) architecture capable of supporting multiple precisions. The vector MAC can perform one 64x64, two 32x32, four 16x16 or eight 8x8 bit signed/unsigned multiply-accumulates using essentially the same hardware as a scalar 64-bit MAC and with only a small increase in delay. The scalar MAC architecture is "vectorized" by inserting mode-dependent multiplexing into the partial product generation and by inserting mode-dependent kills in the carry chain of the reduction tree and the final carry-propagate adder. This is an example of "shared segmentation" in which the existing scalar structure is segmented and then shared between vector modes. The vector MAC is area efficient and can be fully pipelined which makes it suitable for high-performance processors and possibly dynamically reconfigurable processors.

I. INTRODUCTION

The addition of vector capabilites to a processor architecture can provide a significant boost in performance for multimedia type applications [1]. However, the addition of vector capabilities is typically expensive in terms of area. This paper presents a 64-bit fixed-point vector multiply-accumulator (MAC) architecture capable of supporting multiple precisions using essentially the same hardware as a 64-bit scalar MAC with only a small increase in delay. The vector MAC can perform one 64x64, two 32x32, four 16x16 or eight 8x8 signed/unsigned multiply-accumulates. The multiplier is based on a radix-4 Booth encoder and a Wallace carry-save adder (CSA) reduction tree. The final 128-bit carry propagate adder is a modified carry-lookahead (CLA) adder. The scalar MAC architecture is "vectorized" by inserting mode-dependent multiplexing into the partial product generation and by inserting mode-dependent kills in the carry chain of the reduction tree and the final carry-propagate adder. This is an example of "shared segmentation" in which the existing scalar structure is segmented and then shared between vector modes. This method can also be applied to other arithmetic algorithms.

The remainder of this paper is organized as follows. First the basic architecture of the 64-bit vector MAC is presented. Then the details of each part of the scalar 64-bit MAC are presented along with the required modifications necessary to support the vector functionality. Next some existing alternative implementations are discussed for the purposes of comparison. Following that, a VLSI implementation of the vector MAC is presented with a delay and area comparison against the scalar version of the MAC.

II. MULTIPLY-ACCUMULATE (MAC) ARCHITECTURE

II.I. Overview

The block diagram of the vector MAC unit is shown in Fig. 1. The vector MAC unit consists of a modified Booth [2] recoder partial product generator, a Wallace [4] carry-save adder (CSA) reduction tree and a final carry-propagate adder (CPA). The CPA is implemented using a carry-lookahead adder (CLA) consisting of 4-bit carry-lookahead blocks. The accumulator is fed into the Wallace tree and only adds one extra stage of CSA delay. This is a fairly typical MAC design. The *mode[3:0]* control signals determine which of the vector modes the MAC is to operate in. The *unsigned* control signal selects unsigned multiplication. The details of the implementation which facilitate the vector functionality will be covered in the proceeding sections.

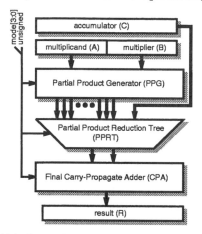

Fig. 1. Multiply-Accumulate (MAC) Architecture Block Diagram

II.II. Scalar Partial Product Generator (PPG)

The Scalar Partial Product Generator (PPG) uses radix-4 Booth recoding which reduces the number of partial products from *n* to *(n/2+1)* for an *n x n* bit multiplication. The extra partial product is for handling signed/unsigned multiplication and is generated when the multiplier is sign extended. The partial products use sign encoding instead of sign extension for handling the negatively weighted most significant bit [3],[6]. This avoids having to sign extend the partial products to the full width of the 128-bit result. The radix-4 Booth algorithm encodes groups of three multiplier bits overlapping by one bit into five possibilities - *selze* ("zero"), *selp1* ("+1"), *selp2* ("+2)", *seln1* ("-1") and *seln2* ("-2).

As mentioned previously, the last partial product is used to handle signed and unsigned numbers. Note that each partial product is appended with two bits, one of which may be a "1" if the previous partial product is being twos-complemented. Hence these extra two bits are the *seln2* and *seln1* Booth selects from the previous partial product and are known as "hot ones" [3].

II.III. Vector Partial Product Generator (VPPG)

The block diagram for the Vector PPG is shown in Fig. 2 and is identical to a scalar PPG except for some additional multiplexing of the multiplicand and some masking applied to the multiplier.

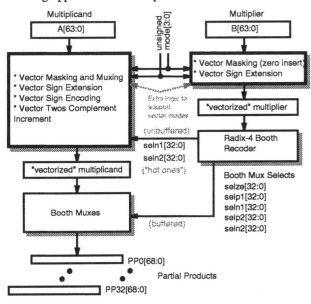

Fig. 2. Vectorized Partial Product Generator Block Diagram

The Vector Partial Product Generator (VPPG) must be able to generate the correct partial products for each of the supported vector modes. Each of the vector modes has a corresponding array of partial products. Each of these arrays of partial products can be overlayed on top of the array of partial products required for the 64-bit scalar mode as shown in Fig. 3, Fig. 4 and Fig. 5. A more detailed example of the partial product array for the 8x8 vector mode is shown in Fig. 6. In this example, each subgroup of partial products corresponds to an 8x8 bit vector element multiplication. Note that the fourth partial product of each subgroup has been truncated by one bit since it overlaps by one bit with the first partial product of the next subgroup. Furthermore, take special note of the last partial product which is used to support unsigned mode for all vector elements instead of having an extra partial product for each vector element.

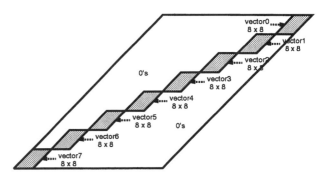

Fig. 3. Simplified Vector Booth Partial Product Array for 64x64 bit multiplication in 8 bit-mode

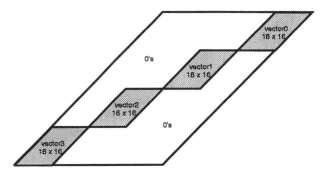

Fig. 4. Simplified Vector Booth Partial Product Array for 64x64 bit multiplication in 16 bit-mode

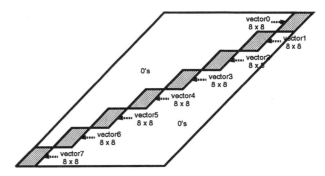

Fig. 5. Simplified Vector Booth Partial Product Array for 64x64 bit multiplication in 32 bit-mode

We will next examine what needs to be done to enable these partial product arrays to be merged. First we will consider changes to the multiplier (*B*) operand path through the Booth recoder and then the multiplicand (*A*) operand path through the Booth muxes.

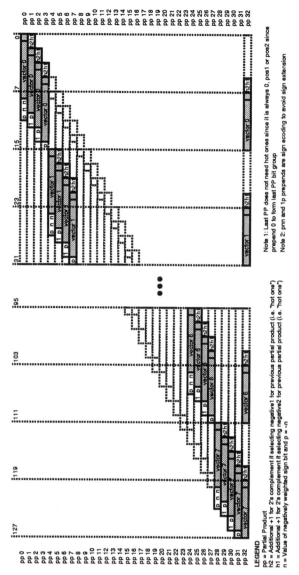

Fig. 6. Vector Booth Partial Product Array for 64x64-bit multiplication in 8-bit vector mode.

II.IV.I. Vector Booth Recoder

In the vector MAC, the Booth recoder is the same as that used for the scalar MAC except that the multiplier operand (*B*) input changes depending on the vector mode. This is illustrated in Fig. 7 which shows the Booth recoder multiplier input for each vector mode.

In the scalar PPG, the first partial product is generated by assuming *B[-1] = 0*. In the vector Booth Recoder, this "0" must be inserted at vector element boundaries for the first partial product of each subgroup. These are the extra "0" bits in Fig. 7. For example, in the 8-bit vector mode, the fifth partial product uses bit triplet *{B[9],B[8],"0"}* instead of *{B[9],B[8],B[7]}* as used in the scalar PPG. The zero insertion is implemented by masking the multiplier operand bits with a signal that is a function of the mode and bit position. Hence an extra two-input nand gate is added to the critical path.

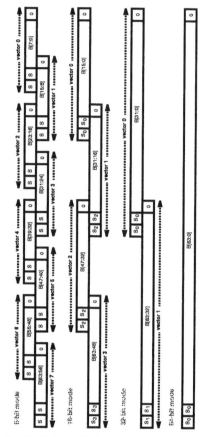

Fig. 7. Vector Booth Recoder Inputs

To support unsigned numbers, the scalar PPG generates an additional partial product since the multiplier operand is extended by two bits to form the bit triplet. Similarly, each vector element of the multiplier operand B must be extended by two bits. This is indicated in Fig. 7 by the additional "sign" bits that are denoted by s_i where i is the vector element index.

Hence the vector Booth PPG must generate an additional partial product for every vector element. The sign bits are different for each vector element and are formed according to equation (1).

$$s_i = \text{MSB} \cdot \sim\text{unsigned} \quad (1)$$

where MSB is the most significant bit of the vector element i.

The sign bits are used to generate the extra partial product for each vector element which are then combined into a single partial product that is twice the width of the operands i.e. 128-bits. This combination can be done because the extra partial products for each vector element do not overlap. They do however cross the vector element boundaries and need to be truncated by 3 bits i.e. the sign encoding bits are truncated at the MSB end. Note that the Booth select for the last partial product is either *selze* ("select zero") or *selp1* ("select plus one") and therefore does not need the "hot ones" bits.

II.V.II. Vector Booth Muxes

The MSB of each partial product generated by the PPG is negatively weighted. Consequently, the partial products must be either sign extended out to the full width of the result or sign encoded as mentioned earlier. In sign encoding, the negatively weighted MSB is replaced by *{p,n,n}* for the first partial product or *{1,p}* for the remaining partial products where n is the MSB or sign-bit of the multiplicand operand and $p=\sim n$ [6]. In the Vector PPG, the sign encoding must be performed at each vector element boundary using the appropriate sign-bit for that vector element. The sign-bit is dependent on the vector mode, the vector element in question and whether signed or unsigned mode is selected.

The final complication in the Vector PPG involves handling the two's complement of the multiplicand when one the Booth selects *seln1* ("select -1") or *seln2* ("select -2") are asserted. The two's complement requires inverting the multiplicand and then adding one to the LSB. To avoid adding extra partial products for the sole purpose of performing the increment, the extra ones are simply appended to the next partial product. Thus the extra bits appended are *{seln2,seln1}* for the previous partial product. This is a standard practice [3]. However, in the vector PPG, these extra "hot one" bits must be inserted at vector element boundaries which vary depending on the mode.

It appears that each vector mode has a unique set of partial products. The naive or non-optimal approach to merging these partial products would be to form the partial products for each mode and select the correct partial product based on the vector mode as shown in Fig. 8. This requires an additional five 64-bit 4:1 multiplexers for each partial product. The Booth recoder is usually the critical path so adding the extra 4:1 mux delay prior to the Booth muxes does not add to the critical path except for the appending of the "hot one" bits *{seln2,seln1}*. These bits are the Booth selects for the previous partial product.

The naive approach can be improved by observing that at most bit positions there are less than four unique data to select from. Infact, at some bit positions there is no additional logic required at all compared to the scalar Booth PPG. This is illustrated in Fig. 9 which shows a selection of partial products and how they overlap for each mode. In the worst case a 4-input mux is required, such as for the sign encoding bits in partial product 30. The remaining bits require either a 3-input mux, a 2-input mux, a 2-input "and" gate or no extra logic at all depending on significance. The mux selects are a function of the mode control signals, significance and partial product number. Similarly, one input to the "and" gate is a function of the mode control signals, significance and partial product number. This "and" gate input simply masks the other input which is the multiplicand. The 2-input and 3-input mux inputs are either tied to ground, the multiplicand, the sign encoding bits, or the *seln1* and *seln2* Booth selects for the previous partial products. A more detailed example of the partial product overlap is given in Fig. 10.

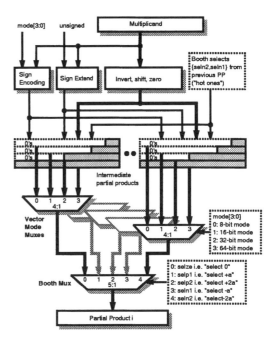

Fig. 8. Naive Vector Partial Product Generation

Note that the last partial product cannot be generated this way and requires full multiplexing as in Fig. 8. This is because the Booth selects for the last partial product are different for each vector element and each mode. The last partial product is also the full width of the result (128 bits) and thus enters the Wallace tree in one of the last two stages.

The theoretical critical path is from the multiplier *B*, through the vector masking, the Booth Recoder, the vector-mode muxes and finally the booth mux. The extra 2-input "nand" gate used for the masking of the multiplier occurs in parallel with part of the Booth recoder logic and therefore does not add delay to the critical path. It is assumed the mask is precalculated to avoid adding to the critical path. The vector mode muxing is designed to allow for the late "seln1" or "seln2" Booth select and inserts the delay of a 2-input mux. The delay of the 2-input mux will occur in parallel to the significant buffering of the Booth selects before they reach the Booth muxes. Assuming these delays are about equal the extra logic does not add an significant delay to the theoretical critical path. The real critical path is of course dependent on the technology and the implementation details.

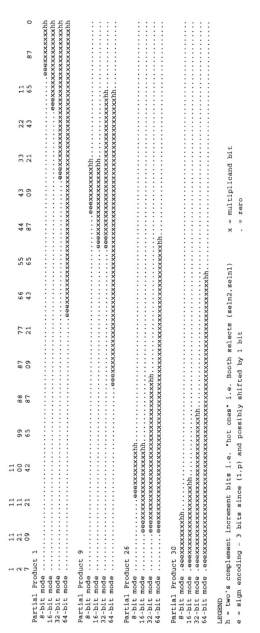

Fig. 9. Example of partial product overlap between vector modes

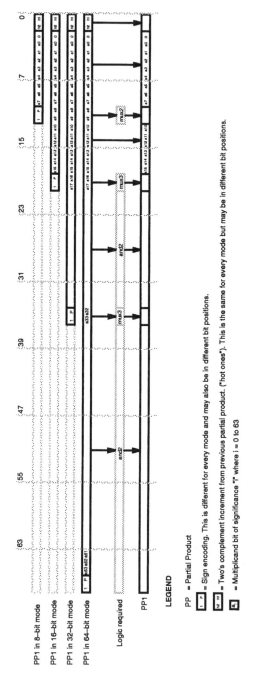

Fig. 10. Example of Partial Product 1 assuming Booth select "*selp2*" ("select positive two") asserted

II.VI. Scalar Partial Product Reduction Tree (PPRT)

The Scalar Partial Product Reduction Tree (PPRT) reduces the set of partial products down to two for the final carry-propagate addition. A Wallace CSA Tree consisting of 3:2 compressors or full-adders (FAs) is used to implement the scalar PPRT [4]. The accumulator is also added into the Wallace tree to avoid an extra carry-propagate addition. The number of 3:2 levels in the Wallace tree is given by equation .

$$\text{Levels} = \lceil (\log N)/(\log 1.5) \rceil \quad (2)$$

where N = total number of partial products

For the MAC the number of partial products is $64/2 + 1 + 1 = 34$. One of the two additional partial products is for supporting signed/unsigned numbers and the other is the accumulator. Hence the Wallace tree has $\lceil \log\langle 34\rangle/\log(1.5)\rceil = \lceil 8.7 \rceil = 9$ levels. The total number of full adders required is 3088.

II.VII. Vector Partial Product Reduction Tree (PPRT)

There are two methods for vectorizing the scalar PPRT. The first method observes that the 16-bit vector mode reduction trees can be created from the 8-bit vector mode reduction trees with the two's complement partial product and accumulator excluded. An additional 4:2 stage is used to reduce these partial products and produce the result for each of the 8-bit vector elements. The vector result for the 8-bit mode is formed by concatenating the vector element results. This is then applied recursively until the reduction tree for the 64-bit mode is created. The vector results for each mode must then be multiplexed together to form the final two partial products. The first method is illustrated in Fig. 11. Note for each mode only one branch of the vector reduction tree is shown fully.

The second method involves killing the carries which cross the vector element boundaries at every level of the CSA tree. The bit positions of the vector boundaries depend on the vector mode. The second method could be applied equally well to a Wallace tree or a Dadda [5] tree and will be elaborated below. The killing of carries in the second method can be achieved by inserting a 2-input "and" gate at each vector element boundary to mask the carry-in input of each corresponding full-adder. Since a CSA does not have a carry-propagate path, the carry-kill "and" gate can be incorporated into the full-adder design such that it does not add any additional delay through the CSA tree as shown in equation (3).

$$\begin{aligned} sum &= (a \oplus b) \oplus (c_{in} \cdot \neg kill) \\ c_{out} &= (a \cdot b) + [(a+b) \cdot (c_{in} \cdot \neg kill)] \end{aligned} \quad (3)$$

A possible realization of the full-adder is illustrated in equation Fig. 12. which clearly shows that the addition of the *kill* signal will not significantly increase delay since the extra gate is in parallel with other terms.

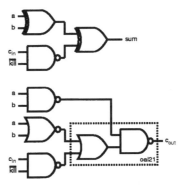

Fig. 12. Full-adder with carry-in kill for Vector PPRT - Method 2

The second method was chosen for the vector MAC because it does not add delay to the critical path and does not require much extra hardware. The first method adds a 4-input mux to the critical path and requires several extra rows of CSAs.

II.VIII. Scalar Final Carry Propagate Adder (CPA)

The final CPA sums the two remaining partial products generated by the Wallace Tree. The scalar MAC uses a standard 128-bit carry-lookahead (CLA) adder comprised of 4-bit CLA blocks. This adder was chosen because it is a fast adder and can easily be modified to create a vector adder. It should be noted that CLA adders comprised of 4-bit CLA blocks are most efficient for data widths that are powers of 4. Thus there are probably better choices for a fast 128-bit adder. For the purposes of this paper this can be overlooked.

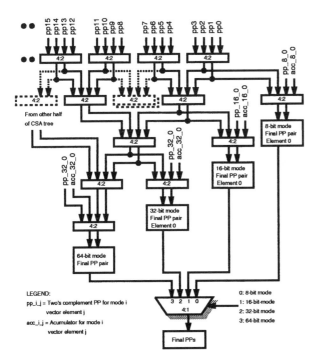

Fig. 11. Vector Partial Product Reduction Tree - Method 1

II.IX. Vector Final carry Propagate Adder (CPA)

There are essentially two methods for vectorizing the final CPA. The first method involves extending the width of the adder by one bit at every potential vector element boundary and conditionally inserting zeros in order to kill the carries. Hence this requires that the adder is extended by the number of vector elements supported by the lowest granularity vector mode. In the case of the 64-bit vector MAC we require an extra 8 bits since for 8-bit mode there are 8 vector elements. The total width of the adder would thus be 136 bits. For each extra significance, there is a pair of inputs. Both inputs are set to zero if a carry kill is required. Alternatively, one of the inputs is set to a one and the other to zero if a carry propagate (no carry kill) is required. The extra bits are then dropped when forming the final result.

The second method involves killing the carries which cross the vector element boundaries that are determined by the vector mode selected. This is the same method used in the Vector Wallace tree. The extra 2-input "and" gate can be combined into the 4-bit CLA blocks without adding additional delay to the critical path. The 4-bit CLA equations with the carry-in kill term are given in equation (4).

$$cout_0 = g_0 + c_{in} \cdot p_0 \cdot (\neg kill) \qquad (4)$$
$$cout_1 = g_1 + g_0 \cdot p_1 + c_{in} \cdot p_0 \cdot p_1 \cdot (\neg kill)$$
$$cout_2 = g_2 + g_1 \cdot p_2 + g_0 \cdot p_1 \cdot p_2 + c_{in} \cdot p_0 \cdot p_1 \cdot p_2 \cdot (\neg kill)$$
$$g_3^0 = g_3 + g_2 \cdot p_3 + g_1 \cdot p_2 \cdot p_3 + g_0 \cdot p_1 \cdot p_2 \cdot p_3$$
$$p_3^0 = p_0 \cdot p_1 \cdot p_2 \cdot p_3$$

where g_3^0 is the block generate from bit 0 to bit 3
and p_3^0 is the block propagate from bit 0 to bit 3

In the scalar 4-bit CLA block, the critical path is from the carry-in c_{in} and the group generate inputs since the group propagates are available much earlier. Therefore the c_{in} and the group generate inputs are typically placed close to the output in the circuit topology. This means that there is less capacitance to discharge when the critical signals finally change. In the vector CLA block the extra kill term is also available early. Consequently, it does not add additional capacitance to discharge provided the critical inputs remain closest to the output. The vector MAC was implemented using the second method since it does not add extra delay to the critical path.

III. Existing Implementation Schemes

There are essentially two existing schemes for implementing a vector multiply-accumulator [6], [7], [8], [9], [10], [11], [12], [13]. The first scheme involves having separate hardware for each vector element of each mode and then muxing the result at the end. Hence to achieve the equivalent of the MAC presented in this paper, the first scheme requires one 64-bit MAC, two 32-bit MACs, four 16-bit MACs and eight 8-bit MACs. This is a brute force scheme and wastes a great deal of area. In addition, the delay of the final 4:1 result mux is inserted into the critical path. Such a design can be fully-pipelined without causing pipeline stalls.

The second scheme involves building wider vector elements out of several of the narrower vector elements and then adding the multiple results together. This can be done consecutively or by recirculating the data back through the unit over more than one cycle. For instance, a *2Nx2N* multiplier can be built out of four *NxN* multipliers by generating four *2N* products which can then be added to form the *4N* product. This scheme is not suited to supporting more than two vector modes. Furthermore, if the data is recirculated back through the MAC unit, then the pipeline feeding the MAC must stall. This is unsuitable for high performance processors since it complicates the instruction scheduling. Typically a vector MAC will use a combination of the above schemes.

IV. Implementation

The vector MAC presented in this paper was implemented in Verilog HDL at a structural level and then synthesized, placed and routed in a 0.13 um bulk-silicon technology. Synopsys Physical Compiler was used for synthesis and initial placement. Cadence QPlace/Warp was used for final placement and routing. A "SPICE-accurate" in-house static timing tool was used to determine propagation delays. In addition, the scalar MAC design was also implemented for direct comparison. The results are shown in Table 1 for both MAC designs with no pipeline registers. There was a 8% increase in delay and a 2% increase in area over the scalar 64-bit MAC.

TABLE 1. Implementation Summary

Design	Delay (ps)	Area (μm^2)
Scalar MAC	2282.45	187 553
Vector MAC	2470.44	191 301

The vector MAC was tested by cycling through all possible combinations of inputs in which the two most-significant bits and two least significant-bits for each vector element were varied while the intermediate bits were all ones or all zeros. In addition, some random pattern generation was used.

V. CONCLUSIONS

In this paper, we presented the design and implementation of a vector multiplier-accumulate (MAC) unit that can perform one 64x64, two 32x32 bit, four 16x16 or eight 8x8 signed/unsigned multiply-accumulates using essentially the hardware as a 64-bit multiplier-accumulator and without significantly more delay. The concept of *"shared segmentation"* is introduced in which the existing scalar hardware structure is segmented and then shared between vector modes. In the case of the MAC, the scalar architecture is "vectorized" by inserting mode-dependent masking into the partial product generation and by inserting mode-dependent kills in the carry chain of the reduction tree and final carry-propagate adder. The *"shared segmentation"* concept can be applied to other arithmetic units such as floating-point addition or multiplication.

FUTURE WORK

We are in the process of designing and implementing other vector arithmetic units by applying the *"shared segmentation"* concept with particular focus on floating-point addition and multiplication. We are also evaluating different interconnection strategies in which an array of such "vectorized" execution units are connected together to form a powerful execution engine suitable for computationally intensive applications such as multimedia and digital signal processing.

REFERENCES

[1] Lee, C.G.; Stoodley, M.G. "Simple vector microprocessors for multimedia applications" Microarch., 1998. MICRO-31. Proc. 31st Annual ACM/IEEE Int. Symp. on , pp. 25-36, 1998.

[2] Booth, "A Signed Binary Multiplication Algorithm", Qt. J. Mech. Appl. Math., vol. 4, pp. 236-240, 1951.

[3] S. Vassiliadis; E.M. Schwarz; B.M. Sung, "Hardwired multipliers with encoded partial products", IEEE Trans. on Computers, Vol. 40, Issue 11 , Nov 1991, pp. 1181 -1197.

[4] C.S.Wallace, "A suggestion for a fast multiplier", IEEE Trans. Electron. Comput., vol. EC-13, pp. 14-17, 1964.

[5] L. Dadda, "Some Schemes For Parallel Multipliers", Alta Freq., 34, pp. 349-356, 1965.

[6] Behrooz Parhami, "Computer Arithmetic - Algorithms and Hardware Designs", pp. 178-180, 191-195, ISBN 0-19-512583-5, 2000.

[7] M.S. Schmookler et al, "A Low-power, High-speed Implementation of a PowerPC™ Microprocessor Vector Extension', Comp. Arith., 1999. Proc. 14th IEEE Symp., 1999.

[8] A.A. Farooqui; V.G. Oklobdzija, "General Data-Path Organization of a MAC unit for VLSI Implementation of DSP Processors", Circ. & Sys., ISCAS '98. Proc. IEEE Int. Symp., Vol. 2, pp. 260 -263, 1998.

[9] W.F. Wong; E. Goto, "Division and Square-Rooting using a split multiplier", Electr. Letters, Vol. 28 No. 18, pp. 1758-1759, Aug 1992.

[10] Y. Liao; D.B. Roberts ,"A High-Performance and Low-Power 32-bit Multiply-Accumulate Unit with Single-Instruction-Multiple-Data (SIMD) Feature", IEEE J. of Solid-State Cir., Vol. 37, No. 7, July 2002.

[11] R.B. Lee, "Multimedia Extensions for General-Purpose Processors", Sig. Proc. Sys., SIPS 97, Nov 1997, pp. 9 -23.

[12] Tang, K.C.; Wu, A.K.M.; Fong, A.S.; Pao, D.C.W., "Integrated partition integer execution unit for multimedia and conventional applications", IEEE Int. Conf. on Electr., Circ. and Sys., 1998, Vol. 2, pp. 103 -107.

[13] Rong Lin, "Trading bitwidth for array size: a unified reconfigurable arithmetic processor design", Int. Symp. on Qality Elec. Design, 2001, pp. 325-330.

Some Optimizations of Hardware Multiplication by Constant Matrices

Nicolas Boullis, Arnaud Tisserand
Arénaire Project (CNRS–ENSL–INRIA–UCBL)
LIP, École Normale Supérieure de Lyon
46 allée d'Italie, F-69364 Lyon, France
`Nicolas.Boullis@ens-lyon.fr,Arnaud.Tisserand@ens-lyon.fr`

Abstract

This paper presents some improvements on the optimization of hardware multiplication by constant matrices. We focus on the automatic generation of circuits that involve constant matrix multiplication (CMM), i.e. multiplication of a vector by a constant matrix. The proposed method, based on number recoding and dedicated common sub-expression factorization algorithms was implemented in a VHDL generator. The obtained results on several applications have been implemented on FPGAs and compared to previous solutions. Up to 40% area and speed savings are achieved.

1 Introduction

Important optimizations of the speed, area and power consumption of circuits can be achieved by using dedicated operators instead of general ones whenever possible. Multiplication by constant is a typical example. Indeed, if one operand of the multiplication is constant, one can use some shifts and additions/subtractions to perform the operation instead of using a full multiplier. This usually leads to smaller, faster and less power-consuming circuits.

Applications involving multiplication by constant are common in signal processing, image processing, control and data communication. Finite impulse response (FIR) filters, discrete cosine transform (DCT) and discrete Fourier transform (DFT), for instance, are central operations in high-throughput systems, and they use a huge amount of such operations. Their optimization widely impacts the performance of the global system that uses them. In [11] there is an analysis of this operation frequency.

The problem of the optimization of multiplication by constant has been studied for a long time. For instance, the famous recoding presented by Booth in [3] can simplify the multiplication by constant operation as well as the full multiplication. This recoding and the algorithm proposed by Bernstein in [1] were widely used on processors without multiplication unit.

The main goal in this problem is the minimization of the computation quantity. The multiplication by constant problem seems to be simple, but its resolution is a hard problem due to its combinatorial properties. This problem can occur in more or less complex contexts. In the case of a single multiplication of one variable by one constant, it seems possible to explore the whole parameter space. But in the case of the multiplication of several variables by several constants, the space to explore is so huge that we have to use heuristics.

A first solution proposed to optimize multiplication by constant was the use of the constant recoding, such as Booth's. This solution just avoids long strings of consecutive ones in the binary representation of the constant. Better solutions are based on the factorization of common sub-expressions, simulated annealing, tree exploration...

Our work deals with constant matrix multiplication (CMM), i.e. one useful form of the multiplication of several variables by several constants. A lot of applications involve such linear operations. This method is based on constants recoding followed by some dedicated common sub-expression factorization algorithms. The proposed method was implemented in a VHDL generator. The generated results for several applications have been implemented on Xilinx FPGAs and compared to other solutions. Some significant improvements have been obtained: up to 40% area saving in the DCT case for instance.

This paper is organized as follows. The problem is presented in Section 2. In Section 3 some related works are presented. Our algorithms are presented in Section 4. The developed generator and the target architectures are discussed in Section 5. Finally, the results of the implementation of some applications and their comparison to other solutions are presented in Section 6.

2 Problem Definition

In this paper, and in the related works, the central problem is the substitution of full multipliers by an optimized sequence of shifts and additions and/or subtractions. We focus on integers but all the results can be extended to fixed-point representations.

All the values are represented using a standard radix-2 notation or two's complement unless it is specified. The notation $x \ll k$ denotes the k-bit left shift of the variable x (i.e. $x \times 2^k$). As we look at FPGA implementations, we assume that shift is just routing and addition and subtraction have the same area and speed cost.

As an example, let us compute p as the product of the input variable x by the constant $c = 111463 = 11011001101100111_2$. The simplest algorithm consists in using the distributiveness of multiplication. There is one addition of x (after some potential shift) for each one in the binary representation of c. In the case $c = 111463$, it leads to 10 additions:

$$111463x = x \ll 16 + x \ll 15 + x \ll 13 + x \ll 12$$
$$+ x \ll 9 + x \ll 8 + x \ll 6 + x \ll 5$$
$$+ x \ll 2 + x \ll 1 + x.$$

The central point in this problem is the minimization of the total number of operations. It can be significantly reduced by using a recoding of the constant and/or subexpression elimination and sharing. The theoretical complexity of this problem seems to be still unknown.

Depending on the target application, this problem can occur at different levels of complexity. It starts with the multiplication of one constant by one variable. After, the multiple constant multiplication (MCM) problem appears with the multiplication of several constants by one variable [17]. In this present work, we deal with a more general version of this problem with the multiplication of one constant matrix by one variable vector: the constant matrix multiplication (CMM).

3 Related Works

There are, at least, four types of methods to address the multiplication by constant problem:

- Direct recoding methods.
- Evolutionary methods.
- Cost-function based search methods.
- Pattern search methods.

3.1 Direct Recoding Methods

In the radix-2 signed digit (SD) representation, the digits belong to the set $\{\bar{1} = -1, 0, 1\}$. A number is said to be in the canonical signed digit (CSD) format if no two non-zero digits are consecutive, then the number of non-zero digits is minimal. Using the CSD format on a n-bit value, the number of non-zero digits is bounded by $(n + 1)/2$ and it tends asymptotically to an average value of $n/3 + 1/9$, as shown in [7]. For our example, using Booth's canonical recoding we have: $111463 = 11011001101100111_2 = 100\bar{1}0\bar{1}0100\bar{1}0\bar{1}0100\bar{1}_2$ and the product $p = c \times x$ reduces to 7 additions/subtractions:

$$111463x = x \ll 17 - x \ll 14 - x \ll 12 + x \ll 10$$
$$- x \ll 7 - x \ll 5 + x \ll 3 - x.$$

The KCM algorithm [5] was specifically designed for LUT-based FPGAs. It decomposes the binary representation of the variable into 4-bit chunks (a radix-16 representation). There is a more general version of this decomposition problem with distributed arithmetic. For instance, in [19] it was used on a DCT operator with an area saving of 17% compared to the direct implementation of the whole computation.

There are some recent works on the use of high-radix recoding. For instance, in [2], they use a radix-8 representation with punctured coefficients with the digits in the set $\{0, \pm 1, \pm 2, \pm 4\}$ instead of the set $\{0, \pm 1, \pm 2, \pm 3, \pm 4\}$. This is a lossy representation, so they have to deal with some additional accuracy requirements. In our case, we want to study this problem for a lossless representation at first, but this approach seems to be interesting.

Another recoding solution was proposed with the use of multiple-radix representations and especially with the double-base number system (DBNS) [6]. In this solution, the authors use both radices 2 and 3 simultaneously, i.e. the values are expressed by $a = \sum_{i,j} a_{i,j} 2^i 3^j$ with $a_{i,j} \in \{0, 1\}$. This multiple-radix representation, sometimes useful in some analog circuits, does not seem to be efficient in the multiplication by constant problem in digital circuits.

3.2 Evolutionary Methods

Some evolutionary methods such as evolutionary graph generation [8] have been proposed to generate arithmetic circuits and especially for constant multipliers. These methods based on genetic algorithms seem to provide very bad results. For instance, in [8] the results are slightly better than the straightforward CSD encoding which is very far from the best known results. Furthermore, it seems that these methods are limited to the problem of multiplication by one constant and have never been used to produce real circuits.

3.3 Cost-Function based Search Methods

The algorithm presented by Bernstein in [1] allows to reuse some intermediate values that are just used once in recoding methods. A more detailed and corrected version of this algorithm can be found in [4]. The algorithm based on a tree exploration defines three kinds of operations: $t_{i+1} = t_i \ll k$, $t_{i+1} = t_i \pm x$ and $t_{i+1} = t_i \ll k \pm t_i$. A cost can be specified for each operation according to the target technology. The cost function used to guide the exploration is the sum of the costs of all the implied operations. This algorithm only shares a few common sub-expressions. For our example $p = c \times x$, this algorithm gives a 5-addition solution:

$$t_1 = ((x \ll 3 - x) \ll 2) - x,$$
$$t_2 = t_1 \ll 7 + t_1,$$
$$p = (((t_2 \ll 2) + x) \ll 3) - x.$$

There are some other cost-function based search methods such as simulated annealing. In [14], this technique was used to produce multiplication by a small set of constants. The same multiplier is used for a small set of different coefficients. This problem is different from our's.

In [15] a greedy algorithm is used to determine a solution with a low total operation cost. A 28% average area saving is achieved. This solution seems to be limited due to local attraction of the greedy algorithm.

3.4 Pattern Search Methods

Most of the pattern search methods are based on the same general idea. The algorithm recursively builds a set of useful constants to be optimized. This set is initialized with the recoded initial constants. The different methods differ in the way they match the common sub-expressions and the way they reuse and share them.

The multiple constant multiplication (MCM) solution presented in [17] performs a tree exploration with selection of matching parts of the SD representation of the constants. This paper is the most cited one, and it presents a lot of details about the algorithm as well as about the comparisons.

In [9] the matches between constants are represented using a graph. The exploration of this graph is used to produce a specific form of FIR filters with a reduced number of adders.

A solution based on an algebraic formulation of the possible matches between constants is presented in [12]. Unfortunately, the authors use random filters for their tests without specifying the coefficients. So it is difficult to compare their results to the other solutions.

In [13] a factorization method based on the selection of the best pair of matching digits is used. This solution can be easily extended to the selection of common parts of words larger than two digits.

We will base our solution on extensions and improvements of the algorithms presented in [10] and [16]. A detailed description of this idea is presented below.

One can notice that among all the abundant bibliography about the multiplication by constant problem there is no general solution to the CMM problem.

4 Proposed Algorithms

4.1 Lefèvre's algorithm

In 2001, Lefèvre proposed a new algorithm to efficiently multiply a variable integer by a given set of integer constants [10]. As a special case, this algorithm can be used to multiply a variable by a single constant.

4.1.1 Definitions

The principle of the algorithm is to keep a list of constants to be optimized, and to find a "pattern" that appears several times in the set of constants. The constants are recoded using the CSD format in the very beginning. A pattern is a sequence of digits in $\{\bar{1}, 0, 1\}$. The number of non-zero digits in the pattern is called its weight.

A pattern P is said to occur in a constant C with a shift α when for each 1 in position k of P, there is a 1 in position $k + \alpha$ in C, and for each $\bar{1}$ in position k of P, there is a $\bar{1}$ in position $k + \alpha$ in C. And a pattern is said to occur negatively when there is a $\bar{1}$ in C for each 1 in P, and a 1 in C for each $\bar{1}$ in P. This last point is one of the main difference between the two papers [10] and [16], Lefèvre's algorithm allows to use more complex patterns and leads to slightly better optimizations.

When two occurrences of two patterns in the same constant match the same non-zero digit of the constant, the two occurrences are said to conflict. For example, in the number $51 = 10\bar{1}010\bar{1}_2$, the pattern $10\bar{1}$ occurs positively with shift 0, negatively with shift 2 and positively with shift 4. The first and third occurrences both conflict with the second one. And the pattern 10001 occurs negatively with shift 0 and positively with shift 2. Those occurrences are overlapping but not conflicting.

On our previous example $p = c \times x$, Lefèvre's algorithm gives a solution with only 4 additions:

$$t_1 = (x \ll 3) - x,$$
$$t_2 = (t_1 \ll 2) - x,$$
$$p = (t_2 \ll 12) + (t_2 \ll 5) + t_1.$$

4.1.2 Description of the algorithm

The principle of the algorithm can be described by the pseudo-code presented in Algorithm 1.

Algorithm 1 : Principle of Lefèvre's algorithm.

While $\begin{pmatrix} \text{there are some patterns with weight} \geq 2 \\ \text{and at least 2 non-conflicting occurrences} \end{pmatrix}$

    choose a pattern with maximal weight and 2 non-conflicting occurrences
    remove the chosen pattern for both occurrences
    add the pattern as a new constant

Then, multiplication by each constant in the final set can be implemented in the usual way: for each 1 ($\bar{1}$) in position p, add (subtract) x shifted by p bits to the left. And then, by rolling back the algorithm, each constant can be computed by shifts and additions/subtractions, with roughly one addition/subtraction for each chosen occurrence of a pattern.

4.2 Extensions and enhancements to Lefèvre's algorithm

Mathematically speaking, Lefèvre's algorithm deals with the multiplication of a number by a constant vector. The first thing to do is to extend it for the multiplication of a vector by a constant matrix. This extension is rather straightforward: we simply replace each constant with a constant vector. Patterns are then replaced by vectors of patterns and shifts are done component-wise. With this algorithm, it is possible to share all kinds of expression.

For example, let us consider the computation of $A = 5x + 5y + z$ and $B = 5x + 5y + 4z$. The algorithm will first share the computation of $5x + 5y$ between A and B. After, it will share $x + y$ in $(5x + 5y) = 4(x+y) + (x+y)$, effectively sharing the multiplication by 5 between x and y.

One point is kept unspecified in Lefèvre's algorithm: which maximal pattern and which occurrences to choose? In his implementation, Lefèvre simply chose the first maximal pattern he found, with the first two occurrences. This solution is probably not the best, so we tried to find something better.

The first idea was to find all the maximal-weight patterns with at least two non-conflicting occurrences, and all their occurrences. And then, we try to choose a set of patterns and a set of, at least two, occurrences such that two chosen occurrences (of the same pattern or of different patterns) do not conflict. The choice is performed in order to maximize the gain in the weight of all the constants; with a constant with i occurrences, we gain $i - 1$ times its weight. As all the chosen patterns have the same maximal weight, we want to maximize the sum, for each pattern, of the number of occurrences diminished by one.

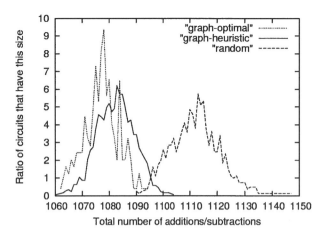

Figure 1. Heuristics Influence Distributions (from the optimization of many large 2D DCT operators).

We tried three different solutions for this. The first one, called "random", and which is the closest to the original algorithm, is to recursively choose, on random, a pattern with 2 non-conflicting occurrences, and to remove everything that conflicts with these occurrences. The second one, called "graph-heuristic", is to recursively choose a pattern with a maximal set of non-conflicting occurrences, and a minimal set of conflicts with the other patterns, and then remove everything that conflicts with these occurrences. And the third one, called "graph-optimal", is to build all the maximal sets of patterns and non-conflicting occurrences, and to choose the best one. This last solution can be very computationally intensive.

We tried to compare those three solutions, by running them several times for the same constant matrix (a huge 2D IDCT operator). The results in Figure 1 show that the "graph-optimal" and "graph-heuristic" are roughly equivalent, and better than the "random", with a tiny advantage to "graph-optimal". The time required to generate these results is less than one minute for "graph-heuristic" and "random", while it can grow to hours for "graph-optimal". Hence, we generally choose the "graph-heuristic" solution, so we can do lots of tries (thanks to its speed), and then choose the best solution.

4.3 Beyond the mathematical optimization

The improvements described above only deal with the minimization of the total number of additions and subtractions. Translated to hardware, this is not enough. Some additions and subtractions should be reordered without changing their total number, thanks to properties such as associativity and commutativity.

First of all, one may want to have a small circuit. When

three numbers a, b and c are added, the order in which they are added influences the size of the adders. For example, if a and b are narrow numbers, while c is wide, computing $(a + b) + c$ leads to a smaller circuit than $(a + c) + b$ or $(b + c) + a$. Hence, for space optimization, it is generally better to add the narrowest numbers first.

On the other hand, one may want to have a fast circuit. The order in which three numbers are added influences also the worst case delay of the circuit. For example, if a and c are available early while b is available late, the result is available earlier if we compute $(a + c) + b$ than $(a + b) + c$ or $(b+c)+a$. Hence, for speed optimization, it is preferable to add the earliest available values first.

Of course, those two kinds of optimization may often conflict. Hence, the user must specify if he or she prefers to optimize for speed or for area. Of course, if the user prefers to optimize speed, and if two solutions are equivalent for speed, the smallest one is chosen.

Moreover, the algorithmic optimization is not enough. We need to generate some real circuits. Hence, we decided to generate some VHDL code. Our VHDL code generator is currently targeted for Xilinx FPGAs. So additions and subtractions are performed on the dedicated carry-propagate adders and subtracters. The generator is able to produce VHDL code for fully parallel circuits, or for digit-serial circuits with radix 2^n for any n.

5 Implementation

Our implementation is mainly in two parts. The first part performs the mathematical optimization, with our extended and enhanced version of Lefèvre's algorithm. This part was written in C++, and is approximately 1500 line long. This part is not a program by itself, but a collection of simple classes that can be easily interfaced with any C or C++ program. Hence, it would be easy, for example, to interface this with a program that computes coefficients for FIR filters. Then the user could simply choose the type of filter and the pass and cut-off frequencies, and the program would compute the coefficients and generate some efficient VHDL code for it.

After the mathematical optimization, everything is implemented as plug-ins. Hence, there are, for example, plug-ins that optimize the order of the additions and subtractions, or plug-ins that generate the output VHDL code. This structure with plug-ins make the whole thing very modular. Hence, if someone wants, for example, to generate some Verilog code, or some assembly language code for a DSP, it is sufficient to write a new output plug-in. Then if someone wants to get pipelined circuits, he or she should write a new pipelining plug-in, and it can then be used with any output plug-in. Those plug-ins are also written in C++, but could be written in C or any language that can be interfaced with C. The collection of plug-ins is currently approximately 2500 line long.

The plug-ins communicate between themselves and the main program with simple interfaces that describe the circuit as a graph. In this representation, vertices represent mathematical values. Hence, there are vertices for input values, for output values, and also for intermediate values. Then there is an arc, from vertex x to vertex y, tagged with $(shift, sign, delay)$, if x shifted $shift$ bits to the left and delayed of $delay$ clock cycles is a positive or negative part (according to $sign$) of y. The $delay$ part will be useful for filters or for pipelined circuits. This representation has the quality of being independent from the desired output.

Let's give an simple example of how this can be used, and what is generated. As a simple example, we will consider building a constant multiplier by 111463. The corresponding source code and generated VHDL are presented Figure 2 and Figure 3 respectively.

```
#include "decompose.h"
int main(int argc, char **argv) {
  int value[] = {111463};
  Expr::initialize(argc, argv);
  CombinationSet<int> work(1);
  work.addLine(value);
  Key<int> *k;
  k = work.findBestPattern();
  while ((k!=NULL)&&(k->weight()>1)) {
    work.applyKey(*k, &cout);
    delete k;
    k = work.findBestPattern();
  }
  work.finish();
  return 0;
}
```

Figure 2. Multiplication by 111463 optimization source code.

6 Results and Comparisons

The synthesis have been performed using Xilinx ISE XST 4.2.03i tools for a Virtex XCV200-5 FPGA. The operators are not pipelined. The area is measured in number of slices (2 LUTs per slice in Virtex devices). The required area compared to the 2352 available slices in a XVC200 device is also reported in parentheses. The delay is expressed in nanoseconds. The number of additions/subtractions and the number of FA cells are computed by our generator (the two last lines of Figure 3); the number of FA cells is only an estimation .

Only a few papers give enough elements to compare our solutions. In [17] and [13], there are useful values for the

```vhdl
library IEEE;
  use IEEE.STD_LOGIC_1164.all;
  use IEEE.STD_LOGIC_arith.all;

entity MULT_111463 is
  port (
    x1 : in  std_logic_vector (7 downto 0);
    y1 : out std_logic_vector (24 downto 0)
  );
end MULT_111463;

architecture struct of MULT_111463 is
  signal t3 : std_logic_vector (22 downto 0);
  signal t2 : std_logic_vector (15 downto 0);
  signal t1 : std_logic_vector (22 downto 0);
begin -- struct
-- t3     = 20641*x1
   t3 <= ( t2(15)&t2(15)&t2(15 downto 0) & "00000" )
        +( t1(22 downto 0) );
-- t2     = 129*x1
   t2 <= ( x1(7)&x1(7)&x1(7)&x1(7)&x1(7)&x1(7)&x1(7)
          &x1(7)&x1(7 downto 0) )
        +( x1(7)&x1(7 downto 0) & "0000000" );
-- t1     = 16513*x1
   t1 <= ( t2(15)&t2(15)&t2(15)&t2(15)&t2(15)&t2(15)
          &t2(15)&t2(15 downto 0) )
        +( x1(7)&x1(7 downto 0) & "00000000000000" );
-- y1     = 111463*x1
   y1 <= ( t1(21 downto 0) & "000" )
        -( t3(22)&t3(22)&t3(22 downto 0) );
end struct;
-- Number of additions: 4
-- Estimated number of FA cells:  61
```

Figure 3. Multiplication by 111463 generated VHDL.

DCT application. Table 1 presents the number of additions/subtractions for the 1D 8-point DCT for several word sizes. Our generator improves the best previous results from 17% up to 44%. Table 2 gives the synthesis results for the corresponding generated operators.

operator	initial	[17]	[13]	our
DCT 8b	300	94	73	56
DCT 12b	368	100	84	70
DCT 16b	521	129	114	89
DCT 24b	789	212	—	119

Table 1. Number of additions/subtractions comparison for some 1D 8-point DCT operators.

We performed some other comparisons on some error-correcting codes from [17] and [13]: the 8×8 Hadamard matrix transform, (16, 11) Reed-Muller, (15, 7) BCH and (24, 12, 8) Golay codes. The comparison with the previous works in [17] and [13] is presented in Table 3 and the corresponding synthesis results are presented in Table 4. These results show that for very simple operators such as a small BCH code, some improvements are still possible. In the case of the 8×8 Hadamard matrix transform, we obtained the same results than the previous work [17], which results

operator	synthesis		generator	
	slices	delay	# ±	# FA
DCT 8b	401 (17%)	19.5	56	739
DCT 12b	647 (27%)	21.7	70	1202
DCT 16b	1085 (46%)	25.7	89	2009
DCT 24b	2106 (89%)	27.9	119	3934

Table 2. Synthesis results of some 1D 8-point DCT generated operators.

may be optimal.

operator	initial	[17]	[13]	our
8×8 Had.	56	24	—	24
(16, 11) R.-M.	61	43	31	31
(15, 7) BCH	72	48	47	44
(24, 12, 8) Golay	76	—	47	45

Table 3. Number of additions/subtractions comparison for some error-correction benchmarks.

operator	synthesis		generator	
	slices	delay	# ±	# FA
8×8 Had.	128 (5%)	11.9	24	240
(16, 11) R.-M.	39 (1%)	12.1	31	86
(15, 7) BCH	461 (19%)	18.2	44	861
(24, 12, 8) Golay	63 (2%)	12.2	45	136

Table 4. Synthesis results for some error-correction benchmarks.

Table 5 presents the synthesis results of the same operator with the three possible optimizations of our generator: none, area or speed. The operator is a 1D 8-point IDCT for 14-bit constants and 8-bit inputs. From the same initial addition/subtraction number the optimizations presented in Section 4.3 lead to significant improvements, 40% for the speed optimization for instance. The generation time for all these operators is around a few seconds on a standard desktop PC.

Some low-pass FIR filters used in [18] and [9] have been synthesized, these results are presented in Table 7. The specifications of the different filters are presented in Table 6. Their coefficients have been generated using the remez Matlab function: remez(#tap,[0 f_p f_s 1],[1 1 0 0]), and rounded to the target width. As the other authors do not specify the way they generate the coefficients, it is not possible to fairly compare our results. There are several function that use the Remez's algorithm in Matlab, and there are many options. Without more details it is not possible to

operator	synthesis		generator	
	slices	delay	# ±	# FA
IDCT	769 (32%)	30.2	81	1382
IDCT area	665 (28%)	19.8	81	1196
IDCT speed	666 (28%)	18.0	81	1241

Table 5. Influence of the generator optimizations on a 1D 8-point IDCT operator.

determine which sets of coefficients have been used in the previous papers.

Filter	f_p	f_s	#tap	width
FIR 1	0.15	0.25	60	14
FIR 2	0.15	0.20	60	16
FIR 3	0.10	0.15	60	14
FIR 4	0.10	0.12	100	18

Table 6. Low-pass FIR filters specifications.

operator	synthesis		generator	
	slices	delay	# ±	# FA
FIR 1	792 (33%)	29.1	86	1464
FIR 2	1075 (45%)	29.3	104	2021
FIR 3	792 (33%)	27.3	88	1509
FIR 4	1978 (84%)	34.9	173	3752

Table 7. Low-pass FIR filters synthesis results.

In Section 4.3, we said that our generator can produce digit-parallel as well as digit-serial circuits using different output plug-ins. Table 8 presents the synthesis results of a 1D 8-point IDCT operator for several solutions: digit-parallel and radix-2, 4, 8, 16, 64 and 256 digit-serial versions. Digit-serial implementations lead to small area and short cycle time operators. But in order to fairly compare digit-serial v.s. digit-parallel solution, we have to compare with pipelined parallel operator.

operator	slices	delay
parallel	614	40
serial radix-2	85	22
serial radix-4	153	36
serial radix-8	194	46
serial radix-16	242	47
serial radix-64	349	47
serial radix-256	446	48

Table 8. Synthesis results for 1D 8-point digit-parallel and digit-serial IDCT operators.

7 Conclusion

A new algorithm for the problem of multiplication by constant was presented. We generalized the previous results by dealing with the problem of the optimization of multiplication of one vector by one constant matrix. This algorithm is based on extensions and enhancements of previous algorithms [10] and [16]. Compared to the best previous results, our solution leads to a significant drop in the total number of additions/subtractions, up to 40%.

We implemented this algorithm in a VHDL generator. Based on a simple mathematical description of the computation, the generator produces an optimized VHDL code for Xilinx FPGAs. At the moment, the generated operators are non-pipelined parallel or digit-serial ones. We will extend our generator to produce pipelined circuits to reach higher clock frequencies.

We will also improve our algorithm in the case of operators that use different time samples of the same inputs such as filters. At the moment, the algorithm only factorizes some common sub-expressions without any temporal aspect.

We also want to extend our algorithm and generator to standard-cell based ASICs. The way to implement the adder/subtracter will widely impact the performance of the complete operator. The optimization required for low-power consumption may also change our solutions.

Another way to explore in the future, is the use of lossy representations such as [2]. In a lot of applications, the models include some approximations and the coefficients quantization. It may be a good idea to allow small perturbations of the coefficients.

Acknowledgments

The authors would like to thank the "Ministère Français de la Recherche" (grant # 1048 CDR 1 "ACI jeunes chercheurs") and the Xilinx University Program for their support.

References

[1] R. Bernstein. Multiplication by integer constants. *Software – Practice and Experience*, 16(7):641–652, July 1986.

[2] P. Boonyanant and S. Tantaratana. FIR filters with punctured radix-8 symmetric coefficients: Design and multiplier-free realizations. *Circuits Systems Signal Processing*, 21(4):345–367, 2002.

[3] A. D. Booth. A signed binary multiplication technique. *Quart. J. Mech. App. Math.*, IV(2):236–240, 1951.

[4] P. Briggs and T. Harvey. Multiplication by integer constants. Technical report, Rice University, 1994.

[5] K. D. Chapman. Fast integer multipliers fit in FPGAs. *EDN Magazine*, May 1994.

[6] V. S. Dimitrov, G. A. Jullien, and W. C. Miller. Theory and applications of the double-base number system. *IEEE Transactions on Computers*, 48(12):1098–1106, 1999.

[7] R. I. Hartley. Subexpression sharing in filters using canonic signed digit multipliers. *IEEE Transactions on Circuits and Systems—II: Analog and Digital Signal Processing*, 43(10):677–688, Oct. 1996.

[8] N. Homma, T. Aoki, and T. Higuchi. Evolutionary graph generation system with transmigration capability and its application to arithmetic circuit synthesis. *IEE Proceedings*, 149(2):97–104, Apr. 2002.

[9] H.-J. Kang and I.-C. Park. FIR filter synthesis algorithms for minimizing the delay and the number of adders. *IEEE Transactions on Circuits and Systems—II: Analog and Digital Signal Processing*, 48(8):770–777, Aug. 2001.

[10] V. Lefèvre. Multiplication par une constante. *Réseaux et Systèmes Répartis, Calculateurs Parallèles*, 13(4-5):465–484, 2001.

[11] D. J. Magenheimer, L. Peters, K. W. Pettis, and D. Zuras. Integer multiplication and division on the HP precision architecture. *IEEE Transactions on Computers*, 37(8):980–990, Aug. 1988.

[12] M. Martínez-Peiró, E. I. Boemo, and L. Wanhammar. Design of high-speed multiplierless filters using a nonrecursive signed common subexpression algorithm. *IEEE Transactions on Circuits and Systems—II: Analog and Digital Signal Processing*, 49(3):196–203, Mar. 2002.

[13] A. Matsuura, M. Yukishita, and A. Nagoya. A hierarchical clustering method for the multiple constant multiplication problem. *IEICE Transactions on Fundamentals of Electronics, Communications and Computer Sciences*, E80-A(10):1767–1773, Oct. 1997.

[14] M. F. Mellal and J.-M. Delosme. Multiplier optimization for small sets of coefficients. In *International Workshop Logic and Architecture Synthesis*, pages 13–22, Grenoble, France, Dec. 1997.

[15] H. T. Nguyen and A. Chatterjee. Number-splitting with shift-and-add decomposition for power and hardware optimization in linear DSP synthesis. *IEEE Transactions on Very Large Scale Integration (VLSI) Systems*, 8(4):419–424, Aug. 2000.

[16] R. Paško, P. Schaumont, V. Derudder, S. Vernalde, and D. Ďuračková. A new algorithm for elimination of common subexpressions. *IEEE Transactions on Computer-Aided Design of Integrated Circuits and Systems*, 18(1):58–68, Jan. 1999.

[17] M. Potkonjak, M. B. Srivastava, and A. P. Chandrakasan. Multiple constant multiplications: Efficient and versatile framework and algorithms for exploring common subexpression elimination. *IEEE Transactions on Computer-Aided Design of Integrated Circuits and Systems*, 15(2):151–165, Feb. 1996.

[18] H. Samueli. An improved search algorithm for the design of multiplierless FIR filters with power-of-two coefficients. *IEEE Transactions on Circuits and Systems*, 36(7):1044–1047, July 1989.

[19] S. Yu and E. E. Swartzlander. DCT implementation with distributed arithmetic. *IEEE Transactions on Computers*, 50(9):985–991, Sept. 2001.

A Less Recursive Variant of Karatsuba-Ofman Algorithm for Multiplying Operands of Size a Power of Two [*]

Serdar S. Erdem [†]
Gebze Yüksek Teknoloji Enstitüsü
Elektronik Mühendisliği Bölümü
Gebze 41400 Kocaeli, Turkey
erdem@gyte.edu.tr

Çetin K. Koç
Oregon State University
Electrical & Computer Engineering
Corvallis, Oregon 97331, USA
koc@ece.orst.edu

Abstract

We propose a new algorithm for fast multiplication of large integers having a precision of 2^k computer words, where k is an integer. The algorithm is derived from the Karatsuba-Ofman Algorithm and has the same asymptotic complexity. However, the running time of the new algorithm is slightly better, and it makes one third as many recursive calls.

1 Introduction

Multi-precision integer arithmetic is used in many applications, including cryptography. Efficient software implementations of multi-precision operations are nededed for several public-key cryptographic systems, for example, RSA, Diffie-Hellman, and Elliptic Curve Digital Signature Algorithms [10, 2, 4, 9]. Among the arithmetic operations, the multi-precision multiplication is one of the most time consuming operations with its $\mathcal{O}(n^2)$ complexity. The Karatsuba-Ofman Algorithm (KOA) is a fast multiplication algorithm for multi-precision numbers with $\mathcal{O}(n^{1.58})$ asymptotic complexity [5, 6, 7]. We modify this algorithm and obtain a less recursive algorithm. However, our algorithm works only if the operand size is a power of two in computer words, bytes, digits, etc. In this paper, we describe KOA, the new algorithm, and give their analyses. The detailed proofs of the analyses are omitted in this paper for brevity, and can be found in [3]. We also give an example of multiplication using the new algorithm and the results of our implementations comparing KOA and the new algorithm.

[*]The reader should note that Oregon State University has filed US and International patent applications for inventions described in this paper.
[†]This work was performed while the first author was with Oregon State University.

2 Multi-Precision Numbers and Operations

In this paper, the variables in bold face denote multi-precision numbers. Let $\mathbf{a}$ be an n-digit number represented in base z. We denote the digits of $\mathbf{a}$ from the most significant to least significant by $\mathbf{a}[n-1], \mathbf{a}[n-2], \cdots, \mathbf{a}[0]$, i.e.,

$$\mathbf{a} = \mathbf{a}[n-1]z^{n-1} + \cdots + \mathbf{a}[1]z + \mathbf{a}[0] \ .$$

Also, $\mathbf{a}^l[k]$ denotes an l-digit number whose jth digit is $\mathbf{a}[k+j]$, i.e.,

$$\mathbf{a}^l[k] = \mathbf{a}[k+l-1]z^{l-1} + \cdots + \mathbf{a}[k+1]z + \mathbf{a}[k] \ .$$

We use the following operations on multi-digit numbers:

- The addition or subtraction of two n-digit numbers produces another n-digit number and an extra bit. This extra bit is a carry bit for addition or a borrow (sign) bit for subtraction. Multi-precision addition and subtraction are relatively easy operations. For further details and implementation, refer to [6, 8].

- Because $z = 2^w$, multiplying a number with z^i is equivalent to shifting the words in its array representation by i positions. The jth word becomes the $(i+j)$th word and the 0th through $(i-1)$th words are filled with zeros.

- We can assign a value to the subarray of a number. The assignment $\mathbf{a}^l[k] := \mathbf{b}$ overwrites the digits of $\mathbf{a}$ in our notation. The digits $\mathbf{a}[k+i]$ for $i = 0, \cdots, l-1$ are replaced with the digits $\mathbf{b}[i]$ for $i = 0, \cdots, l-1$.

We can also define more complex operations for multidigit numbers using our notation. For example, the operation

$$(c, \mathbf{t}^l[k]) := \mathbf{a}^l[k'] + \mathbf{b}^l[k'']$$

adds the l-digit numbers $\mathbf{a}^l[k']$ and $\mathbf{b}^l[k'']$ derived from $\mathbf{a}$ and $\mathbf{b}$. It then stores the result in $\mathbf{t}^l[k]$ and the carry bit in c. More explicitly, the following code segment is performed:

$$c := 0$$
$$\text{for } i = 0 \text{ to } l - 1$$
$$(c, \mathbf{t}[k+i]) := \mathbf{a}[k'+i] + \mathbf{b}[k''+i] + c$$
$$\text{endfor}$$

3 Karatsuba-Ofman Algorithm (KOA)

The classical multi-precision multiplication algorithm multiplies every digit of a multiplicand by every digit of the multiplier and adds the result to the partial product. It has $\mathcal{O}(n^2)$ complexity, where n is the operand size (number of digits). KOA is an alternative multi-precision multiplication method [5]. KOA has $\mathcal{O}(n^{1.58})$ complexity and thus it multiplies large numbers faster than the classical method. KOA is a recursive algorithm and follows a divide and conquer strategy.

Let $\mathbf{a}$ and $\mathbf{b}$ be two n-digit numbers in radix z where n is even. We can split them in two parts as

$$\mathbf{a} = \mathbf{a_L} + \mathbf{a_H} z^{n/2} \ , \ \mathbf{b} = \mathbf{b_L} + \mathbf{b_H} z^{n/2} \ ,$$

where $\mathbf{a_L} = \mathbf{a}^{n/2}[0]$, $\mathbf{b_L} = \mathbf{b}^{n/2}[0]$, $\mathbf{a_H} = \mathbf{a}^{n/2}[n/2]$, and $\mathbf{b_H} = \mathbf{b}^{n/2}[n/2]$. This means $\mathbf{a_L}$ and $\mathbf{b_L}$ are the numbers represented by the low order digits (the first $n/2$ digits), while $\mathbf{a_H}$ and $\mathbf{b_H}$ are the numbers represented by the high order digits (the last $n/2$ digits). We can write $\mathbf{t} = \mathbf{a} \cdot \mathbf{b}$ in terms of the half-sized numbers $\mathbf{a_L}$, $\mathbf{b_L}$, $\mathbf{a_H}$, and $\mathbf{b_H}$ as

$$\begin{aligned}\mathbf{t} &= \mathbf{a} \cdot \mathbf{b} \\ &= (\mathbf{a_L} + \mathbf{a_H} z^{n/2})(\mathbf{b_L} + \mathbf{b_H} z^{n/2}) \\ &= \mathbf{a_L b_L} + (\mathbf{a_L b_H} + \mathbf{a_H b_L})z^{n/2} + \mathbf{a_H b_H} z^n \ .\end{aligned}$$

Thus, we can compute the product $\mathbf{t}$ from 4 half-sized products $\mathbf{a_L b_L}$, $\mathbf{a_L b_H}$, $\mathbf{a_H b_L}$, and $\mathbf{a_H b_H}$. On the other hand, following the idea of KOA, we can use the equality

$$\mathbf{a_L b_H} + \mathbf{a_H b_L} = \mathbf{a_L b_L} + \mathbf{a_H b_H} + (\mathbf{a_L} - \mathbf{a_H})(\mathbf{b_H} - \mathbf{b_L})$$

in the above equation and obtain

$$\begin{aligned}\mathbf{t} &= \mathbf{a_L b_L} + [\mathbf{a_L b_L} + \mathbf{a_H b_H} + \\ & \quad (\mathbf{a_L} - \mathbf{a_H})(\mathbf{b_H} - \mathbf{b_L})]z^{n/2} + \mathbf{a_H b_H} z^n \ . \quad (1)\end{aligned}$$

The equation above shows that only 3 half-sized multiplications are sufficient to compute $\mathbf{t}$ instead of 4. These products are $\mathbf{a_L b_L}$, $\mathbf{a_H b_H}$ and $(\mathbf{a_L} - \mathbf{a_H})(\mathbf{b_H} - \mathbf{b_L})$. We obtain this decrease in the number of products at the expense of more additions and subtractions.

KOA computes a product from 3 half-sized products using Eq. (1). In the same fashion, KOA computes each of these half-sized products from 3 quarter-sized products. This process goes recursively. When the products get very small (for example, when their operands reduce to one digit), the recursion stops and these small products are computed by the classical method.

The following recursive function implements KOA. We assume that the inputs can be split into lower and higher order digits evenly in each recursion. As a consequence, the input size n is required to be a power of two. Of course, one can also write a general KOA function which splits its inputs approximately when the input size is an odd number.

function: KOA($\mathbf{a}, \mathbf{b}$: n-word number; n : integer)
$\mathbf{t}$: $2n$-digit number
$\mathbf{a_L}, \mathbf{a_M}, \mathbf{a_H}$: $(n/2)$-digit number
low, mid, high : n-digit number
/*** When the input size is one digit ***/
Step 1:   if $n = 1$ then return $\mathbf{t} := \mathbf{a}[0] \cdot \mathbf{b}[0]$
/*** Generate 3 pairs of half-sized numbers ***/
Step 2: $\mathbf{a_L} := \mathbf{a}^{n/2}[0]$
Step 3: $\mathbf{b_L} := \mathbf{b}^{n/2}[0]$
Step 4: $\mathbf{a_H} := \mathbf{a}^{n/2}[n/2]$
Step 5: $\mathbf{b_H} := \mathbf{b}^{n/2}[n/2]$
Step 6: $(s_a, \mathbf{a_M}) := \mathbf{a_L} - \mathbf{a_H}$
Step 7: $(s_b, \mathbf{b_M}) := \mathbf{b_H} - \mathbf{b_L}$
/*** Multiply the half-sized numbers ***/
Step 8:   low := $KOA(\mathbf{a_L}, \mathbf{b_L}, n/2)$
Step 9:   high := $KOA(\mathbf{a_H}, \mathbf{b_H}, n/2)$
Step 10: mid := $KOA(\mathbf{a_M}, \mathbf{b_M}, n/2)$
/*** Combine the subproducts ***/
Step 11: $\mathbf{t}$:= low + (low + high + $s_a s_b$ mid)$z^{n/2}$ + highz^n
Step 12: return $\mathbf{t}$

In Step 1, we check if $n = 1$. If the input operands are 1-digit, we multiply the inputs and return the result. If not, we continue with the remaining steps. In Steps 2 through 5, $(n/2)$-digit numbers $\mathbf{a_L}$, $\mathbf{b_L}$, $\mathbf{a_H}$ and $\mathbf{b_H}$ are generated from the lower and higher order digits of the inputs. In Steps 6 and 7, we obtain $\mathbf{a_M}$, $\mathbf{b_M}$, s_a and s_b using the subtraction operations as described below

$$\begin{aligned} s_a &= sign(\mathbf{a_L} - \mathbf{a_H}) \ , \ \mathbf{a_M} = |\mathbf{a_L} - \mathbf{a_H}| \ , \\ s_b &= sign(\mathbf{b_H} - \mathbf{b_L}) \ , \ \mathbf{b_M} = |\mathbf{b_H} - \mathbf{b_L}| \ . \end{aligned}$$

The terms $\mathbf{a_M}$, $\mathbf{b_M}$, s_a and s_b are the magnitudes and the signs of the results of the subtractions in Steps 6 and 7. Clearly, $\mathbf{a_M}$ and $\mathbf{b_M}$ are $n/2$ digits as $\mathbf{a_L}$, $\mathbf{b_L}$, $\mathbf{a_H}$ and $\mathbf{b_H}$. In Steps 8, 9 and 10, we multiply these $n/2$-digit numbers by recursive calls. Here we have

$$\begin{aligned}\text{low} &= \mathbf{a_L b_L} \ , \\ \text{high} &= \mathbf{a_H b_H} \ , \\ \text{mid} &= |\mathbf{a_L} - \mathbf{a_H}||\mathbf{b_H} - \mathbf{b_L}| \ . \end{aligned}$$

Finally, in Step 11, we find the product $t = a \cdot b$ using Eq. (1). We substitute **low** into $a_L b_L$, **high** into $a_H b_H$ and $s_a s_b$**mid** into $(a_L - a_H)(b_H - b_L)$. The last subsitution is due to the fact that

$$s_a s_b \mathbf{mid} = (s_a |a_L - a_H|)(s_b |b_H - b_L|)$$
$$= (a_L - a_H)(b_H - b_L) .$$

4 Efficient Implementation of KOA

In the previous section, we presented a naive implementation of KOA in order to illustrate the algorithm. Here, we present an efficient implementation of KOA which is more suitable for computer arithmetic.

function: $KOA(\mathbf{a}, \mathbf{b} : n\text{-digit number}; n : \text{integer})$
$\mathbf{t} : 2n$-digit number
$\mathbf{a_L}, \mathbf{a_M}, \mathbf{a_H} : (n/2)$-digit number
$\mathbf{mid} : n$-digit number
/*** When the input size is one digit ***/
Step 1:   if $n = 1$ then return $\mathbf{t} := \mathbf{a} \cdot \mathbf{b}$
/*** Generate 3 pairs of half-sized numbers ***/
Step 2: $\mathbf{a_L} := \mathbf{a}^{n/2}[0]$
Step 3: $\mathbf{b_L} := \mathbf{b}^{n/2}[0]$
Step 4: $\mathbf{a_H} := \mathbf{a}^{n/2}[n/2]$
Step 5: $\mathbf{b_H} := \mathbf{b}^{n/2}[n/2]$
Step 6: $(s_a, \mathbf{a_M}) := \mathbf{a_L} - \mathbf{a_H}$
Step 7: $(s_b, \mathbf{b_M}) := \mathbf{b_H} - \mathbf{b_L}$
/*** Multiply the half-sized numbers ***/
Step 8: $\mathbf{t}^n[0] := KOA(\mathbf{a_L}, \mathbf{b_L}, n/2)$
Step 9: $\mathbf{t}^n[n] := KOA(\mathbf{a_H}, \mathbf{b_H}, n/2)$
Step 10: $\mathbf{mid} := KOA(\mathbf{a_M}, \mathbf{b_M}, n/2)$
/*** Combine the subproducts ***/
Step 11a: if $s_a = s_b$ then
          $(c, \mathbf{mid}) := \mathbf{t}^n[0] + \mathbf{t}^n[n] + \mathbf{mid}$
          else
          $(c, \mathbf{mid}) := \mathbf{t}^n[0] + \mathbf{t}^n[n] - \mathbf{mid}$
Step 11b: $(c', \mathbf{t}^n[n/2]) := \mathbf{t}^n[n/2] + \mathbf{mid}$
Step 11c: $\mathbf{t}^{n/2}[3n/2] := \mathbf{t}^{n/2}[3n/2] + c' + c$
Step 12: return $\mathbf{t}$

This new implementation first differs from the previous one in Steps 8 and 9. The product $a_L b_L$ and $a_H b_H$ are respectively stored into the lower and the higher halves of $\mathbf{t}$, i.e., $\mathbf{t}^n[0]$ and $\mathbf{t}^n[n]$, instead of usingthe variables **low** and **high**. It is clear that Steps 8 and 9 give

$$\mathbf{t} = \mathbf{low} + \mathbf{high} z^n = a_L b_L + a_H b_H z^n .$$

The result above is a part of the computation performed in Step 11. Thus, with the help of Steps 8 and 9, we save some storage space in Step 11, since we do not use the variables **low** and **high**. Step 11 is accomplished in three substeps. We compute $a_L b_L + a_H b_L = a_L b_L + a_H b_H + s_a s_b |a_L - a_H||b_L - b_H|$ in Step 11a. We store the result into n-digit variable **mid** and 1-bit carry into c. For this computation, we add $\mathbf{t}^n[0]$ and $\mathbf{t}^n[n]$ containing $a_L b_L$ and $a_H b_H$. Also, if $s_a = s_b$, we add $\mathbf{mid} = |a_L - a_H||b_L - b_H|$ to the sum, if not, we subtract it from the sum. This is because if $s_a = s_b$, we have $s_a s_b = 1$, otherwise, $s_a s_b = -1$. In Step 11b and 11c, we perform the computation $\mathbf{t} = \mathbf{t} + (c, \mathbf{mid}) z^{n/2}$. Because $\mathbf{t} = a_L b_L + a_H b_H z^n$ and $(c, \mathbf{mid}) = a_L b_L + a_H b_H + s_a s_b |a_L - a_H||b_L - b_H|$, the computations in Steps 11a, 11b, and 11c are equivalent to Step 11 of KOA implementation in § 3.

5 Complexity of KOA

KOA function contains several multi-digit additions and subtractions. The operands of these operations need to be read from the memory and their results need to be written back to the memory. We take the memory read and write operations into account in addition to the arithmetic operations. An n-digit addition or subtraction requires $2n$-digit memory read and n-digit memory write operations. Table 1 gives the number of arithmetic and read/write operations in KOA function.

Steps	Operation	Read	Write
6, 7	n	$2n$	n
8, 9, 10	recursions		
11a	$2n$	$4n$	$2n$
11b	n	$2n$	n
Total	$4n$	$8n$	$4n$

Table 1. The complexity of KOA with $n > 1$.

We do not perform any computations in Steps 2 through 5, because a_L, a_H, b_L and b_H are just the copies of the lower and higher halves of the inputs. In practice, we can avoid the copy operations by using pointers for the lower and higher halves of the inputs.

Also, we view Step 11c as a single digit addition and neglect its cost. This is because we assume that the addition of a multi-digit number with a carry only affects the least significant digit of the number and does not cause a carry propagation through the higher order digits. We can justify this assumption in software implementations where a digit is usually stored into a 32-bit word, i.e., the base $z = 2^{32}$. Adding a carry to a digit produces another carry with $1/z = 2^{-32}$ probability.

Let $T(n)$ denote the complexity of KOA function. It

can be given as

$$T(n) = 3T(n/2) + 4n + 8n + 4n$$
$$= 3T(n/2) + 16n. \quad (2)$$

The solution if this recurrence is the asympotic complexity $T(n) = \mathcal{O}(n^{1.58})$, see, for example, [1].

We are also interested in computing the total number of recursive calls made in KOA. Let $R(n)$ denote the number of recursive calls with input size $n = 2^k$, where k is an integer. The initial call makes 3 recursive calls with $n/2$-digit inputs. These 3 recursive calls each leads to $R(n/2)$ recursions. Thus, we have the recursion

$$R(n) = 3 + 3R(n/2). \quad (3)$$

Taking $R(1) = 1$, we find the solution of this recursion easily as

$$R(n) = 3 + 9 + \ldots + 3^k + 3^k = 3(3^k - 1)/2.$$

6 New Algorithm KOA2^k

In this section, we present a new algorithm derived from KOA to multiply numbers of size a power of two in digits. We name this algorithm as KOA2^k due to the restriction in its input size. Let $\mathbf{a}$ and $\mathbf{b}$ be the input operands to be multiplied.

Let $\mathbf{a}$ and $\mathbf{b}$ be two n-digit numbers, and k be a positive integer such that 2^k divides n. We define

$$\mathbf{sumP}_k = \sum_{i=0}^{2^k-1} \mathbf{P}_{k,i} z^{i(n/2^k)},$$

where $\mathbf{P}_{k,i} = \mathbf{a}^m[im]\mathbf{b}^m[im]$ and $m = n/2^k$. It is clear that if 2^k divides n, then

$$\mathbf{sumP}_k, \mathbf{sumP}_{k-1}, \cdots, \mathbf{sumP}_1, \mathbf{sumP}_0$$

are all defined. The last term $\mathbf{sumP}_0$ is the most important one, since

$$\mathbf{sumP}_0 = \sum_{i=0}^{2^0-1} \mathbf{P}_{0,i} z^{i(n/2^0)} = \mathbf{P}_{0,0} = \mathbf{a} \cdot \mathbf{b}.$$

The goal of KOA2^k is to find $\mathbf{sumP}_0$ which is equal to the product $\mathbf{a} \cdot \mathbf{b}$. The outline of KOA$2^k$ is given below.

- Restrict the operand size n to a power of 2. The recursion depth is $\log_2 n$. Furthermore, $\mathbf{sumP}_k$ is defined for all recursion levels k from 0 to $\log_2 n$.

- Compute $\mathbf{sumP}_{\log_2 n}$ in terms of the operands. We show how to accomplish this step in Proposition 1.

- Compute $\mathbf{sumP}_{k-1}$ from $\mathbf{sumP}_k$ iteratively to obtain $\mathbf{sumP}_0$, which is the final result. We give the iteration relation in Proposition 2

- During the computations, the term $\mathbf{sumP}_k$ needs to be stored. The size of this multi-digit number is given in Proposition 3.

We now give 3 propositions whose proofs are given in [3].

Proposition 1 Let $\mathbf{a}$ and $\mathbf{b}$ be two n-digit numbers where $n = 2^{k_0}$ for some integer k_0. We have

$$\mathbf{sumP}_{\log_2 n} = \mathbf{sumP}_{k_0} = \sum_{i=0}^{n-1} \mathbf{a}[i] \cdot \mathbf{b}[i] z^i.$$

$\square$

Proposition 2 Let $\mathbf{a}$ and $\mathbf{b}$ be two n-digit numbers such that 2^k divides n for some integer $k \geq 0$. Then, the term $\mathbf{sumP}_{k-1}$ is related to $\mathbf{sumP}_k$ in the following way:

$$\mathbf{sumP}_{k-1} = (1 + z^m)\mathbf{sumP}_k +$$
$$\sum_{i=0}^{2^{k-1}-1} s_a(i) s_b(i) \mathbf{mid}(i) z^{(2i+1)m},$$

where $m = n/2^k$ and

$$\mathbf{mid}(i) = |\mathbf{a}^m[2im] - \mathbf{a}^m[(2i+1)m]|$$
$$|\mathbf{b}^m[(2i+1)m] - \mathbf{b}^m[2im]|,$$
$$s_a(i) = sign(\mathbf{a}^m[2im] - \mathbf{a}^m[(2i+1)m]),$$
$$s_b(i) = sign(\mathbf{b}^m[(2i+1)m] - \mathbf{b}^m[2im]).$$

$\square$

Proposition 3 Let $\mathbf{a}$ and $\mathbf{b}$ be two n-digit numbers such that 2^k divides n for some integer $k \geq 0$. Then, the term $\mathbf{sumP}_k$ is of $n + m$ words where $m = n/2^k$. $\square$

The discussion suggests a new algorithm in which the input size n must be a power of two. The algorithm computes $\mathbf{t} = \mathbf{sumP}_k$ iteratively, until $\mathbf{t} = \mathbf{sumP}_0$ is obtained.

function: KOA2^k($\mathbf{a}, \mathbf{b}$: n-digit number; n : integer)
$\mathbf{t}$: $2n$-digit number
m : integer
$\mathbf{a_M}$: m-digit number /*** $\max(m) = n/2$ ***/
$\mathbf{mid}$: $2m$-digit number
/*** When the input size is one digit ***/
Step 1:   if $n = 1$ then return $\mathbf{a}[0] \cdot \mathbf{b}[0]$
/*** Compute $\mathbf{sumP}_{\log_2 n}$ ***/
Step 2: $\mathbf{t} := \sum_{i=0}^{n-1} \mathbf{a}[i] * \mathbf{b}[i] z^i$
/*** Compute $\mathbf{sumP}_{k-1}$ ***/
          for $k = \log_2 n$ downto 1

Step 3: $\quad m = n/2^k$
$\quad\quad\quad \mathbf{t} := \mathbf{t}(1 + z^m)$
$\quad\quad\quad$ for $i = 0$ to $2^{k-1} - 1$
Step 4: $\quad\quad (s_a, \mathbf{a_M}) := \mathbf{a}^m[2im] - \mathbf{a}^m[(2i+1)m]$
Step 5: $\quad\quad (s_b, \mathbf{b_M}) := \mathbf{b}^m[(2i+1)m] - \mathbf{b}^m[2im]$
Step 6: $\quad\quad \mathbf{mid} := \text{KOA}2^k(\mathbf{a_M}, \mathbf{b_M}, m)$
Step 7: $\quad\quad \mathbf{t} := \mathbf{t} + s_a s_b \mathbf{mid} z^{(2i+1)m}$
$\quad\quad\quad$ endfor
$\quad\quad$ endfor
Step 8: $\quad$ return $\mathbf{t}$

7 Efficient Implementation of KOA2^k

In the previous section, we presented a naive implementation of KOA2^k in order to illustrate its properties. In this section, we present an efficient implementation which is more suitable for computer arithmetic. The algorithm computes $\mathbf{sumP}_k$ and stores it into the digits of $\mathbf{t}$ from $\mathbf{t}[\alpha]$ to $\mathbf{t}[2n-1]$ such that $\mathbf{t}[\alpha + i] = \mathbf{sumP}_k[i]$. Since $\mathbf{sumP}_k$ is of $n+m$ digits, we have $\alpha = 2n - (n+m) = n - m$. When $k = 0$, we have $\mathbf{sumP}_k = \mathbf{sumP}_0$, $m = n/2^k = n$ and $\alpha = n - m = 0$. The algorithm computes $\mathbf{sumP}_0 = \mathbf{a} \cdot \mathbf{b}$ and stores it to the digits from $\mathbf{t}[0]$ to $\mathbf{t}[2n-1]$.

function: $KOA2^k(\mathbf{a}, \mathbf{b} : n\text{-digit number}; n : \text{integer})$
$\mathbf{t}$: $2n$-digit number
α, m : integer
$\mathbf{a_M}$: m-digit number /*** $\max(m) = n/2$ ***/
$\mathbf{mid}$: $2m$-digit number
/*** When the input size is 1 digit ***/
Step 1: $\quad$ if $n = 1$ then return $\mathbf{a}[0] \cdot \mathbf{b}[0]$
/*** Compute $\mathbf{sumP}_{\log_2 n}$ ***/
$\quad\quad \alpha := n - 1$
Step 2a: $(\mathbf{C}, \mathbf{S}) := \mathbf{a}[0] \cdot \mathbf{b}[0]$
Step 2b: $\mathbf{t}[\alpha] := \mathbf{S}$
$\quad\quad$ for $i = 1$ to $n - 1$
Step 2c: $(\mathbf{C}, \mathbf{S}) := \mathbf{a}[i] \cdot \mathbf{b}[i] + \mathbf{C}$
Step 2d: $\mathbf{t}[\alpha + i] := \mathbf{S}$
$\quad\quad$ endfor
Step 2e: $\mathbf{t}[\alpha + n] := \mathbf{C}$
/*** Compute $\mathbf{sumP}_{k-1}$ ***/
$\quad\quad$ for $k = \log_2 n$ downto 1
$\quad\quad\quad m = n/2^k \quad \alpha = n - m$
Step 3a: $\mathbf{t}^m[\alpha - m] := \mathbf{t}^m[\alpha]$
Step 3b: $(c, \mathbf{t}^n[\alpha]) := \mathbf{t}^n[\alpha] + \mathbf{t}^n[\alpha + m]$
Step 3c: $\mathbf{t}^m[\alpha + n] := \mathbf{t}^m[\alpha + n] + c$
$\quad\quad\quad$ for $i = 0$ to $2^{k-1} - 1$
Step 4: $\quad (s_a, \mathbf{a_M}) := \mathbf{a}^m[2im] - \mathbf{a}^m[(2i+1)m]$
Step 5: $\quad (s_b, \mathbf{b_M}) := \mathbf{b}^m[(2i+1)m] - \mathbf{b}^m[2im]$
Step 6: $\quad \mathbf{mid} := KOA2^k(\mathbf{a_M}, \mathbf{b_M}, m)$
$\quad\quad\quad$ if $s_a = s_b$ then
Step 7a: $\quad (c, \mathbf{t}^{2m}[\alpha + 2im]) := \mathbf{t}^{2m}[\alpha + 2im] + \mathbf{mid}$
Step 7b: $\quad \text{propagate}(\mathbf{t}[\alpha + 2im + 2m], c)$
$\quad\quad\quad$ else
Step 7c: $\quad (b, \mathbf{t}^{2m}[\alpha + 2im]) := \mathbf{t}^{2m}[\alpha + 2im] - \mathbf{mid}$
Step 7d: $\quad \text{propagate}(\mathbf{t}[\alpha + 2im + 2m], b)$
$\quad\quad\quad$ endfor
$\quad\quad$ endfor
Step 8: $\quad$ return $\mathbf{t}$

In Step 1, we multiply the inputs and return the result if $n = 1$. Otherwise, we continue with the remaining steps. The steps in this new implementation correspond to the steps in the previous implementation, however, they are divided into substeps. In Step 2, we compute

$$\mathbf{sumP}_{\log_2 n} = \sum_{i=0}^{n-1} \mathbf{a}[i] \cdot \mathbf{b}[i] z^i \ .$$

The result is stored into the digits of $\mathbf{t}$ from $\mathbf{t}[\alpha]$ to $\mathbf{t}[2n-1]$. Since $k = \log_2 n$, we have $m = n/2^k = 1$ and $\alpha = n - m = n - 1$. The product $\mathbf{a}[i] \cdot \mathbf{b}[i]$ for $i = 0, \cdots, n-1$ yields the two-digit result $(\mathbf{C}, \mathbf{S})$ such that $\mathbf{C}$ and $\mathbf{S}$ are the most and least significant digits, respectively. Since $\mathbf{a}[i] \cdot \mathbf{b}[i]$ is multiplied with z^i, we add $\mathbf{S}$ to $\mathbf{t}[\alpha + i]$ and $\mathbf{C}$ to $\mathbf{t}[\alpha + i + 1]$.

In Steps 3 to 7, we obtain $\mathbf{sumP}_{k-1}$ from $\mathbf{sumP}_k$. These steps are in a loop running from $k = \log_2 n$ to $k = 1$. Inside the loop, we have $m = n/2^k$ and $\alpha = n - m$.

When Step 3 starts, the digits of $\mathbf{t}$ from $\mathbf{t}[\alpha]$ to $\mathbf{t}[2n-1]$ represent $\mathbf{sumP}_k$. In Step 3, we add $\mathbf{sumP}_k$ to the m-digit shifted copy of itself to find $(1+z^m)\mathbf{sumP}_k$, and then, store the result into the digits $\mathbf{t}[\alpha - m]$ to $\mathbf{t}[2n-1]$.

The magnitudes and the signs of the result of the subtractions in Steps 4 and 5 are $\mathbf{a_M}$, $\mathbf{b_M}$, $sign_a$, and $sign_b$. Here $\mathbf{a_M}$ and $\mathbf{b_M}$ are m-digit numbers. We multiply them by a recursive call and obtain the $2m$-digit number $\mathbf{mid}$ in Step 6.

When Step 7 starts, the digits of $\mathbf{t}$ from $\mathbf{t}[\alpha - m]$ to $\mathbf{t}[2n-1]$ represent the multi-digit number $(1+z^m)\mathbf{sumP}_k$. In Step 7, we add $s_a s_b \mathbf{mid} z^{(2i+1)m}$ to this number. If $s_a = s_b$ and $s_a s_b = 1$, we add $\mathbf{mid} z^{(2i+1)m}$, following Steps 7a and 7b. Otherwise if $s_a s_b = -1$, we subtract $\mathbf{mid} z^{(2i+1)m}$, following Steps 7c and 7d. Since $\mathbf{mid}$ is multiplied by $z^{(2i+1)m}$, the least significant digit of $\mathbf{t}$ involving the operations in Step 7 is $\mathbf{t}[\alpha - m + (2i+1)m] = \mathbf{t}[\alpha + 2im]$. We add (subtract) $\mathbf{mid}$ to (from) the consecutive $2m$ digits of $\mathbf{t}$ in Step 7a (7c), starting from $\mathbf{t}[\alpha + 2im]$. Then, we propagate the resulting carry (borrow) through the higher order digits of $\mathbf{t}$ in Step 7b (7d), starting from $\mathbf{t}[\alpha + 2im + 2m]$. The function $\text{propagate}(t[k], c)$ is given as follows:

$\quad$ while$(c > 0)$
$\quad\quad (c, \mathbf{t}[k]) := \mathbf{t}[k] + c$
$\quad\quad k := k + 1$

The function propagate($t[k], c$) adds (subtracts) a carry (borrow) to (from) the kth digit of t and propagates it through the higher order digits.

8 Complexity of KOA2^k

A detailed (step by step) complexity analysis of KOA2^k function is performed in [3], and the results are summarized in Table 2 below. We neglect the cost of addition with a single carry and subtraction with a single borrow. Thus, Steps 3c, 7b and 7d do not take place in Table 2.

Steps	Operation	Read	Write
		$nT(1)$	
		$2(n-1)$	$n-1$
3a		$n-1$	$n-1$
3b	$n \log_2 n$	$2n \log_2 n$	$n \log_2 n$
4	$\frac{n}{2} \log_2 n$	$n \log_2 n$	$\frac{n}{2} \log_2 n$
5	$\frac{n}{2} \log_2 n$	$n \log_2 n$	$\frac{n}{2} \log_2 n$
6	recursions		
7a,7c	$n \log_2 n$	$2n \log_2 n$	$n \log_2 n$
Total	$nT(1) + 12n \log_2 n + 5n - 5$		

Table 2. The complexity of KOA2^k with $n > 1$.

The single digit multiplications in Step 2, $\mathbf{a}[i] \cdot \mathbf{b}[i]$ for $i = 0, \cdots, n-1$, cost $nT(1)$ where $T(1)$ denotes the cost of multiplying two digits, including the cost of reading the operands and writing the result. Also, a single digit read and a 2-digit addition are performed in Step 2c in order to read C and add it to $\mathbf{a}[i] \cdot \mathbf{b}[i]$ in a loop iterating $n-1$ times. Thus, we have $2(n-1)$ additions and $n-1$ reads in Step 2c.

We have $m = n/2^k$ assignments in a loop iterating from $k = \log_2 n$ to 1 in Step 3a. This makes a total of $\sum_{k=1}^{\log_2 n} (n/2^k) = n-1$ assignments. We also add the n-digit numbers in the same loop in Step 3b, which costs a total of $n \log_2 n$ additions.

Steps 4 to 7 are in two loops. The outer loop iterates $\log_2 n$ times while the inner loop iterates 2^{k-1} times. Steps 4 and 5 perform operations on m-digit numbers. Thus, $m 2^{k-1} \log_2 n = (n/2) \log_2 n$ operations are needed to perform in Steps 4 and 5 each. Step 7 performs operations on $2m$-digit numbers. Thus, we perform $2m 2^{k-1} \log_2 n = n \log_2 n$ operations in Step 7.

Step 6 makes a recursive call with m-digit input and is embedded in two loops: The inner loop iterates 2^{k-1} times, while the outer loop iterates from $k = \log_2 n$ to 1. Therefore, the complexity of $KOA2^k$ function, denoted as $T(n)$, can be given as

$$T(n) = \sum_{k=1}^{\log_2 n} 2^{k-1} T(n/2^k) + Total(n),$$

where $Total(n)$ is the number operations, reads and writes given in the last row of Table 2, which is equal to

$$Total(n) = nT(1) + 12n \log_2 n + 5n - 5.$$

As shown in [3], the above recursion can be simplified as

$$T(n) = 3T(n/2) + 12n + 5. \quad (4)$$

This recurrence is similar to the recurrence in Eq. (2). The asympotic complexity of KOA2^k is also $\mathcal{O}(n^{1.58})$ as KOA. However, since $12n + 5 < 16n$ for $n > 2$, the running time of KOA2^k is better than KOA, i.e., the constant in front of the order is smaller.

Similarly, we compute the number of recursive calls made by KOA2^k. It makes 2^{k-1} recursive calls with m-digit inputs in a loop iterating from $k = \log_2 n$ to 1 in Step 6. Thus, we have the following recurrence:

$$R(n) = \sum_{k=1}^{\log_2 n} 2^{k-1} + \sum_{k=1}^{\log_2 n} 2^{k-1} R(n/2^k)$$
$$= n - 1 + \sum_{k=1}^{\log_2 n} 2^{k-1} R(n/2^k),$$

where $n \geq 1$ and $R(1) = 0$. It turns out that this recursion can also be simplified as

$$R(n) = 1 + 3R(n/2), \quad (5)$$

as shown in [3]. The solution of this recurrence is given as

$$R(n) = (3^k - 1)/2.$$

In § 5, we found the total number of recursive calls in KOA function as $R(n) = 3(3^k - 1)/2$. We conclude that KOA2^k makes one third as many recursive calls as KOA, as we have claimed.

9 A Multiplication Example by KOA2^k

We will multiply the hexadecimal numbers $\mathbf{a} = F3D1$ and $\mathbf{b} = 6CA3$ using KOA2^k. The operand size and the base is given as $n = 4$ and $z = 16$. Let $\mathbf{a}[i]$ and $\mathbf{b}[i]$ denote the ith digits of $\mathbf{a}$ and $\mathbf{b}$, respectively.

Step 1: Since $n > 1$, we continue with the remaining steps.

Step 2: We need to compute

$$t := \text{sumP}_{\log_2 n} = \sum_{i=0}^{n-1} a[i]b[i]z^i \ .$$

The individual multiplications are

$a[0] \cdot b[0] = 1 \cdot 3 = 03$ $a[1] \cdot b[1] = D \cdot A = 82$
$a[2] \cdot b[2] = 3 \cdot C = 24$ $a[3] \cdot b[3] = F \cdot 6 = 5A$

Since multiplication by z means 1-digit shift, the sum sumP_2 is computed as

```
                     0  3
                  8  2
               2  4
       +    5  A
       ─────────────────
t = sumP₂ =  5  C  C  2  3
```

Iteration: We have $k = \log_2 n$ down to 1 and $m = n/2^k$.

Step 3 (1st Iteration): The computation of $t := t(1+z^m)$ for $m = 1$ is accomplished as

```
            5  C  C  2  3
      +  5  C  C  2  3
      ────────────────────
  t = 6  2  8  E  5  3
```

Steps 4, 5 and 6 (1st Iteration): We need to compute the terms $s_a(i)s_b(i)\text{mid}(i)$, where $i = 0,\cdots,2^{k-1}-1$ for $k = 2$ as

$$\begin{aligned} s_a(0)s_b(0)\text{mid}(0) &= (a[0]-a[1])(b[1]-b[0]) \\ &= (1-D)(A-3) \\ &= -54 \ , \\ s_a(1)s_b(1)\text{mid}(1) &= (a[0]-a[1])(b[1]-b[0]) \\ &= (3-F)(6-C) \\ &= 48 \ . \end{aligned}$$

Step 7 (1st Iteration): We compute

$$t := t + s_a s_b \text{mid} z^{(2i+1)m} \ ,$$

where $i = 0,\cdots,2^{k-1}-1$ for $k = 2$ and $m = 1$ as follows

```
            6  2  8  E  5  3
       -           5  4
       +       4  8
       ────────────────────
t = sumP₁ = 6  7  0  9  1  3
```

Step 3 (2nd Iteration): We compute $t := t(1+z^m)$ for $m = 1$ as

```
            6  7  0  9  1  3
      +  6  7  0  9  1  3
      ──────────────────────
  t = 6  7  7  0  1  C  1  3
```

Steps 4, 5, and 6 (2nd Iteration): We compute the terms $s_a(i)s_b(i)\text{mid}(i)$ for $i = 0,\cdots,2^{k-1}-1$ and $k = 1$. Since $k = 1$, we have only one term for $i = 0$, which is $s_a(0)s_b(0)\text{mid}(0)$, and computed as

$$\begin{aligned} &= (a^2[0]-a^2[2])(b^2[2]-b^2[0]) \\ &= (D1-F3)(6C-C3) \\ &= 74E \ . \end{aligned}$$

Step 7 (2nd Iteration): We compute

$$t := t + s_a s_b \text{mid} z^{(2i+1)m} \ ,$$

where $i = 0,\cdots,2^{k-1}-1$ for $k = 1$ and $m = 2$ as follows

```
            6  7  7  0  1  C  1  3
       +              7  4  E
       ─────────────────────────
t = sumP₀ = 6  7  7  6  A  1  3
```

We obtain the result at the end of Step 7 as $t = 67776A13$ which is the product $t = \text{sumP}_0 = a \cdot b = F3D1 \cdot 6CA3$.

10 Implementation Results

In order to compare their practical implementations, we have written assembly language programs for KOA and KOA2^k and obtained timings on a 350-MHz Pentium PC running Windows 2000 operating system with 256 megabytes of memory. The timing results (in milliseconds) are summarized in Table 3.

Operand (bits)	Threshold (words)	KOA (ms)	KOA2^k (ms)	Speedup %
1024	16	0.0278	0.0272	2.2
1536	12	0.0575	0.0548	4.7
2048	16	0.0895	0.0854	4.6
3072	12	0.1809	0.1702	5.9
4096	16	0.2788	0.2656	4.7
8192	16	0.8638	0.8142	5.7

Table 3. Timings of KOA and KOA2^k.

During the multiplication of two large operands using KOA or KOA2^k, recursive calls are made to multiply smaller operands. When the operand size becomes equal to or less than a particular threshold, no more recursive calls are made. Instead, the operands are multiplied using the classical multiplication method. This is because neither KOA nor KOA2^k can outperform the standard multiplication method with small operands. We experimentally obtained the optimum threshold our in platform as 12 or 16 computer words. Table 3 also lists the threshold values in words in the second column.

The third column of Table 3 lists the speedup in percentage of KOA2^k with respect to KOA. We have obtained approximately 5% speedup when the operands are larger than 1024 bits. Note that the speedup for 1536-bit operands is more than the speedup for 2048-bit operands. Similarly, the speedup for 3072-bit operands is more than the speedup for 4096-bit operands. This shows that KOA2^k performs better than KOA for small threshold values.

References

[1] T. H. Cormen, C. E. Leiserson, and R. L. Rivest. *Introduction to Algorithms*. MIT Press, 1990.

[2] W. Diffie and M. E. Hellman. New directions in cryptography. *IEEE Transactions on Information Theory*, 22:644–654, November 1976.

[3] S. S. Erdem. *Improving the Karatsuba-Ofman Multiplication Algorithm for Special Applications*. PhD thesis, Department of Electrical and Computer Engineering, Oregon State University, November 2001.

[4] IEEE. P1363: Standard specifications for public-key cryptography. Draft Version 13, November 12, 1999.

[5] A. Karatsuba and Y. Ofman. Multiplication of multidigit numbers by automata. *Soviet Physics-Doklady*, 7:595–596, 1963.

[6] D. E. Knuth. *The Art of Computer Programming, Volume 2, Seminumerical Algorithms*. Addison-Wesley, Third edition, 1998.

[7] Ç. K. Koç. High-Speed RSA Implementation. Technical Report TR 201, RSA Laboratories, 73 pages, November 1994.

[8] A. Menezes, P. Van Oorschot, and S. Vanstone. *Handbook of Applied Cryptography*. CRC Press, 1997.

[9] National Institute for Standards and Technology. Digital Signature Standard (DSS). FIPS PUB 186-2, January 2000.

[10] R. L. Rivest, A. Shamir, and L. Adleman. A method for obtaining digital signatures and public-key cryptosystems. *Communications of the ACM*, 21(2):120–126, February 1978.

Session 3:
Division

Chair: Paolo Montuschi

Revisiting SRT Quotient Digit Selection*

Peter Kornerup
Dept. of Mathematics and Computer Science
University of Southern Denmark, Odense, Denmark
E-mail: kornerup@imada.sdu.dk

Abstract

The quotient digit selection in the SRT division algorithm is based on a few most significant bits of the remainder and divisor, where the remainder is usually represented in a redundant representation. The number of leading bits needed depends on the quotient radix and digit set, and is usually found by an extensive search, to assure that the next quotient digit can be chosen as valid for all points (remainder, divisor) in a set defined by the truncated remainder and divisor, i.e., an "uncertainty rectangle".

This paper presents expressions for the number of bits needed for the truncated remainder and divisor, thus eliminating the need for a search through the truncation parameter space for validation. It also presents simple algorithms to properly map truncated negative divisors and remainders into non-negative values, allowing the quotient selection function only to be defined on the smaller domain of non-negative values.

1. Introduction

The SRT class of division algorithms is characterized by the use of redundant representations for the quotient, and most often as well for the remainder. Since the invention in the late fifties simultaneously by D. Sweeney, J.E. Robertson [12] and K.D. Tocher [15], and the introduction of the use of redundant representations for the remainders by D.E. Atkins [1], these methods have been extensively studied and implemented in processors. The famous "Pentium™ bug", where certain anomalies in the behavior of the floating point divide instruction were discovered, turned out to be caused by a few incorrect entries in the table employed by the quotient digit determination algorithm used for the radix 4 SRT implementation [5].

Due to the redundancy in the quotient digit set, there are overlaps between digit selection regions in the Robertson diagram, allowing a choice between two digit values.

Hence even if the information on the remainder and divisor is incomplete (representing an uncertainty interval), it may be possible to choose one of the alternative quotient digit values. By allowing such a relaxed quotient digit determination, it is possible to base the quotient digit selection on leading digits of the divisor and of the remainder in a redundant representation. The determination of the truncation parameters, i.e., how many digits of the remainder and divisor will be needed, has been extensively studied since the paper by Atkins, e.g., in [14, 16, 2, 7, 10] to list a few. These all use extensive searches to check the validity of a given set of parameters, recently [11] reduced the search to four pairs of truncation parameters. To cite [10] "*It is not possible to determine the optimal choices of δ and f analytically, as several factors are involved in making these choices.*" (δ and f here being the number of fractional digits needed in the truncated divisor resp. remainder.) It is, however, well-known that there is a simple lower bound on δ, and [4] has a lower bound on f, and [11] upper bounds on δ and f.

Here it is shown that, given a value of δ satisfying the bound mentioned above, it is indeed possible to determine analytically the other parameter f, such that a valid quotient digit selection function can be specified, eliminating the need for any further checking by search. Also, simple algorithms are presented, to properly map negative remainders and divisors into non-negative values, allowing the quotient selection function only to be defined on the smaller domain of non-negative values.

Section 2 introduces the fundamentals of SRT division and the notation used, together with certain bounds used in the quotient digit selection. Section 3 then develops the theory leading to the determination of the truncation parameters, and thus the specification of the quotient digit selection function. A few examples then illustrate the results. Then in Section 4 simple, constant time algorithms are developed for mapping truncated divisors and redundant remainders into their non-negative equivalents, for employing the digit selection function exclusively on non-negative arguments. Finally Section 5 concludes with comments on the extensibility of the results to other truncation procedures.

---

*Work supported by the Danish Natural Science Research Council, grant no. 21-00-0679

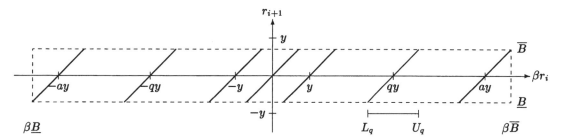

Figure 1. Robertson diagram for SRT division

2. Fundamentals of SRT Division

SRT is not really a specific kind of division, rather it is a class of division methods, characterized by the following:

- The divisor is normalized.
- A redundant symmetric quotient digit set is used.
- Quotient digits selected by a few leading digits of remainder and divisor.
- The remainders may be in a redundant representation.

Let β be the quotient radix and $D = \{-a, \cdots, 0, \cdots, a\}$ the quotient digit set with $\beta/2 \leq a \leq \beta - 1$, and define the *redundancy factor*

$$\rho = \frac{a}{\beta - 1} \quad \text{with} \quad \tfrac{1}{2} < \rho \leq 1 \quad (1)$$

where $\rho = 1$ corresponds to a maximally redundant digit set.

Let x be the dividend and y the positive divisor, r_i the remainder (with $r_0 = x$) and q_i the digit selected in the ith step. The purpose of the *digit selection function*, $\sigma(r_i, y)$, is to select the next quotient digit q_{i+1}, while keeping the new remainder $r_{i+1} = \beta r_i - q_{i+1} y$ bounded, say with bounds $\underline{B}$ and $\overline{B}$,

$$\underline{B} \leq r_{i+1} \leq \overline{B}. \quad (2)$$

Let the selection interval $[L_q, U_q]$ be the interval for βr_i, $\beta \underline{B} \leq \beta r_i \leq \beta \overline{B}$, for which it is possible to chose $q_{i+1} = q$ while keeping the updated remainder $r_{i+1} = \beta r_i - qy$ bounded by (2), i.e., for $L_q \leq \beta r_i \leq U_q$

$$\underline{B} \leq r_{i+1} = \beta r_i - qy \leq \overline{B}$$

must hold, corresponding to the Robertson diagram in Figure 1.

From the diagram it is seen that

$$L_q = qy + \underline{B} \quad \text{and} \quad U_q = qy + \overline{B}, \quad (3)$$

in particular (3) must hold for $q_{i+1} = \pm a$, the extremal digit values, hence

$$L_{-a} = -ay + \underline{B} \quad \text{and} \quad U_a = ay + \overline{B}.$$

But as seen from Figure 1, $\beta \underline{B} = L_{-a}$ and $\beta \overline{B} = U_a$, hence it follows that

$$\underline{B} = -\rho y \quad \text{and} \quad \overline{B} = \rho y$$

and from (3) we find

$$L_q = (q - \rho)y \quad \text{and} \quad U_q = (q + \rho)y. \quad (4)$$

To assure that at least one digit value can be chosen for any r_i, every value of βr_i must fall in at least one digit selection interval, i.e., it is necessary that

$$U_{q-1} \geq L_q,$$

hence by (4) we must require that $(q - 1 + \rho)y \geq (q - \rho)y$, or $\rho \geq \tfrac{1}{2}$ which is always satisfied since $a \geq \beta/2$. Actually by (1) recalling that we require $y > 0$

$$U_{q-1} - L_q = (2\rho - 1)y > 0, \quad (5)$$

thus consecutive selection intervals overlap, such that there are values of r_i for which in general there is a choice between two digit values (and possibly three for $\rho = 1$).

In summary, provided that $-\rho y \leq r_i \leq \rho y$, then the selection function can deliver at least one digit value q_{i+1} such that the new remainder satisfies $-\rho y \leq r_{i+1} \leq \rho y$. However, for $i = 0$, the dividend is used as the first remainder, $r_0 = x$, thus we must require that x and/or y are normalized such that $-\rho y \leq x \leq \rho y$, or $-\rho \leq \frac{x}{y} \leq \rho$. Any scaling applied for this normalization must then be used to correct the final quotient remainder pair.

Observation 1 *With quotient radix $\beta \geq 2$, quotient digit set $\{-a, \cdots, 0, \cdots, a\}$, and $\rho = \frac{a}{\beta - 1}$, there exists a digit selection function $\sigma(r_i, y)$ that delivers a next quotient digit q_{i+1} such that the next remainder*

$$r_{i+1} = \beta r_i - q_{i+1} y \quad \text{for} \quad i = 0, 1, \cdots$$

satisfies

$$-\rho y \leq r_{i+1} \leq \rho y,$$

provided that the dividend x, equal to the initial remainder r_0, and divisor $y > 0$ are normalized such that $-\rho \leq \frac{x}{y} \leq \rho$.

To simplify the analysis of the digit selection, we are assuming that $y > 0$, we shall see later that this restriction is not necessary.

3. Digit Selection

Due to the overlap $U_{q-1} - L_q = (2\rho - 1)y > 0$ it is not necessary to know the exact value of the remainder r_i to be able to select a correct next digit q_{i+1}. The digit selection intervals are conveniently illustrated in a *P-D diagram*[1] or *Taylor diagram* as in Figure 2, showing the intervals as functions of the divisor y, assumed normalized $\frac{1}{2} \leq y < 1$, and the shifted remainder βr_i. We shall here assume that the remainder updating and digit selection takes place in binary arithmetic.

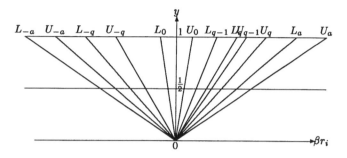

Figure 2. P-D diagram for digit selection with normalized divisor.

For any given fixed value of the divisor y, it is now possible to choose partition points, $S_q(y)$, in the selection intervals $[L_q(y), U_q(y)]$, or rather in the stricter overlap intervals $S_q(y) \in [L_q(y), U_{q-1}(y)]$, such that the selection function $\sigma(r_i, y)$ returning q_{i+1} may be defined by

$$\beta r_i \geq 0: \quad S_q(y) \leq \beta r_i < S_{q+1}(y) \Rightarrow q_{i+1} = q.$$
$$\beta r_i < 0: -S_{q+1}(y) \leq \beta r_i < -S_q(y) \Rightarrow q_{i+1} = -q, \quad (6)$$

using the symmetry around the y-axis, allowing us to restrict the analysis to $q \geq 0$.

To simplify the following discussion, we shall often assume that y is fixed and drop the argument y in the notation of selection intervals $[L_q(y), U_q(y)]$ and partition points $S_q(y)$. However, these can be pictured as functions of y in Figure 3.

Due to the overlap between selection intervals, the partition points can be chosen such that it is sufficient to check a few of the leading digits of the (possibly redundant) value of βr_i. Let $\widehat{\beta r_i}$ denote a truncated value of βr_i, and $\text{ulp}(\widehat{\beta r_i})$ denote the unit in the last place of the truncated value, say $\text{ulp}(\widehat{\beta r_i}) = 2^{-t}$. Define the truncation error, ϵ_r, by

$$\beta r_i = \widehat{\beta r_i} + \epsilon_r$$

then for various binary representations of r_i we have:

---

[1]From **P**artial remainder, **D**ivisor diagram, where "partial remainder" is a traditional notation.

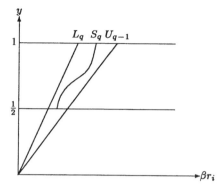

Figure 3. $S_q(y)$ **as a function of** y

2's complement: $\quad 0 \leq \epsilon_r < \text{ulp}(\widehat{\beta r_i})$
2's compl. carry-save: $\quad 0 \leq \epsilon_r < 2\text{ulp}(\widehat{\beta r_i})$
borrow-save[2]: $\quad -\text{ulp}(\widehat{\beta r_i}) < \epsilon_r < \text{ulp}(\widehat{\beta r_i})$

as illustrated below, where τ is the number of integer bits, which we shall not be concerned with here.

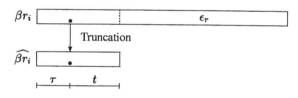

It is essential to note that truncation is assumed to take place on the redundant representation, before conversion into non-redundant representation. Thus it is not necessary to convert the full-length remainder into non-redundant.

Similarly, considering only a few leading digits of the divisor, let $\widehat{y}$ denote the truncated value of y with $\text{ulp}(\widehat{y}) = 2^{-u}$ for $u \geq 1$, and truncation error δ defined by $y = \widehat{y} + \delta$, where $0 \leq \delta < \text{ulp}(\widehat{y})$ for y in 2's complement.

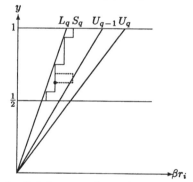

Figure 4. $S_q(y)$ **as a function of truncated divisor** $\widehat{y}$

The function $S_q(y)$ now becomes a stair-case function, delimiting rectangles within which a particular quotient digit can be chosen, as pictured in Figure 4. We shall assume that $S_q(y)$ is chosen as far to the left as possible.

---

[2]Often also denoted signed-digit

The stair-case function $S_q(y)$ is now determined by a set of constants $\widehat{S}_q(\widehat{y})$, corresponding to the various truncated values $\widehat{y}$. These constants are assumed specified to the same accuracy as $\widehat{\beta r_i}$ (i.e., $\text{ulp}(\widehat{\beta r_i}) = 2^{-t}$). For fixed $\widehat{y} = k2^{-u}$, where $\text{ulp}(\widehat{y}) = 2^{-u}$, $\widehat{S}_q(\widehat{y})$ can then be written as an integer multiple $s_{q,k}2^{-t}$ of $\text{ulp}(\widehat{\beta r_i}) = 2^{-t}$.

The dotted rectangle in Figure 4 located with lower left-hand corner at $(\widehat{S}_q(\widehat{y}), \widehat{y})$, for y in non-redundant 2's complement and βr_i redundant carry-save 2's complement, shows the set of points $(\beta r_i, y)$ satisfying

$$\widehat{S}_q(\widehat{y}) \leq \beta r_i < \widehat{S}_q(\widehat{y}) + 2\text{ulp}(\widehat{\beta r_i})$$
$$\text{and} \quad \widehat{y} \leq y < \widehat{y} + \text{ulp}(\widehat{y}) \quad (7)$$

where the point $(\widehat{S}_q(\widehat{y}), \widehat{y})$ and the truncations have to be chosen such that the next quotient digit $q_{i+1} = q$ can be selected for any point in the rectangle (7).

For the shifted remainder βr_i in borrow-save, $\widehat{S}_q(\widehat{y})$ would just have to be chosen as the midpoint of the lower edge (two-sided error), but for the following analysis we will assume that the representation is carry-save.

Using (4) for $q > 0$ the rectangle has to be to the right of the line $L_q(y) = (q - \rho)y$, yielding the following condition on the upper left-hand corner:

$$L_q(\widehat{y} + \text{ulp}(\widehat{y})) = (q - \rho)(\widehat{y} + \text{ulp}(\widehat{y})) \leq \widehat{S}_q(\widehat{y}). \quad (8)$$

These rectangles are overlapping, since they are of width $2\text{ulp}(\widehat{\beta r_i})$, but are positioned at a horizontal spacing of $\text{ulp}(\widehat{\beta r_i})$. Since $\widehat{S}_q(\widehat{y})$ is chosen as small as possible, for any point in the rectangle overlapping from the left, the digit value $q - 1$ must be chosen. Thus the midpoint of the bottom edge must be to the left of the line $U_{q-1}(y)$, yielding this additional condition:

$$\widehat{S}_q(\widehat{y}) + \text{ulp}(\widehat{\beta r_i}) \leq U_{q-1}(\widehat{y}) = (q - 1 + \rho)\widehat{y}. \quad (9)$$

But the lower right-most corner must also be to the left of the line $U_q(y)$, hence we must also require

$$\widehat{S}_q(\widehat{y}) + 2\text{ulp}(\widehat{\beta r_i}) \leq U_q(\widehat{y}) = (q + \rho)\widehat{y}. \quad (10)$$

It is easy to see that for $t \geq 1$ (which we shall see later is always the case), the upper bound on $\widehat{S}_q(\widehat{y})$ obtained from (9) is smaller than or equal to the bound found from (10). Thus combining conditions (7) and (9) to determine the size and position of the rectangles, we must require

$$(q - \rho)(\widehat{y} + \text{ulp}(\widehat{y})) \leq \widehat{S}_q(\widehat{y}) \leq (q - 1 + \rho)\widehat{y} - \text{ulp}(\widehat{\beta r_i}).$$

But $\widehat{S}_q(\widehat{y})$ has to be an integer multiple of $\text{ulp}(\widehat{\beta r_i}) = 2^{-t}$, hence defining $\widehat{S}_q(\widehat{y}) = s_{q,k}2^{-t}$ we must require:

$$\lceil 2^{t-u}(q - \rho)(k + 1) \rceil = s_{q,k} \leq \lfloor 2^{t-u}(q - 1 + \rho)k - 1 \rfloor \quad (11)$$

for $q > 0$, using $\text{ulp}(\widehat{y}) = 2^{-u}$ and defining $\widehat{y} = k2^{-u}$, for integer k, $2^{u-1} \leq k < 2^u$.

Recall that we required $\widehat{\beta r_i} \geq 0$ and thus it should also be possible to choose $q = 0$, but obviously then $\widehat{S}_0(\widehat{y}) = 0$ for all $\widehat{y}$, or $s_{0,k} = 0$ for all k, $2^{u-1} \leq k < 2^u$. Note that the right-most bounds on the uncertainty rectangles for $q = 0$ are implicitly chosen by the choice of $\widehat{S}_q(\widehat{y})$ for $q = 1$.

Without restrictions on t, u and ρ, there is only the integer term -1 which can be moved in and out of the floor and ceiling functions, but reorganizing terms then condition (11) for $q > 0$ can be written as:

$$\lceil 2^{t-u}(q - \rho)k + 2^{t-u}(q - \rho) + 1 \rceil$$
$$\leq \lfloor 2^{t-u}(q - \rho)k + 2^{t-u}(2\rho - 1)k \rfloor, \quad (12)$$

where the ceiling and floor expressions are linear functions of k:

$$\lceil Ak + B \rceil \leq \lfloor (A + C)k \rfloor, \quad (13)$$

with $A \geq 0$, $B \geq 1$ and $C > 0$ for $q \geq 1$. Clearly it is necessary that $Ck \geq B$ for this condition to be satisfied, but it is easily seen that $Ck - B \geq 1$ is a sufficient condition, since then there is at least one integer between the ceiling and floor expressions. Hence if the condition

$$2^{t-u}((2\rho - 1)k - (q - \rho)) \geq 2,$$

holds for the minimal value $k = 2^{u-1}$ and the maximal value $q = a$, then this is sufficient for (12) to hold.

Thus the stronger condition derived from $Ck - B \geq 1$

$$2^{-t} \leq \left((\rho - \tfrac{1}{2}) - (a - \rho)2^{-u}\right)/2 \quad (14)$$

may be used to find values of t, $2^{-t} = \text{ulp}(\widehat{\beta r_i})$ and u, $2^{-u} = \text{ulp}(\widehat{y})$, for which (12) is satisfied for all $q \in \{1, \cdots, a\}$ and all k such that $2^{u-1} \leq k \leq 2^u - 1$, i.e., $\tfrac{1}{2} \leq \widehat{y} < 1$. Note, however, that the solutions to (14) need not be optimal in the sense that t is minimal, since it might be sufficient to require $Ck \geq B$, i.e., only require

$$2^{-t} \leq \left((\rho - \tfrac{1}{2}) - (a - \rho)2^{-u}\right), \quad (15)$$

which would allow a solution for t which is one smaller than the solution to (14). We shall below return to the choice between the two conditions.

Obviously, the right-hand side of both (14) and (15) must be strictly positive for solutions to exist for t, hence we want a u satisfying

$$2^{-u} < \frac{\rho - \tfrac{1}{2}}{a - \rho}, \quad (16)$$

provided that $a > \rho$, or $\beta > 2$, since $\beta = 2$ is the only case where $\rho = a (= 1)$. As seen in Example 3 below, the case $\beta = 2$ can be handled separately. Hence (16) is a sufficient condition on u for all $\beta > 2$, $q \in \{1, \cdots, a\}$ and all k such that $2^{u-1} \leq k \leq 2^u - 1$.

Returning to (13) and the choice between (14) and (15) to determine a minimal value of t, we need the following lemma:

Lemma 2 *Given constants $A \geq 0$, $C > 0$, then there exist integers $k_0 \leq k_1$ such that the inequality*

$$\lceil Ak + B \rceil \leq \lfloor (A+C)k \rfloor \qquad (17)$$

holds for all $k \geq k_0$, provided that:

i) $\lceil Ak + B \rceil = \lfloor (A+C)k \rfloor$ *for $k_0 \leq k \leq k_1 - 1$,*
and ii) $\lceil Ak_1 + B \rceil < \lfloor (A+C)k_1 \rfloor$.

Proof: Define $\Delta(x) = \lfloor (A+C)x \rfloor - \lceil Ax + B \rceil$. Since $\Delta(x) \leq (A+C)x - (Ax + B) = Cx - B$ and $\Delta(x)$ is integral, it follows that $\Delta(x) \leq \lfloor Cx - B \rfloor$. If on the other hand $\lfloor Cx - B \rfloor = n$ then $\Delta(x) \geq n - 1$ (there are at least $n-1$ integers between $Ax + B$ and $(A+C)x$), thus

$$\lfloor Cx - B \rfloor - 1 \leq \Delta(x) \leq \lfloor Cx - B \rfloor. \qquad (18)$$

Hence $\Delta(x) - 1 \leq \lfloor Cx - B \rfloor - 1 \leq \lfloor Cy - B \rfloor - 1 \leq \Delta(y)$ for $y \geq x$, thus:

$$y \geq x \Rightarrow \Delta(y) \geq \Delta(x) - 1. \qquad (19)$$

$\Delta(x)$ may not always increase with x as shown here

where $\Delta(k) = 0$, but $\Delta(k+1) = \Delta(k+2) = -1$.

Since $C > 0$, by (18) there exists a minimal k_0 such that $\Delta(k) \geq 0$ for $k \geq k_0$ until eventually there is a minimal $k_1 \geq k_0$ such that $\Delta(k_1) \geq 1$. By (19) the lemma then has been proven. $\square$

We can now combine the previous discussion with the lemma, into the following result:

Theorem 3 (SRT digit selection constants)
For radix β SRT division for $\beta > 2$ with digit set $D = \{-a, \cdots, a\}$ and $\rho = \frac{a}{\beta - 1}$, the selection constants $\widehat{S_q}(\widehat{y}) = s_{q,k} 2^{-u}$ can be determined for $1 \leq q \leq a$ and $\widehat{y} = k \cdot \mathrm{ulp}(\widehat{y})$ as

$$s_{q,k} = \lceil 2^{t-u}(q - \rho)(k+1) \rceil$$

for $k = 2^{u-1}, \cdots, 2^u - 1$, using truncation parameters $\mathrm{ulp}(\widehat{S_q}(\widehat{y})) = \mathrm{ulp}(\widehat{\beta r_i}) = 2^{-t}$ and $\mathrm{ulp}(\widehat{y}) = 2^{-u}$, where u has to satisfy

$$2^{-u} < \frac{\rho - \frac{1}{2}}{a - \rho}.$$

To determine t for given u, let t_0 be the smallest t satisfying

$$2^{-t} \leq (\rho - \tfrac{1}{2}) - (a - \rho) 2^{-u},$$

and let

$$\Delta(u, t, k) = \lfloor 2^{t-u}(a - 1 + \rho)k - 1 \rfloor - \lceil 2^{t-u}(a - \rho)(k+1) \rceil,$$

then

$$t = \begin{cases} t_0 & \text{if } \Delta(u, t_0, 2^{u-1}) \geq 1 \\ t_0 & \text{if } \Delta(u, t_0, k) = 0 \text{ for } k = 2^{u-1}, \cdots, k_1 - 1, \\ & \text{and } \Delta(u, t_0, k_1) \geq 1 \\ t_0 + 1 & \text{otherwise.} \end{cases}$$

Proof: The expression for $s_{q,k}$ is from (11), and the condition on u from (16) was shown to be sufficient for $q > 0$, by using $\max q = a$, and $\min k = 2^{u-1}$, for all values of t.

Rewriting (11) we found that u and t must satisfy the condition

$$\lceil Ak + B \rceil \leq \lfloor (A+C)k \rfloor, \qquad (20)$$

or equivalently $\Delta(u, t, k) \geq 0$, for $q = a$ and for all k, $2^{u-1} \leq k < 2^u$, where

$$\begin{aligned} A &= 2^{t-u}(q - \rho) \geq 0 \\ B &= 2^{t-u}(q - \rho) + 1 \\ C &= 2^{t-u}(2\rho - 1) > 0. \end{aligned}$$

Using Lemma 2 the two first choices for t imply that (11) holds for all $k \geq k_0 = 2^{u-1}$. As we saw before $Ck - B \geq 0$ for all q, $1 \leq q \leq a$ translates into the condition

$$2^{-t_0} \leq (\rho - \tfrac{1}{2}) - (a - \rho) 2^{-u}.$$

If the conditions for the first two choices for t fail, then the stronger condition

$$2^{-t} \leq \left((\rho - \tfrac{1}{2}) - (a - \rho) 2^{-u}\right)/2,$$

corresponding to $t = t_0 + 1$, implies that $Ck - B \geq 1$ for $k = k_0 = 2^{u-1}$ and $q = a$, and again by the lemma this implies that condition (11) holds for all $k \geq k_0$, and then also for all q, $1 \leq q \leq a$. $\square$

Example 1 (*Minimally redundant radix 4 SRT*)
Let $\beta = 4$ with minimally redundant digit set $D = \{-2, -1, 0, 1, 2\}$, hence $a = 2$ and $\rho = \frac{2}{3}$, and from (16) we derive $u \geq 4$. Choosing $u = 4$ by Theorem 3, $t = t_0$ has to be the smallest solution to

$$\begin{aligned} 2^{-t} &\leq (\rho - \tfrac{1}{2}) - (a - \rho) 2^{-u} \\ &= (\tfrac{2}{3} - \tfrac{1}{2}) - (2 - \tfrac{2}{3}) 2^{-4} = \tfrac{1}{12}, \end{aligned}$$

hence $t_0 = 4$.

For $k = 2^{u-1} = 8, 9, 10$, $\Delta(4,4,k) = 0$ but $\Delta(4,4,11) = 1$. Thus by the second choice for t in Theorem 3, $t = t_0 = 4$. Hence we can compute the values of the constants $\widehat{S}(\widehat{y}) = s_{q,k} 2^{-u}$ for $q > 0$ as

$$s_{q,k} = \lceil 2^{t-u}(q-\rho)(k+1) \rceil = \lceil (q - \tfrac{2}{3})(k+1) \rceil,$$

resulting in the following table of comparison constants

$k = 16\widehat{y}$	8	9	10	11	12	13	14	15
$16\widehat{S}_1$	3	4	4	4	5	5	5	6
$16\widehat{S}_2$	12	14	15	16	18	19	20	22

Utilizing the definitions (4) of the functions $L_q(y)$ and $U_q(y)$ we have

$q =$	-2	-1	0	1	2
$L_q(y)$	$\tfrac{-8}{3}y$	$\tfrac{-5}{3}y$	$\tfrac{-2}{3}y$	$\tfrac{1}{3}y$	$\tfrac{4}{3}y$
$U_q(y)$	$\tfrac{-4}{3}y$	$\tfrac{-1}{3}y$	$\tfrac{2}{3}y$	$\tfrac{5}{3}y$	$\tfrac{8}{3}y$

which together with the table of $\widehat{S}_q$ yields the P-D diagram for the first quadrant in Figure 5. In practice the left part of the diagram (for $\widehat{\beta r_i} < 0$) is not needed as we shall see later, when utilizing symmetries in the uncertainty rectangles. □

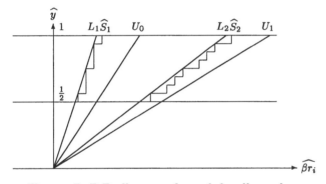

Figure 5. P-D diagram for minimally redundant radix 4 SRT

Example 2 Continuing the previous example, if instead of choosing the minimal $u = 4$ the value $u = 5$ is chosen, we find

$$\begin{aligned} 2^{-t} &\leq (\rho - \tfrac{1}{2}) - (a - \rho) 2^{-u} \\ &= (\tfrac{2}{3} - \tfrac{1}{2}) - (2 - \tfrac{2}{3}) 2^{-5} \\ &= \tfrac{1}{6} - \tfrac{1}{24} = \tfrac{1}{8}, \end{aligned}$$

having the minimal solution $t = t_0 = 3$. However, for $k = 2^{u-1} = 16$ we find for $\Delta(u, t_0, k)$

$$\begin{aligned} \Delta(5,3,16) &= \lfloor 2^{-2} \cdot \tfrac{5}{3} \cdot 16 - 1 \rfloor - \lceil 2^{-2} \cdot \tfrac{4}{3} \cdot 17 \rceil \\ &= \lfloor \tfrac{17}{3} \rfloor - \lceil \tfrac{17}{3} \rceil \\ &= 5 - 6 = -1, \end{aligned}$$

thus by Theorem 3 it is necessary to increase $t = t_0 + 1 = 4$, hence $(u,t) = (5,4)$ is also a valid pair of truncation parameters, but obviously not as good as $(4,4)$ found in the previous example. As a check we find $\Delta(5,4,16) = \lfloor \tfrac{37}{3} \rfloor - \lceil \tfrac{34}{3} \rceil = 12 - 12 = 0$ and $\Delta(5,4,17) = \lfloor \tfrac{79}{6} \rfloor - \lceil \tfrac{36}{3} \rceil = 13 - 12 = 1$. □

Example 3 *(Radix 2 SRT)*
For $\beta = 2$ and digit set $\{-1, 0, 1\}$ the condition on u in Theorem 3 cannot be used since $a = \rho = 1$. However, for $u = t = 1$ and $k = 1$, corresponding to $2^{u-1} = 1 = k < 2^u = 2$, it is easily seen that (11) is satisfied for $q = 1$.

Thus only one bit of the fraction part of y is needed for $\widehat{y}$, and this bit is always 1 since $\tfrac{1}{2} \leq y < 1$, implying that the quotient selection is independent of y. Note that $\mathrm{ulp}(\widehat{\beta r_i}) = 2^{-1} = \tfrac{1}{2}$. From (4) the lower and upper bounds are then found to be

$q =$	-1	0	1
$L_q(y)$	$-2y$	$-y$	0
$U_q(y)$	0	y	$2y$

and it now easy to see that $\widehat{S}_0 = -\tfrac{1}{2}$ and $\widehat{S}_1 = 0$, since the uncertainty rectangle has height $\tfrac{1}{2}$ and width 1. The resulting full P-D diagram is shown in Figure 6. □

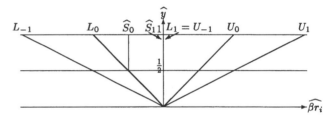

Figure 6. P-D diagram for radix 2 SRT with redundant remainder

4. Exploiting Symmetries

A simple way of implementing the digit selection function of (6) is to compare the value of $\widehat{\beta r_i}$ against selection bounds $\widehat{S}_q(\widehat{y})$ corresponding to the stair-case function. This can be performed by a set of comparators, using a priority encoder to determine the position of the largest value of q for which the comparison returns true. By exclusively accepting non-negative values of the remainder, the number of such comparators can be almost halved, changing the sign of the digit value found if the remainder was originally negative.

Another often used implementation is to use a table, indexed by the leading truncated digits of the remainder and of the divisor. To minimize the number of bits to be used for the table index, obviously the leading digits of the redundant remainder $\widehat{\beta r_i}$ must be converted to non-redundant representation. Beyond this assimilation of redundancy, it is advantageous also here to exploit the symmetry so that look-up is based on non-negative remainders.

However, it is not quite trivial to see how to change the sign of a negative remainder, since we have to make sure that the "uncertainty" rectangle is mapped correctly from the left half-plane onto the right half-plane of Figure 4. We will start with the case of remainders in 2's complement carry-save representation, as illustrated in Figure 7.

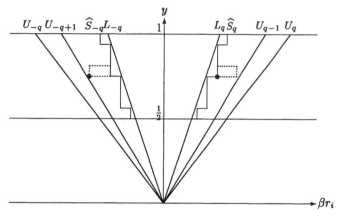

Figure 7. Mapping negative remainders to positive

Observe that the left rectangle is determined by the point $(\widehat{\beta r_i}, \widehat{y})$ in its lower **left-hand** corner. The symmetric point given by $(-\widehat{\beta r_i}, -\widehat{y})$ is the lower **right-hand** corner of the symmetric rectangle to the right. But we need the coordinates of the lower **left-hand** corner of it, which is located at the point $(-\widehat{\beta r_i} - 2\text{ulp}(\widehat{\beta r_i}), \widehat{y})$.

Fortunately, there is a very simple way of performing this mapping when $\widehat{\beta r_i}$ is represented in 2's complement carry-save. Recall that negation of a 2's complement, carry-save number is performed by inverting the bit-patterns of the "save" part as well as the "carry" part, and adding a unit in the least significant position of **both**. Hence by just inverting all the bits of the encoding of $\widehat{\beta r_i}$ (i.e., forming the 1's complement, and **not** adding a unit to both), we obtain precisely the effect of subtracting the correction $2\text{ulp}(\widehat{\beta r_i})$, and thus the correct mapping of the left rectangle into the right.

Now, of course, to know whether to invert or not, it is necessary to know the sign of $\widehat{\beta r_i}$, but $\widehat{\beta r_i}$ also has to be converted into non-redundant 2's complement representation (be compressed). However, it is possible in parallel to add the carry- and save-parts of $\widehat{\beta r_i}$, as well as adding the inverted bit-patterns. The sign of the former addition then determines which of the two sums is to be the final (non-negative) compressed result, to be used as index to a table look-up, or as input to some other quotient determi-

nation logic. Hence the delay of the mapping is just that of a selector, i.e., a constant time overhead, following the adders. From the literature e.g., [8, 9, 3], it appears like this mapping previously has been performed by first adding (compressing) the carry- and save parts, conditionally followed by a 2's complementation, essentially with a delay corresponding to an extra addition.

By analogy it is also possible to map the rectangle corresponding to a negative divisor $\widehat{y}$ from the lower half-plane to the upper, by simply inverting the bits of $\widehat{y}$, which implies that there is no need to perform a time consuming negation of the 2's complement divisor y initially in the division algorithm. If $y < 0$ then the bits of $\widehat{y}$ just have to be inverted.

If the remainder is represented in borrow save, the situation is also very simple. Here the rectangle of uncertainty is determined by the mid-point of the lower edge of the rectangle. This implies that the mapping is obtained by negating $\widehat{\beta r_i}$, which is the trivial constant time operation of simply interchanging the negatively and positively weighted bit-vectors of the encoding. Here it is sufficient with two parallel subtractions, followed by a selection of the non-negative result, to realize the compression as well as the mapping.

Observation 4 *For the purpose of SRT quotient digit selection, truncated negative remainders and/or divisors can be mapped in constant time into the first quadrant by:*

2's complement: *Complement all bits of the encoding bit-vector(s),*

Borrow-save: *Interchange the positively and negatively weighted bit-vectors,*

thereby assuring that the "uncertainty rectangles" are mapped into their non-negative equivalents. The selected quotient digit q_{i+1} changes its sign according to the mappings performed, while the updating of the shifted remainder remains $r_{i+1} = \beta r_i - q_{i+1} y$, employing the signed divisor y and remainder r_i.

5. Conclusions

It has been shown, that it is indeed possible to analytically define the truncation parameter t for the shifted remainder, $2^{-t} = \text{ulp}(\widehat{\beta r_i})$, given a value of the divisor truncation parameter u satisfying a certain bound, e.g., the minimal such value of u. Thus the quotient digit selection function can be defined without the need to extensively check the validity of some chosen parameters.

We have here used standard truncation, just discarding digits below a certain position. In some of the previous papers, e.g., [16, 10, 2], also different truncations of the "save" and "carry"-parts in carry-save representations have been analyzed, and similarly for different truncations of the positive and negative parts in borrow-save represented remainders. It is also possible to reduce the truncation error by including a carry-bit from a few extra positions beyond the truncation point, as suggested in [16]. Another (equivalent) possibility is to apply the digit-parallel PN-recoding from [6]. Such extra (user specified) truncation parameters could also be included in the analysis presented here, but will of course complicate it.

We have also shown an algorithm for constant time mapping of the truncated remainders $\widehat{\beta r_i}$ into their non-negative, non-redundant equivalents for quotient digit determination, by simply selecting the non-negative result of two parallel additions (compressions).

6. Acknowledgements

The author appreciates the detailed comments provided by the reviewers, in particular those of reviewer #3, for comments leading to improvements of Lemma 2 and Theorem 3.

References

[1] D. E. Atkins. Higher-Radix Division Using Estimates of the Divisor and Partial Remainders. *IEEE Transactions on Computers*, C-17:925–934, 1968. Reprinted in [13].

[2] N. Burgess and T. Williams. Choices of Operand Truncation in the SRT Division Algorithm. *IEEE Transactions on Computers*, 44(7):933–938, July 1995.

[3] S.C. Chung. Simplification of Lookup Table. US Patent #5.777.917, July 1998.

[4] L. Ciminiera and P. Montushi. Higher Radix Square Rooting. *IEEE Transactions on Computers*, 39(10):1220–1231, October 1990.

[5] T. Coe and P. Tang. It takes Six Ones to Reach a Flaw. In S. Knowles and W. H. McAllister, editors, *Proc. 12th IEEE Symposium on Computer Arithmetic*. IEEE Computer Society, 1995.

[6] M. Daumas and D.W. Matula. Further Reducing the Redundancy of Notation Over a Minimally Redundant Digit Set. *Journal of VLSI Signal Procesing*, 33(1/2):7–18, 2003.

[7] M. Ercegovac and T. Lang. *Division and Square Root: Digit-Recurrence Algorithms and Implementations*. Kluwer Academic Publishers, 1994.

[8] J. Fandrianto. Algorithms for High Speed Shared Radix 4 Division and Radix 4 Square-Root. In *Proc. 8th IEEE Symposium on Computer Arithmetic*, pages 73–79. IEEE Computer Society, 1987.

[9] J. Fandrianto. Algorithms for High Speed Shared Radix 8 Division and Radix 8 Square Root. In *Proc. 9th IEEE Symposium on Computer Arithmetic*, pages 68–75. IEEE Computer Society, 1989.

[10] S. Oberman and M. Flynn. Minimizing the Complexity of SRT Tables. *IEEE Transactions on VLSI systems*, 6(1):141–149, March 1998.

[11] B. Parhami. Precision Requirements for Quotient Digit Selection in High-Radix Division. In *Proc. 35-th Asilomar Conference on Circuits, Systems and Computers*, pages 1670–1673. IEEE Press, 2001.

[12] J. Robertson. A New Class of Digital Division Methods. *IRE Transactions on Electronic Computers*, EC-7:218–222, 1958. Reprinted in [13].

[13] E. E. Swartzlander, editor. *Computer Arithmetic, Vol I*. Dowden, Hutchinson and Ross, Inc., 1980. Reprinted by IEEE Computer Society Press, 1990.

[14] G. Taylor. Radix 16 SRT Dividers with Overlapped Quotient Selection Stages. In *Proc. 7th IEEE Symposium on Computer Arithmetic*, pages 64–71. IEEE Computer Society, 1985.

[15] K. Tocher. Techniques of Multiplication and Division for Automatic Binary Computers. *Quarterly Journal of Mechanics and Applied Mathematics*, 11:364–384, 1958.

[16] T. E. Williams and M. Horowitz. SRT division diagrams and their usage in designing custom integrated circuits for division. Technical Report CSL-TR-87-326, Stanford University, 1986.

SRT Division Algorithms As Dynamical Systems

Mark McCann* and Nicholas Pippenger[†]
Department of Computer Science
The University of British Columbia
Vancouver, British Columbia, Canada, V6T 1Z4
mccann@cs.ubc.ca nicholas@cs.ubc.ca

Abstract

SRT division, as it was discovered in the late 1950s represented an important improvement in the speed of division algorithms for computers at the time. A variant of SRT division is still commonly implemented in computers today. Although some bounds on the performance of the original SRT division method were obtained, a great many questions remained unanswered. In this paper, the original version of SRT division is described as a dynamical system. This enables us to bring modern dynamical systems theory, a relatively new development in mathematics, to bear on an older problem. In doing so, we are able to show that SRT division is ergodic, and is even Bernoulli, for all real divisors and dividends. With the Bernoulli property, we are able to use entropy to prove that the natural extensions of SRT division are isomorphic by way of the Kolmogorov-Ornstein Theorem. We demonstrate how our methods and results can be applied to a much larger class of division algorithms.

1 Introduction

Since the discovery of the first radix-2 SRT division algorithm, the use of the term "SRT division" has expanded to include a wide variety of higher radix non-restoring division algorithms that are loosely based on the original. For example, there is the infamous implementation of a radix-4 SRT division algorithm in the first release of the Pentium™ CPU that has become widely known as the "Pentium™ Bug." One major difference between this implementation of radix-4 SRT division and the original radix-2 SRT division is that the former produces a constant number of quotient bits per step, while the latter produces a variable number. Modern implementations of SRT division use carry-save adders to perform additions and subtractions in constant time. Earlier implementations, however, used carry-propagate adders with delays that grow with the word length. Therefore, the primary goal of the early investigators was to reduce the number of uses of the costly adder. In the late 1950's, Sweeney [3], Robertson [16], and Tocher [20] independently made the observation that whenever a partial remainder is in the range $(-\frac{1}{2}, \frac{1}{2})$, there will be one or more leading zeros that can be shifted through in a very short amount of time (usually one cycle) thereby reducing the use of the adder. Although the aforementioned have received most of the credit for the algorithm which is named after them, it can be argued that Nadler described an equivalent algorithm in a 1956 paper [12]. The description of higher-radix SRT division which is the basis for modern SRT division is generally attributed to Atkins [1], but this is not the version of division that we will be concerned with in this paper.

Although what is considered to be "costly" for a division algorithm has changed, it is still interesting and important to understand the behaviour of successive partial remainders on average for a given divisor. Surprisingly, some of the most basic questions that one might have concerning the behaviour of partial remainders for even simple radix-2 SRT division have remained unanswered for over forty years. The difficulty that early investigators experienced in answering such questions was mainly due to a lack of necessary mathematical tools and results. During that past thirty years, the field of "dynamical systems theory" or "ergodic theory" has come into existence in mathematics and has been greatly developed. In this paper we show how to apply some of what is now known in dynamical systems theory to the earliest version of SRT division. In doing so, we are able to prove several previously unknown properties for simple SRT division. The results are quite general and lend themselves to be adapted to other division algorithms. For the remainder of this paper, the term SRT division will refer to the original algorithm unless otherwise stated.

The SRT division algorithms analyzed by Freiman [4] and Shively [19] are the same, but the authors differ in what they take to be a step of the algorithm: Freiman defines a step to be the operations from one use of the adder to the next, while Shively defines it to be the operations from one normalizing shift (of a single place) to the next. The following definitions are consistent with Shively's:

(a) n represents the number of iterations performed in the algorithm.

(b) p_0 is the dividend (or initial partial remainder) normal-

---

*The work reported here was supported by an NSERC Research Grant.
[†]The work reported here was supported by an NSERC Research Grant and a Canada Research Chair.

ized so that $p_0 \in [\frac{1}{2}, 1)$.

(c) $p_i \in (-1, 1)$, $i \in \mathbb{N}$, is the partial remainder after the ith step.

(d) D is the divisor normalized to $[\frac{1}{2}, 1)$.

(e) $q_i \in \{-1, 0, 1\}$ ($i \in \{0, \ldots, n-1\}$) is the quotient digit generated by the ith step.

(f) $Q_n = \sum_{i=0}^{n-1} \frac{q_i}{2^i}$ is the "rounded off" quotient generated after n steps of the algorithm.

Given the above definitions, after n steps of the division algorithm, we would like it to be true that

$$p_0 = DQ_n + \varepsilon(n)$$

where $\varepsilon(n)$ is a term that goes to zero as n goes to infinity.

A recurrence relation for the SRT division algorithm can be stated as

$$p_{i+1} = \begin{cases} 2p_i & : |p_i| < \frac{1}{2} \\ 2(p_i - D) & : |p_i| \geq \frac{1}{2} \text{ and } p_i \geq 0 \\ 2(p_i + D) & : |p_i| \geq \frac{1}{2} \text{ and } p_i < 0, \end{cases}$$

and

$$q_i = \begin{cases} 0 & : |p_i| < \frac{1}{2} \\ 1 & : |p_i| \geq \frac{1}{2} \text{ and } p_i \geq 0 \\ -1 & : |p_i| \geq \frac{1}{2} \text{ and } p_i < 0. \end{cases}$$

By observing that

$$p_{i+1} = \begin{cases} 2(p_i - (0)D) & : |p_i| < \frac{1}{2} \\ 2(p_i - (1)D) & : |p_i| \geq \frac{1}{2} \text{ and } p_i \geq 0 \\ 2(p_i - (-1)D) & : |p_i| \geq \frac{1}{2} \text{ and } p_i < 0, \end{cases}$$

we can rewrite the definition of p_{i+1} as

$$p_{i+1} = 2(p_i - q_i D).$$

After n steps have been completed, we have

$$p_n = 2^n p_0 - 2^n q_0 D - 2^{n-1} q_1 D - \cdots - 2^1 q_{n-1} D,$$

and then after dividing by 2^n and solving for p_0 we find that

$$p_0 = \frac{p_n}{2^n} + \frac{q_0 D}{2^0} + \frac{q_1 D}{2^1} + \cdots + \frac{q_{n-1} D}{2^{n-1}}$$

$$= D \sum_{i=0}^{n-1} \frac{q_i}{2^i} + \frac{p_n}{2^n} = DQ_n + \frac{p_n}{2^n}.$$

Now let $\varepsilon(n) = p_n/2^n$ and let $Q^* = \lim_{n\to\infty} Q_n$. Since $|p_n| < 1$, in the limit as n goes to infinity

$$p_0 = DQ^*.$$

The quotient bits being generated are not in a standard binary representation, but it is a simple matter to convert the answer back to standard binary on-the-fly without using any expensive operations.

Table 1 shows an example of using the SRT division algorithm to divide 0.67 by 0.75. The steps that produce non-zero quotient bits have been shown. In this example, after six uses of the adder, the quotient ($0.89\bar{3}$) has been determined to four digits of precision.

Table 1. SRT division example

p_0	$= 0.67$	$= 0.67$		
p_1	$= 2(0.67 - D)$	$= -0.16$	Q_0	$= 1$
p_4	$= 2(2^2(-0.16) + D) =$	0.22	Q_3	$= 0.875$
p_7	$= 2(2^2(0.22) - D) =$	0.26	Q_6	$= 0.890625$
p_9	$= 2(2^1(0.26) - D) =$	-0.46	Q_8	$= 0.89453125$
p_{11}	$= 2(2^1(-0.46) + D) =$	-0.34	Q_{10}	$= 0.893554688$
p_{13}	$= 2(2^1(-0.34) + D) =$	0.14	Q_{12}	$\doteq 0.893310547$

Now, with this simple system of division in hand, we might want to ask certain questions about its performance. For example, we could ask "How many bits of precision are generated per iteration of the algorithm on average?" To answer this question, we must look at the magnitude of $|Q^* - Q_n| = |p_n/2^n|$. The number of bits of precision on the nth step is then $n - \log_2 p_n$. In the worst case, p_n is close to 1, and therefore we get at least one bit of precision per iteration of the algorithm, regardless of the values of D or p_0. Of course, a designer of actual floating-point hardware probably wants to know the expected performance based on the expected values of p_n. To answer the many variants of this type of question, it is clear that we must know something about the distribution of partial remainders over time. The remainder of this paper is devoted to extending what is known about the answer to this type of question as it relates to SRT division and its variants.

2 SRT Division as a Dynamical System

The example in table 1 makes it clear that keeping track of the signs of successive partial remainders is irrelevant in determining how many times the adder will be used for a particular calculation. For this reason, we only need to consider the magnitudes of successive partial remainders. We now give a reformulation of SRT division that will allow us to look at division as a dynamical system.

Definition 1 (SRT Division Transformation). For $D \in [\frac{1}{2}, 1)$, we define the function $T_D : [0, 1) \to [0, 1)$ as

$$T_D(x) = \begin{cases} 2x & : 0 \leq x < \frac{1}{2} \\ 2(D - x) & : \frac{1}{2} \leq x < D \\ 2(x - D) & : D \leq x < 1. \end{cases}$$

This transformation of the unit interval represents the successive partial remainders that arise as SRT division is carried out by a divisor D on a dividend x. D is normalized to $[\frac{1}{2}, 1)$. The dividend x is normalized to $[\frac{1}{2}, 1)$ initially, while each of the successive partial remainders $T_D^n(x)$ ($n \in \mathbb{N}$) subsequently ranges through $[0, 1)$.

By using the characteristic function for a set Δ defined as
$$1_\Delta(x) = \begin{cases} 1 &: x \in \Delta \\ 0 &: x \notin \Delta, \end{cases}$$
we can rewrite T_D as
$$T_D(x) = 2x \cdot 1_{[0,\frac{1}{2})}(x) + 2(D-x) \cdot 1_{[\frac{1}{2},D)}(x) + \\ 2(x-D) \cdot 1_{[D,1)}(x). \quad (1)$$

If we plot (1) on the unit interval, we obtain a very useful visualization of our transformation. Figure 1 shows the plot of $T_{0.75}(x)$ combined with a plot of the successive partial remainders that arise while dividing 0.67 by 0.75. The heavy solid lines represent the transformation $T_{0.75}$, while the abscissa of the thin vertical lines represent successive partial remainder magnitudes. This is the same system that was presented earlier in table 1. Notice that a vertical line in the interval $[\frac{1}{2}, D)$ corresponds to a subsequent flip in the sign of the next partial remainder.

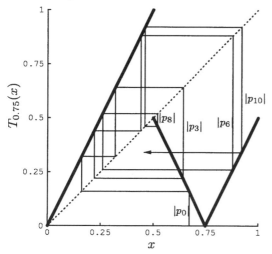

Figure 1. Following partial remainder magnitudes graphically for $D = 0.75$ and $p_0 = 0.67$.

Figure 1 shows an example of following the trajectory of a single partial remainder for a particular divisor. After ten applications of the $T_{0.75}$, there is not any obvious regular pattern, although we expect to see one eventually since the quotient is rational in this case*. Of course, most numbers are not rational and we can deduce that for most numbers, the transformation will never exhibit a repeating pattern. In figures 2 and 3, we see that a very small change in the value of the initial partial remainder quickly produces large differences in the observed behaviour of the subsequent partial remainders. Our system appears to be chaotic (it certainly has sensitive dependence on initial conditions and is topologically transitive), and, if this the case, we will gain little understanding by studying the trajectories of individual partial remainders. The logical next step is to study the behaviour of distributions of points over the whole interval.

---

*With redundant representations, rational numbers can have aperiodic representations, though we do not expect this to happen.

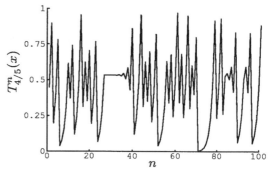

Figure 2. The result of applying $T_{4/5}$ to $x = \frac{\pi}{7}$ one hundred times.

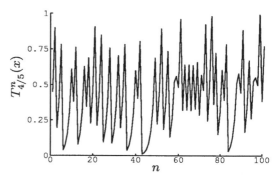

Figure 3. The result of applying $T_{4/5}$ to $x = \frac{\pi}{7} + 0.00001$ one hundred times.

The area of understanding the behaviour of ensembles of points under repeated transformation is the realm of dynamical systems theory. For the remainder of this paper, we assume a certain amount of familiarity with the fundamentals of dynamical systems theory (or ergodic theory), which requires some basic understanding of measure theory. We will include a few helpful background material definitions, but mostly we will provide references. A very good introduction to the study of chaotic systems is Lasota and Mackey's book *Chaos, Fractals, and Noise* [7]. For a more detailed introduction to ergodic theory (along with the necessary measure theory needed to understand this paper), Peter Walters's book *An Introduction to Ergodic Theory* [21] and Karl Petersen's book *Ergodic Theory* [15] are highly recommended.

Definition 2 (Probability Space). If $\mathcal{B}$ is a σ-algebra on subsets of a set X and if m is a measure on $\mathcal{B}$ where $m(X) = 1$, then the triple $(X, \mathcal{B}, m)$ is called a *probability space*. (See [21, pp. 3–9] and [7, pp. 19–31] for a good overview of basic measure theory and Lebesgue integration.)

Definition 3 (Perron-Frobenius operator). For a probability space $(X, \mathcal{B}, m)^\dagger$, the *Perron-Frobenius* operator $P : L^1 \to L^1$ associated with a non-singular transformation $T : X \to X$ is defined by

---

$\dagger$For a probability space $(X, \mathcal{B}, m)$, the L^1 space of $(X, \mathcal{B}, m)$ is the set of $f : X \to \mathbb{R}$ satisfying $\int_X |f(x)|\, dm < \infty$.

$$\int_B Pf(x)\,\mathrm{d}m = \int_{T^{-1}(B)} f(x)\,\mathrm{d}m, \quad \text{for } B \in \mathcal{B}.$$

For a piecewise monotonic $C^{2\ddagger}$ transformation T with n monotonic pieces, we can give an explicit formula for the Perron-Frobenius operator. Let $A = \{A_1, A_2, \ldots, A_n\}$ be the partition of X which separates T into n pieces. For $i \in \{1, \ldots, n\}$, let $t_i(x)$ represent the natural extension of the ith C^2 function $T(x)|_{A_i}$. The Perron-Frobenius operator for T is then

$$Pf(x) = \sum_{i=1}^{n} \left| \frac{\mathrm{d}}{\mathrm{d}x} t_i^{-1}(x) \right| f(t_i^{-1}(x)) \cdot 1_{t_i(A_i)}(x).$$

In particular, for T_D (as in (1)),

$$\begin{aligned} Pf(x) = &\tfrac{1}{2} f(\tfrac{1}{2}x) \cdot 1_{[0,1)}(x) \\ &+ \tfrac{1}{2} f(D - \tfrac{1}{2}x) \cdot 1_{(0,2D-1]}(x) \\ &+ \tfrac{1}{2} f(D + \tfrac{1}{2}x) \cdot 1_{[0,2-2D)}(x). \end{aligned} \quad (2)$$

With (2) we can show precisely what happens to an initial distribution of points (described by an integrable function) after they are repeatedly transformed under T_D. Figures 4 and 5 show what happens to two different initial distribution of points after five applications of the Perron-Frobenius operator associated with $T_{3/5}(x)$. By the fifth application, the distributions look remarkably similar. One might guess that they are both approaching the same final distribution. This situation is in marked contrast to the chaotic behaviour observed in figures 2 and 3.

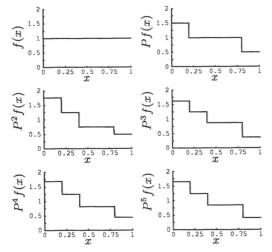

Figure 4. The result of applying the Perron-Frobenius operator P associated with $T_{3/5}$ to $f(x) = 1$ **six times**.

$\ddagger C^2$ denotes the set of all functions with two continuous derivatives.

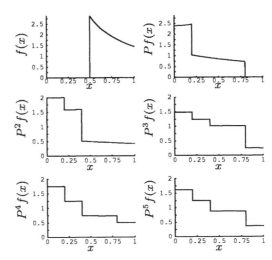

Figure 5. The result of applying the Perron-Frobenius operator P associated with $T_{3/5}$ to

$$f(x) = \frac{1}{\log 2} \int_{1/2}^{1} \frac{\mathrm{d}x}{x} \text{ six times.}$$

Definition 4 (Stationary Distribution). Let $(X, \mathcal{B}, m)$ be a probability space, let P be the Perron-Frobenius operator associated with a non-singular transformation $T : X \to X$, and let L^1 denote the L^1 space of $(X, \mathcal{B}, m)$. If $f \in L^1$ is such that $Pf \stackrel{\circ}{=} f$,[§] then f is called a *stationary distribution* of T.

A practical use of the Perron-Frobenius operator is in deriving and verifying the equations of stationary distributions for given divisors. As an example of this we verify the correctness of a previously known stationary distribution for $D \in [\tfrac{3}{4}, 1)$. An exact equation for the stationary distribution when $D \in [\tfrac{3}{4}, 1)$ was first given by Freiman [4] and is restated by Shively [19] as

$$f(x) = \frac{1}{D} 1_{[0,2D-1)}(x) + \frac{1}{2D} 1_{[2D-1,1)}(x). \quad (3)$$

To verify that this is a stationary distribution function, we begin by applying the Perron-Frobenius operator as given in (2) to (3) and verifying that $Pf(x) \stackrel{\circ}{=} f(x)$. So then, applying P to f we get

$$\begin{aligned} Pf(x) = &\tfrac{1}{2} \left(\tfrac{1}{D} 1_{[0,2D-1)}(\tfrac{1}{2}x) + \right. \\ &\quad \left. \tfrac{1}{2D} 1_{[2D-1,1)}(\tfrac{1}{2}x) \right) 1_{[0,1)}(x) \\ + &\tfrac{1}{2} \left(\tfrac{1}{D} 1_{[0,2D-1)}(D - \tfrac{1}{2}x) + \right. \\ &\quad \left. \tfrac{1}{2D} 1_{[2D-1,1)}(D - \tfrac{1}{2}x) \right) 1_{(0,2D-1]}(x) \\ + &\tfrac{1}{2} \left(\tfrac{1}{D} 1_{[0,2D-1)}(D + \tfrac{1}{2}x) + \right. \\ &\quad \left. \tfrac{1}{2D} 1_{[2D-1,1)}(D + \tfrac{1}{2}x) \right) 1_{[0,2-2D)}(x). \end{aligned}$$

Assuming that $D \in [\tfrac{1}{2}, 1)$, and observing that $x \in [0, 1)$,

---

[§]The $\circ$ symbol will be used to indicate that a given relation holds except possibly on a set of measure zero.

$$Pf(x) = \tfrac{1}{2}\left(\tfrac{1}{D}1_{[0,4D-2)}(x)+\right.$$
$$\left.\tfrac{1}{2D}1_{[4D-2,1)}(x)\right)1_{[0,1)}(x)$$
$$+\tfrac{1}{2}\left(\tfrac{1}{D}1_{(2-2D,1)}(x)+\right.$$
$$\left.\tfrac{1}{2D}1_{(0,2-2D]}(x)\right)1_{(0,2D-1]}(x)$$
$$+\tfrac{1}{2}\left(\tfrac{1}{2D}1_{[0,2D-1)}(x)\right)1_{[0,2-2D)}(x).$$

Finally, assuming that $D \in [\tfrac{3}{4}, 1)$, we have

$$Pf(x) = \tfrac{1}{2D}1_{[0,1)}(x) + \tfrac{1}{2D}1_{(2-2D,2D-1]}(x)+$$
$$\tfrac{1}{4D}1_{(0,2-2D]}(x) + \tfrac{1}{4D}1_{[0,2-2D)}(x)$$
$$= \tfrac{3}{4D}1_{[0,2-2D)}(x) + \tfrac{1}{4D}1_{(0,2-2D]}(x)$$
$$+ \tfrac{1}{2D}1_{[2-2D,2D-1)}(x)$$
$$+ \tfrac{1}{2D}1_{(2-2D,2D-1]}(x) + \tfrac{1}{2D}1_{[2D-1,1)}(x)$$
$$\stackrel{\circ}{=} \tfrac{1}{D}1_{[0,2D-1)}(x) + \tfrac{1}{2D}1_{[2D-1,1)}(x) = f(x).$$

In the case of variable quotient-bits-per-cycle algorithms such as the original SRT division, one of the primary uses of a formula for the distribution of partial remainders is for calculating the *shift average* for a given divisor. The shift average is the average number uses of the shift register (single shift or multiplication by two) between uses of the adder. Under the assumption that a register shift is a much faster operation than using the adder, the shift average gives a useful characterization of the expected performance of our algorithm for a given divisor. With (3), we know the fraction of bits that require the use of the adder. To calculate the average number of zero bits generated between non-zero bits (bits requiring use of the adder), we take the reciprocal of the fraction of bits that require the adder. We calculate the shift average for a divisor $D \in [\tfrac{3}{4}, 1)$ to be

$$s(D) = \left(1 - \tfrac{1}{2D}\right)^{-1} = \tfrac{2D}{2D-1}. \quad (4)$$

Unfortunately, since we have not proven that the stationary distributions from SRT division are unique, we have no way of knowing whether or not the shift average calculation in (4) is correct. To prove that all stationary distributions are unique, we need to show that T_D is ergodic for all $D \in [\tfrac{1}{2}, 1)$. Freiman [4] shows that T_D is ergodic for rational D, but we extend this result to real D. In the next section we give a summary of our results from [10] that let us show that all T_D are Bernoulli. It is known that having the Bernoulli property implies ergodicity. Ergodicity is defined as follows:

Definition 5 (Ergodic [7]). Let $(X, \mathcal{B}, m)$ be a probability space and let a nonsingular transformation $T : X \to X$ be given. Then T is *ergodic* if for every set $B \in \mathcal{B}$ such that $T^{-1}(B) = B$, either $m(B) = 0$ or $m(X \setminus B) = 0$.

3 Bernoulli Property

Our central result, which we present in this section, is that the class of transformations of the interval that characterizes SRT division for all real divisors D has the property that each transformation T_D is Bernoulli. Although the basic concept of a Bernoulli shift (the things to which transformations having the Bernoulli property are isomorphic) is not difficult, a complete definition requires enough auxiliary concepts from measure theory (concepts not used anywhere else in this paper) that we chose to refer the interested reader to [14, 15, 18, 21]. Neither an understanding of Bernoulli shifts, nor a formal definition of what it means to be Bernoulli is required to follow our presentation. Having said this, we should mention informally the connection between Bernoulli shifts and transformations having the Bernoulli property.

The transformation T_D is an non-invertible endomorphism of the unit interval. This means that from a given partial remainder we can predict all future partial remainders, but we cannot uniquely predict past partial remainders. There is a natural way (called the natural extension) to make our transformation invertible (an automorphism) on a larger space. Specifically, each non-invertible transformation T_D having the Bernoulli property has an extension to an automorphic transformation, isomorphic to a two-sided Bernoulli shift [15, pp. 13,276]. From the way that entropy for a transformation is defined, the entropy for an automorphic Bernoulli transformation associated with a non-invertible Bernoulli transformation is the same as the entropy for the non-invertible Bernoulli transformation. By proving that all transformations T_D are Bernoulli, and by proving that entropy of each T_D is the same, we will be able to conclude that the natural extensions of SRT division algorithms are isomorphic to each other for all divisors.

Due to space limitations, we are unable to present our proofs in this paper. Instead, we will summarize our results and mention some of the the external theorems that were central to our proofs. The omitted proofs and more results can be found in [10].

Our ability to prove that SRT division is Bernoulli, largely depends on the following two theorems of Bowen [2]:

Theorem 1 (of Bowen [2]). *Let T be a piecewise C^2 map of $[0,1]$, μ be a smooth T-invariant probability measure, and $\lambda = \inf_{0 \leq x \leq 1} |f'(x)| > 1$. If the dynamical system (T, μ) is weak-mixing, then the natural extension of (T, μ) is Bernoulli.*

Theorem 2 (of Bowen [2]). *With T and μ as in Theorem 1, (T, μ) will be weak-mixing if T is expanding.*

With these theorems, we are only left to show that SRT division has the expanding property.

Definition 6 (of Bowen [2], Expanding). We will say that a transformation T on an interval is *expanding* if it has the property that $\sup_{n>0} \mu(T^n U) = 1$ for all open intervals U with $\mu(U) > 0$, where μ is any normalized measure that is absolutely continuous with respect to Lebesgue measure.

The definition for expanding allows us to focus on the behaviour of intervals of points, rather than having to worry about arbitrary sets of points with non-zero measure. In [10], we are able to show that that that the following sequence of subsets of intervals is expanding:

$U_1 = U$ and

$$U_{i+1} = \begin{cases} T_D(U_i) & : U_i^\circ \subseteq [0, \tfrac{1}{2}) \text{ or } U_i^\circ \subseteq [\tfrac{1}{2}, 1) \\ T_D(U_i \cap [0, \tfrac{1}{2})) & : U_i^\circ \not\subseteq [0, \tfrac{1}{2}) \text{ and } U_i^\circ \not\subseteq [\tfrac{1}{2}, 1) \\ & \text{and } m(U_i \cap [0, \tfrac{1}{2})) \geq m(U_i \cap [\tfrac{1}{2}, 1)) \\ T_D(U_i \cap [\tfrac{1}{2}, 1)) & : U_i^\circ \not\subseteq [0, \tfrac{1}{2}) \text{ and } U_i^\circ \not\subseteq [\tfrac{1}{2}, 1) \\ & \text{and } m(U_i \cap [0, \tfrac{1}{2})) < m(U_i \cap [\tfrac{1}{2}, 1)). \end{cases}$$

$\{U_i\}_{i=1}^\infty$ is constructed by throwing away the smaller piece whenever an interval would be split in two by T_D. Since each $U_i \in \{U_i\}_{i=1}^\infty$ is a subset of the same set of interval that arise by just repeatedly applying T_D to the same initial U, showing that $\{U_i\}_{i=1}^\infty$ eventually fills the entire interval X implies that T_D is expanding. The sequence $\{U_i\}_{i=1}^\infty$ greatly simplifies the analysis because intervals never become fragmented.

In order to show that T_D satisfies the other requirements of Bowen's theorems, we must show that there exists a measure μ, which is invariant under T_D. We were able to establish that all T_D have invariant measures from the following theorem of Lasota and Yorke [8]:

Theorem 3 (of Lasota and Yorke [8]). *Let $(X, \mathcal{B}, m)$ be a probability space and let $T : X \to X$ be a piecewise C^2 function such that $\inf |T'| > 1$. If P is the Perron-Frobenius operator associated with T, then for any $f \in L^1$, the sequence $(\frac{1}{n} \sum_{k=0}^{n-1} P^k f)_{n=1}^\infty$ is convergent in norm to a function $f^* \in L_1$. The limit function f^* has the property that $Pf^* = f^*$ and consequently, the measure $d\mu^* = f^* dm$ is invariant under T.*

From the definition of T_D, we see that T_D is C^2 and that $\inf_{0 \leq x \leq 1} |T_D'(x)| = 2 > 1$ since $|T_D'(x)| = 2$ for all x for which the derivative is defined. By Theorem 3 we can conclude that there exists a smooth T_D-invariant probability measure μ. Given that T_D is expanding, we see that Theorem 2 holds. Hence, (T_D, μ) is weak-mixing and, by Theorem 1 (T_D, μ) is Bernoulli.

We mention here that the *natural extensions* of (T, μ) are the associated automorphic transformations that we alluded to at the beginning of this section. See Petersen [15, p. 13] for an exact definition.

Knowing that all T_D are Bernoulli is a very useful property because we can use entropy as a complete invariant to show isomorphism amongst the two-sided Bernoulli shifts associated with T_D that have the same entropy. This comes from the contribution of Ornstein to the Kolmogorov-Ornstein Theorem.

Theorem 4 (of Kolmogorov [5, 6] and Ornstein [13]). *Two Bernoulli shifts are isomorphic if and only if they have the same entropy.*

In general it can be very difficult to calculate the entropy for a class of transformations straight from the definition of entropy (see Walters, [21, pp. 75-87] for a definition of the entropy of a transformation). Even with the many standard formulas that have been derived for calculating entropy, a great number of systems found in practice are not covered. Simple SRT division is one such dynamical system that is not easy to calculate the entropy for from results found in standard textbooks on ergodic theory. Fortunately, a result by Ledrappier [9] does allow us to calculate the entropy for simple SRT division. With Ledrappier's results, we are able to show that the following formula for entropy holds:

$$h(T_D) = \int \log |T_D'| \, d\mu = \log 2 \int d\mu = \log 2.$$

This formula was shown by Rohlin [17] to hold for a smaller class of transformations, which does not include the T_D associated with SRT division.

With the above results, we have established isomorphism amongst the automorphic transformations (or natural extensions) associated with simple SRT division transformations by an application of the Kolmogorov-Ornstein Theorem. The key to obtaining this result was being able to show that T_D has Bowen's expanding property. In Section 4, we extend the results of this section to a more general type of SRT division.

4 Multi-Threshold SRT Division

A simple optimization to the original SRT division algorithm, at least with the historical concern of avoiding additions and subtractions in mind, is the inclusion of additional divisors to increase the shift average. In this section, we give a summary of our proof that all such division algorithms with reasonable assumptions on the separation of the divisor multiples have the expanding property. We begin by giving a precise definition of the class of "multi-threshold" SRT transformations.

Definition 7. Let $\alpha \in \mathbb{R}^n$ be such that
(a) $0 < \alpha_1 < \alpha_2 < \ldots < \alpha_n$, and
(b) For all $x, D \in [\tfrac{1}{2}, 1)$, there exists $i \in \{1, \ldots, n\}$ such that $|\alpha_i D - x| < \tfrac{1}{2}$.

We define $\mathfrak{A}_n$ to be the set of all $\alpha \in \mathbb{R}^n$, satisfying conditions (a) and (b). Also, $\mathfrak{A} = \bigcup_{n \in \mathbb{N}} \mathfrak{A}_n$.

Definition 8 (Peaks and Valleys). Given an $\alpha \in \mathfrak{A}_{n \geq 2}$, the point of intersection between two lines $f(x) = 2(x - \alpha_i D)$ and $g(x) = 2(\alpha_{i+1} D - x)$ will be called a *peak* and is denoted by $\psi_i = (\tfrac{1}{2} D(\alpha_{i+1} + \alpha_i), D(\alpha_{i+1} - \alpha_i))$. For convenience, we will refer to the abscissa as $\psi_i^x = \tfrac{1}{2} D(\alpha_{i+1} + \alpha_i)$, and to the ordinate as $\psi_i^y = D(\alpha_{i+1} - \alpha_i)$. The point of intersection of the two lines $f(x) = 2(\alpha_i D - x)$ and $g(x) = 2(x - \alpha_i D)$ is $(\alpha_i D, 0)$ and will be called a *valley*.

Definition 9. For a $D \in [\tfrac{1}{2}, 1)$ and $\alpha \in \mathfrak{A}$, define the transformation $T_{D, \alpha}(x) : [0, 1) \to [0, 1)$. For $\alpha \in \mathfrak{A}_1$, we get the familiar transformation

$$T_{D, \alpha}(x) = \begin{cases} 2x & : 0 \leq x < \tfrac{1}{2} \\ |2(D - x)| & : \tfrac{1}{2} \leq x < 1. \end{cases}$$

For $\alpha \in \mathfrak{A}_2$,

$$T_{D,\alpha}(x) = \begin{cases} 2x & : 0 \leq x < \frac{1}{2} \\ |2(\alpha_1 D - x)| & : \frac{1}{2} \leq x < \psi_1^x \\ |2(\alpha_2 D - x)| & : \frac{1}{2} \leq x \text{ and } \psi_1^x \leq x < 1. \end{cases}$$

For $\alpha \in \mathfrak{A}_{n \geq 3}$,

$$T_{D,\alpha}(x) = \begin{cases} 2x & : 0 \leq x < \frac{1}{2} \\ |2(\alpha_1 D - x)| & : \frac{1}{2} \leq x < \psi_1^x \\ |2(\alpha_i D - x)| & : \frac{1}{2} \leq x \text{ and } \psi_i^x \leq x < \psi_{i+1}^x \\ |2(\alpha_n D - x)| & : \frac{1}{2} \leq x \text{ and } \psi_{n-1}^x \leq x < 1. \end{cases}$$

Definition 10. Define $\mathfrak{M}_n = \{T_{D,\alpha} : D \in (\frac{1}{2}, 1], \alpha \in \mathfrak{A}_n\}$ and define $\mathfrak{M} = \bigcup_{n \in \mathbb{N}} \mathfrak{M}_n$. We call $\mathfrak{M}_n$ the set of all *n-threshold SRT division transformations* and we call $\mathfrak{M}$ the set of *multi-threshold SRT division transformations*.

Condition (b) in Definition 7 guarantees that the division algorithm generates a new quotient bit every step. In order to restrict the set of $\mathfrak{M}$ begin considered, we make the following definition to place restrictions on the relative distance between divisor multiples:

Definition 11 (Separation). For $\alpha \in \mathfrak{A}_n$, we define the *separation* in α as

$$separation(\alpha) = \max_{i \in \{1, \ldots, n-1\}} \frac{\alpha_{i+1}}{\alpha_i}.$$

If $separation(\alpha) = r$, we say that "the divisor multiples in α are separated by at most a factor of r."

Table 2 shows an example dividing 0.67 by 0.75 using multi-threshold SRT division with $\alpha = (0.75, 1, 1.25)$. This example is performing the same calculation as in table 1, but it has computed the dividend with twice as many digits of precision with the same effective number of uses of the adders. We say "effective" because in multi-threshold SRT division, there are several adders working in parallel. In a real implementation of multi-threshold SRT division, the values for α must be carefully chosen so that not too much overhead is required to select a good partial remainder. There is also a tradeoff between the amount of overhead in choosing a good partial remainder and the precision to with which a good partial remainder is selected.

Table 2. Multi-threshold SRT division example

p_0	$= 0.67$	$= 0.67$	Q_0	$= 1$
p_1	$= 2(0.67 - \alpha_2 D)$	$= -0.16$		
p_4	$= 2(2^2(-0.16) + \alpha_1 D)$	$= -0.155$	Q_3	$\doteq 0.90625$
p_7	$= 2(2^2(-0.155) + \alpha_1 D)$	$= -0.115$	Q_6	$\doteq 0.89453125$
p_{11}	$= 2(2^3(-0.115) + \alpha_3 D)$	$= 0.035$	Q_{10}	$\doteq 0.893310547$
p_{16}	$= 2(2^4(0.035) - \alpha_1 D)$	$= -0.005$	Q_{15}	$\doteq 0.893333435$
p_{24}	$= 2(2^7(0.005) + \alpha_1 D)$	$= -0.155$	Q_{23}	$\doteq 0.893333346$

By constructing a sequence of intervals similar to what was done for simple SRT division, we are able to prove that a multi-threshold SRT division transformation $T_{D,\alpha} \in \mathfrak{M}$ is expanding when $separation(\alpha) \leq \frac{5}{3}$. By making use of Theorems 1, 2 and 3 from the previous section, it is straightforward to show that $T_{D,\alpha} \in \mathfrak{M}$ is Bernoulli when $separation(\alpha) \leq \frac{5}{3}$.

The calculation for entropy in multi-threshold SRT division follows the same method used for single divisor SRT division. Ledrappier's results extend in the same way to multi-threshold SRT division, and we are able to show that the entropy of any $T_{D,\alpha} \in \mathfrak{M}$ with $separation(\alpha) \leq \frac{5}{3}$ is $\log 2$.

In [10] we proved that for all $T_{D,\alpha} \in \mathfrak{M}$, if $separation(\alpha) \leq \frac{5}{3}$, then $T_{D,\alpha}$ is Bernoulli. In [10] we prove that for $T_{D,\alpha} \in \mathfrak{M}_{n \geq 4}$, if $separation(\alpha) > \frac{5}{3}$, then for each $D \in [\frac{1}{2}, 1)$, there exist uncountably many α for which $T_{D,\alpha}$ is not ergodic. We also show that for $T_{D,\alpha} \in \mathfrak{M}_3$, with $separation(\alpha) \geq \frac{9}{5}$ there are examples of non-ergodic systems and similarly for $T_{D,\alpha} \in \mathfrak{M}_2$, $separation(\alpha) > 3$ there are non-ergodic systems. We do not provide the proofs here, but instead present figure 6 which demonstrates the existence of a large class of non-ergodic $T_{D,\alpha} \in \mathfrak{M}_{n \geq 4}$ when $separation(\alpha) > \frac{5}{3}$. In this example, $n = 4$, $D = \frac{11}{16}$, $\alpha = (\frac{37}{66}, \frac{21}{22}, 1, \frac{59}{33})$, and $separation(\alpha) = \frac{5}{3} + \frac{5}{51}$. The thick lines represent $T_{D,\alpha}$ and the coarse dashed line represents the necessary separation restriction on α to guarantee that $T_{D,\alpha}$ is ergodic. Partial remainders in the set $A = [\frac{11}{48}, \frac{13}{48}] \cup [\frac{22}{48}, \frac{26}{48}] \cup [\frac{44}{48}, 1)$ are mapped back to A by $T_{D,\alpha}$. This means that $T_{D,\alpha}$ is not ergodic, and therefore not Bernoulli.

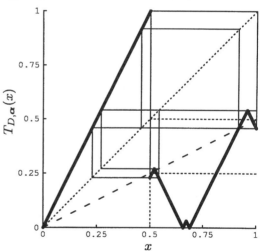

Figure 6. An example of a non-ergodic system for $T_{D,\alpha} \in \mathfrak{M}_{n \geq 4}$.

5 Conclusions

The original question that inspired this paper was "Is simple SRT division ergodic for all real divisors?" In pursuing the answer to this problem, we discovered that not only is simple SRT division ergodic for all divisors, but it is also Bernoulli. Having established a Bernoulli property, and having calculated the entropy for our transformations, we were able to use the Kolmogorov-Ornstein theorem to conclude that our transformations are equivalent to each other in the sense that their natural extensions are isomorphic. In proving these important properties for simple SRT division, we made extensive use of more general results from dynamical systems theory. Consequently, our results were shown to be easily extensible to more general division systems. In general, it is difficult to prove that a particular class of transformations are ergodic or Bernoulli. Our results have provided an effective means of proving both of these properties for a large class of SRT-like division algorithms.

From the standpoint of understanding an algorithm's expected performance, it is necessary to know that when a stationary distribution is found, it is unique. Having established the uniqueness of stationary distributions, the next step is to find the actual stationary distribution for as wide a class of transformations as possible. In section 3, we verified a known expression for the stationary distribution function for T_D where $D \in [\frac{3}{4}, 1)$. In addition, many of the stationary distribution functions have been classified by Shively and Freiman for $D \in [\frac{3}{5}, \frac{3}{4}]$, although the derivations are not nearly as simple as for $D \in [\frac{3}{4}, 1)$. It turns out that things become very complicated when $D \in [\frac{1}{2}, \frac{3}{5}]$. In his thesis [19], Shively shows many interesting properties for the stationary distribution functions in this region. For example, he shows that there are many different intervals of D where there are an infinite number of different stationary distribution equations. As such, the graph of the shift average for $D \in [\frac{1}{2}, \frac{3}{5}]$ is far from complete and appears to have a complex pattern (from the few points that have been plotted in this region). This is surprising considering the simplicity of the underlying transformation. A better understanding of this final region of simple SRT division would be an interesting goal to pursue.

In the work of Freiman [4], it was first shown that the shift average for $D \in [\frac{3}{5}, \frac{3}{4}]$ is constantly 3, which can be easily shown to be the maximum possible shift average. This property was then used by Metze [11] to obtain a version of SRT division that has an expected shift average of 3 for all divisors. Another area to pursue would be to explore shift averages for multi-threshold SRT division and, if other plateaus are found, they could possibly be used to obtain higher expected shift averages for all possible divisors. Undoubtedly, obtaining a complete understanding of the stationary distribution functions for multi-threshold division would be even more difficult than it is for simple SRT division. It is possible that such results in this area could lead to improvements in modern SRT division. Related to this, it would be interesting to attempt to extend the results of this paper to modern SRT division.

References

[1] D. E. Atkins. Higher-radix division using estimates of the divisor and partial remainders. *IEEE Trans. on Computers*, C-17(10), Oct. 1968.

[2] R. Bowen. Bernoulli maps of the interval. *Israel J. Math.*, 28(1-2):161–168, 1977.

[3] J. Cocke and D. W. Sweeney. High speed arithmetic in a parallel device. Technical report, IBM, February 1957.

[4] C. V. Freiman. Statistical analysis of certain binary division algorithms. *Proc. IRE*, 49:91–103, 1961.

[5] A. N. Kolmogorov. A new invariant for transitive dynamical systems. *Dokl. Akad. Nauk SSSR*, 119:861–864, 1958.

[6] A. N. Kolmogorov. Entropy per unit time as a metric invariant of automorphisms. *Dokl. Akad. Nauk SSSR*, 124:754–755, 1959.

[7] A. Lasota and M. C. Mackey. *Chaos, fractals, and noise*. Springer-Verlag, New York, second edition, 1994. Stochastic aspects of dynamics.

[8] A. Lasota and J. A. Yorke. On the existence of invariant measures for piecewise monotonic transformations. *Trans. Amer. Math. Soc.*, 186:481–488 (1974), 1973.

[9] F. Ledrappier. Some properties of absolutely continuous invariant measures on an interval. *Ergodic Theory Dynamical Systems*, 1(1):77–93, 1981.

[10] M. McCann. S-R-T division algorithms as dynamical systems. Master's thesis, University of British Columbia, April 2002.

[11] G. Metze. A class of binary divisions yielding minimally represented quotients. *IRE Trans. on Electronic Computers*, EC-11:761–764, Dec. 1961.

[12] M. Nadler. A high-speed electronic arithmetic unit for automatic computing machines. *Acta Tech. (Prague)*, (6):464–478, 1956.

[13] D. S. Ornstein. Bernoulli shifts with the same entropy are isomorphic. *Advances in Math.*, 4:337–352, 1970.

[14] D. S. Ornstein. *Ergodic theory, randomness, and dynamical systems*. Yale University Press, New Haven, Conn., 1974. James K. Whittemore Lectures in Mathematics given at Yale University, Yale Mathematical Monographs, No. 5.

[15] K. Petersen. *Ergodic theory*. Cambridge University Press, Cambridge, 1983.

[16] J. E. Robertson. A new class of digital division methods. *IRE Trans. Electronic Computers*, EC-7:218–222, Sept. 1958.

[17] V. A. Rohlin. Exact endomorphisms of a Lesbesgue space. *Amer. Math. Soc. Transl.*, 39, 1964.

[18] P. Shields. *The theory of Bernoulli shifts*. The University of Chicago Press, Chicago, Ill.-London, 1973. Chicago Lectures in Mathematics.

[19] R. Shively. *Stationary distribution of partial remainders in S-R-T digital division*. PhD thesis, University of Illinois, 1963.

[20] K. D. Tocher. Techniques of multiplication and division for automatic binary computers. *Quart. J. Mech. Appl. Math.*, 11:364–384, July–September 1958.

[21] P. Walters. *An introduction to ergodic theory*. Springer-Verlag, New York, 1982.

A New Iterative Structure for Hardware Division: the Parallel Paths Algorithm

Eric Rice   Richard Hughey

Department of Computer Engineering
University of California, Santa Cruz, CA 95064
E-mail: elrice,rph@soe.ucsc.edu

Abstract

This paper presents a new approach to hardware division—the parallel paths algorithm. In this approach, prescaling allows the division recurrence to be implemented by three processes which can be calculated in parallel during iterations. While two of the processes must complete in a single iteration, the third—which includes the most expensive division operations—can be calculated over multiple iterations. Iteration latency is determined by the slowest of the three paths, and in many cases can be limited to that of carry-save addition and latching. A radix-4 implementation of the algorithm is shown to achieve better performance than other commonly used methods while requiring a modest increase in area.

Index Terms: Computer arithmetic, hardware division, prescaling, linear convergence.

1 Introduction

Digit-recurrence algorithms used in special-purpose division hardware involve several steps, the most problematic of which is quotient selection. In basic SRT division [1, 2], quotient selection represents half or more of iteration latency [3]. As a result, considerable effort has been made to reduce its impact on division latency.

Two strategies are of particular interest to speed quotient selection. The first is to overlap multiple SRT stages in a single iteration, making use of multiple speculative calculations that can be selected among quickly when a key result becomes available. This overlap can involve quotient selection (used in the Ultra Sparc64 processor [4]), partial remainder formation (used in the Hal Sparc64 processor [5]), or both [6]. These are currently the best approaches for accelerating division without significant increase in area.

A second strategy to reduce the impact of quotient selection is to prescale the divisor close to unity. When prescaled with sufficient precision, this allows quotient selection to be performed by simply rounding or truncating the partial remainder [7]. While prescaling has been proposed for low-radix algorithms, the overhead associated with prescaling prevents them from being as efficient as overlapped SRT. High-radix prescaling algorithms proposed by Ercegovac et. al. [8] and by Wong and Flynn [9] lead to faster division than overlapped SRT but require significant increases in area.

A prescaling algorithm proposed by Ercegovac and Lang [10] uses prescaling in a different way. Although it does not quite achieve sufficient compensation for the prescaling overhead, it is of interest because of its similarity to our proposed algorithm in that (1) prescaling is performed to greater precision than needed by the iteration radix, and (2) this added precision is used to allow quotient selection to be calculated over multiple iterations (this is fixed at two iterations in the cited paper).

This paper presents a new strategy that can eliminate quotient selection from the critical path altogether. In this approach, prescaling allows the division recurrence to be implemented by three separate processes which can be calculated in parallel during iterations. Iteration latency is determined by the slowest of the three paths, and in many cases can be limited to that of carry-save addition and latching. We call this approach the parallel paths algorithm.

This paper is organized as follows. After introducing basic terms and concepts in Section 2, Section 3 describes the Parallel Paths algorithm. A radix-4 implementation is described in Section 4, followed by a comparison of the implementation with other efficient algorithms in Section 5.

2 Division Basics and Notation

We address the problem of calculating $Q=a/b$, where the standard recurrence for an iterative division algorithm is:

$$P_{i+1} = P_i - bq_i. \qquad (1)$$

for partial remainder P_i (where $P_0=a$) and partial quotient q_i. Calculating equation (1) involves several steps:

- SEL: Select next quotient digit q_i.
- DRV: Drive selected digit (encoded) to array of full adders that will be used for partial remainder update.

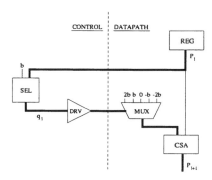

Figure 1. Basic radix-4 SRT algorithm.

- **MUX:** Use driven signals to multiplex appropriate (precalculated) bq_i.
- **CSA:** Perform partial remainder update via array of carry-save adders.
- **REG:** Latch new partial remainder.

Figures 1 and 2 illustrate how these basic blocks are used in a basic radix-4 SRT algorithm and a radix-4 SRT algorithm using overlapped radix-2 stages. In both cases, SEL is performed via table lookup based on b and the most significant bits of P_i after a short carry-propagate addition (CPA).

2.1 Prescaling algorithms

In a prescaling algorithm, a scaling factor M—an estimate of $1/b$—is used to scale a and b. The division problem then becomes:
$$Q = \frac{a}{b} = \frac{Ma}{Mb}.$$
and the recurrence becomes:
$$P_{i+1} = P_i - q_i(Mb) \quad (2)$$
$$= (P_i - q_i) + q_i(1 - Mb), \quad (3)$$
where $P_0 = Ma$.

The reformulation in equation (3) shows the relationship between the precision of the scaling factor and the number of quotient bits that can be retired in each iteration by rounding or truncating. Since $(P_i - q_i)$ can have an arbitrarily large number of leading zeros (we could use a very over-redundant quotient digit set and select $q_i = P_i$), the magnitude of $q_i(1 - Mb)$ determines the maximum number of bits that can be retired.

3 The Parallel Paths Algorithm

The parallel paths algorithm is based on equation (3), but aims at minimizing iteration latency rather than maximizing the number of bits that can be retired per iteration. The strategy is developed by studying the two terms of equation (3) and noticing two important differences between them.

The first difference is their complexity. While $(P_i - q_i)$ can be calculated quickly for some quotient selection methods (described below), adding $q_i(1 - Mb)$ requires the combined latency of SEL→DRV→MUX→CSA, the entire standard digit-recurrence iteration.

The second difference between the terms of equation (3) is the relative magnitudes of the two terms. While subtracting q_i in $(P_i - q_i)$ zeroes the leading fractional bits and significantly affects P_{i+1}, the magnitude of $q_i(1 - Mb)$ is determined by the accuracy of the scaling factor M.

The new approach uses this last fact to address the high latency of calculating q_i and adding $q_i(1 - Mb)$ to the partial remainder. By prescaling with sufficient precision, $q_i(1 - Mb)$ can be made small enough so that it will not affect quotient selection in one or more subsequent iteration(s). This allows postponing its introduction, spreading its latency over multiple iterations.

While this leads to partial remainders that are not fully adjusted with respect to previous quotient digits, it is easy to see the validity of the process:
$$\begin{aligned} Q &= \sum_{j=1}^{i} q_j + \frac{P_{i+1}}{Mb} \\ &= \sum_{j=1}^{i} q_j + \frac{P_i - q_i(Mb)}{Mb} \\ &= \sum_{j=1}^{i} q_j + \frac{(P_i - q_i)}{Mb} + \frac{q_i(1 - Mb)}{Mb}. \end{aligned}$$
Since $q_i(1 - Mb)$ will eventually be added to the partial remainder where its value will be divided by Mb, it does not matter that it has been postponed.

An iteration of the new approach involves three parallel processes. This is because addition of $q_i(1 - Mb)$ is divided into two parts. While SEL→DRV→MUX can be distributed over multiple iterations, its subsequent addition to the partial remainder (CSA) must occur within a single iteration since each iteration will introduce new partial product(s) to the partial remainder. The three processes are thus:

- **SUB** Calculation of $(P_i - q_i)$. (Despite its name and function, an appropriate SUB will generally not involve subtraction, as discussed below.)

- **SEL→DRV→MUX** Preparation of the partial products for partial remainder update. This requires selection of q_i, driving such to the CSA array, and multiplexing between $q_i(1 - Mb)$ values.

- **CSA** Partial remainder update through carry-save addition.

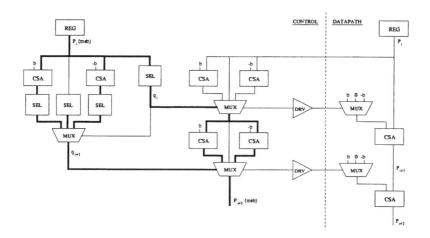

Figure 2. Radix-4 SRT using overlapped radix-2 stages (with overlapped quotient selection and overlapped partial remainder formation).

Both SUB and CSA update the partial remainder in each iteration and are performed simultaneously. SUB (based on q_i) adjusts the high-order bits of P_i, and CSA (based on q_{i-k} for k postponed stages) adjusts the lower-order bits. Since the latency of SEL→DRV→MUX can be 'hidden' by sufficiently postponing the introduction of partial products, iteration latency is often limited by the maximum of the SUB and CSA latencies.

Surprisingly, SUB is not the limiting factor when the right SEL is used. For example, by calculating quotient digits q_i via carry propagate addition (CPA) of the most significant bits of P_i followed by truncation, $(P_i - q_i)$ is simply the fractional result of the CPA plus the bits of P_i not included in the CPA (Figure 3). This SEL method thus allows $(P_i - q_i)$ to be calculated without first calculating q_i, and requires less time than a CSA.

Figure 4 illustrates how the parallel paths strategy can reduce iteration latency through increased prescaling precision and postponement of partial products. For every two additional bits of prescaling precision, each partial product can be postponed one additional iteration, since its magnitude relative to the 'local' partial remainder will be the same (ensuring the same bounds). The figure shows how this affects iteration latency for assumed latencies of $2\ t_{xor}$ for CSA, $6\ t_{xor}$ for SEL→DRV→MUX, and a SUB requiring $2\ t_{xor}$ or less. Four possible places for adding the partial products $q_i(1-Mb)$ are shown, along with corresponding prescaling requirements and iteration latencies. Note that the 2nd, 3rd and 4th latencies are determined by the latency of SEL→DRV→MUX divided by the number of iterations the partial products are postponed, but that latency cannot be reduced further due to CSA latency. (Also, note that the latency estimates given are for illustrative purposes only, as are the two leading zeros in partial products.)

Designing a good implementation of the parallel paths algorithm is a complicated optimization problem. In addition to many of the same parameter choices available in previous digit-recurrence algorithms, there is the added structural choice of how many stages to postpone introduction of partial products. The three-way race inherent in iterations leads to different priorities for evaluating a given design, including the requirement for a fast SUB and increased importance of CSA latency. There is also a tradeoff between prescaling costs and iteration costs, since extra prescaling precision often leads to reduced iteration latency.

While an important purpose of this paper is to introduce the parallel paths strategy (rather than a particular implementation) our exploration of this design space has led to several promising designs, including the following radix-4 algorithm that provides improved performance over the best currently used algorithms without significant increase in area.

4 Parallel Paths Implementation: a Radix-4 Example

To develop a radix-4 parallel paths algorithm based on $q_i \in \{-2, \ldots, 2\}$, we begin with a simpler related algorithm based on $q_i \in \{0, \ldots, 4\}$, and then convert it to the preferred digit set (preferred primarily because allowing $q_i = 3$ requires calculating a non-redundant $3(1-Mb)$ for efficient partial product formation).

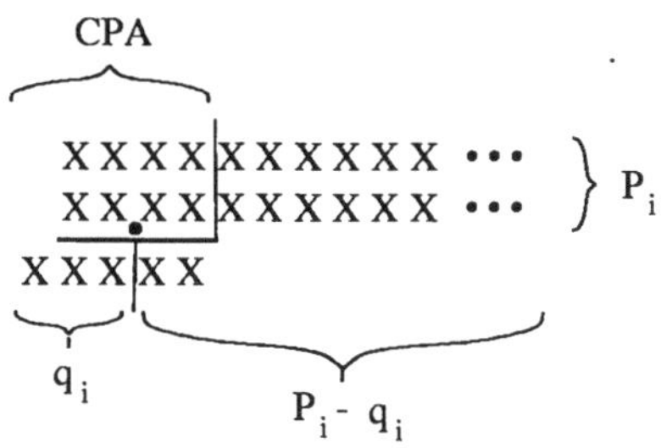

Figure 3. Implementing SEL via a CPA of the most significant bits of P_i allows fast calculation of $P_i - q_i$. (Each 'X' represents one bit.)

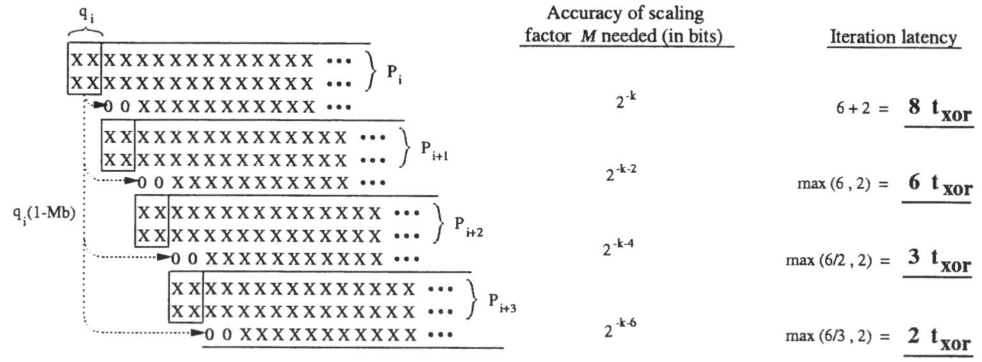

Figure 4. Reducing iteration latency in a radix-4 algorithm with the parallel paths algorithm.

4.1 Algorithm based on $q_i \in \{0, \ldots, 4\}$

The basic recurrence of the $q_i \in \{0, \ldots, 4\}$ algorithm is shown in Figure 5(a). The SEL and SUB functions are implemented by a CPA of P_i that incorporates two fractional bits (shaded region), with the integer result becoming q_i (implementing SEL). The fractional result (digits A and B) plus the bits of P_i not incorporated in the CPA, becomes $P_i - q_i$ (implementing SUB). P_{i+1} is formed through CSA addition of the less significant bits of P_i and $q_{i-2}(1-Mb)$, with digits A and B being forwarded directly to P_{i+1}.

Algorithm bounds are ensured by requiring that each partial product have four leading fractional zeros relative to the partial remainder to which it is added (Figure 5(b)). This ensures that the bits of P_{i+1} which determine q_{i+1} (shaded region) have a maximum value of 4.75, so that $q_{i+1} \leq 4$. Note that for the parallel paths algorithm one must assume that a worst-case partial product will be added to each P_i when checking bounds, since it is not based on q_i.

A bit-wise analysis of the CPA as done in Figure 5(b) is necessary. For example, one could instead calculate the maximum allowable P_i:

$$(P_{max} - q_{max} + [q(1-Mb)]_{max}) \times radix = P_{max}$$
$$(P_{max} - 4 + 1/16) \times 4 = P_{max}$$
$$P_{max} = 5.25,$$

but while this is in fact the value of P_{i+1} in Figure 5(b), a smaller P_i could cause CPA overflow. For example:

```
 1  1  1  1  0  0  0  0 ...
    1 .0  1  0  0  0  0 ...
```

would lead to the $q_i = 5$. While it would be possible to accomodate this by selecting $q_i = 4$ and leaving the remaining '1' for P_{i+1}, such would require a more complex (slow) SEL. Fortunately, as Figure 5(b) shows, the present implementation avoids partial remainders that would cause such CPA overflow.

For perspective, there are many ways to design a CPA-based radix-4 algorithm. For example, if SEL were implemented using a CPA that incorporated 3 fractional bits of P_i (instead of 2) just three leading fractional zeros would be needed in partial products (Figure 6(a)). In fact, just two leading zeros would be needed if the 3rd fractional bit of the partial product were also incorporated in the CPA (Fig-

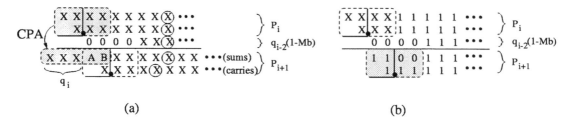

Figure 5. (a) Recurrence for $q_i \in \{0,\ldots,4\}$ algorithm. (The circled X's show the inputs and outputs of one full adder.) (b) Demonstrating algorithm bounds.

Figure 6. Two other CPA-based radix-4 schemes. Different forms of CPA (shaded regions) require different number of leading fractional zeros in partial products to ensure $q_{i+1} \leq 4$.

ure 6(b)). Note that the CPA could not also incorporate the 2nd fractional bit of the partial product (i.e., so that partial products would require just 1 leading fractional zero) because then the CPA could produce two carries, leading to $q_i=5$.

While each of the SEL methods in Figure 6 would require increased latency, one could compensate for this by postponing partial products a third iteration. Using the SEL shown in Figure 6(b) would then require the same prescaling accuracy as in the proposed implementation, but SEL→DRV→MUX would have the latency of an additional CSA to complete.

Iteration latency of the proposed algorithm is designed to be based on CSA latency. By postponing introduction of partial products two iterations, SEL→DRV→MUX has at least 2 t_{FA} to finish without affecting iteration latency, and 2.5 t_{FA} if the CSA is designed to allow one input to arrive late (which we will assume).

4.2 Converting the algorithm to $q_i \in \{-2,\ldots,2\}$

Converting the algorithm to $q_i \in \{-2,\ldots,2\}$ can be accomplished by inserting a constant '1' in the fifth fractional position of each partial product and later subtracting it when it 'migrates' to an integer position (Figure 7). Because the constant represents a '2' when it migrates out in a partial quotient, subtracting it neatly converts $q_i \in \{0,\ldots,4\}$ to $q_i \in \{-2,\ldots,2\}$ [1]. (Note that while the figure indicates that there is a '1' in each constant position, this digit becomes '0' when the corresponding partial product is negative.)

Adding a constant to each $q_i(1-Mb)$—rather for example than adding all of them to P_0—is necessary to ensure that partial remainders remain positive, since we need $q_i \in \{0,\ldots,4\}$ before subtracting 2. Thus we cannot 'save' one bit of prescaling precision by adding the constants to P_0. (For cases as in Figure 6(a) where the number of leading fractional zeros is odd—i.e. where the first non-zero position will represent a '1' when it migrates to an integer position—we can minimize prescaling requirements by adding half of each constant to the partial product and half to P_0.)

Prescaling requirements for this algorithm can be easily calculated. Since the 5th fractional digit of each partial product is used for the constant '1', each $q_i(1-Mb)$ requires 5 leading fractional zeros (if positive) or 5 leading fractional ones (if negative) relative to the partial remainder to which it is added. Since the radix point of this partial remainder has shifted four places from the partial remainder from which q_i is selected, we need $-2^{-(4+5)} < q_i(1-Mb) < 2^{-(4+5)}$, leading to:
$$-2^{-10} < (1-Mb) < 2^{-10}$$
$$-2^{-10} < b(1/b - M) < 2^{-10}.$$

Since the value of b can approach 2, this seems to imply that we need:
$$-2^{-11} < (1/b - M) < 2^{-11}.$$

---

[1] As pointed out by Ercegovac, the process of adding and later subtracting these constants implements a sort of rounding—the previous two partial products cause the fractional bits of each partial remainder to be overvalued by 5/8. Without this or some other form of rounding, it would be theoretically impossible to design an algorithm based on truncation without using an overredundant quotient digit set [3].

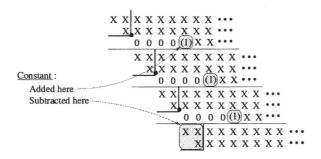

Figure 7. A $q_i \in \{0,\ldots,4\}$ radix-4 algorithm is converted to a $q_i \in \{-2,\ldots,2\}$ algorithm by adding a constant to each partial product and later subtracting it from each quotient digit.

However, because most methods for reciprocals produce estimates that are considerably more accurate for $b \approx 2$ than $b \approx 1$ (the curve $y = 1/x$ is flatter at $b \approx 2$), we can use methods aimed at just 2^{-10} reciprocal accuracy.

Based on analysis in [11] (though eventually found empirically) we found a linear approximation that produces an 11-bit redundant reciprocal estimate after a short multiply-accumulate:
$$M = C_0 - C_1(b - p),$$
where p is b truncated after four fractional bits. This method requires two lookup tables for C_0 and C_1 of sizes ($2^4 \times 11$) and ($2^4 \times 7$), respectively. In the next section we describe how M can be calculated using iteration hardware.

4.3 Implementation Details

Figure 8 shows the hardware blocks needed for iterations of the proposed implementation, excluding the scaling factor lookup tables and hardware for on-the-fly conversion of partial quotients. Two copies of iteration hardware are used because of the sensitivity of the parallel paths algorithm to latching latency.

There are several subtle features of the implementation. To ensure maximum efficiency, REG C must be latched 1 t_{xor} later than the other registers (its output will be the late input to the CSA).

Primary iteration hardware can be used to calculate M and the prescaled operands $(1 - Mb)$ and Ma. To see how this works, consider a generic multiply-add:
$$xy + z,$$
where x has fewer digits than y. By storing x in the most significant positions of REG B and selecting y in the Q-MUX, each iteration will process two radix-4 digits of x—multiplying each by y and adding the results to the partial remainder. Since iteration hardware left-shifts partial remainders 2 places after each CSA, the various partial products $x_i y$ end up correctly aligned, and the digits of x are slowly shifted out of REG B. (Note that at the beginning we need a right-shifted y in Q-MUX to compensate for these shifts.) By loading a right-shifted z into P_0 at the beginning (adjacent to x), iteration hardware will calculate a redundant $(xy + z)$ in REG B.

Based on this method, the choices of inputs to the two MUXes at the top of Figure 8 allow iteration hardware to calculate:

1) $C_0 - C_1(b - p) = M$
2) $1 - Mb$
3) Ma

before beginning division iterations.

5 Comparison with Other Algorithms

We now compare the radix-4 parallel paths algorithm presented above with several of the best competing algorithms for double precision division. In particular, our comparison includes:

- Basic radix-4 SRT.
- Radix-16 SRT using two overlapped radix-4 stages. (One of the best approaches currently in use.)
- Radix-512 very high radix algorithm (VHR) of Ercegovac et. al. (We do not include higher radices of this approach due to their large area requirements.)

In Table I we evaluate area and latency requirements for these algorithms, relying primarily on the methodology and results of Montuschi et. al. [12], where parameterized area and latency estimates are provided for common hardware blocks. We use their results for basic radix-4 SRT and radix-16 overlapped SRT without modification. For each algorithm, we also evaluate latency when a faster MUX/REG design is used, as proposed in [13]. (Due to lack of data, we assume the same areas when using this design.)

Figure 9 plots speed versus area for these algorithms. As can be seen, the parallel paths algorithm is slightly faster

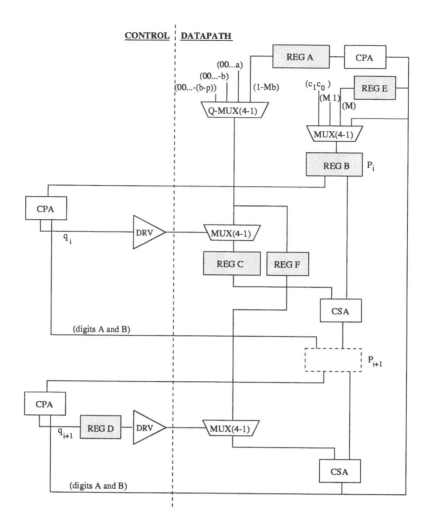

Figure 8. Primary hardware blocks needed for iterations.

Table I. Estimated component latency and area for parallel paths and Radix-512 VHR algorithms.

	Radix-4 Parallel Paths 2 copies	R-512 VHR
scaling factor M	50	150
REG A	.6(66)	.6(68)
Q-MUX	.8(72)	.8(68)
REG E	.6(11)	–
MUX (for REG B)	.8(72)	–
REG B	1.2(72)	1.2(68)
SEL	–	–
DRV	–	–
MUX (part. product)	.8(72)(2)	.8(68)(6)
REG C	.6(72)	–
REG F	.6(72)	–
CSA	72(2)	2.3(68)(3)
REG D	–	–
O-T-Fly	1.45(56)	1.45(56)
CPA	2.4(66)	2.4(68)
TOTAL AREA (A_{FA})	883	1367

	Radix-4 Parallel Paths 2 copies	R-512 VHR
scaling factor M	4	1
CSA(1-Mb)	3	1
CPA(1-Mb),CSA(Ma)	3	1
q_i's	15	7
finish	2	1
total cycles	26	11
cycle time	4/2.75	7.5/6.25
TOTAL LATENCY (t_{FA})	104/71.5	82.5/70

	R-4 SRT	R-16 Overlapped SRT
TOTAL AREA (A_{FA})	300	550
total cycles	28	15
TOTAL LATENCY (t_{FA})	170/135	120/101.25

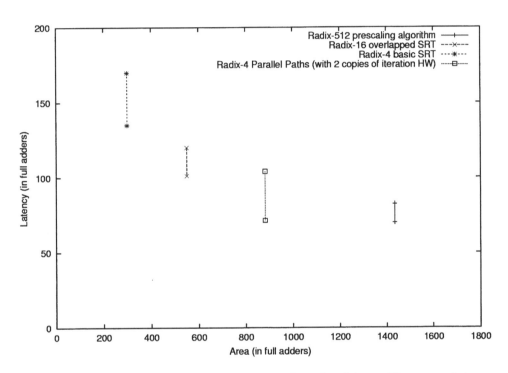

Figure 9. Speed versus area for several of the best division algorithms. The upper latency shown for each algorithm is based on component estimates from [12]; the reduced latencies assume use of a fast MUX/REG as described in [13].

than overlapped SRT when the slower registers are used, and significantly faster when using the faster registers.

5.1 Higher Precision Division

For higher-precision division, the parallel paths approach increases its advantage over the SRT algorithms. This is because, for example, if the division must be carried out to twice as many bits, the SRT algorithms will require fully twice as many iterations to perform the division. The parallel paths algorithms, on the other hand, while requiring twice as many quotient-digit producing iterations (representing 50-60% of its double precision latency) will see very little increase in latency in the other 40+% of the algorithm.

For *very* high-precision division, another variation of the Parallel Paths algorithm can attain speeds approaching 1 bit every $0.25t_{FA}$, though (not unexpectedly) this algorithm requires considerable area.

6 Conclusions/Future Work

This paper introduced the parallel paths algorithm, a new hardware division strategy that allows the latency of quotient selection and partial product formation to be 'hidden' by distributing their calculation over multiple iterations. While still looking for an optimal design—the parallel paths approach is in fact a class of algorithms—we examined a specific radix-4 implementation that can achieve an estimated speedup of 1.4 over radix-16 overlapped SRT while requiring a factor of 1.6 increase in area. These results indicate that the parallel paths strategy offers a new option in terms of area/performance tradeoff. Future work will include more detailed design and simulation, with eventual implementation in silicon.

7 Acknowledgments

This work was supported in part by NSF grant EIA-9905322 and an ARCS foundation scholarship.

References

[1] J. E. Robertson, "A new class of digital division methods," *IRE Trans. Electronic Computers*, vol. 7, pp. 218–222, Sept. 1958.

[2] K. D. Tocher, "Techniques of multiplication and division for automatic binary computers," *Quarterly J. Mech. Appl. Math.*, vol. 11, pp. 364–384, 364–384 1958.

[3] M. D. Ercegovac and T. Lang, *Division and Square Root: Digit-recurrence algorithms and applications*. Dordrecht, The Netherlands: Kluwer Academic Publishers, 1994.

[4] J. A. Prabhu and G. B. Zyner, "167 MHz radix-8 divide and square root using overlapped radix-2 stages," in *Proc. Int. Conf. Computer Design*, pp. 155–162, IEEE, 1995.

[5] T. E. Williams and M. A. Horowitz, "A zero-overhead self-timed 160-ns 54-b CMOS divider," *IEEE Journal of Solid-State Circuits*, vol. 26, pp. 1651–1661, Nov. 1991.

[6] D. Harris and M. A. Horowitz, "SRT division architectures and implementations," in *Proc. Int. Conf. Computer Design*, pp. 18–25, IEEE, 1997.

[7] A. Svoboda, "An algorithm for division," in *Proc. 9th Symp. Inform. Processing Machines*, pp. 25–34, 1963.

[8] M. D. Ercegovac, T. Lang, and P. Montuschi, "Very high radix division with selection by rounding and prescaling," *IEEE Trans. Computers*, vol. 43, pp. 909–918, Aug. 1994.

[9] D. Wong and M. Flynn, "Fast division using accurate quotient approximations to reduce the number of iterations," *IEEE Trans. Computers*, vol. 41, pp. 981–995, Aug. 1992.

[10] M. D. Ercegovac and T. Lang, "A division algorithm with prediction of quotient digits," in *Proc. Int. Conf. Computer Design*, pp. 51–56, IEEE, 1985.

[11] P.-M. Seidel, "High-speed redundant reciprocal approximation," *Integration, the VLSI Journal*, vol. 28, pp. 1–12, 1999.

[12] P. Montuschi and T. Lang, "Boosting very-high radix division with prescaling and selection by rounding," *IEEE Transactions on Computers*, vol. 50, pp. 13–27, Jan. 2001.

[13] F. Klass, C. Amir, A. Das, K. Aingaran, C. Truong, R. Wang, A. Mehta, R. Heald, and G. Yee, "A new family of semidynamic and dynamic flip-flops with embedded logic for high-performance processors," *IEEE Journal of Solid-State Circuits*, vol. 34, pp. 712–716, May 1999.

Prescaled Integer Division

David W. Matula, Alex Fit-Florea
Department of Computer Science and Engineering
Southern Methodist University
{matula,alex}@engr.smu.edu

Abstract

We describe a high radix integer division algorithm where the divisor is prescaled and the quotient is postscaled without modifying the dividend to obtain an identity $N = Q^* \times D + R^*$ with the quotient Q^* differing from the desired integer quotient Q only in its lowest order high radix digit. Here the "oversized" partial remainder R^* is bounded by the scaled divisor with at most one additional high radix digit selection needed to reduce the partial remainder and augment the quotient to obtain the desired integer division result $N = Q \times D + R$ with $0 \leq R \leq D - 1$.

We present a high radix multiplicative version of this algorithm where a $k \times p$ digit base β rectangular aspect ratio multiplier allows quotient digit selection in radix β^{k-1} with a cost of only one $k \times p$ digit multiply per high radix digit, plus the fixed pre- and post-scaling operation costs. We also present a Booth radix 4 additive version of this algorithm where appropriately compressed representation of the partial remainder with Booth digits $\{-2, -1, 0, 1, 2\}$ allows successive quotient digit selection from the leading partial remainder digit without the iterative table lookups required in SRT division.

1 Introduction

Digit serial division algorithms generating quotient digits in a higher radix in most significant digit first order are well known. The SRT division procedure characterized by Atkins in [1] iteratively employs table lookup for determining the next quotient digit and an adder to update the partial remainder. Using a short-by-long rectangular aspect ratio multiplier the digit serial division procedure may be implemented in a much higher radix using multiplication of the partial remainder by a short reciprocal of the divisor with rounding to determine the next quotient digit. Updating the remainder is then accomplished employing another short-by-long multiply. See [3] and [5] for an extensive list of references on these methods.

Prescaled division is an enhanced version of digit serial division introduced by Svoboda in [9] that has been followed with a rich literature noted in [5]. The process of prescaled division scales both the numerator and denominator by a predetermined short reciprocal of the divisor. Division of the scaled numerator by the scaled divisor yields the identical quotient to any precision with a remainder that is implicitly also scaled by the short reciprocal. The efficiency advantage of prescaled division does not apply to integer division as the necessary additional work to effectively "descale" the scaled remainder costs as-much-as or more than the savings gained.

We present an alternative methodology for prescaled integer division of a dividend N by a divisor D that provides both the quotient Q and remainder R, achieving almost the same efficiency as that of a straightforward prescaling procedure providing the same quotient with a scaled remainder. Our algorithm involves *divisor prescaling* and *quotient postscaling*. That is, the (unscaled) dividend is divided by the scaled divisor resulting in a "reduced" integer quotient and an "oversized" remainder R^*. Subsequent scaling of the reduced quotient by the short reciprocal yields an identity $N = Q^*D + R^*$. The oversized remainder R^* may then be processed by a traditional integer division method (noting that essentially only one high radix digit need be determined) to augment Q^* to obtain the integer quotient Q and determine the final integer remainder R satisfying $N = Q \times D + R$ with $0 \leq R \leq D - 1$.

The general theory of our prescaled integer division method is outlined in Section 2. Our main result in Section 3 is a prescaled integer division algorithm employing a $k \times p$ digit rectangular aspect ratio multiplier that can be proven to determine the quotient and remainder of $2p$ by p digit division in $\left(\lceil \frac{p}{k-1} \rceil + 3\right) k \times p$ digit multiplications.

In Section 4 we describe a version of prescaled integer division employing an adder for partial remainder computation where quotient digit selection in Booth radix 4 is available from the leading digit of the partial remainder. Both the divisor prescaling and the quotient postscaling operations are executable as 3-1 additions with one initial table lookup to determine the short divisor reciprocal value to be employed as the scale factor.

Section 5 provides a conclusion and further directions for development of this procedure.

2 Scaling Integer Division

Consider integer division with the divisor D being a p-digit base β integer normalized to the range $\beta^{p-1} \leq D \leq \beta^p - 1$ and the dividend N being a $2p$-digit integer satisfying $0 \leq N \leq D\beta^p - 1$. The result of *integer division* of N by D is an integer *quotient* Q and an integer *remainder* R which are uniquely determined by

$$N = Q \times D + R, \text{ with } 0 \leq R \leq D - 1. \quad (1)$$

The bounds on N and D imply $0 \leq Q \leq \beta^p - 1$ with both Q and R being p-digit base β integers.

Digit serial division develops Q by the dependent sequential steps

- next quotient digit selection
- partial remainder computation

Both of these steps can cost significant delay. When a rectangular aspect ratio $k \times p$ digit multiplier is available, the next quotient digit can be determined in the high radix β^{k-1} by a $k \times p$ digit multiplication by a predetermined short reciprocal and rounding. The next partial remainder is then computable by a $k \times p$ digit multiplication. When only an adder is available to support digit serial division, the SRT algorithm employs table lookup for quotient digit selection and subtraction of a selected digit multiple of the multiplicand for remainder update.

A much simpler division process occurs when the divisor is a value very close to a power of the base. For the decimal divisor $D = 100445568$, it is evident that the next quotient digit may be taken as the leading radix 100 digit of the partial remainder. Thus when the divisor is near a power of the base the steps of partial remainder computation and next quotient digit selection are effectively merged into a single operation, taking only about one half the time per high radix quotient digit generated.

Straightforward prescaled integer division scales a divisor and dividend by a predetermined integer scale factor $\bar{D}$ selected to generate a *scaled divisor* $D\bar{D}$ very close to a power of the base. Scaling both the integer divisor N and dividend D by $\bar{D}$ yields

$$N\bar{D} = Q(D\bar{D}) + R\bar{D} \text{ with } 0 \leq R\bar{D} \leq D\bar{D} - \bar{D} \quad (2)$$

Integer division of $N\bar{D}$ by $D\bar{D}$ provides a more efficient procedure to determine the identical integer quotient Q, but then concurrently provides a scaled remainder value $R\bar{D}$. This form of prescaled division is not applicable to integer division when both the integer quotient Q and integer remainder R must be determined.

We herein present an alternative prescaling algorithm that returns both the quotient Q and remainder R of integer division with nearly the same efficiency as a straightforward prescaling algorithm that would generate the same quotient with a scaled remainder.

Given N and D as in (1), we shall determine Q and R in terms of three intermediate integer parameters q, r, and R^* that are defined as follows. Integer division of Q by $\bar{D}$ uniquely specifies an integer *reduced quotient* q and an integer *reduced remainder* r by the equation

$$Q = q\bar{D} + r, \quad 0 \leq r \leq \bar{D} - 1. \quad (3)$$

Substituting equation (3) into equation (1) we obtain

$$N = qD\bar{D} + (rD + R).$$

Letting $R^* = rD + R$, we obtain from the bounds $0 \leq r \leq \bar{D} - 1$ and $0 \leq R \leq D - 1$ that $0 \leq R^* \leq D\bar{D} - 1$. Then $N = q \times (D\bar{D}) + R^*$ with $0 \leq R^* \leq D\bar{D} - 1$ means that the integers q and R^* can be determined by integer division of the (unscaled) dividend N by the scaled divisor $D\bar{D}$.

We now wish to find Q and R from

$$Q = q\bar{D} + r,$$

$$R = R^* - rD.$$

Note that both Q and R can be found from a $k \times p$ digit multiply given that we can also determine r having found q and R^* from the integer division of N by $(D\bar{D})$. Observe that

$$R^*\bar{D} = r(D\bar{D}) + R\bar{D}.$$

Since $rD\bar{D} \leq R^*\bar{D} < (r+1)D\bar{D}$ and since for an appropriately chosen $\bar{D}$ we can obtain $\frac{D\bar{D}}{\beta^{p+k-1}} = 1 + \epsilon$ with $|\epsilon| < \frac{1}{\beta^{k-1}}$, the leading digit $r' = \left\lfloor \frac{R^*\bar{D}}{\beta^{p+k-1}} \right\rfloor$ either equals r or is very close to r. Using r' we may compute a trial quotient, by $q\bar{D} + r'$ and trial remainder by $R^* - r'D$, with a final small quotient and remainder adjustment as necessary if $R^* - r'D$ is not in the range $[0, D-1]$. The process is formalized in the next section.

3 Multiplicative Prescaled Integer Division

For multiplicative prescaled integer division we assume support of a $k \times p$ digit base β integer multiplier generating a $k + p$ digit product. Multiplicative prescaled division is particularly effective when the aspect ratio $p : k$ of the larger multiplicand size p to the smaller multiplier size k is in the range from $4 : 1$ to $10 : 1$. To obtain maximum benefit from the use of the $k \times p$ digit multiplier, it is critical that the scale factor used to prescale the divisor have both length and accuracy corresponding to the shorter size k of the multiplier, allowing division by the prescaled divisor to be performed with digit selection in radix β^{k-1}.

For $1 \leq k < p$ with integer divisor D satisfying $\beta^{p-1} \leq D < \beta^p$, let $\beta^{k-1} \leq \bar{D} \leq \beta^k - 1$ and $0 \leq S \leq D - 1$ be determine by

$$\beta^{k+p-1} = \bar{D}D - S.$$

Here $\bar{D} = \left\lceil \frac{\beta^{k+p-1}}{D} \right\rceil$ is termed an *integer-normalized k-digit short reciprocal* of the integer divisor D and $S = \bar{D}D - \beta^{k+p-1}$ is the residue of $\bar{D}D$ modulo β^{k+p-1}. Equivalently, $\bar{D}$ is the quotient of β^{k+p-1} divided by D selected so as to have the negative remainder $-(D-1) \leq -S \leq 0$.

Prescaled integer division of the integer dividend N by the integer divisor D proceeds as follows

- The divisor D is prescaled by a k-digit short reciprocal yielding the *scaled divisor* $D\bar{D}$.

- Integer division of N by $D\bar{D}$ is performed with digit selection radix β^{k-1}, determining a reduced quotient, oversized remainder pair q, R^* satisfying

$$N = q(D\bar{D}) + R^*, \text{ with } 0 \leq R^* \leq D\bar{D} - 1.$$

- The reduced quotient q is scaled by $\bar{D}$ with the divisor "descaled" from $D\bar{D}$ to D yielding the identity

$$N = (q\bar{D})D + R^*$$

- A final quotient digit adjustment is performed to reduce the remainder from the oversized range $0 \leq R^* < D\bar{D} - 1$ to a remainder R satisfying $0 \leq R < D - 1$.

The process is illustrated by the following numeric example.

Example: [Prescaled Integer Division]
Problem: For the 12×6 digit decimal integer division arguments

$N = 365748375204$ (12 digit dividend)
$D = 784731$ (6 digit divisor)

find the 6 digit integer quotient Q and non negative remainder R with $0 \leq R \leq D - 1$ such that
$N = Q \times D + R$.

Note: We assume our multiplier is a 3×6 digit decimal multiplier, so $p = 6$, $k = 3$, and we pursue prescaled integer division in radix $10^{k-1} = 100$.

Step 1: Determine the $k = 3$ digit short reciprocal
$\bar{D} = \left\lceil \frac{10^8}{784731} \right\rceil = 128$.

Step 2: Determine the 9 digit scaled divisor and 6 digit residue

$D \times \bar{D} = 784731 \times 128 = 100445568$
$S = 445568$

Step 3: Perform digit serial division of the dividend by the the scaled divisor determining $\left\lceil \frac{p}{k-1} \right\rceil - 1 = 2$ digits in the high radix 100. Note that digit selection is obtained directly from the leading digits of each partial remainder with the partial remainder computations comprising multiplications of the radix 100 digit by the 6 digit residue $S = 445568$ in the 3×6 digit multiplier.

```
                         36 41.
   1 00 44 55 68  | 36  57 48 37 52 04.          first digit selection
   36 × 44 55 68    (36) 16 04 04 48
                         41 44 33 04             second digit selection
   41 × 44 55 68       (41)18 26 82 88
                         26 06 21 16
```

Step 4: Perform divisor descaling with quotient scaling,

$365748\,375204$ $= 3641 \times 100445568 + 26062116$
(divisor descaling:) $= 3641 \times (128 \times 784731) + 26062116$
(quotient scaling:) $= (3641 \times 128) \times 784731 + 26062116$
$= 466048 \times 784731 + 26062116$

Step 5: Reformat digit serial division and determine a low order digit incrementation by one step of the short reciprocal division procedure applied to the "oversized" remainder.

```
                    +33.         (quotient adjustment:)
                    46 60 48.    = 46 60 81
   78 47 31  | 36 57 48 37 52 04.    (digit augmentation selection:)
   128×         26 06 21 16    = 033 35 95 08 48
   33 × 78 47 31   25 89 61 23    (partial remainder adjustment)
                    16 59 93
```

Step 6: Determine (confirm) positive remainder and provide quotient in non redundant form.
$Q = 466081$,
$R = 165993$, $0 \leq R < 784731$.

The digit selection procedure for the first phase of Prescaled Integer Division utilized in Step 3 of the example is identical to digit selection in p *by* p digit prescaled division. It follows that $\left(\lceil \frac{p}{k-1} \rceil - 1\right)$ high radix digits are determined at a cost of just one $k \times p$ digit multiplication per digit. The remainder $R^* = 26062116$ in Step 5 in general satisfies $|R^*| < \beta^{p+k-1} + S$, so multiplication by the short reciprocal determines a $p + 2k - 1$ digit product with leading k digit tuple magnitude no greater than the short reciprocal. Here we obtain $033 \leq 128$.

The digit selection multiplication in Step 5 can be performed as a $k \times p$ digit multiplication in the form $\bar{D} \times \left\lfloor \frac{R^*}{\beta^{k-1}} \right\rfloor = 128 \times 260621 = \underline{033}359488$, with $R^* \bmod \beta^{k-1}$, (here being 16) then added to the remainder in Step 6. Since the augmented digit selected in Step 5 is in the range $[-\bar{D}, \bar{D}]$, the partial remainder adjustment can be supported by a $k \times p$ digit multiplication, with the resulting next partial remainder output in Step 5 possibly needing further adjustment to obtain absolute value less than the divisor D.

Algorithm 1 (Prescaled Integer Division)
Stimulus:
A high radix β^{k-1} with $(k-1) \geq 2$, a precision p with $p \geq k+1$, a p-digit integer divisor D with $\beta^{p-1} \leq D \leq \beta^p - 1$, a $2p$-digit integer dividend N with $0 \leq N \leq D\beta^p - 1$.
Response:
An integer quotient Q with $0 \leq Q \leq \beta^p - 1$, an integer remainder R with $0 \leq R \leq D - 1$, where Q, D satisfy $N = QD + R$.
Method:
$Q := 0$;
$\bar{D} := \left\lceil \frac{\beta^{p+k-1}}{D} \right\rceil$; {short reciprocal}
$R = N$; {initiate partial remainder}
$S = D\bar{D} - \beta^{p+k-1}$; {residue of scaled dividend}
$n := \lceil p/(k-1) \rceil$; {number of high radix digits in Q}
for $i := n - 1$ **downto** 1 **do**
  (1) $digit := R \,\textbf{div}\, \beta^{i(k-1)+p}$; {leading digit selection}
  (2) $R := R - digit \times \beta^{i(k-1)+p}$; {leading digit deletion}
  (3) $R := R - digit \times S \times \beta^{(i-1)(k-1)}$; {partial remainder determination}
  (4) $Q := Q \times \beta^{k-1} + digit$; {quotient accumulation}
od;
(5) $Q := Q \times \bar{D}$; {quotient scaling}
(6) $digit := (\bar{D} \times R) \,\textbf{div}\, \beta^{(k-1)+p}$; {digit augmentation selection}
(7) $R := R - digit \times D$; {partial remainder adjustment}
(8) $Q := Q + digit$; {quotient accumulation}
(9) **while** $R < 0$ **do**
   $Q := Q - 1$;
   $R := R + D$
od;
(10) **while** $R \geq D$ **do**
   $Q := Q + 1$;
   $R := R - D$
od;

Lemma 1 *Algorithm 1 for Prescaled Integer Division performs $2p$ by p digit base β integer division in the high radix β^{k-1}, with $k - 1 \geq 2$, employing $(\lceil \frac{p}{k-1} \rceil + 3)$ $k \times p$ digit multiplications.*

Outline of Proof: A single $k \times p$ digit base β multiplication in Statement (3) determines both the next partial remainder and the next high radix quotient digit, yielding a total of $(\lceil \frac{p}{k-1} \rceil - 1)$ $k \times p$ digit multiplications in the **for** loop. The prescaling of the divisor determining S and the postscaling of the quotient in Statement (5) contributes two more $k \times p$ digit multiplications. The remainder scaling in Statement (6) for low order digit augmentation selection and the partial remainder adjustment in Statement (7) can be performed by two more $k \times p$ digit multiplications for the total of $(\lceil \frac{p}{k-1} \rceil + 3)$.

For correctness of Algorithm 1 note initially that $R = N < D\bar{D}\beta^{(n-1)(k-1)}$. Using induction for $i = n-1, n-2, \ldots, 1$, assume $|R| < D\bar{D}\beta^{i(k-1)}$. Consider the case $R \geq 0$. The rational $\frac{\beta^{p+k-1}R}{D\bar{D}}$ determines the digit string for the rest of the infinitely precise quotient, and since $\bar{D} \times R$ exceeds the value by less than one part in β^{k-1}, the high radix digit determined in Statement (1) is either identical to the leading digits of the string or is 1 unit larger. If identical, the next partial remainder R satisfies $0 \leq R < D\bar{D}\beta^{(i-1)(k-1)}$, and if the high radix digit is 1 unit larger, the next partial remainder is negative with $0 < |R| < D\bar{D}\beta^{(i-1)(k-1)}$. The case for $R < 0$ follows by symmetry, so by induction we terminate the **for** loop with $N = Q' \times D + R'$, where $|R'| \leq D\bar{D} - 1$.

As noted in the paragraph before Algorithm 1, the final high radix digit determined in Statement (6) is in the range $[-\bar{D}, \bar{D}]$. Although this digit can exceed β^{k-1}, it does not exceed β^k in magnitude and the remainder adjustment can be supported by a $k \times p$ digit multiplier, followed possibly by some further additive adjustments to obtain $0 \leq R \leq D - 1$.

Note that determining a full $p \times p$ digit multiplication with a $k \times p$ digit multiplier employs $\lceil \frac{p}{k} \rceil$ $p \times k$ digit multiplications. For $1 << k << p$, it follows that the number of $p \times k$ digit multiplications employed for prescaled integer division to determine Q, R satisfying $N = Q \times D + R$ is not much greater than the number of multiplies to compute $Q \times D + R$ to verify the answer. For example, employing a 12×128 bit multiplier for a 128×128 bit product utilizes 11 short-by-long multiplies, and finding the quotient and remainder by prescaled integer division as given by Algorithm 1 utilizes 15 short-by-long multiplies.

In summary the $2p$ by p digit prescaled integer division algorithm prescales only the initial divisor, and then postscales the resulting *reduced quotient* and *oversized remainder* of this modified division problem. The postscaling of the oversized remainder is employed to find a single "oversized" high radix digit to adjust the quotient and compute an appropriately bounded remainder. The foundation of the method with more details on determining the final high radix digit for quotient adjustment is included in the following theorem.

Theorem 2 *For $2 \leq k \leq (p-1)$, assume given a*

- *p-digit divisor D with $\beta^{p-1} \leq D \leq \beta^{p-1}$,*
- *2p-digit dividend N with $0 \leq N \leq D\beta^{p-1}$,*
- *k-digit short reciprocal $\bar{D}$ and residual S satisfying $\bar{D}D = \beta^{p+k-1} + S$ with $0 \leq S \leq D - 1$,*
- *$[p-(k-1)]$-digit reduced quotient q and oversized remainder R^* satisfying $N = q \times (\bar{D}D) + R^*$, with $0 \leq R^* \leq \bar{D}D - 1$.*

If $0 \leq R^* \leq D-1$, then the p-digit quotient $Q = q \times \bar{D}$ and remainder $R = R^*$ satisfies $N = QD + R$ with $0 \leq R \leq D - 1$.
Else if $D \leq R^* \leq \bar{D}D - 1$, then the quotient Q and remainder R given by
$Q = q \times \bar{D} + \left\lfloor \frac{R^* \times \bar{D} - S}{\beta^{p+k-1}} \right\rfloor$,
$R = R^* - (\left\lfloor \frac{R' \times \bar{D} - S}{\beta^{p+k-1}} \right\rfloor \times D)$,
satisfies $N = QD + R$ with $-D \leq R \leq D - 1$.

4 Additive Prescaled Integer Division

Our additive prescaled integer division algorithm will use a slightly different procedure for determining the postscaled quotient and corresponding remainder.

For the radix 4 we will assume our short reciprocal $\bar{D}$ is a 3-digit radix 4 integer. We divide N by the scaled divisor $D\bar{D}$ so as to obtain 3 fractional digits giving

$$N = \frac{Q'}{64}(D\bar{D}) + \frac{R'}{64}$$

Letting $\frac{Q'\bar{D}}{64} = Q'' + \frac{j}{64}$ we obtain

$$N = Q''D + \frac{jD + R'}{64}.$$

Now $jD + R'$ must be divisible by 64, yielding an integer $R'' = \frac{jD+R'}{64}$ where $|R''| < 2D$. Thus $N = Q''D + R''$ where an augmentation of Q'' by at most a value from $\{-1, 1, 2\}$ yields $N = QD + R$ with $0 \leq R \leq D - 1$. Division radix 4 requires generation of $\lfloor \frac{p+1}{2} \rfloor$ digits in the Booth radix 4 quotient representation with digits $\{-2,-1,0,1,2\}$, which for $p = 64$ is 33 digits. Since the number of quotient digits is relatively large, the generation of an additional 3 fractional digits with quotient and remainder adjustment by adding a fractional part times D back into the remainder does not significantly increase the number of additive steps on a relative basis.

Our additive algorithm assumes $\bar{D}$ (normalized to $(1,2]$) is chosen so that $1 \leq D\bar{D} < 1\frac{1}{6}$ for a radix 4 divisor D normalized to $\frac{1}{2} \leq D < 1$, the above Table shows that a 3 digit $\bar{D}$ may be chosen so that $1 \leq D\bar{D} < 1\frac{1}{6}$. Our partial remainder computation

Divisor Interval	Short Reciprocal $\bar{D}$	Scaled Interval $D\bar{D}$
$[\frac{1}{2}, \frac{4}{7})$	2.00	$[1, \frac{8}{7})$
$(\frac{4}{7}, \frac{2}{3})$	$\frac{7}{4} = 2.\bar{1}0$	$(1, \frac{7}{6})$
$(\frac{2}{3}, \frac{3}{4})$	$\frac{4}{3} = 1.20$	$(1, \frac{9}{8})$
$[\frac{3}{4}, \frac{4}{5})$	$\frac{11}{8} = 1.12$	$(\frac{33}{32}, \frac{11}{10})$
$(\frac{4}{5}, \frac{8}{9})$	$\frac{4}{5} = 1.10$	$(1, \frac{10}{9})$
$(\frac{8}{9}, 1)$	$\frac{9}{8} = 1.02$	$(1, \frac{9}{8})$

is performed so that when the leading digit of the partial remainder is 1 or 2, then all subsequent digits are in the set $\{-1, 0, 1, 2\}$ constituting a fractional part in the range $\left(-\frac{1}{3}, \frac{2}{3}\right)$. Similarly when the partial remainder is -1 or -2, all subsequent digits are in the set $\{-2, -1, 0, 1\}$ constituting a fractional part in the range $\left(-\frac{2}{3}, \frac{1}{3}\right)$. This provides for partial remainder computation of $R := 4(R - dD)$, where $d \in \{-2, -1, 0, 1, 2\}$ is chosen to maintain the condition that $(R - dD)$ is in the range $\left(-\frac{2}{3}, \frac{2}{3}\right)$. The following example illustrates this additive integer prescaled division procedure for integer division of an 8 digit radix 4 dividend by a 4 digit radix 4 divisor.

[Additive Prescaled Integer Division]
Problem: For the 8×4 digit radix 4 integer division arguments

$N = 13222012_4$ (8 digit dividend)
$D = 2211_4$ (4 digit divisor)

find the 4 digit integer quotient Q and remainder R with $0 \leq R \leq D - 1$ such that

$N = Q \times D + R$.

We assume our adder is able to perform signed digit summations and compress the remainder to either the digit set $\{-2, -1, 0, 1\}$ or $\{-1, 0, 1, 2\}$

Step 1: Determine the three digit short reciprocal according to Table 1.
$\bar{D} = 2\bar{1}0_4$ since $\frac{2211_4}{4^4}$ is in the interval $(\frac{4}{7}, \frac{2}{3})$.

Step 2: Determine the 7 digit scaled divisor $D\bar{D}$
$D \times \bar{D} = 2211 \times 2\bar{1}0 = 1102200_4 = 10201\bar{1}0_4$

Step 3: Perform digit serial division of the dividend by the the scaled divisor determining 5 quotient digits including 3 fractional digits radix 4. Note that digit selection is obtained from the leading digit of the new partial remainder. If the leading digit is not zero and the second digit has an opposite sign or is zero, the bias of the digit set is changed from $\{-2, -1, 0, 1\}$ to $\{-1, 0, 1, 2\}$ or from $\{-1, 0, 1, 2\}$ to $\{-2, -1, 0, 1\}$ in compressing the next partial remainder. The positive dividend becomes the initial partial remainder with compression to $\{-1, 0, 1, 2\}$. The leading two digits of each partial remainder are underlined to indicate the control.

$$\begin{array}{r}
2\bar{1}.\bar{1}1\bar{2} \\
1\,0\,2\,0\,1\,\bar{1}\,0 \overline{\smash{)}2\bar{1}2\,2\,2\,0\,1\,2\,.\,0\,0\,0} \quad \in \{-1,0,1,2\} \\
\underline{2\,1\,0\,0\,2\,\bar{2}} \\
\underline{1\,\bar{1}\,2\,1\,2\,1\,2} \quad \in \{-2,-1,0,1\} \\
\bar{1}\,0\,\bar{2}\,0\,\bar{1}\,1\,0 \\
\underline{1\,0\,1\,1\,1\,2\,0} \quad \in \{-2,-1,0,1\} \\
\bar{1}\,0\,\bar{2}\,0\,\bar{1}\,1\,0 \\
\underline{\bar{1}\bar{1}\,1\,1\,1\,0\,0} \quad \in \{-1,0,1,2\} \\
1\,0\,2\,0\,1\,\bar{1}\,0 \\
\underline{\bar{2}\,1\,1\,0\,1\,0\,0} \quad \in \{-2,-1,0,1\} \\
\bar{2}\,\bar{1}\,0\,0\,\bar{2}\,2\,0 \\
\underline{2\,1\,0\,2\,2\,0} \quad \in \{-1,0,1,2\}
\end{array}$$

Step 4: Additively perform divisor descaling with quotient scaling,
$$\begin{aligned}
2\bar{1}22012 &= 2\bar{1}.\bar{1}1\bar{2} \times 10201\bar{1}0 + 210.22 \\
&= 2\bar{1}.\bar{1}1\bar{2} \times (2\bar{1}0 \times 2211) + 210.22 \\
&= (2\bar{1}.\bar{1}1\bar{2} \times 2\bar{1}0) \times 2211 + 210.22 \\
&= (1\bar{1}0\bar{1}2.\bar{1}2) \times 2211 + 210.22
\end{aligned}$$

Step 5: Add fractional quotient digits times divisor into remainder.
$$\begin{aligned}
R &= \quad .\bar{1}2 \times 2211 + 210.22 \\
&= \bar{1}\bar{1}\bar{1}.12 + 210.22 \\
&= 100
\end{aligned}$$

Step 6: Determine (confirm) positive remainder and provide quotient in non redundant form.
$$\begin{aligned}
Q &= \quad 2332_4 \\
R &= \quad 100_4, \quad 0 \leq R \leq D-1.
\end{aligned}$$

Observation 2 *The additive Prescaled Integer Division procedure as described in conjunction with the example yields a quotient Q and remainder $0 \leq R \leq D-1$ satisfying $N = QD + R$.*

The procedure described may be improved so that the partial remainder need not be fully compressed to obtain the next quotient digit. It can be shown that a short reciprocal table can be determined requiring no more than 3 digits in $\bar{D}$ where the scaled divisor $D\bar{D}$ is restricted to the narrower range $[1, 1\frac{1}{15})$. This allows partial compression of the redundant partial remainder where effectively only a couple leading digits of the partial remainder need be compressed to obtain a quotient digit that will preserve the sufficiently compressed range of the redundant partial remainder. The determination of appropriate reciprocal values is aided by the methodology in [8]. The possibility of using only a limited number of leading digits of the redundant remainder to determine the next quotient digit follows from the methodology in [5].

By using higher precision table lookup of the reciprocal to bound the size of $D\bar{D}$ in a much smaller neighborhood of unity, it is possible with partial compression of the remainder to perform additive integer prescaled division with higher radices such as radix 8 or 16. Employing efficient partial compression techniques such as described in [2, 4, 6, 7] might allow implementations where a radix 8 or 16 digit could be determined each cycle.

5 Conclusion

We have demonstrated a new divisor-prescaling quotient-postscaling digit serial division algorithm that can be applied to integer division. The procedure allows each quotient to be computed at a rate of one high radix quotient digit per $k \times p$ digit multiplication for the multiplicative version and one high radix quotient digit per partially compressed addition for the additive version, with the remainder available at an additional cost essentially independent of the number of digits generated in the quotient. Fine tuning of the additive version and some implementation studies of the algorithms are planned to determine the overall feasibility of these integer division algorithms compared to current implementations of existing integer division methods.

References

[1] D. E. Atkins, "Higher-Radix-Division Using Estimates of the Divisor and Partial Remainders" *IEEE Transactions on Computers*, vol. 17, no. 10, pp. 925-934, 1968.

[2] M. Daumas and D. W. Matula, "Further reducing the Redundancy of a Notation Over a Minimally Redundant Digit Set", *J. VLSI*, vol. 33, pp. 7-18, 2003 (see also: M. Daumas and D. W. Matula, "Recoders for Partial Compression and Rounding", *Research Report 97-01 Laboratoire de l'Informatique du Parallelisme*, Lyon, France, 1997).

[3] M. D. Ercegovac and T. Lang, *Division and Square Root*, Norwell, Mass.: Kluwer Academic, 1994.

[4] P. Kornerup, "Digit-Set Conversion: Generalizations and Applications" *IEEE Transactions on Computers*, vol. 43, no. 5, pp. 622-629, 1994.

[5] L. A. Montalvo, K. K. Parhi, and A. Guyot, "New Svoboda-Tung Division", *IEEE Transactions on Computers*, vol. 47, no. 9, pp. 1014-1020, 1998.

[6] A. N. Nielsen, D. W. Matula, C. N. Lyu and G. Even, "An IEEE Compliant Floating Point Adder that Conforms with the Pipelined Packet Forwarding Paradigm", *IEEE Transactions on Computers*, vol. 49, no. 4, pp. 33-47, 2000.

[7] P.-M. Seidel "High Speed Redundant Reciprocal Approximation" *3rd Real Numbers and Computers Conference*, Paris, France, pp. 219-229, 1998.

[8] D. Das Sarma and D. W. Matula, "Measuring the Accuracy of ROM Reciprocal Tables.", *IEEE Transactions on Computers*, vol. 43, no. 8, pp. 932-940, 1994.

[9] A. Svoboda, "An Algorithm for Division", *Information Processing Machines*, no. 9, pp. 25-32, 1963.

Session 4:
Floating Point

Chair: Andrew Beaumont-Smith

Hardware Implementations of Denormalized Numbers

Eric M. Schwarz
IBM Server Division
2455 South Rd., MS:P310
Poughkeepsie, NY 12601 USA
eschwarz@us.ibm.com

Martin Schmookler
IBM Server Division
11400 Burnett Road
Austin, TX 78758 USA
martins@us.ibm.com

Son Dao Trong
IBM Server Division

Boeblingen, Germany
daotrong@de.ibm.com

Abstract

Denormalized numbers are the most difficult type of numbers to implement in floating-point units. They are so complex that some designs have elected to handle them in software rather than hardware. This has resulted in execution times in the tens of thousands of cycles, which has made denormalized numbers useless to programmers. This does not have to happen. With a small amount of additional hardware, denormalized numbers and underflows can be handled close to the speed of normalized numbers. This paper will summarize the little known techniques for handling denormalized numbers. Most of the techniques discussed have only been discussed in filed or pending patent applications.

1. Introduction

The IEEE 754 binary floating-point standard [1] defines a set of normalized numbers and four sets of special numbers. The special numbers are Not-a-Numbers(NaNs), infinities, zeros, and denormalized numbers which are sometimes referred to as subnormals or denormals. Operations on the first three special numbers require no computation. The only type of special number which requires computation for an arithmetic operation is denormals. Normalized numbers are represented by the following:

$$X = (-1)^{X_s} * 1.X_f * 2^{X_e - bias}$$

where X is the value of the normalized number, X_s is the sign bit, X_f is the fractional part of the significand, X_e is the exponent, and $bias$ is the bias of the format (127, 1023, and 16383, for single, double and quad). Denormals are represented by the following:

$$X = (-1)^{X_s} * 0.X_f * 2^{1-bias} \quad , \quad X_e = 0, \; X_f \neq 0$$

There is no implied bit, and the exponent is not equal to $X_e - bias$, but instead has to be forced up by 1, to $Emin$ which is equal to -126, -1022, and -16382 depending on the format. Typically the dataflow of a Floating-Point Unit (FPU) is optimized for normalized numbers since they are the most common but there must be some mechanism to handle denormals.

There have been many variations as to how much of the denormalization handling should be done in hardware versus software. Some implementations force all denormalized input to software while others handle easy cases in hardware. Several SPARC implementations support gross underflow in hardware but force other underflow cases to software[2, 3]. Some implementations take either a trap to software or stall in hardware when a denormalized operand is detected and transform the denormal to a normalized number with a greater exponent range. The Motorola G4 vector unit traps on denormal inputs and underflow results in Java mode, while it also has a non-compliant mode where it forces denormal inputs and results to zero. The problem with this technique is that it stalls dispatching instructions and does not solve all the execution issues of denormals and underflow. This paper will present techniques for handling denormal input as well as underflow cases which require "denormalization" in hardware.

There are two obvious areas where denormals must be handled, 1) as input to an arithmetic operation, and 2) when an intermediate result underflows and traps are disabled, a denormal number needs to be produced. First, architecture variations will be discussed which affect the handling of denormals. Handling denormal input will be discussed in the following four sections. Section 3 will discuss using tagging of register files to indicate denormal input and its effect on loads and store instructions. Section 4, 5, and 6 discuss denormal input for the operations of addition, multiplication, and fused multiply-add. The following two sections address the handling of an intermediate result which underflows. Section 7 discusses denormalizing an intermediate result, and Section 8 addresses how to prevent denormalization from being needed. Section 9 presents a case study, the Power4 FPU and shows how it combined several of these techniques to implement denormals in hardware with very little area and keeping most of the execution in hardware.

2. Architecture Variations

The handling of denormal numbers is dependent on both the instruction set architecture and constraints set by the overall processor microarchitecture. The important attributes of the architecture include the different precisions which are supported, how they are represented internally in the registers, and the operations which can be performed on each of these data types.

For some architectures, all data types are converted internally to the widest supported format. For example, the Intel IA-32 architecture [4] as well as some other architectures [5] support single, double and double-extended formats in memory, but all data is automatically converted to the double-extended format when loaded from memory. This 80-bit format includes 15 bits of exponent and 64 bits of significand. Single and double denormals are normalized during this conversion. Similarly, the PowerPC architecture [6] supports single and double in memory, but converts all single format data to normalized double format in the floating-point registers (FPRs). When storing data to memory, both of these architectures must also denormalize some data values that have a lower precision than the internal format. An actual implementation can vary from this architectural model. Some PowerPC implementations, for example, provide extra bits called tags with each register which allow denormals and sometimes other special values to be quickly recognized during instruction processing. Use of these tags would allow a denormal single to be loaded into the FPRs unnormalized. The constraints of the microarchitecture, such as timing considerations of load and store operations, how instructions are dispatched to each unit, whether registers are renamed, and so on, may affect such design choices.

Despite the similarities described above for handling denormal data from memory, the IA-32 and PowerPC differ in how they handle results which correspond to denormal values. The IA-32 architecture supports both single and double precision instructions when the results are normalized. However, the precision control only affects the fraction length. The exponent range for all operations is the same as for double extended operations. Thus, results may differ from an architecture which fully implements both single and double precision as specified in the IEEE 754 standard [1], although this difference would usually result in greater accuracy. Therefore, the IA-32 arithmetic operations can only produce a denormal result in its double extended range. Single and double denormals are only produced when storing such values to memory. The PowerPC, on the other hand, when executing single precision instructions, must round all denormal results at the proper boundary. This requires that it first produce an unnormalized result with its exponent clamped at $Emin$, and then convert it to the normalized double format. If tags are used which permit unnormalized single precision representations in the FPRs, then conversion after rounding can be avoided.

The Intel IA-64 [7] illustrates yet another variation in which all precisions are represented only in the widest format, which is 82 bits. It includes a 17 bit exponent field and a 64 bit significand. Like the IA-32, the significand includes an integer bit and a 63 bit fraction. However, the IA-32 disallows unnormalized representations other than true double extended denormals. The IA-64, on the other hand, represents the integer bit, or implied bit, so that single and double denormals which are loaded from memory or which result from single or double operations are represented in the registers with unnormalized significands and with the exponent set to $Emin$ of the corresponding precision. Thus, a scheme for avoiding normalization corresponds to non-architected representations in the PowerPC and in some implementations of IA-32, but corresponds to architected representations in the IA-64. Thus, the IA-64 provides full IEEE compliance for single and double precision. A separate range control bit is provided which either sets the exponent range to 17 bits, or to the range which corresponds to the specified precision.

The IBM zSeries architecture[8] (which is the new name for the 64-bit version of S/390[9]) illustrates a very different choice for internal representation of the supported precisions. Both single and double precision data are represented in the FPRs with the same format as in memory. Thus, single precision data only occupy the left half of the 64-bit registers, and all denormals must be represented in the denormalized format within these registers. However, since the execution dataflow is optimized for double precision, the single format data is converted to an internal format which supports six different data types. There are single, double, and quadword data types for both Binary Floating-Point (BFP) and Hexadecimal Floating-Point (HFP). The architecture allows instructions defined for any of these data types to operate on the contents of any of the FPRs, whether it makes sense or not. Therefore, creating tags for special operand values, as in the PowerPC, is not too useful for this architecture.

Predetermining whether data is denormalized and creating tags has limited value even for architectures such as the PowerPC. It may solve some problems while creating new ones. In some cases, tagging can save hardware and eliminate critical paths since the detection of denormalized operands is done prior to arithmetic calculation. But then creating the tags themselves may present difficulties. On the other hand, even without tagging it is possible to have implementations which can handle denormalized operands at speed. In the next section, more details concerning tags are discussed. Then the following sections describe handling of denormal operands for various operations without use of tags. The case study in section 9 then illustrates how tags are used in the Power4 FPU and in related implementations.

3. Register Files with Tags

In the previous section, for architectures such as the PowerPC, we showed that tagging the data that is loaded into the FPRs can simplify the conversion of single denormals to double format. For normal data, this conversion only requires adding three exponent bits corresponding to the complement of the exponent high order bit, and padding zeros to the low order bits of the fraction. But for denormalized operands, the data either needs to be normalized or else must be represented in a non-architected format. This format might consist of just a tag bit and an unnormalized significand which must be dealt with in later operations. The tag bit is needed to distinguish this value from a normalized double precision number having the same exponent.

If normalization is to be done at the time the data is loaded, then either special hardware must be added for counting the number of leading zeros and then shifting the fraction, or else one must pass the data through the floating-point adder dataflow, making use of existing hardware for doing these tasks. The FPRs may have separate write ports just for data loaded from memory, so using the adder dataflow would be difficult. In many cases, tagging does not eliminate having to normalize the data before instruction processing, but merely delays it until the data is needed and the adder dataflow can more easily be used.

Once tagging is added, it can be exploited to help simplify instruction processing in other ways. A tag could be used not just for single precision denormals but for double precision denormals as well in determining the value of the implied bit. Also, if processing of an instruction requires that the operands be first normalized, then subsequent instructions must be prevented from being issued to the unit. It is important to recognize these cases quickly, and tag bits can help in their detection. Additional tag bits can also be used to quickly detect other special values, such as infinities and NaNs. If instruction issuing needs to be stalled for prenormalizing operands, then it is also desirable to distinguish denormals from zeros, which do not require prenormalization. Another use for tagging, which was mentioned in the previous section, is to avoid renormalizing single precision denormals after they are rounded. Also, if they are later stored to memory in single format, denormalizing them could also be avoided.

There are some costs and disadvantages to using tags also. First, there is the extra circuitry and delay to determine the tags. This may be much more significant if the fraction must also be examined to distinguish zeros from denormals. If tags are also used for double precision denormals, then all instructions must also produce correct tags. This could complicate instructions such as convert-to-integer, since the architecture may allow floating-point instructions to operate on the result, although it would be nonsense. There is also the cost of providing the tag bits to all of the registers, but that is not significant.

Another disadvantage is that multiple formats may be used for the same data values. Double precision store instructions may have to normalize unnormalized single precision data, while single precision store instructions may have to denormalize data which is normalized but in the single denormal range. All double precision instructions may become more difficult to execute when an operand may be an unnormalized single denormal. In some PowerPC processors, all unnormalized operands are prenormalized before the instruction is executed, while in some other PowerPC processors, prenormalization is done only in special cases.

Also, design verification becomes more difficult, because testcases may need to verify correct processing for each format of a particular data value.

As previously noted, the use of tag bits may take several forms. One method is to merely set a tag bit for data from a single precision load, or possibly for the result from any single precision instruction. When the data is later read from the FPR, the tag and eight exponent bits can be used to determine if the data is a single precision denormal in a non-architected form. This form of tagging requires no extra delay when loading data from memory. If more time is available during the load operation, the exponent can be examined for all zeros, and the tag bit could then correspond to the implied bit itself. This method is used in several implementations. The implied bit is determined during all load, arithmetic and conversion operations, and stored explicitly with the data in the FPR. The exponent LSB is also set to one, so that the exponent corresponds to the proper *Emin*. This simplifies the execution of subsequent arithmetic operations for both single and double precision. Considerably more circuitry and delay is encountered if the fraction is also examined to set tags which distinguish zeros and infinities from denormals and NaNs. In the IA-64 architectural model, the exponent is also forced to all zeros when the data is zero. In the Power4 implementation of the PowerPC, the exponent is not forced for these values, but the tags override the exponent contents when the data is used.

4. Denormal Input for Addition

Floating-point addition of denormal operands is more complex than for normalized operands. Floating-point addition involves an exponent comparison and difference calculation, alignment, conditional complementation, addition of significands, normalization, and rounding. If there are denormal input operands the exponent difference calculation will be off by one and this needs to be corrected. The exponent difference drives the aligner and is timing critical. Typically implementations compute the exponent difference, D, and $D-1$, and $D+1$.

$$\begin{aligned}
Sum &= A + B \\
A &= (-1)^{A_s} * (a_0 + A_f) * 2^{A_e + \overline{a_0} - bias} \\
D &= A_e + \overline{a_0} - (B_e + \overline{b_0}) \\
&= A_e - B_e + \overline{a_0} - \overline{b_0} \\
&= A_e - B_e + z \quad, z \in \{-1, 0, +1\}
\end{aligned}$$

A late detection of operands equal to denormals then selects between the exponent differences. An alternative implementation is to add a stage to the aligner to perform a late correction shift based on which operand is denormal.

Also implied ones of the significands must be correct too. Since the critical path is not the significand input, this is not a problem. The most timing critical path is the exponent difference calculation and the implied bit should be correct by the end of this calculation. One significand is shifted by the aligner while the other is not needed until the carry propagate addition. Therefore the significand can easily be corrected for an implied bit.

5. Denormal Input for Multiplication

Floating-point multiplication typically involves a Booth decode, partial product generation (Booth multiplexing), a counter tree, a carry propagate addition, normalization and rounding. Multiplication can be performed with a Booth decode which reduces the number of partial products or as direct bit by bit multiplication. For floating-point multiplication the product, P, is calculated for the multiplicand, X, and the multiplier, Y, as shown by the following [10]:

$$\begin{aligned}
X &= x_0 + \sum_{i=1}^{n-1} x_i * 2^{-i} \\
Y &= y_0 + \sum_{j=1}^{n-1} y_j * 2^{-j} \\
P &= x_0 * y_0 + x_0 * \sum_{j=1}^{n-1} y_j * 2^{-j} + y_0 * \sum_{i=1}^{n-1} x_i * 2^{-i} \\
&\quad + \sum_{i=1}^{n-1} \sum_{j=1}^{n-1} x_i y_j * 2^{-(i+j)}
\end{aligned}$$

$$\begin{aligned}
P &= loc1 + loc2 + P' \quad, \quad where \\
P' &= \sum_{i=1}^{n-1} \sum_{j=1}^{n-1} x_i y_j * 2^{-(i+j)} \\
loc1 &= x_0 * (y_0 + \sum_{j=1}^{n-1} y_j * 2^{-j}) \\
loc2 &= y_0 * \sum_{i=1}^{n-1} x_i * 2^{-i}
\end{aligned}$$

Williams [11] separated the partial products that are dependent on an implied one as denoted by $loc1$ and $loc2$ (for leading ones correction), from the other partial products denoted by P'. For a 53 bit direct multiplication (non-Booth) there were 52 partial products representing P' and just two others ($loc1$ and $loc2$) dependent on the implied ones of the multiplier and multiplicand. Williams noted that in a counter tree it is common to have a few inputs that can have delayed arrival times since they have less counters in their path. Also, some counter designs are tapered having varying propagation delay based on input and output. Thus, it is possible to delay the two late arriving partial products while the exponent is examined for all zeros to see if the implied bit should be a one or a zero.

A similar technique is used in the next zSeries FPU which uses a Booth radix-4 multiplier [12]. Rather than adding in a leading ones correction, leading zero correction terms are subtracted from the partial product array.

$$X = x_0 + \sum_{i=1}^{n-1} x_i * 2^{-i}$$

$$Y = y_0 + \sum_{j=1}^{n-1} y_j * 2^{-j}$$

$$Y = \sum_{j=1}^{\lfloor \frac{n-1}{2} \rfloor + 1} W_j * 4^{-j}$$

$$W_j \; \epsilon \; \{-2, -1, 0, +1, +2\}$$

$$P = \sum_{j=1}^{\lfloor \frac{n-1}{2} \rfloor + 1} W_j * X * 4^{-j}$$

$$X' = 1 + \sum_{i=1}^{n-1} x_i * 2^{-i}$$

$$X = X' - \overline{x_0}$$

$$P = \sum_{j=1}^{\lfloor \frac{n-1}{2} \rfloor + 1} W_j * X' * 4^{-j} - Y * \overline{x_0}$$

$$lzc1 = -Y * \overline{x_0}$$

Two terms could be used to correct for the multiplier and multiplicand if two late inputs are available in the counter tree and the extra counter area is not a concern. Note for each added term there will be one additional 3:2 counter. A second solution is to correct the Booth decode (W_1 term) prior to creating the partial product dependent on the multiplier's implied bit. This would necessitate a delayed partial product but would not add an additional partial product. Only one correction term would be needed to correct for the multiplicand's implied bit. Actually $W_1(y_0 = 0)$ and $W_1(y_0 = 1)$ are calculated in parallel and multiplexed after y_0 is known.

Thus, there are multiple ways to correct for denormal input into a multiplier. Additional partial products are needed with delayed inputs. In implementations where the counter tree can accept more rows without adding stages, this type of design is non-timing critical. It adds area of 1 or 2 - 3:2 counters but this is small in comparison to the overall counter tree area.

6. Denormal Input for Fused Multiply/Add

Several architectures including the PowerPC floating-point architecture are optimized for the fused multiply-add operation. This operation is a multiply-add or multiply-subtract instruction, e.g. A*C + B or A*C - B, where the AC product is not rounded before the addition or subtraction. In all implementations, the pipeline structure is optimized to exploit this operation, even though it may increase the latency of Add, Subtract and Multiply instructions. For Add and Subtract, the C operand is set to 1.0, and for Multiply, the B operand is set to zero. Alignment of operands is one function whose design is significantly different for this architecture.

The usual method of aligning operands for Add and Subtract is to compare the operand exponents and then shift the operand with the smaller exponent to the right. In a multiply-add operation, the B operand is the only operand aligned and complemented. The alignment is to any position up to a little over 53 bits greater than the AC product or to the least significant bit of the product. To minimize latency, the alignment of B is done at the same time as the product is developed, then merged with the product in the final carry save adder.

If the addend, B, is denormalized the main problem is correcting the shift amount. The significand can easily be corrected before reaching the aligner. The exponent correction is timing critical and will probably require multiple adders to compute the exponent difference based on any of three operands being denormalized.

$$Sum = B + A * C$$
$$B = (-1)^{B_s} * (b_0 + B_f) * 2^{B_e + \overline{b_0} - bias}$$
$$P = (-1)^{P_s} * P_m * 2^{(A_e + \overline{a_0} + C_e + \overline{c_0} - bias) - bias}$$
$$D = B_e + \overline{b_0} - (A_e + \overline{a_0} + C_e + \overline{c_0} - bias)$$
$$= B_e - A_e - C_e + bias + \overline{b_0} - \overline{a_0} - \overline{c_0}$$
$$= B_e - A_e - C_e + bias + z \;\;, z \; \epsilon \; \{-2, -1, 0, +1\}$$

Note that there is no need to consider both A and C denormalized ($z = -2$) since this will severely underflow and thus will not have to be aligned with B even if it is a denormal too.

If the multiplier or multiplicand are denormalized, the exponent shift amount calculation is affected and the significand calculation needs to be adjusted. Either of the two techniques mentioned in the previous section can be used to correct the significand.

A multiply-add dataflow can also have difficulty with underflow if underflow traps are enabled. The result written is a normalized significand rounded and with a rebiased exponent prior to invoking the exception handler. The problem is that it is difficult to produce 53 bits of significance for the case of a denormalized addend and when the product is less than the least significant bit of the addend. The dataflow does not separate the addend from the product properly and instead incorrectly concatenates the addend with a couple guard bits to the product. To handle this case some implementations like Power3 and Power4 prenormalize input denormals while others trap to software when they detect the "disjoint case" such as the latest zSeries processor [13].

7. Denormalization

Once an intermediate result completes normalization it can be determined whether the operand underflows. If the underflow trap is disabled, then the intermediate result needs to be aligned to an exponent equal to $Emin$ and then rounded. This alignment and subsequent rounding operation is called denormalization. The problem with denormalization is that by the time underflow is detected it is too late in the pipeline to utilize the normalizer. The data either needs a denormalization unit, or somehow needs to wrap back to the top of the pipeline avoiding other instructions, or somehow needs to avoid denormalization altogether as discussed in the next section.

There have been several designs that assumed denormalization units such as by AMD [14, 15]. Basically, rather than stalling the FPU pipeline, an intermediate result requiring denormalization would be sent to another unit. The complication with this type of an implementation is that it requires an out-of-order execution design since subsequent operations would pass the instruction requiring denormalization. There would have to be a checkpoint ordering buffer to re-order instructions coming from both the FPU and the denormalization unit. This buffering is already available in an out-of-order execution design. However, the denormalization unit requires a large right shifter.

An alternative to adding a dedicated unit for denormalization is to utilize the existing shifters in the FPU. In a non-pipelined design it is simple to feed data back to the top of the pipeline to utilize the aligner or normalizer for denormalization [16]. But in a pipelined design there needs to be a mechanism to squash or reorder subsequent instructions following the instruction requiring denormalization. Schwarz [17] shows a mechanism for feeding back an intermediate result to an early stage in the pipeline if it is detected that there are no other instructions in the pipeline. In the case there is another instruction in the pipeline, all instructions are flushed from the pipeline and the one requiring denormalization is re-issued in non-pipelined mode. If the second execution requires denormalization, the instruction is guaranteed to have the FPU pipeline to itself and be able to perform denormalization. No results are saved from the first execution and if another processor in the configuration happens to write over the instruction or data, then the second execution may no longer require denormalization. This technique was used on the 1998 S/390 G5 FPU [18] and only added a little control logic since there was already a mechanism for conditionally issuing an instruction in a pipelined or non-pipelined manner.

Other similar techniques to this proposal have been used. Another FPU detected underflow early enough in the pipeline to stall a subsequent instruction from reaching the normalizer. And then it fed back the normalizer output back to its input to effectively right shift the intermediate result. The key to this technique is detecting possible underflow in an early stage and forcing stalls to separate instructions.

Another variation that the authors have seen of this technique is to detect underflow very late and provide a small shifter at the bottom of the dataflow. Underflow is detected in the normalizer. The pipeline is stalled at this point until denormalization is complete. The latch feeding the rounder also has a feedback path with a small multiplexor which enables a hold of the latch or up to a 4 bit shift. The data is right shifted 4 bits each cycle until it reaches the point where the exponent is equal to $Emin$. Then the stall is released and the data rounded properly to complete denormalization. This technique only requires a very small shifter and not much detection logic. It does create stalls in the pipeline which can be timing critical and therefore is not implemented frequently. The denormalization process for double precision can require up to 13 cycles but this is much less than trapping to software. The main drawback is the late detect of a stall can cause timing critical control signals. This type of technique was implemented in the 1998 S/390 G5 FPU to handle quad precision denormal numbers since the feedback paths in the dataflow were only double precision.

8. Preventing Denormalization

Denormalization can be prevented. The trick is to prevent normalizing past the radix point of a denormalized result. Urano [19] shows the simple technique of comparing the shift amounts for a denormal result versus the shift amount from a leading zero anticipator, and selecting the least shift amount. Goshtein and Khlobytev [20] also show a design of creating the two shift amounts in parallel, and they go on to suggest two units for the implementation. One unit supports the normalized dataflow with limited shifts while the other dataflow is slow and supports the maximum shift amounts. Grushin and Vlasenko [21] also suggest creating both shift amounts but they go into a reduction of the equation of the shift amount. They reduce the comparison and selection of the lesser shift amount into one equation. All of these techniques create a separate shift amount for denormals and for complete normalization, and have different techniques for choosing the lesser of the shift amounts.

Some high speed designs get rid of the choice between two shift amounts and combine the maximum shift amount of a denormal back earlier in execution. Naffziger and Beraha [22] determine the bit of the LZA which corresponds to the position of most significant bit of a denormal and force this bit to a one. All bits are examined in parallel and basically a decode of the maximum shift amount is done and used to force the LZA bit to a one. Bjorksten, Mikan, and Schmookler [23] in Power3 create a vector corresponding to the denormal maximum shift amount, using a monotonic mask. It has ones in every bit starting with the most significant bit of a denormal. This denormal vector is logically ORed with a monotonic LZA vector, and the resulting vector is used to encode the shift amount. The following shows a similar method without using a monotonic mask:

$$\begin{aligned} V_i &= \overline{P_{i-2}}Z_{i-1}\overline{Z_i} + \overline{P_{i-2}}G_{i-1}\overline{G_i} \\ &\quad + P_{i-2}G_{i-1}\overline{Z_i} + P_{i-2}Z_{i-1}\overline{G_i} \\ U_i &= (i = (E_{product} - E_{min})) \\ M_i &= V_i + U_i \\ Shift &= LZD(M) \end{aligned}$$

where V is a commonly used LZA vector which examines three bits in parallel, P is a bit propagate using an exclusive OR function, G is a bit generate, Z is a bit zero term, U is a vector of the maximum a denormal can be shifted, and + represents the logical OR function, and juxtaposition represents a logical AND function. M is the combination of the LZA vector and denormal vector, and the resulting shift amount should be based on a Leading Zero Detect of M. The shift amount calculation can be made a little simpler with a monotonic mask.

Handlogten [24] in the PowerPC A50 moved the logical ORing process one step further back. Handlogten ORs the denormal vector with both the carry and sum input to the LZA. And then just the LZA output and the resulting LZD is used to create the shift amount.

9. Case study of Power4

The Power4 FPU design illustrates the use of several techniques for handling denormal operands and results. Early in the program, performance considerations forced several key decisions regarding how denormals would be handled. Each of these decisions required that certain mechanisms be provided to handle special cases. However, we then expanded on each of these mechanisms to further simplify the design or reduce critical timing paths, without significantly affecting performance.

The first key decision was that denormal data from memory would be loaded into dedicated write ports of the FPRs without first normalizing it. This would avoid the delay and area for counting leading zeros in the fraction and then shifting it. Therefore at least one tag bit was needed to help identify a single precision denormal value. It was determined that we would have enough time while transmitting data from cache to also determine whether the exponent field was all zeros or all ones, and whether the fraction was all zeros. So, three tag bits were added, along with an integer bit corresponding to the implied bit. The tag bits allowed all special values to be quickly identified for special handling during execution of arithmetic instructions.

The second decision was that some mechanism would be needed for normalizing denormal operands for some special case arithmetic operations. The previous processor, the Power3, had attempted to eliminate stalling instruction issuing based on data values. It successfully eliminated stalls based on unusual results such as for denormal values. However, there were several rare cases involving a denormal addend with the fused multiply-add instructions which were too difficult to handle without first normalizing it. The Power3 prenormalizes the addend just for those cases, passing it through the pipeline, utilizing the leading zero anticipator (LZA) and normalizer and then returning it to the top of the pipeline before executing the operation. During this *prenormalization stall*, the instruction queue is prevented from issuing instructions. From this experience, it was decided that Power4 would also need a *prenormalization stall*. However, since denormal operands are very rare, it was also decided that all operands would be prenormalized for both single and double precisions instructions. For the multiply-add instructions, which may have three denormal operands, they are pipelined so that each additional denormal operand takes only three more cycles. Prenormalization simplifies the execution of most instructions, and the effect of the stalls on performance is negligible. The tag bits used in Power4 enable denormal operands to be detected more quickly, thereby allowing the stall signal to be sent out earlier to prevent the next instruction from entering the pipeline.

Prenormalization of a double precision operand results in an intermediate exponent which is below E_{min}, and thus requires another exponent bit. In Power4, two internal exponent bits are added. This allows for the product of two denormals when the underflow trap is enabled, and also avoids ambiguity in the alignment shift count which could otherwise wrap past zero or all ones. Although these extra bits are only needed in the dataflow, Power4 adds them also to the contents of the FPRs. When data is loaded from memory, these bits are determined along with the tag bits. All arithmetic results must also produce the 13 bit exponent.

A third important decision in the design was that a short stall would also be allowed when various unusual results are produced. These include cases where very large normalization shifts might be needed, denormal results occur, or rebiasing of the exponent is needed for trapped underflow or overflow. Power3 was able to avoid stalls for these cases but with difficulty. In a multiply-add operation, a 108-bit LZA and normalizer is needed, and the normalizer must also be limited when the intermediate exponent is near $Emin$. Detection of a *possible* unusual result would cause a *back-end stall* two cycles before the actual stall, thus allowing the instruction queue to halt and the upper stages of the pipeline to also halt. The stall would allow the output data to recycle back through the last two stages, which are the normalizer and the rounder. Thus, most stalls are only two cycles. If a denormal result is needed, the data is normalized the first trip through the pipeline but is not rounded. During the stall, it is then sent back to the normalizer but aligned 65 bit positions to the right. The low order bits of the intermediate exponent, which is smaller than $Emin$, provide the proper shift amount for the normalizer. The stall also allows the LZA and normalizer to be much smaller. Even with stalls occurring at times when the result is normal, the two-cycle delay does not have a very significant effect on performance.

It is possible to have both types of stalls in progress within the pipeline at the same time. An instruction may begin a *prenormalization stall* in stage1. It may advance up to the fourth stage when the previous instruction reaches stage5 and begins a *back-end stall*. The *prenormalization stall* is then stopped until the *back-end stall* is completed.

There is one other interesting mechanism that is used for denormals which has not yet been described. Single precision denormal values may be held in the FPRs with the significand either normalized or unnormalized. If a double precision store to memory is to be executed and it is unnormalized, then a *prenormalization stall* is taken to normalize it. However, if a *single* precision store needs to be executed and it is normalized, then it needs to be denormalized. Rather than taking the data through the pipeline and denormalizing it, the alignment shifter is used. All data for stores use the Add operand input in the multiply-add dataflow. Since there are no other operands entering the multiplier, constants are forced into the exponents which correspond to those operands. If the sum of those constants is $Emin$, then the aligner will shift the significand to the right a distance equal to the difference of $Emin$ and the exponent of the normalized operand.

10. Conclusion

Implementing denormalized numbers in hardware is possible with a small amount of additional hardware. The usefulness of tagging has been discussed and its utility is partially dependent on the architecture of the processor. Denormalized input operands can be handled for multiply and add operations by performing multiple corrections of the exponent difference calculation for alignment and by correcting the multiplication result by adding correction terms. An underflow condition with traps disabled requires a denormalized result. Denormalization can be handled in a denormalization unit or it can be prevented by stopping the normalizer from shifting past the radix point of denormalized number. This is accomplished by modifying the leading zero anticipate logic to prevent an indication of more than this radix point.

Also, shown is a case study for the new Power4 FPU which handles denormalized numbers in hardware. It uses a combination of tagging and prenormalization to prevent denormalization and to handle denormal input. The result is a processor which executes denormalized operands with only a few additional cycles over the execution of normalized operands.

References

[1] "IEEE standard for binary floating-point arithmetic, ANSI/IEEE Std 754-1985," The Institute of Electrical and Electronic Engineers, Inc., New York, Aug. 1985.

[2] R. Yu and G. Zyner. "167 MHz Radix-4 Floating Point Multiplier," In *Proc. of Twelfth Symp. on Comput. Arith.*, pp. 149-154, Bath, England, July 1995.

[3] A. Naini, A. Dhablania, W. James, and D. D. Sarma. "1-GHz HAL SPARC65 dual floating point unit with RAS features," In *Proc. of Fifteenth Symp. on Comput. Arith.*, pp. 173-183, Vail, Colorado, June 2001.

[4] Intel Corporation. "IA-32 Intel Architecture Sofware Developer's Manual Volume 1 Basic Architecture," ftp://download.intel.com/ design/ Pentium4/ manuals/ 24547008.pdf, 1997.

[5] M. D. V. Dyke-Lewis and W. Meeker. "Method and apparatus for performing fast floating point operations," *U.S. Patent No. 5,966,085*, p. 7, Oct. 12, 1999.

[6] C. May, E. Silha, R. Simpson, and H. Warren, editors. *"The PowerPC Architecture: a specification for a new family of RISC processors,"* Morgan Kaufman Publishers, Inc., San Francisco, CA, 2002.

[7] Intel Corporation. "Intel Itanium Architecture Sofware Developer's Manual Volume 1 Application Architecture," ftp://download.intel.com/ design/ Itanium/ Downloads/ 24531703s.pdf, Dec. 2001.

[8] "z/Architecture Principles of Operation," Order No. SA22-7832-1, available through IBM branch offices, Oct. 2001.

[9] "Enterprise Systems Architecture/390 Principles of Operation," Order No. SA22-7201-5, available through IBM branch offices, Sept. 1998.

[10] S. Vassiliadis, E. M. Schwarz, and D. J. Hanrahan. "A general proof for overlapped multiple-bit scanning multiplications," *IEEE Trans. Comput.*, 38(2):172–183, Feb. 1989.

[11] T. Williams. "Method and apparatus for multiplying denomalised binary floating point numbers without additional delay," *U.S. Patent No. 5,347,481*, p. 16, Sep. 13, 1994.

[12] C. A. Krygowski and E. M. Schwarz. "Floating-point multiplier for de-normalized inputs," *U.S. Patent Application No. 2002/0124037 A1*, p. 8, Sep. 5, 2002.

[13] G. Gerwig, H. Wetter, E. M. Schwarz, and J. Haess. "High performance floating-point unit with 116 bit wide divider," In *Proc. of Sixteenth Symp. on Comput. Arith.*, Spain, June 2003.

[14] S. Gupta, R. Periman, T. Lynch, and B. McMinn. "Normalizing pipelined floating point processing unit," *U.S. Patent No. 5,267,186*, p. 10, Nov. 30, 1993.

[15] S. Gupta, R. Periman, T. Lynch, and B. McMinn. "Normalizing pipelined floating point processing unit," *U.S. Patent No. 5,058,048*, p. 11, Oct. 15, 1991.

[16] M. P. Taborn, S. M. Burchfiel, and D. T. Matheny. "Denormalization system and method of operation," *U.S. Patent No. 5,646,875*, p. 8, Jul. 8, 1997.

[17] E. Schwarz, B. Giamei, C. Krygowski, M. Check, and J. Liptay. "Method and system for executing denormalized numbers," *U.S. Patent No. 5,903,479*, p. 6, May 11, 1999.

[18] E. M. Schwarz and C. A. Krygowski. "The S/390 G5 floating-point unit," *IBM Journal of Research and Development*, 43(5/6):707–722, Sept./Nov. 1999.

[19] M. Urano and T. Taniguchi. "Method and apparatus for normalization of a floating point binary number," *U.S. Patent No. 5,513,362*, p. 15, Apr. 30, 1996.

[20] V. Y. Gorshtein and V. T. Khlobystov. "Multiplication apparatus and methods which generate a shift amount by which the product of the significands is shifted for normalization or denormalization," *U.S. Patent No. 5,963,461*, p. 22, Oct. 5, 1999.

[21] A. I. Grushin and E. S. Vlasenko. "Computer methods and apparatus for eliminating leading non-significant digits in floating point computations," *U.S. Patent No. 5,732,007*, p. 34, May 24, 1998.

[22] S. D. Naffziger and R. G. Beraha. "Method and apparatus for bounding alignment shifts to enable at speed denormalized result generation in an FMAC," *U.S. Patent No. 5,757,687*, p. 13, May 26, 1998.

[23] A. A. Bjorksten, J. D.G. Mikan, and M. S. Schmookler. "Fast floating point results alignment apparatus," *U.S. Patent No. 5,764,549*, p. 9, Jun. 9, 1998.

[24] G. H. Handlogten. "Method and apparatus to perform pipelined denormalization of floating-point results," *U.S. Patent No. 5,943,249*, p. 10, Aug. 24, 1999.

Representable correcting terms for possibly underflowing floating point operations

Sylvie Boldo & Marc Daumas
E-mail: Sylvie.Boldo@ENS-Lyon.Fr & Marc.Daumas@ENS-Lyon.Fr
Laboratoire de l'Informatique du Parallélisme
UMR 5668 CNRS – ENS de Lyon – INRIA
Lyon, France

Abstract

Studying floating point arithmetic, authors have shown that the implemented operations (addition, subtraction, multiplication, division and square root) can compute a result and an exact correcting term using the same format as the inputs. Following a path initiated in 1965, many authors supposed that neither underflow nor overflow occurred in the process. Overflow is not critical as this kind of exception creates persisting non numeric quantities. Underflow may be fatal to the process as it returns wrong numeric values with little warning. Our new conditions guarantee that the correcting term is exact when the result is a number. We have validated our proofs against Coq automatic proof checker. Our development has raised many questions, some of them were expected while other ones were surprising.

1 Introduction

It was recognized in 1991 [2] that the most widely used algebraic operations of floating point arithmetic can return an exact correcting term provided the same floating point format is used for the inputs, the result and the correcting term. First results on a similar subject were probably presented in 1965 [11, 16] as techniques to enhance precision for additions and accumulations.

As the IEEE 754 and 854 standards [21, 4] spread the use of correct rounding modes including rounding to the nearest floating point value, rationales were studied for the four mentioned operations leading to a generic theorem such as the one below.

Result 1 (Adapted from Bohlender *et al*, 1991)
Let $\mathbb{F}$ be the set of real numbers represented with the defined floating point arithmetic and let $\oplus, \ominus, \otimes, \oslash, \circ(\sqrt{\cdot})$ be the implemented addition, subtraction, multiplication, division and square root rounded to the nearest floating point value. Provided neither underflow nor overflow precludes a dependable behavior, an exact correcting term that is known to belong to $\mathbb{F}$ can be defined as follows from inputs x and y and the result,

Result	Correcting term
$x \oplus y$	$x + y - x \oplus y$
$x \ominus y$	$x - y - x \ominus y$
$x \otimes y$	$x \times y - x \otimes y$
$x \oslash y$	$x - (x \oslash y) \times y$
$\circ(\sqrt{x})$	$x - (\circ(\sqrt{x}))^2$

The correcting term is still in $\mathbb{F}$ if we use a directed rounding mode for the multiplication $\otimes$ and the division $\oslash$.

The fact that many authors have overlooked consequences of an overflow or an underflow in their related work [6, 13, 2, 19], may have arisen from an inadequacy of the existing formalisms to build a tight condition for underflow to be harmless. We will see in this text that an error may occur even when neither inputs nor the result are subnormal numbers.

A sufficient condition to be able to define an exact correcting term is that the exact error lies above the gradual underflow threshold as we will see in the conclusion of this text. This is not a necessary condition. We will present examples and counter examples in the text to show that all our conditions are both sufficient and tight.

In the process of giving tight conditions for Result 1 to be correct even when underflow occurs, we have isolated a very surprising situation. In an intuitive deduction, the rounded result r is the most significant part of the exact result and the correcting term e is a remainder. It is easy to jump to a conclusion that

$|e|$ must be significantly smaller than $|r|$ in additions, subtractions and multiplications, and smaller than $|x|$ in divisions and square roots. We have exhibited cases were $|e|$ is close to $|x|$. Exploring the directed rounding modes, we have found cases where $|e|$ is significantly larger than $|x|$.

All theorems presented in this text have been developed with a strong focus on genericity. The radix, the number of significant digits of the mantissa, the underflow threshold and the rounding mode used are parameters possibly set in the hypotheses. Jumping to the conclusions of this text, we have no restriction on the radix provided it is an integer greater than or equal to 2 and no restriction on the number of digits of the mantissa provided once again it is greater than or equal to 2. The tie breaking mechanism when rounding to the nearest has no effect on our theorems and can be even, odd or use any combination. However, precise rounding to a nearest floating point number is necessary for additions and square roots.

Following Section 2, discussing our formalism and rounding, we present results with increasing difficulty on multiplications (Section 3), additions and subtractions (Section 4), divisions (Section 5), and square roots (Section 6). As our question is easily connected to the correct behavior of the remainder (FPREM) operation defined by the IEEE standard, we have built a machine-checked proof of this fact in the last section of this text, answering a question raised in 1995 [15] (Section 7). All the proofs can be downloaded through the Internet at the following address.

`http://www.ens-lyon.fr/~sboldo/coq/`

2 Floating point representation

Numbers are represented with pairs (n, e) that stand for $n \times \beta^e$ where β is the radix of the floating point system. We use both an integral signed mantissa n and an integral signed exponent e for sake of simplicity. As we use integral mantissa and exponent, components have to be slightly adapted to get the usual IEEE standard **exponent** and the usual IEEE standard **mantissa** stored in computers.

The underflow exponent, a constant, is the lowest exponent $-e_{min}$ available. We do not set an upper bound on the exponent as overflows are easily detected by other means. Mantissas are stored in signed-magnitude format using p digits. We define a **representable** pair (n, e) such that $|n| < \beta^p$ and $e \geq -e_{min}$.

The above definition is not sufficient to identify one unique pair (n, e) for a represented quantity. A representable pair is **normal** if $\beta \cdot |n| \geq \beta^p$ and it is **subnormal** if $\beta \cdot |n| < \beta^p$ and $e = -e_{min}$. Each represented number has one unique representation, the **machine** pair, that is either normal or subnormal.

In some sense, our internal representation is similar to the one proposed in the late 1950s [1]. However, our approach is very different as computers only manipulate unique machine representations. This formalism is used to model the numbers but not to compute on them. Additional representations are used to state conditions and prove theorems.

In all the following theorems, we use any representation. We do not have to use the machine representation. Our theorems are valid with subnormals but no example use them explicitly.

Our formalism was introduced in [5] for developments using Coq proof environment [9]. Other formalisms of floating point arithmetic are in use with PVS [15, 10], HOL [3, 8] or ACL2 [18]. Using Curry Howard isomorphism, Coq and HOL rely on a very small inference engine to check the correctness of proofs. Although Coq and HOL lack many automatic techniques implemented in PVS or ACL, they allow users to safely explore properties of a system.

Most available general purpose processors have long been compliant with the IEEE 754 standard on floating point arithmetic. It means that they implement precise rounding for the four arithmetic operations: addition, multiplication, division and square root. The result obtained for any of these computer operations is the one given by using a user-chosen rounding function on the result of the exact mathematical operation. The standard specifies four rounding functions: rounding to the nearest with the even tie breaking rule, rounding up, down or toward 0. The rounding functions and the arithmetic operators are defined and used in our formalism.

3 Multiplication

Concerning the question raised in this paper, the multiplication is most probably the simplest operation. This fact was recognized early and IBM S/370 has a special instruction to produce the rounded product and the exact correcting term [12].

An algorithm has long been developed and tested to compute these pairs with only IEEE standard operations [6, 13] although it relies on some developments on the addition presented Section 4. The task of producing the two quantities is much simpler with a computer that provides a full accuracy fused multiply and accumulate operator such as with Intel IA64 [14].

The following theorem is based on some early development of our project.

Theorem 2 (RepMult_gen in FroundMult) *Let $\otimes$ be the implemented multiplication rounded to a nearest floating point value or with a directed rounding mode. Given inputs x and y such that $x \otimes y$ is neither an infinity nor a NaN, the correcting term*

$$x \times y - x \otimes y$$

is representable if and only if there exist two representable pairs (n_x, e_x) and (n_y, e_y) representing x and y such that

$$e_x + e_y \geq -e_{min}.$$

Proof sketch adapted from [5]: The product of two p-digit mantissas produces a $2p$-digit mantissa that can be split into two p-digit floating point pairs whatever the rounding mode used.

The condition $e_x + e_y \geq -e_{min}$ is sufficient as it means that the extended mantissa is larger than the underflow threshold. It is also necessary. Suppose that there are no representable pairs for x and y such that $e_x + e_y \geq -e_{min}$, we conclude that $x \times y \mod \beta^{-e_{min}}$ is different from zero. Such difference cannot be represented as it is below the underflow threshold.

□

Counter example against weaker conditions: We use the two following pairs representing the numbers $9 \times 2^{-\lfloor \frac{e_{min}}{2} \rfloor - 1}$ and $11 \times 2^{-\lceil \frac{e_{min}}{2} \rceil}$. Our notation uses $p = 4$ digits and an arbitrary value for e_{min}.

$$(1001_2, -\left\lfloor \frac{e_{min}}{2} \right\rfloor - 1)_2 \quad \text{and} \quad (1011_2, -\left\lceil \frac{e_{min}}{2} \right\rceil)_2$$

The exact product $99 \times 2^{-e_{min}-1}$ rounds to the nearest floating point pair $96 \times 2^{-e_{min}-1} = 12 \times 2^{-e_{min}+2}$ represented by the pair $(1100_2, -e_{min} + 2)_2$. The exact correcting term is $-3 \times 2^{-e_{min}-1}$ and cannot be represented. Still neither inputs nor the result is a subnormal pair.

4 Addition and subtraction

Authors have long exhibited two different situations in the production of an exact representable correcting term for additions and subtractions. If the exponents of the inputs are close enough, the exact result can be written with a $2p$-digit mantissa. In this case, the rounded value and the error can be stored with representable pairs whatever the rounding mode being either to a nearest or directed.

If the exponents of the inputs are too far away one from another, we have to make sure that the rounded result is the largest input in magnitude. This fact is obtained only when rounding to a nearest floating point value if the operations are precisely rounded. It was proved in some early part of our development [5] by adapting the proof of the correctness of the algorithm published in [19] to obtain the correcting term.

Theorem 3 (errorBoundedPlus in ClosestPlus) *Let $\oplus$ and $\ominus$ be the implemented addition and subtraction rounded to a nearest floating point value. Given inputs x and y, for each result $x \oplus y$ and $x \ominus y$ that is neither an infinity nor a NaN, the correcting terms*

$$x + y - x \oplus y \quad \text{and} \quad x - y - x \ominus y$$

are representable.

Counter example against weaker conditions: We present now an example where the double rounding of Texas Instruments' TMS 320 C3x or Intel's IA32 introduces an unrepresentable error. Let the radix be $\beta = 2$ and the precision $p > 4$ arbitrary large. We assume that the extended precision less than doubles the number of digits in the mantissa. We compute the sum of the two normalized numbers $(-(2^{p-1} + 1), p + 1)_2$ and $(2^p - 3, 0)_2$.

The first input has value $-2^{2p} - 2^{p+1}$ and the exact result is $-2^{2p} - 2^p - 3$. As the extended precision is limited, the last two bits of the result are lost and the first rounding returns $-2^{2p} - 2^p$. This result is next rounded to -2^{2p} and the error is $-2^p - 3$ that cannot be represented.

We obtain the same unrepresentable correcting term with the same inputs using the IEEE standard directed rounding mode to $+\infty$.

5 Division

Theorem 4 exhibits two conditions for the correcting term to be representable. The first condition (1) is expected. The second condition (2) is very new and deals with a situation that only occurs with some of the directed rounding modes.

Theorem 4 (errBoundedDiv in FroundDivSqrt.v) *Let $\oslash$ be the implemented division rounded to a nearest floating point value or with a directed rounding mode. Given inputs x and y, whenever $x \oslash y$ is neither an infinity nor a NaN, the correcting term*

$$x - (x \oslash y) \times y$$

is representable if and only if there exist two representable pairs (n_y, e_y) and (n_q, e_q) representing y and $x \oslash y$ such that
$$e_y + e_q \geq -e_{min}, \quad (1)$$
and
$$|x \oslash y| \neq \beta^{-e_{min}} \quad \text{or} \quad \frac{\beta^{-e_{min}}}{2} \leq \left|\frac{x}{y}\right|. \quad (2)$$

Proof: We will prove that the exact remainder $r = x - (x \oslash y) \times y$ computed with appropriate care is representable so that the two conditions are sufficient. We first define $q = x \oslash y$ with any rounding mode. We assume that there exist some representations (n_q, e_q) of q and (n_y, e_y) of y that satisfy (1) and let (n_x, e_x) be a representation of x.

Let (n'_q, e'_q) be the machine representation of q, that means that
$$n_q \times \beta^{e_q} = n'_q \times \beta^{e'_q}$$
and we can define the **unit in the last place** function $\text{ulp}(q) = \beta^{e'_q}$. We know from previous results that
$$e'_q \leq e_q \quad (\texttt{FcanonicLeastExp})$$
and
$$\left|\frac{x}{y} - q\right| < \text{ulp}(q) \quad (\texttt{RoundedModeUlp}).$$

We will show that the floating point pair (n_r, e_r) with
$$\begin{aligned}
n_r &= n_x \times \beta^{e_x - \min(e_x, e_q + e_y)} \\
&\quad - n_q \times n_y \times \beta^{e_q + e_y - \min(e_x, e_q + e_y)} \\
e_r &= \min(e_x, e_q + e_y)
\end{aligned}$$
is a representable pair for r. We easily check that (n_r, e_r) is a representation of $x - qy$. To prove that r is representable, we consider two cases.

First, if $e_q + e_y \leq e_x$, then $e_r = e_q + e_y$ and $e_r \geq -e_{min}$ from (1). We check that
$$\begin{aligned}
|n_r| &= |r| \times \beta^{-e_r} \\
&= |x - qy| \times \beta^{-e_q - e_y} \\
&\leq |y| \times \left|\frac{x}{y} - q\right| \times \beta^{-e_q - e_y} \\
&< |n_y| \times \text{ulp}(q) \times \beta^{-e_q} \\
&< |n_y| \times \beta^{e'_q - e_q} \\
&< |n_y|
\end{aligned}$$
and finally, $|n_r| < \beta^p$.

Second, if $e_x < e_q + e_y$ then $e_r = e_x$ and $e_r \geq -e_{min}$. So $|n_r| < \beta^p$ is the only question left to finish our proof. We examine three cases depending on $|n'_q|$. If $|n'_q| = 0$, then $q = 0$ and $r = x$. If $|n'_q| > 1$, we check that
$$\begin{aligned}
\left|\frac{x}{y} - q\right| &< \text{ulp}(q) \\
|n'_q| \times |x - qy| &< |n'_q| \times \text{ulp}(q) \times |y| \\
&< |q| \times |y| \\
&\leq |x| + |x - qy|
\end{aligned}$$
and finally $(|n'_q| - 1) \times |x - qy| \leq |x|$. Since $|n'_q| \geq 2$, we deduce that $|x - qy| \leq |x|$ and then $|n_r| \leq |n_x| < \beta^p$.

The last case lies when $|n'_q| = 1$. Since (n'_q, e'_q) is a machine representation, it is subnormal and $q = \pm\beta^{-e_{min}}$. We deduce $\beta^{-e_{min}}/2 \leq |x/y|$ from hypothesis (2) and $|x/y| < 2 \times \beta^{-e_{min}}$ since we used a rounding mode, so that the successor of the rounded result bounds the real value. We conclude that
$$\frac{|q|}{2} \leq \left|\frac{x}{y}\right| < 2|q| \quad \text{and} \quad \left|\left|\frac{x}{y}\right| - |q|\right| \leq \left|\frac{x}{y}\right|.$$

As $\frac{x}{y}$ and q have the same sign (properties `RleRoundedR0` and `RleRoundedLessR0` of rounding modes independent of the operation),
$$\left|\frac{x}{y} - q\right| \leq \left|\frac{x}{y}\right| \quad \text{and} \quad |x - qy| \leq |x|.$$

That ends the proof.

$\square$

Counter examples against weaker conditions: We will show that both hypotheses (1) & (2) are tight with an example when one of the hypotheses is not satisfied.

- The case where (1) is not satisfied follows an expected path. Let the radix be $\beta = 2$ and the precision $p = 4$ with
$$\begin{aligned}
x &= 1001_2 \times 2^{-e_{min}+3} \\
y &= 1101_2 \times 2^{-\lfloor \frac{e_{min}}{2} \rfloor}.
\end{aligned}$$
The division is rounded to the nearest, that is towards $-\infty$ here. We get
$$q = 1011_2 \times 2^{-1 - \lceil \frac{e_{min}}{2} \rceil}.$$
and $e_q + e_y = -e_{min} - 1$.
The correcting term $x - qy$ is
$$r = -1101011_2 \times 2^{-e_{min}-1},$$
that cannot be represented.

- The case where (2) is not satisfied is more surprising with radix $\beta = 2$ and

$$x = 2^{-e_{min}+p} - 2^{-e_{min}+1}$$
$$y = 2^{p+1}.$$

The exact division

$$x/y = \frac{2^{-e_{min}}}{2} - 2^{-e_{min}-p}$$

is rounded towards $+\infty$ and we get

$$q = 2^{-e_{min}}.$$

The correcting term $x - qy$ is

$$-(2^p + 1) \times 2^{-e_{min}}$$

that cannot be represented and that is surprisingly larger than x. A relatively larger correcting term arises rounding towards $+\infty$ with $x = 2^{-e_{min}+p}$ and $y = 2^{2p+1}$. Such a situation has never been presented in the past but it can be related to the switch from relative error to absolute error in the bounds of residuals presented in [7].

We have proved in theorems errorBoundedDivClosest and errorBoundedDivToZero, that hypothesis (2) is always true when rounding to a nearest floating point or towards zero.

We will discuss again in Section 7 on the choice of the method used to prove the theorems on the division. The proof published in [2] is based on the usual division algorithm and it uses properties that are not known *a priori* in a proof checking environment.

6 Square root

It is common knowledge that square root extractions are similar to division algorithms in many ways. As a consequence, we have written proofs of this section with the skeleton of the preceding ones. However the inequalities use distinct mathematical properties. Merging the two theorems into a single one is impossible unless we use a system able to prove that a large set of inequalities leads to our goal without human assistance. Such a system does not exist now.

As stated in the following theorem only one condition (3) has to be satisfied. On the other hand, the correcting term can be defined only when the operator precisely rounds the result to a nearest floating point number.

Theorem 5 (errBoundedSqrt in FroundDivSqrt.v)
Let $\circ(\sqrt{\cdot})$ be the implemented square root operation rounded to a nearest floating point value. Given the input x, whenever $\circ(\sqrt{x})$ is neither an infinity nor a NaN, the correcting term

$$x - \circ(\sqrt{x}) \times \circ(\sqrt{x})$$

is representable if and only if there exist a representable pair (n_q, e_q) representing $\circ(\sqrt{x})$ such that

$$2\,e_q \geq -e_{min}. \quad (3)$$

Proof: Once again, we prove that the exact remainder $r = x - \circ(\sqrt{x}) \times \circ(\sqrt{x})$ computed with appropriate care is representable so that the condition is sufficient. We first define $q = \circ(\sqrt{x})$ rounded to a nearest floating point value. We assume that there exists some representation (n_q, e_q) of q that satisfies (3) and let (n_x, e_x) be a representation of x.

Let (n'_q, e'_q) be the machine representation of q.

$$|\sqrt{x} - q| \leq \frac{\text{ulp}(q)}{2} \quad (\text{ClosestUlp}).$$

We will show that the floating point pair (n_r, e_r) with

$$n_r = n_x \times \beta^{e_x - \min(e_x, 2\,e_q)}$$
$$- n_q^2 \times \beta^{2\,e_q - \min(e_x, 2\,e_q)}$$
$$e_r = \min(e_x, 2\,e_q)$$

is a representable pair for r. We easily check that (n_r, e_r) is a representation of $x - q^2$. To prove that r is representable, we consider two cases.

First, if $2\,e_q \leq e_x$, then $e_r = 2\,e_q$ and $e_r \geq -e_{min}$ from (3). We check that

$$|n_r| = |x - q^2| \times \beta^{-2\,e_q}$$
$$= |\sqrt{x} - q| \times |\sqrt{x} + q| \times \beta^{-2\,e_q}$$
$$\leq \frac{\text{ulp}(q)}{2} \times (|\sqrt{x} - q| + 2\,|q|) \times \beta^{-2\,e_q}$$
$$\leq \frac{\text{ulp}(q)}{2} \times \left(\frac{\text{ulp}(q)}{2} + 2\,|q|\right) \times \beta^{-2\,e_q}$$
$$\leq \frac{1}{4} \times \beta^{2\,e'_q - 2\,e_q} + |n'_q| \times \beta^{2\,e'_q - 2\,e_q}$$
$$\leq \frac{1}{4} + |n'_q|$$
$$\leq \frac{1}{4} + (\beta^p - 1)$$
$$\leq \beta^p - \frac{3}{4}$$

and finally, $|n_r| < \beta^p$.

Second, if $e_x < 2\,e_q$ then $e_r = e_x$ and $e_r \geq -e_{min}$. So $|n_r| < \beta^p$ is only left to finish our proof. We examine two cases depending on (n'_q, e'_q) is normal or subnormal.

In the case where (n'_q, e'_q) is normal, we use the fact that $\text{ulp}(q) \leq |q| \times \beta^{1-p}$ (`FulpLe2`) with $p \geq 2$, and that $q \geq 0$ (`RleRoundedR0`), so

$$q \leq \sqrt{x} + \frac{\text{ulp}(q)}{2}$$
$$\leq \sqrt{x} + \frac{1}{2} \times \beta^{1-p} \times q$$
$$q \leq \frac{\sqrt{x}}{1 - \frac{\beta^{1-p}}{2}}$$
$$q^2 \leq x \times \left(\frac{1}{1 - \frac{\beta^{1-p}}{2}}\right)^2$$
$$\leq 2\,x$$

As q^2 and x have the same sign and $q^2 \leq 2\,x$, we have $|x - q^2| \leq x$ and then again $|n_r| \leq |n_x| < \beta^p$.

The case where (n'_q, e'_q) is subnormal cannot happen with any of the *common* single and double precision floating-point formats and it was expectedly dismissed by previous authors. But if the radix is $\beta = 2$, the precision is $p = 5$ and the underflow exponent is $-e_{min} = -3$, the machine representation of the square root of 1, the number represented by $(1,0)$ is subnormal as it is $(1000_2, -3)$.

Again, such a system would be extremely silly but neither automatic proof checkers nor higher order logic have *common sense*. We would need to define constraints such as the one presented in the IEEE 854 standard [4, § 3.2] to explain that 1 should not be a subnormal number. As a specification should be as concise as possible to remain trusted and since we are able to present a proof without additional definitions, we prefer to limit ourselves to the 17 definitions of our generic specification and no constraint at all but $\beta \geq 2$ and $p \geq 2$.

In this case, $e'_q = -e_{min}$ and

$$\begin{aligned}
|n_r| &= |x - q^2| \times \beta^{-e_x} \\
&= |\sqrt{x} - q| \times |\sqrt{x} + q| \times \beta^{-e_x} \\
&\leq \frac{\beta^{-e_{min}}}{2} \times (|\sqrt{x} - q| + 2\sqrt{x}) \times \beta^{-e_x} \\
&\leq \frac{\beta^{-e_{min}}}{2} \times \left(\frac{\beta^{-e_{min}}}{2} + 2\sqrt{x}\right) \times \beta^{-e_x} \\
&\leq \frac{1}{4}\beta^{-2\,e_{min}-e_x} + \sqrt{n_x}\sqrt{\beta^{e_x}}\beta^{-e_{min}-e_x} \\
&\leq \frac{1}{4} + \sqrt{n_x} \times \sqrt{\beta^{-2e_{min}-e_x}} \\
&\leq \frac{1}{4} + n_x \leq \beta^p - \frac{3}{4}
\end{aligned}$$

and finally, $|n_r| < \beta^p$, so that the property still holds in this case never studied before.

$\square$

Counter examples against weaker conditions: We will show once again that both hypotheses (3) and the rounding mode being to the nearest are sufficient and tight with an example when one hypothesis is not satisfied.

- Let the radix be $\beta = 2$ and the precision be $p = 4$. The case where (1) is not satisfied follows an expected path. We distinguish whether e_{min} is even or not. In the first case (e_{min} is even), with

$$\begin{aligned}
x &= 1010_2 \times 2^{-e_{min}+1} \\
q &= 1001_2 \times 2^{-1-\frac{e_{min}}{2}},
\end{aligned}$$

we check that $2\,e_q = -e_{min} - 2$ and the exact correcting term $x - q^2$ is

$$-2^{-e_{min}-2},$$

that cannot be represented in our floating-point format.

In the second case (e_{min} is odd), we end up to the same conclusion with

$$\begin{aligned}
x &= 1010_2 \times 2^{-e_{min}+3} \\
q &= 1101_2 \times 2^{\frac{-1-e_{min}}{2}} \\
x - q^2 &= -1001_2 \times 2^{-e_{min}-1}.
\end{aligned}$$

- To prove that precise rounding to a nearest is necessary for the square root operation, we focus on the number $x = 2^{2p+2} - 6 \times 2^{p+1}$ where the radix is $\beta = 2$ and the precision $p > 4$ is arbitrary large. A representation of x is $(2^p - 3, p+2)_2$.

The square root of x can be expressed in the form $\sqrt{x} = 2^{p+1} - 3 - \alpha \times 2^{-p}$ with α between $\frac{9}{16}$ and $\frac{9\sqrt{2}}{4}$ from Taylor-Lagrange inequality. Precise rounding would answer $2^{p+1} - 4$ represented by the pair $(2^p - 2, 2)_2$ but some hardware that discards $\alpha \times 2^{-p}$ may round the result to $2^{p+1} - 2$ and return the representable pair $(2^p - 1, 2)_2$.

In the later case, the correcting term $x - q^2$ is $-2 \times 2^{p+1} - 4 = -(2^p + 1) \times 4$ and it cannot be represented.

We have just presented the case of rounding to $+\infty$. A counter example rounding to $-\infty$ or toward 0, is obtained with $x = 2^{2p+2} + 6 \times 2^{p+2}$.

7 Remainder

IEEE 754 and 854 standards define the remainder operation that shall be implemented for a system to comply to the standards. Given two inputs x and y, we define n that is the rounding of x/y to the nearest integer value with the even tie breaking rule and the result is defined as

$$r = x - ny.$$

Both documents state that the remainder can always be represented using the same floating point format as the inputs. The best way to prove this assertion is to look at the common stepwise binary implementation of the Euclidean division. As the quotient is computed bit-wise, most significant bit first, the remainder always fit in the same format as the inputs. This invariant was also used in [2] to prove the theorem on divisions and the authors noticed that this invariant is not true for the stepwise square root extraction.

We would have to describe the Euclidean division with much details and properties to port this proof to an automatic proof checker. This is the reason why the question of the remainder being exact has never been answered with an automatic proof checker before [15]. We present in the following theorem a more elementary proof of this fact.

Theorem 6 (errBoundedRem in FroundDivSqrt.v)
Given inputs x and $y \neq 0$, and n the rounded value of x/y to a nearest integer the remainder

$$x - n \times y$$

is representable if it is neither an infinity nor a NaN

Proof: We define a rounded value of real r to an integer as integer n such that

$$\forall n' \in \mathbb{Z}, \quad |n - r| \leq |n' - r|.$$

We will in fact solely use the property that if n is a rounded value of real r then $|n - r| \leq 1/2$.

We prove the theorem in the case where x and y are non-negative machine pairs. Other cases are handled similarly. Then $r =_\mathbb{R} x - n \times y$ is such that $e_r = \min(e_x, e_y)$.

First, if $e_y \leq e_x$ then $e_r = e_y \geq -e_{min}$ and

$$\begin{aligned}
|n_r| &= |x - n \times y|\beta^{-e_y} \\
&= |y| \times \left|n - \frac{x}{y}\right|\beta^{-e_y} \\
&= |n_y| \times \left|n - \frac{x}{y}\right| \\
&\leq |n_y| \times 1/2 \\
&\leq |n_y| \\
&< \beta^p.
\end{aligned}$$

Second, if $e_x < e_y$ then $e_r = e_x \geq -e_{min}$. We then have three subcases depending on the value of n.

If $n = 0$ then $r = x$ is representable. If $n = 1$, then $r =_\mathbb{R} x - y$ is representable because of Sterbenz's theorem [20]. Indeed as $n = 1$, we have $|1 - x/y| \leq 1/2$ and we deduce that $1/2 \leq x/y \leq 3/2$.

The other cases are impossible. As $x/y - n < 1$, we have $0 \leq x/y < n + 1$ so $n \geq 0$. And as x and y are machine pairs and $e_x < e_y$, we know that $x < y$ (FcanonicPosFexpRlt) so $x/y < 1$ and $n \leq 1$. So we have an integer n such that $n \geq 0$, $n \leq 1$ and n is neither 0 nor 1.

□

8 Conclusion

The urge for the development of automatic proof checking is evident from adventures of published proofs such as the ones described in [17]. We have proved and checked in this document old and new properties on floating point arithmetic. Some of the new properties are almost part of the common knowledge of the community of users of floating point arithmetic. However, validating them has made it possible to detect strange, very uncommon and counter-intuitive cases of failure. Should designers have decided to implement the functionality advocated in [2], such cases might have nurtured dormant bugs.

Although the properties shown in this document have been validated with an automatic proof checker, we do not regard them as unshakable truths. We have so far validated a large number of properties based on our specification and we have been able to connect many of these properties with results published in the literature. Working with an automatic proof checker, we like to compare it to a stubborn but helping colleague that will review our proof and will tirelessly ask for details.

As a conclusion, we regard these properties as highly trusted. They incurred a significant amount of testing not reported in the process of the proof. Most proofs have first been written and approved as a pen and paper proofs before being checked with the computer. The most uncommon achievement of this work is probably the ability to extend our highly trusted properties to any radix, any precision and to digital signal processing circuits implementing a floating point arithmetic slightly different from the IEEE standard.

9 Acknowledgments

This work has benefited from the continuous help of Laurent Théry and Yves Bertot at the INRIA, in Sophia Antipolis. We have heavily used, modified and relied on the first specification of floating point arithmetic in Coq reported in [5] and partially funded by the AOC action of the INRIA.

The authors would also like to thank Peter Markstein for his help in promoting their work and his discussions on intermediate level properties of floating point arithmetic already in use in the development of mathematical libraries.

Finally, we thank the referees for they many comments and suggestions to enhance our work and its readability.

References

[1] R. L. Ashenhurst and N. Metropolis. Unnormalized floating point arithmetic. *Journal of the ACM*, 6(3):415–428, 1959.

[2] G. Bohlender, W. Walter, P. Kornerup, and D. W. Matula. Semantics for exact floating point operations. In P. Kornerup and D. Matula, editors, *Proceedings of the 10th Symposium on Computer Arithmetic*, pages 22–26, Grenoble, France, 1991.

[3] V. A. Carreño. Interpretation of IEEE-854 floating-point standard and definition in the HOL system. Technical Report Technical Memorandum 110189, NASA Langley Research Center, 1995.

[4] W. J. Cody, R. Karpinski, et al. A proposed radix and word-length independent standard for floating point arithmetic. *IEEE Micro*, 4(4):86–100, 1984.

[5] M. Daumas, L. Rideau, and L. Théry. A generic library of floating-point numbers and its application to exact computing. In *14th International Conference on Theorem Proving in Higher Order Logics*, pages 169–184, Edinburgh, Scotland, 2001.

[6] T. J. Dekker. A floating point technique for extending the available precision. *Numerische Mathematik*, 18(3):224–242, 1971.

[7] J. Demmel. Underflow and the reliability of numerical software. *SIAM Journal on Scientific and Statistical Computing*, 5(4):887–919, 1984.

[8] J. Harrison. A machine-checked theory of floating point arithmetic. In Y. Bertot, G. Dowek, A. Hirschowitz, C. Paulin, and L. Théry, editors, *12th International Conference on Theorem Proving in Higher Order Logics*, pages 113–130, Nice, France, 1999.

[9] G. Huet, G. Kahn, and C. Paulin-Mohring. The Coq proof assistant: a tutorial: version 7.2. Technical Report 256, Institut National de Recherche en Informatique et en Automatique, Le Chesnay, France, 2002.

[10] C. Jacobi. Formal verification of a theory of IEEE rounding. In *14th International Conference on Theorem Proving in Higher Order Logics*, pages 239–254, Edinburgh, Scotland, 2001. supplemental proceedings.

[11] W. Kahan. Further remarks on reducing truncation errors. *Communications of the ACM*, 8(1):40, 1965.

[12] A. H. Karp and P. Markstein. High precision division and square root. *ACM Transactions on Mathematical Software*, 23(4):561–589, 1997.

[13] S. Linnainmaa. Software for doubled precision floating point computations. *ACM Transactions on Mathematical Software*, 7(3):272–283, 1981.

[14] P. Markstein. *IA-64 and elementary functions: speed and precision*. Prentice Hall, 2000.

[15] P. S. Miner. Defining the IEEE-854 floating-point standard in PVS. Technical Report Technical Memorandum 110167, NASA Langley Research Center, 1995.

[16] O. Møller. Quasi double-precision in floating point addition. *BIT*, 5(1):37–50, 1965.

[17] J. Rushby and F. von Henke. Formal verification of algorithms for critical systems. In *Proceedings of the Conference on Software for Critical Systems*, pages 1–15, New Orleans, Louisiana, 1991.

[18] D. M. Russinoff. A mechanically checked proof of IEEE compliance of the floating point multiplication, division and square root algorithms of the AMD-K7 processor. *LMS Journal of Computation and Mathematics*, 1:148–200, 1998.

[19] J. R. Shewchuk. Adaptive precision floating-point arithmetic and fast robust geometric predicates. In *Discrete and Computational Geometry*, volume 18, pages 305–363, 1997.

[20] P. H. Sterbenz. *Floating point computation*. Prentice Hall, 1974.

[21] D. Stevenson et al. An American national standard: IEEE standard for binary floating point arithmetic. *ACM SIGPLAN Notices*, 22(2):9–25, 1987.

High Performance Floating-Point Unit with 116 bit wide Divider

Guenter Gerwig, Holger Wetter, Eric M. Schwarz, Juergen Haess
IBM Server Division
ggerwig@de.ibm.com, hwetter@de.ibm.com, eschwarz@us.ibm.com, jhaess@de.ibm.com

Abstract

The next generation zSeries floating-point unit is unveiled which is the first IBM mainframe with a fused multiply-add dataflow. It supports both S/390 hexadecimal floating-point architecture and the IEEE 754 binary floating-point architecture which was first implemented in S/390 on the 1998 S/390 G5 floating-point unit. The new floating-point unit supports a total of 6 formats including single, double, and quadword formats implemented in hardware. The floating-point pipeline is 5 cycles with a throughput of 1 multiply-add per cycle. Both hexadecimal and binary floating-point instructions are capable of this performance due to a novel way of handling both formats. Other key developments include new methods for handling denormalized numbers and quad precision divide engine dataflow. This divider uses a radix-4 SRT algorithm and is able to handle quad precision divides in multiple floating-point and fixed-point formats. The number of iterations for fixed-point divisions depend on the effective number of quotient bits. It uses a reduced carry-save form for the partial remainder, with only 1 carry bit for every 4 sum bits, to save area and power.

1. Introduction

This paper describes a future floating-point unit (FPU) of a high performance microprocessor which is optimized for commercial workloads. The FPU implements two architectures: Binary Floating-Point (BFP) which is compliant with the IEEE 754 Standard [1], and Hexadecimal Floating-Point (HFP) as specified by IBM S/390 Architecture[2] which is now called z/Architecture[3]. There are a total of 6 formats supported which include single, double, and quadword formats for the two architectures as shown in the following table:

Format	bits	sign	exponent	significand	bias
BFP short	32	1	8	24	127
BFP long	64	1	11	53	1023
BFP quad	128	1	15	113	16383
HFP short	32	1	7	24	64
HFP long	64	1	7	56	64
HFP quad	128	1	7	112	64

Unlike many other microprocessors, zSeries microprocessors implement quad precision operations in hardware, and this includes support for both HFP and BFP architectures.

Prior zSeries floating-point units have included the 1996 G3 FPU [4], the 1997 G4 FPU [5, 6], the 1998 G5 FPU [7, 8], the 1999 G6 FPU and the 2000 z900 FPU [9]. Most are remaps of the G5 FPU with extensions for 64-bit integers. The G4 FPU has an aggressive cycle time and can complete a multiply or add in about 3 cycles with a throughput of 1 per cycle. The G5 FPU is the first FPU to implement both BFP and HFP architectures in hardware on one pipeline. The G5 FPU design is based on the G4 FPU so it has the same latency for HFP instructions. BFP instructions involve translating the operands to HFP format, performing the arithmetic operation including rounding and then converting back to BFP format. So, BFP operations take 5 or 6 cycles of latency with a throughput of only one BFP instruction every two cycles.

The G5 FPU was designed with only one year between its announcement and that of the G4 FPU. So, the BFP arithmetic implementation is not optimized for speed, but instead for simplicity. With a longer development schedule for the next zSeries FPU, there were a few new goals: 1) optimize for BFP, 2) optimize for multiply-add, and then 3) optimize for HFP. The first goal was chosen due to the increase of new workloads on zSeries, particularly workloads utilizing Linux. These applications are typically written in Java or C++ and, especially those written in Java, rely on BFP even in commercial applications.

Thus, the primary goal was to create a high performance implementation much like the pSeries workstations. One key element of pSeries floating-point units is that the dataflow supports a fused multiply-add which effectively yields two operations per cycle. Since this type of design is optimal for BFP architectures, a decision was made to base our design on the Power4 design.

The Power4 floating-point unit has a 6 stage binary multiply-add dataflow. It uses tags in the register file to identify denormalized data. It has only 2 data formats, BFP single and double with double format retained in the register file. The major enhancements of our new zSeries FPU to the Power4 design are:

1. Two architectures are supported (HFP and BFP) which results in 6 formats versus only 2 formats of BFP, and 200 different instructions are implemented directly in hardware.

2. The pipeline is reduced to 5 cycles.

3. Denormalized number handling is supported without tags or prenormalization.

4. The normalizer and LZA are expanded to full width.

5. Division and square root are implemented with a quad precision radix-4 SRT algorithm.

These items will be detailed in the remainder of this paper. First, implementing two architectures in one dataflow will be discussed. Then, the overall dataflow will be described along with particular enhancements including handling of denormalized operands and the divide implementation.

2. Dual Architectures

The first machine to implement both BFP and HFP architectures in hardware is the 1998 IBM S/390 G5 processor[7]. A hexadecimal dataflow is used which requires binary operands to be converted to hexadecimal operands before they are operated on. The HFP instructions are capable of performing one add or one multiply per cycle with a latency of about 3 cycles. The BFP instructions can only be pipelined one instruction every other cycle and the latency is 5 or 6 cycles due to the extra conversion cycles and rounding cycle.

The problem with optimizing the dataflow for both HFP and BFP architectures centers on the choice of an internal bias. HFP architecture has a bias of the form 2^{n-1} whereas BFP has a bias of the form $(2^{n-1} - 1)$. To choose one of the biases as the internal bias and to convert to the other format requires shifting the significands and adding constants to the exponent. To avoid a conversion cycle, a separate internal representation

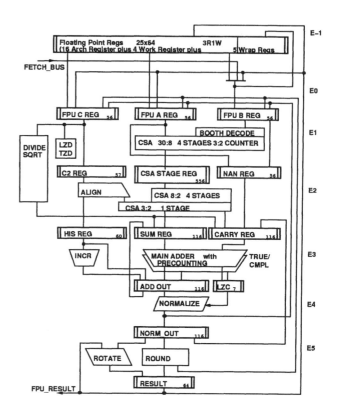

Figure 1. Main Fraction Dataflow of FPU

and bias was chosen for both architectures as shown by the following:

$$X_{BFP_i} = (-1)^{X_s} * (1 + X_f) * 2^{X_e - bias_{Bi}}$$
$$bias_{Bi} = 2^{n-1} - 1 = 32767$$
$$X_{HFP_i} = (-1)^{X_s} * X_f * 2^{X_e - bias_{Hi}}$$
$$bias_{Hi} = 2^{n-1} = 32768$$

This results in no conversion cycles and the dataflow is optimized for both architectures. This requires two different shift amount calculations since the biases differ and the implied radix points differ, but this is a very small amount of hardware.

3. Dataflow Overview

Figure 1 shows the fraction dataflow. At the top of the figure there is the Floating-Point Register file (FPR) with 16 registers of 64 bits each. There are also 5 wrap registers to hold data for loads. Loads are staged through the 5 wrap registers and the dataflow. Loads can be bypassed from any stage in the pipeline to a dependent instruction by using the wrap registers. This eliminates wiring congestion in the FPU dataflow stack and instead localizes it to the register file. When a read of an operand occurs, the data can come from the architected register file, the wrap registers, or a wrap back path from the dataflow, or from memory. In one cycle three registers of 64 bits can be read and one register can be written.

The dataflow is a three operand dataflow, which has a fused multiply and add data structure. One multiplier operand and the addend always come from the FPRs, while the 2nd operand may come from memory. In the starting cycle (labeled E0), the A,B and C registers are loaded with the correct formatting applied, such as zeroing the low order bits of a short precision operand. For binary formats the 'implied one' bit is assumed to be always '1'. If a denormalized number is detected afterwards, this is corrected in the multiplier and/or the aligner logic.

In the first execution cycle (E1), the shift amount for the alignment is calculated (considering potential denormalized operand cases). Also, the multiplication is started with Booth encoding and the first 4 stages of 3:2 counters of the Wallace tree. If there is an effective subtraction, the addend is stored inverted in the C2 register.

In the second execution cycle (E2), the alignment uses the previous calculated shift amount. In the multiplier, the next 4 stages of 3:2 counters reduce the tree to two partial products. These partial products with the aligned addend go through the last 3:2 counter to build the 'sum' and 'carry' of the multiply and add result. To balance the paths for the timing, the propagate and generate logic is performed also in this cycle. The propagate and generate bits are stored in a register instead of the sum and carry bits. A potential high part of the aligner output is stored in the high-sum register (HIS reg).

In the third execution cycle (E3), the main addition takes place. There is a 'True' and a 'Complement' Adder to avoid an extra cycle for recomplementation. Essentially, both $A-B$ and $B-A$ are calculated and the result is selected based on the carry output of the true adder. The number of leading zero bits is calculated using a zero digit count (ZDC) as described in [4]. This algorithm performs a zero digit count on 16 bit block basis of SUM and $SUM+1$. When the carries are known the result is selected among the digits. The aligner bits which did not participate in the add are called the high-sum and they feed an incrementer in this cycle. At the end of this cycle there is a multiplexor which chooses between high-sum and high-sum plus one and also chooses whether to shift the result by 60 bits. If the high-sum is non-zero, the high-sum and upper 56 bits of the adder output are chosen to be latched. If instead the high-sum is zero, only the bits of the adder output are latched. Also the leading zero count is stored in the LZC register.

In the fourth execution cycle (E4), the normalization is done. The stored leading zero count is used directly to do the normalization. No correction is necessary, since the LZC is precise. For hex formats, only the two low order bits of the leading zero count are not used to get the normalized hex result. Additionally, the sticky bits are built according to the format.

In the fifth execution cycle (E5), the rounding and reformatting is done. For hex operands no rounding is needed, but the operands will pass this cycle anyway. Since there is a feedback path from the normalizer to the A, B, and C registers, this does not cost performance.

4. Denormalized Input

The architecture supported is a CISC type architecture and supports both register and memory operands as input to arithmetic instructions. This makes it very difficult to tag input operands in a timely manner. Since data from memory arrive late there is no time to check whether it is denormalized or normalized. The check requires examining the exponent to see if it is all zeros which would require for short and long operands an 8 way and an 11 way NOR function. Detecting denormalized input is instead calculated in the first cycle of execution.

In an implementation of hexadecimal floating-point the operands are 56 bits wide which requires 29 partial products for a radix-2 Booth encoding. 7 levels of 3:2 counters can only handle a maximum of 28 partial products, so 29 must take 8 levels. Additional correction terms do not add significant delay. Our new FPU assumes that the BFP operands are normalized and have an implied one. It corrects the multiplier Y, prior to creating the most significant partial product. The Booth decode term is calculated for both an implied one and implied zero and then selected once the implied bit is determined. This partial product gates into a delayed partial product in the counter tree. The multiplicand, X is corrected by subtracting a term $lzc1$ from the counter tree [10] as shown below (W_j are the booth scans):

$$X = x_0 + \sum_{i=1}^{n-1} x_i * 2^{-i}$$

$$Y = y_0 + \sum_{j=1}^{n-1} y_j * 2^{-j}$$

$$Y = \sum_{j=1}^{\lfloor \frac{n-1}{2} \rfloor + 1} W_j * 4^{-j}$$

$$W_j \in \{-2,-1,0,+1,+2\}$$

$$P = \sum_{j=1}^{\lfloor \frac{n-1}{2} \rfloor + 1} W_j * X * 4^{-j}$$

$$X' = 1 + \sum_{i=1}^{n-1} x_i * 2^{-i}$$

$$X = X' - \overline{x_0}$$

$$P = \sum_{j=1}^{\lfloor \frac{n-1}{2} \rfloor + 1} W_j * X' * 4^{-j} - Y * \overline{x_0}$$

$$lzc1 = -Y * \overline{x_0}$$

The addend can be corrected prior to being aligned into the counter tree. The only difficulty is in correcting the alignment for a denormalized input operand. To do this the alignment is calculated for the exponent difference, D, and $D+1$, and $D-1$, and is selected late based on which operands are denormalized. Thus, denormalized input can be handled without stalling the pipeline or trapping to software. There is one rare case that does trap to software which will be discussed in the next section.

5. Alignment Limitations

The dataflow width is limited to an addend of 56 bits plus 4 guard bits and a product field which is aligned with the adder of 112 bits and 4 guard bits for a total of 176 bit wide dataflow. There are certain cases of unnormalized and denormalized numbers which are difficult to handle in this dataflow. To understand the cases better a case by case study is shown detailing whether a case can be handled by the hardware directly or whether some type of intervention is needed.

In regards to the alignment of the addend with the product, the radix point of the product is fixed in the dataflow. The radix point of the addend is right shifted to achieve the proper fraction alignment prior to the addition of the two.

The BFP arithmetic on this dataflow has to consider the larger dataflow width required by HFP. The dataflow is partitioned as follows for BFP data:

```
|Addend field       | Product field           |
|<---- 60 bits--->|<-- 116 bits ---------->|
|1.cccc...cGGGGGGG|xx.pppp......pGGGGGGGGGG|

| Radix1            | Radix2
```

There are two possible radix points, Radix1 and Radix2, which are used as reference points for the possible right shifts that may be required. Radix1 is the radix point for the addend data. Radix2 is the radix point for the product data. The G bits represent extra guard bits for addend and product fields.

Case 1: Normalized Addend and Normalized Product: This is the most straight forward case. To calculate the right shift amount SA the following equations are used to achieve the correct result S (D is the exponent difference):

$$S = [(1.f_A x 2^{E_A - Bias}) x (1.f_B x 2^{E_B - Bias})] + (1.f_C x 2^{E_C - Bias})$$

$$S = (1.f_A x 1.f_B) x 2^{(E_A + E_B - Bias) - Bias} + (1.f_C x 2^{E_C - Bias})$$

$$D = (E_A + E_B - Bias) - E_C$$

$$SA = E_A + E_B - E_C + K,$$

$$\text{where } Constant\ K = 59 + 2 - Bias$$

If the calculation of the shift amount yields a negative result then the addend C, is not shifted at all and any carry out of the adder is not allowed to propagate since the guard bits, G, are zero. A shift amount which is greater than $54 + 106 = 160$ will result in the sticky calculation being an effective OR of the addend C. The information will be used later in the Rounder.

Case 2: Normalized Addend and Denormalized Product: This case doesn't pose any extra difficulty. If A and/or B are denormalized then there is the possibility that the product may be denormalized. As long as C is normalized, then there are enough bits of precision maintained to form a result consistent with the BFP Architecture since the exponent of the denormalized product will be less than the normalized addend.

Case 3a: Denormalized Addend and Normalized Product: In this case, C is denormalized and P is normalized. The exponent of the resulting sum will be equal to that of the product, P, and the dataflow is sufficient.

Case 3b: Denormalized Addend, Denormalized Product and Underflow Trap Disabled: In this case, C is denormalized and P is denormalized. Since the Underflow Trap is disabled, the result will be rounded to a denormalized number or zero in accordance with the rounding mode and the value of the two guard bits.

Case 3c: Denormalized Addend, Denormalized Product and Underflow Trap Enabled: Since the Underflow Trap is enabled, the result will need to be normalized assuming an unbounded exponent range. This will require at least 53 bits of precision. However, it is possible for the values of the addend and product fields to be disjoint since we don't right shift the product field with respect to the addend field. If both the addend and the product are denormalized, then the result is denormalized and the value of Radix1 will be will be 2^{-1022}. If the value of Radix2 is less than $2^{-1022-60-2} = 2^{-1084}$, then the dataflow is disjoint and is not sufficient. This case is implemented in low level software called millicode. A pseudo-exception is taken by hardware and the millicode handler executes a series of instructions to calculate the correct result. Since we have an underflow condition and the underflow trap is enabled, this case will end in an architectural exception anyhow and so there is no real performance degradation. This is the only case implemented in millicode.

6. Divide

There is an extra wide, 116 bit divider dataflow, in which the mantissa of the quotient and the remainder is calculated using an SRT algorithm. There are numerous different divide instructions and formats to be supported. Not only does the divider support the 6 floating-point formats, but it also performs integer divides on operands 32, 64, or 128 bits wide, which may be signed or unsigned.

6.1. SRT-Algorithm

SRT is a frequently used method for implementing divide and square root on modern microprocessors. It is named after Sweeney, Robertson and Tocher, who independently proposed the algorithm [13, 14]. SRT is an iterative algorithm that retires one digit of the quotient in every iteration. After each iteration, the new partial remainder is calculated by multiplying the previous partial remainder with the radix of the algorithm and subtracting a multiple of the divisor.

This operation is formally described as follows:

$$P_{i+1} = r * P_i - q_{i+1} * D$$

where P represents the partial remainder, q represents the quotient digit guess, D the divisor, and r the radix of the algorithm. The final quotient is the weighted sum of all quotient digits.

The value for the actual quotient digit q_{i+1} is estimated and therefore the partial remainder P_{i+1} could be negative. This can be compensated for by allowing negative values for q_{i+1} too. So, errors in the actual partial remainder can be corrected in later iterations. The convergence of the algorithm requires the following condition be met:

$$|P_{i+1}| < (q_{max} * D)/(r-1)$$

The visual representation of the above equation can be shown in a so called P-D plot. The ranges of q_{i+1} can be seen in Figure 3.

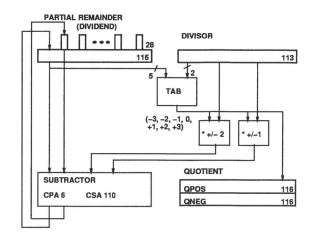

Figure 2. Dataflow structure of divider

Our implementation uses a radix-4 algorithm with a maximally redundant digit set [11, 12]. This reduces the cost of the quotient estimate table lookup at the expense of an increase to the range of quotient digits. Since a full width carry-propagate adder (CPA) would not fit into the required cycle time, a redundant form of the partial remainder is used which allows carry-save adders (CSA) to be used. The increased range of quotient digits has very little effect on a carry-save implementation. The implemented equations are shown by the following:

$$\begin{aligned} P_{s_{i+1}} + P_{c_{i+1}} &= 4(P_{s_i} + P_{c_i}) - q_{i+1} * D \\ q_{i+1} &\in \{-3, -2, -1, 0, +1, +2, +3\} \\ q_{i+1} &= q_{i+1,1} + 2q_{i+2,2} \\ P_{s_{i+1}} + P_{c_{i+1}} &= 4(P_{s_i} + P_{c_i}) - q_{i+1,1} * 1D - q_{i+1,2} * 2D \end{aligned}$$

where P_s and P_c are the partial remainder in a sum and carry redundant form, and $q_{i+1,1}$ and $q_{i+1,2}$ are the quotient digit guesses separated into a guess of 1 and a guess of 2 where each can take on the values -1, 0, or +1.

6.2. Structure of Divider Dataflow

Figure 2 shows the 6 main elements of the divide macro, which contains the divide dataflow:

Divisor Register:
This is a simple register with the maximum width of 113 bits for a BFP quad precision, and is where the divisor is stored for the subtractions.

Table Lookup:
This table consists of a relatively small part of combinatorical logic. It needs the five most significant bits of the partial remainder and the two most significant bits of the divisor, after the implied one. Figure 3 is a

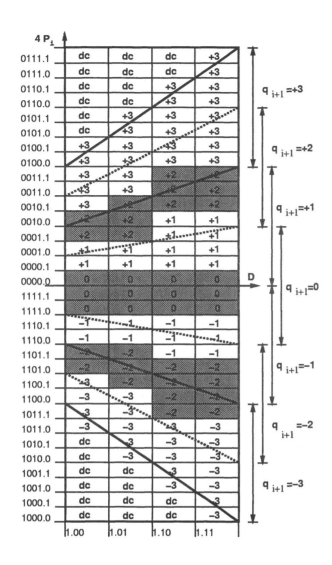

Figure 3. Divide Table

remainder. When the partial remainder is negative, we have to add divisor multiples and when it is positive we have to subtract. The $q_{i+1,1}$ term represents divisor multiples of one and and the $q_{i+1,2}$ represents divisor multiples of two. These terms are added together with the partial remainder.

Partial Remainder Register and Subtractor:
The partial remainder register width of 116 bit is defined by the HFP quadword format width (112) plus one hex guard digit. The register consists of a sum part of 116 bits and a carry part of 28 bits. The 6 high order sum bits must be explicit without a corresponding carry because they are used in the table lookup. The most significant carry bit starts at position 6 and only every fourth carry bit is stored. This is possible since the subtractor does not use a full 4:2 reduction, but uses one stage of 3:2 reduction (CSAs) and one stage of CPAs with a width of 4 bits. On the high order side, one CPA with 6 bit width is needed to deliver an explicit value to the table. These 4 bit wide CPAs in the low order range save latches, area and power and do not cost cycle time, since the 6 bit CPA is needed anyhow in the high order range.

Quotient Register:
The quotient register has a width of 116 bits and consists of two parts: the Q_{POS} and the Q_{NEG} registers. When q_{i+1} is positive, it is stored in Q_{POS}, and when q_{i+1} is negative, it is stored in Q_{NEG}. The pointer, which defines which digit of Q is written, is controlled by a counter in the control logic.

6.3. Execution and performance of floating-point divides

After the operands have been loaded into the dividend and divisor registers, the divide iterations can start. When an operand is denormalized, a normalization in the main dataflow is needed in advance. The required number of divide iterations depends on the data format (single, double, quad). After the divider has completed enough iterations, the sum and carry parts of the remainder and the quotient are moved out into the main dataflow before the main adder. There, they are added to get the explicit value of remainder and quotient. Afterwards the quotient is normalized and rounded, using the main dataflow of the floating-point unit.

The following table shows the required cycles for execution of IEEE floating-point divide instructions:

combination of a P-D plot and the actual implemented lookup table. It illustrates the shifted partial remainder ranges in which a quotient digit can be selected without violating the bounds on the next partial remainder. It can be seen that the table is asymmetric concerning +/-. This is due to the fact that the partial remainder has a redundant form which causes an additional error. Because of this, the high order bits of the partial remainder can be to small (by one ulp), but never be to large. A symmetrical lookup would also be possible, but then we would need one more bit of the partial remainder or the divisor.

Divisor Multiple Generation:
Before selecting the multiples of one or two, the divisor is inverted depending on the sign bit of the partial

Action \ format	single	double	quad
Load Operands	3	3	15
Divide Loops	14	28	58
Readout Remainder/Quotient	4	4	4
Calculate Quotient	1	1	1
Normalize	1	1	1
Round	1	1	1
Write back	1	1	2
Total Latency	25	39	82
Pipelined Latency	20	34	77

6.4. Execution and performance for integer divides

The integer operands are made positive and normalized in the main dataflow. Afterwards the operands are loaded into the dividend and divisor registers, as for floating-point operands. The difference in handling is the pointer, to which q_i within the quotient registers is written.

The number of effective bits n_Q in the quotient can be calculated in advance, when the effective bits n_V and n_D of the dividend and the divisor are known. These values are gained during the normalization process. For a 64 bit integer division the following equations are valid:

$$n_{Q0} = n_V - n_D \quad for \ V_{norm} < D_{norm}$$
$$n_{Q1} = n_V - n_D + 1 \quad for \ V_{norm} \geq D_{norm}$$

Since we gain two bits per cycle, the number of effective quotient bits n_{QE} to be calculated is rounded up to the next even number. The start pointer P_{Start} and the stop pointer P_{Stop} for a 64 bit integer divide are given by:

$$P_{Start} = 64 - n_{QE}; \quad P_{Stop} = 64$$

The following table shows the required cycles for execution of integer divide instructions:

Action	cycles
Load and concatenate	4
Normalize	5
Divide Loops	1-32
Readout Remainder/Quotient	5
Invert Sign (potential)	5
Write back	5
Total Latency	30 - 61
Pipelined Latency	25 - 56

For integer divides the number of divide iterations depend purely on the effective number of quotient bits. Additionally, there are some base cycles which have a dependency on the operand width too. In classical benchmarks, it often occurs that, the divide result has a small number of effective bits; this improves the performance of integer divides considerably.

7. Physical Implementation

The fraction dataflow has been implemented in a bit stack approach. The A,B and C registers have a width of 56 bits. This is widened during alignment and multiplication. The adder, normalizer, and rounder are 116 bits wide. The output of the rounder is reformatted to a width of 64 (with exponent). The layout has a folded form. On the top of Figure 4 are the architectural floating-point registers with A, B, and C registers below. On the bottom is the normalizer. The exponent dataflow is in a stack on the right of the A, B, and C fraction registers.

The divider is also implemented in a stack approach, whereby the divide-table is combinatorial logic which occupies a very small area on the left hand side of the divider macro. Since the interconnection of the divide engine to the main fraction dataflow is not timing critical, this can be located away from the main dataflow and is shown in the right upper corner of the layout. The fraction dataflow is on the left hand side. On the right are the synthesized control logic macros. For each execution pipeline there is one separate control macro. The macros on the bottom contain some miscellaneous logic, which is not related to the floating-point function.

The divider macro is completely designed in standard inverting CMOS logic. Although it has been implemented as a full custom macro, extensive use of a standard cell library has been made in order to keep layout effort small. As a power saving feature, most parts of the floating-point unit can be turned off completely when not in use. For enhanced testability, each of the master-slave latches is accompanied by an additional scan latch. Adding this extra scan latch to the scan chain configuration results in an increased transition fault coverage. The floating-point unit occupies an area of 3.76 mm^2. The divider macro occupies 0.22 mm^2, which is about 6 % of the FPU. It has been fabricated in IBM's 0.13 micron CMOS SOI technology. At a supply voltage of 1.15V and a temperature of 50° C it supports a clock frequency significantly greater than 1 GHz.

8. Summary

A new zSeries floating-point unit has been shown which, for the first time, is based on a fused multiply-

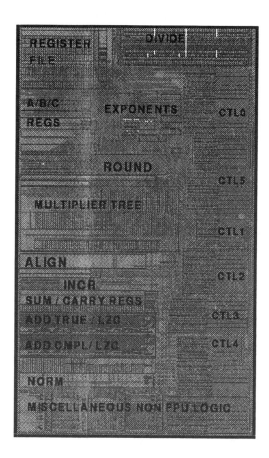

Figure 4. Layout of Floating-Point-Unit

add dataflow capable of supporting two architectures. Both binary and hexadecimal floating-point instructions are supported for a total of 6 formats. The floating-point unit is capable of performing a multiply-add instruction for hexadecimal or binary every cycle with a latency of 5 cycles. This has been accomplished by a unique method of representing the two architectures with two internal formats with their own biases. This has eliminated format conversion cycles and has optimized the width of the dataflow. Though, this method creates complications in the alignment of the addend and product which have been shown in detail. Denormalized numbers are almost exclusively handled in hardware except for one case which is destined for an underflow exception handler anyway.

Also, a fast divide algorithm is used which is capable of supporting a quad precision width and achieves 2 result bits per cycle. The number of divide iterations depends on the effective number of quotient bits, which improves the performance significantly. For the redundant expression of the partial remainder only each fourth carry bit is used, which saves around 80 latches compared to a conventional carry-save approach.

The new zSeries floating-point unit is optimized for both hexadecimal and binary floating-point architecture. It is versatile supporting 6 formats, and it is fast supporting a multiply-add per cycle.

References

[1] "IEEE standard for binary floating-point arithmetic, ANSI/IEEE Std 754-1985," The Institute of Electrical and Electronic Engineers, Inc., New York, Aug. 1985.

[2] "Enterprise Systems Architecture/390 Principles of Operation," Order No. SA22-7201-7, available through IBM branch offices, July 2001.

[3] "z/Architecture Principles of Operation," Order No. SA22-7832-1, available through IBM branch offices, Oct. 2001. available through IBM, Oct. 2001.

[4] G. Gerwig and M. Kroener. "Floating-Point-Unit in standard cell design with 116 bit wide dataflow," In *Proc. of Fourteenth Symp. on Comput. Arith.*, pages 266–273, Adelaide, Austraila, April 1999.

[5] E. M. Schwarz, L. Sigal, and T. McPherson. "CMOS floating point unit for the S/390 parallel enterpise server G4," *IBM Journal of Research and Development*, 41(4/5):475–488, July/Sept. 1997.

[6] E. M. Schwarz, B. Averill, and L. Sigal. "A radix-8 CMOS S/390 multiplier," In *in Proc. of Thirteenth Symp. on Comput. Arith.*, pages 2–9, Asilomar, CA, July 1997.

[7] E. M. Schwarz and C. A. Krygowski. "The S/390 G5 floating-point unit," *IBM Journal of Research and Development*, 43(5/6):707–722, Sept./Nov. 1999.

[8] E. Schwarz, R. Smith, and C. Krygowski. "The S/390 G5 floating point unit supporting hex and binary architectures," In *Proc. of Fourteenth Symp. on Comput. Arith.*, pages 258–265, Adelaide, Austraila, April 1999.

[9] E. M. Schwarz, M. A. Check, C. Shum, T. Koehler, S. Swaney, J. MacDougall, and C. A. Krygowski. "The microarchitecture of the IBM eServer z900 processor," *IBM Journal of Research and Development*, 46(4/5):381–396, July/Sept. 2002.

[10] C. A. Krygowski and E. M. Schwarz. "Floating-point multiplier for de-normalized inputs," *U.S. Patent Application No. 2002/0124037 A1*, page 8, Sep. 5, 2002.

[11] M. D. Ercegovac and T. Lang. "Division and Square Root: digit-recurrence algorithms and implementations," *Kluwer*, Boston, 1994.

[12] D. I. Harris, S. F. Obermann, and M. A. Horowitz. "SRT Division Architectures and Implementations," In *Proc. of Thirteenth Symp. on Comput. Arith.*, pages 18–25, Asilomar, California, July 1997.

[13] K. D. Tocher. "Techniques of multiplication and division for automatic binary computers," Quarterly J. Mech. Appl. Math., vol.11, pt.3, pp.364-384, 1958.

[14] J. E. Robertson "A new class of digital division methods," IRE Transactions on Electronic Computers, vol. EC-7, pp.218-222, Sept. 1958.

The Case For a Redundant Format in Floating Point Arithmetic

Hossam A. H. Fahmy
hfahmy@arith.stanford.edu

Michael J. Flynn
flynn@arith.stanford.edu

Computer Systems Laboratory, Stanford University, USA

Abstract

This work uses a partially redundant number system as an internal format for floating point arithmetic operations. The redundant number system enables carry free arithmetic operations to improve performance. Conversion from the proposed internal format back to the standard IEEE format is done only when an operand is written to memory. A detailed discussion of an adder using the proposed format is presented and the specific challenges of the design are explained. A brief description of a multiplier and divider using the proposed format is also presented. The proposed internal format and arithmetic units comply with all the rounding modes of the IEEE 754 floating point standard. Transistor simulation of the adder and multiplier confirm the performance advantage predicted by the analytical model.

1. Introduction

Addition is the most frequent arithmetic operation in numerically intensive applications. Multiplication follows closely and then division and other elementary functions. This work presents several techniques to improve the effectiveness of floating point arithmetic units in general but with a focus on addition.

A partially redundant number system was previously proposed [5] for use as an internal format within the floating point unit and the associated registers. The format is based on the single and double precision formats of the ANSI/IEEE standard [1]. However, in the proposed format each group of 4 bits of the significand are represented redundantly as a 5 bit signed digit number in $\{-15, \cdots, +15\}$ using two's complement form. The fifth bit (extra bit) represents a negative value with the same weight as the least significant bit of the next higher group. This is shown for the string of bits $a4, a3, a2, a1, a0$ in Fig. 1. This extra bit, $a4$, is saved in the register to the right of the least significant bit in the next higher group and to the left of $a3$. As with IEEE formats, the significand is always positive so there is no

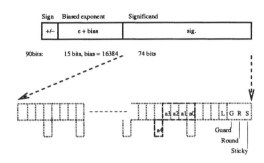

Figure 1. The proposed signed digit format for floating point numbers.

need for the extra bit in the most significant digit. The number is also always normalized in the proposed format. Denormalized IEEE numbers are normalized in the conversion process upon loading into the register file. Each group of 5 bits represents one base 16 digit and therefore, the exponent is applied to base 16 rather than base 2 as is used in the normal IEEE format. The proposed format is in the form, $(-1)^{sign} first\,digit.remaining\,digits \times 16^{exp-bias}$. The guard, round and sticky (GRS) bits are saved in the register file with the unrounded result. The result is then correctly rounded according to the IEEE standard's rules in the following operation when it is used or saved to the memory. This deferred rounding technique moves the rounding computation off the critical path and allows it to be overlapped with the exponent difference calculation in the adder.

In general, SD numbers allow carry free addition by using redundant number representations. Eliminating the carry propagation significantly reduces the latency of arithmetic operations. The conversion from binary to SD form is trivial since the binary format is usually a valid SD representation. However, converting a SD number back into a non-redundant form involves a carry propagation. SD numbers are not commonly used in arithmetic circuits since the SD to binary conversion requires a carry propagation. The proposed system efficiently hides this time delay by overlapping the SD number to binary conversion with the memory store operations, thus removing it from the critical path.

0-7695-1894-X/03 $17.00 © 2003 IEEE

More details of the conversion to and from the proposed internal format to the IEEE format were presented in our earlier work [5].

The format presented above with its specific use of a base 16 signed digits is obviously a special case of more general signed digits [13] where another base or even a mix of different bases [15] can be used. The case of using another base is discussed briefly in section 4 below. The methodology of thinking about the algorithms and trade-offs discussed in this work apply to the general case as well. The specific format presented above is what we implemented in transistors to prove our claims regarding the speed improvement. We deemed this specific format as a practical compromise to give enough speed improvement with a reasonable increase in the area of the register file. The issue of the optimal redundant format to use will obviously depend on the requirements on speed, area and power consumption. That issue is beyond the scope of the current work.

In the following sections, the presented format is used to build efficient arithmetic circuits. Section 2 explains the details of the floating point addition unit. Two design challenges due to the redundant format, namely the leading digit detection and the rounding, are discussed in section 3. Section 4 describes the analytical time delay modeling of the addition unit and discusses the rationale for postponing the rounding and for using a hexadecimal based exponent. The multiplication, division and other elementary functions are computed using the units presented in section 5. Then section 6 provides the simulation results for both the adder and the multiplier. Finally, in section 7 conclusions of this work are presented.

2. Floating point addition

In the current state-of-the-art high performance floating point adders, two-path algorithms are used with integrated rounding similar to the designs proposed by Farmwald [7] and Quach [17]. These adders perform both addition and subtraction. An effective subtraction occurs when both operands are of the same sign and the required operation is a subtraction or when the operands differ in their signs and the operation is an addition. In the case of effective subtraction and an exponent difference of zero or one, a few of the leading digits of the result might become zero. For this case of leading zero digits, there is a need for a left shift of the result for normalization. The two-path algorithm separates this specific case into a path with a specialized left shifter while the general case of operands passes through the regular path. The special path is called the cancellation path (where the leading digits are possibly canceled) or the close path (where the exponents are close to each other) while the regular path is called the far path. The adder design presented here is following a similar approach and is using a

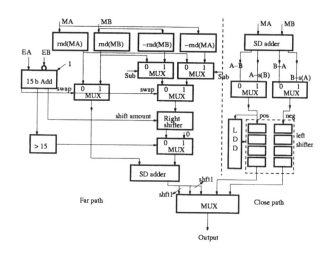

Figure 2. Block diagram of the two-path adder.

two-path algorithm as shown in the block diagram of Fig. 2. In the presented design, the cancellation path is used only in the case of an effective subtraction with an exponent difference of zero or an effective subtraction with an exponent difference of one and a cancellation of some of the leading digits occurring in the result. In all other cases, the far path is used.

The far path of the proposed adder is similar to the far path of other algorithms presented in the literature. The unique aspects of the proposed adder are: first, the use of a hexadecimal base for the exponents; second, the location of the rounding logic in parallel with the exponent difference and third the use of signed digit numbers in the significand. The hexadecimal base of the exponents makes the right shifting for alignment of the two operands a shift to a 4-bit boundary only. So, instead of using an n-way shifter in the conventional adders an $\lceil n/4 \rceil$-way shifter is used here. Such a reduction in the complexity of the shifter reduces its time delay as discussed below. The parallel execution of the rounding logic with the exponent difference logic takes the rounding away from the critical path of the adder. It is possible to simultaneously round and negate the number and this is what is done in the presented design to prepare the operand for the SD (signed digit) adder.

A signal indicating an effective subtraction selects the operand or its negative and a signal indicating which of the operands has a larger exponent allows for swapping them. Then, the operand with the smaller exponent is shifted to the right for alignment and the bits that are flowing out of the shifter are used to calculate the guard, round and sticky bits. Only the least significant bits of the exponent difference are used to indicate the shift amount. If the exponent difference is large enough to completely shift out the smaller operand a zero is forced as the second operand into the adder.

The result of the SD adder may need a normalization

shift by one digit to the right for the case of effective addition and an overflow. The result may otherwise need a normalization shift by one digit to the left for the case of an effective subtraction and cancellation of the Most Significant Digit (MSD). This cancellation of only the MSD can occur even when the exponent difference is larger than one. The SD adder block produces the result and three signals indicating the need for no shift, a shift to the left or a shift to the right. The guard, round and sticky bits are calculated speculatively dependent on the shifting possibilities. The multiplexer unit responsible for choosing between the far and cancellation paths receives those different signals and speculative results and chooses the final result among them in case it chooses the far path.

In the cancellation path the exact exponent difference is not calculated but the least significant bit of each of the two exponents is examined. If the two exponent bits are found to be identical the difference of the exponents is speculated to be zero and the direct subtraction of the operands is needed. If, on the other hand, the two exponent bits are not identical the difference is assumed to be one and the subtracter should produce a result equal to one operand minus the other operand shifted by one bit to the right. The direct subtraction in the case of zero exponent difference may lead to a negative result if the significand of the second operand is larger than the significand of the first operand. To remedy for this negative result in conventional adders, the sign of the floating point result is flipped and the bits representing the result are negated. In the presented design, the subtracter produces the result and its negative and then chooses one of them at the end depending on the sign of the result.

So, assuming the two operands to be labeled A and B, all the possible combinations are produced in the presented design: $A - B$, $A - shift(B)$, $B - A$ and $B - shift(A)$. Then, depending on the speculation of the exponent difference, either the direct subtraction or the one involving a shift is chosen. Since a complete calculation of the exponent difference does not occur in the cancellation path, the rounding is done in conjunction with the significand subtraction. A round digit is computed for each operand and is used within a signed digit subtracter to perform the subtraction. A Leading Digit Detector (LDD) is used to calculate the shift amount needed to normalize the result. This shift amount is applied to two shifters one shifting the result and the other shifting the negative of the result. The sign of the leading digit is detected and the correct sign for the floating point result is decided. The result and its negation as well as a signal indicating the sign of the leading digit are forwarded to the multiplexer unit selecting between the cancellation path and the far path. This unit then makes the decision on the path to choose and appropriate result from each path.

3. Challenges: The leading digit detection and rounding

In an effective subtraction in the close path one or more of the leading digits in the result may become zero. Then, in order to normalize the result, the leading non-zero digit must be detected and the result must be normalized by left shifting the significand by the number of leading zeros. All floating point adders include circuits to either detect or predict the position of the leading non-zero digit after the subtraction is performed. The prediction circuits like the work of Bruguera and Lang [2] or the work of Quach and Flynn [16] operate on the adder's operands in parallel with the significand addition. It is to note that in both schemes the original operands are not redundant while the prediction circuits are working on a redundant representation because the prediction is done before the result of the adder is available. In a redundant number there are several patterns that evaluate to a string of leading zeros. A prediction circuit must then involve a technique for matching all such patterns and taking appropriate actions. If, on the other hand, a detection scheme was used on the non-redundant result of an adder, there will be no need for complicated pattern matching. Since in the proposed design for the floating point adder signed numbers are used for the inputs and output, even a detection scheme must perform some pattern matching.

A few possible patterns become the hard cases in detecting the first non-zero digit. In fact, the leading zeros may be expressed directly as zeros or indirectly as leading insignificant digits: a leading 1 followed by -15s or a leading -1 followed by 15s. The leading 1 (-1) can be converted to a zero and borrowed into the neighbor -15 (15) digit position as a 16 (-16). Since $16 - 15 = 1$ ($-16 + 15 = -1$), the zero propagation may continue into lower significance digits. The following example illustrates how leading non-zero digits may be leading insignificant digits. Assuming $|l| < 15$,

$$\begin{array}{cccccccc} 1 & -15 & -15 & \cdots & -15 & l & m & \cdots = \\ 0 & 0 & 0 & \cdots & 1 & l & m & \cdots \\ -1 & 15 & 15 & \cdots & 15 & l & m & \cdots = \\ 0 & 0 & 0 & \cdots & -1 & l & m & \cdots \end{array}$$

Another pattern is $100\cdots00 - ve = 011\cdots10 + ve$. This pattern and its dual $(-1)00\cdots00 + ve$ are what causes a fine adjustment in the case of the previous work [16, 2]. The fine adjustment is basically to indicate that the location of the leading digit should be shifted by one position. We can mentally think of detecting the leading zeros and the leading insignificant digits as the first step followed by a fine adjustment step. In the fine adjustment step if the leading digit is 1 and is not followed by another positive digit but by 0 then we must detect the sign of the remainder of the number. If that sign is negative, a fine adjustment is needed. The dual case holds for a leading -1.

The need to detect special cases for $+1$ followed by -15 or -1 followed by $+15$ can be eliminated by the use of some recoding techniques similar to what was presented in the work on recoders for partial compression [3, 4]. In the section discussing rounding and leading digit detection, Daumas and Matula state [4]: "Partial compression also realizes virtually all the benefits of leading digit deletion." The word virtually is important; partial compression does not provide a solution for the fine adjustment cases. In fact, from a complexity and time delay point of view, getting the exact bit location of the leading one is essentially the same as doing a carry propagation. The fine adjustment is hence equivalent to transforming the redundant representation into a non-redundant one. As described by Quach and Flynn [16], parallel addition and leading one prediction are both problems of bit pattern detection. They also identified sticky bit computation as the third problem in the category of bit pattern detection.

What is proposed in this work is to perform only the coarse adjustment of finding the leading digit by eliminating any leading zeros or insignificant digits. The main advantage of using a signed digit number system is to eliminate the carry propagation from the critical path. Thus, introducing another circuit similar in complexity (fine adjustment) instead of the carry propagation in the critical path will defeat the purpose. Hence, the fine adjustment is left to the rounding unit and a signed sticky digit (similar to what Matula and Nielsen [10, 12] proposed) is used there. The rounding occurs in parallel with the exponent difference and is not sequentially after the addition, so it is out of the critical path.

Based on the work of Daumas and Matula [3, 4], two recodings are defined to delete the leading insignificant digits. The N-recoding is where a negative one could be added to the digits and the P-recoding is where a positive one could be added. More specifically, for two consecutive digits of the result,

$$\cdots \quad s_{i_3} \quad s_{i_2} \quad s_{i_1} \quad s_{i_0} \quad s_{i-1_3} \quad s_{i-1_2} \quad \cdots$$
$$s_{i_4} \quad \quad \quad s_{i-1_4}$$

The N-recoding is defined as reseting s_{i-1_4} and s_{i_0} to 0 if they were both 1. This is mathematically correct since s_{i-1_4} has a negative value. This recoding can create digits that are equal to -16, however, this is not important since the recoded format is only used within the leading digit detector (LDD) circuits and even with such "out of bound" digits the position of the leading digit can be correctly estimated. The N-recoding when applied to the case of repeated -15 would be as follows assuming that the digit l is positive:

digits	k	-15	$\cdots$	-15	l	$\cdots$
equiv.	$k_3 k_2 k_1 k_0$	0001	$\cdots$	0001	$l_3 l_2 l_1 l_0$	$\cdots$
bits	1	1	$\cdots 1$	l_4	$\cdots$	$\cdots$
result	$k-1$	0	$\cdots$	1	l	$\cdots$

If the digit k is equal to 1, then all the insignificant 1 followed by negative digits have been eliminated by this N-recoding. If the digit l is negative then the result of the recoding will not be $1\,l\cdots$ but rather l_4 will be canceled effectively giving a result of $0\,(16+l)\cdots$. This feature is of value since it ensures that the number is truly normalized. A leading digit of 1 with a negative fractional part is not the normalized format. The condition for the N-recoding to change the bits is $s_{i_0} = s_{i-1_4} = 1$ and hence its output is given by $s_{i_0}^n = s_{i_0}\bar{s}_{i-1_4}, s_{i-1_4}^n = \bar{s}_{i_0}s_{i-1_4}$ while the remaining bits of the digit pass unchanged.

The P-recoding on the other hand is defined to eliminate the case of insignificant leading -1 followed by positive digits. Referring to the two consecutive digits above, if $s_{i_0} = 1$ and $s_{i-1_4} = 0$ then we can split s_{i-1_4} to 1 and -1, add the 1 to s_i and keep the -1 with s_{i-1} as its new s_{i-1_4}. This split is to occur only if s_{i-1} is not exactly equal to zero in order to prevent the new s_{i-1} from becoming -16. Applying P-recoding to the case of repeated digits of 15 the result is:

digits	0	-1	15	$\cdots$	15	l	$\cdots$
equiv.	0	1111	1111	$\cdots$	1111	$l_3 l_2 l_1 l_0$	$\cdots$
bits	1	0	0	$\cdots$	l_4	$\cdots$	$\cdots$
result	0	0	0	$\cdots$	-1	l	$\cdots$

In this, the digit l is assumed negative ($l_4 = 1$) and the whole result is negative at the end. If l is positive then one more digit would become zero and the result will be $0\,(l-16)\cdots$. Note that even in this case, the sign of the result is still negative since $|\,l\,| \leq 15$. In general due to the choice of base and possible values in this number system, any number is of the sign of its leading non-zero digit [14]. The N and P-recodings do not alter that.

The condition mentioned above for the P-recoding to change the bits is $s_{i_0} = 1, s_{i-1_4} = 0$ and $\bar{z}_{i-1} = 1$, where $\bar{z}_{i-1}$ is an indicator to show if the digit $i-1$ is not zero. Let $u_i = s_{i_0}\bar{s}_{i-1_4}\bar{z}_{i-1}$, then if $u_i = 1$ the output of the P-recoding for digit i should be $s_i + 1$ instead of s_i. Obviously, u_i could be added to s_i or, better, it could be used as a select line in a multiplexer which has s_i and $s_i + 1$ as its inputs. This u_i signal also affects the most significant bit of the lower adjacent digit s_{i-1}. If $u_i = 0$ then the output of the P-recoding for this bit, $s_{i-1_4}^p$, is determined just by the outcome of the multiplexer choosing between s_{i-1} and $s_{i-1} + 1$ depending on u_{i-1}. If, on the other hand, $u_i = 1$ then the two possible cases of u_{i-1} need to be analyzed.

$u_{i-1} = 0$: $s_{i-1_4}^p = 1 = \bar{s}_{i-1_4}$ (remember that $s_{i-1_4} = 0$ for $u_i = 1$).

$u_{i-1} = 1$: then two conditions are possible:

sign bit of $s_{i-1} + 1$ is 0: Then, as above, $s_{i-1_4}^p = 1$.

sign bit of $s_{i-1} + 1$ is 1: This means that due to the added 1 an overflow occurred which has a value of $+1$. That positive overflow is canceled out by the -1 resulting from the P-recoding splitting of the original 0 in s_{i-1_4}, thus $s^p_{i-1_4} = 0$.

Hence in all the cases, if $u_i = 1$ the resulting $s^p_{i-1_4}$ bit is the inverse of the bit coming out of the multiplexer choosing between s_{i-1} and $s_{i-1} + 1$.

As is the case for the N-recoding an "out of bound" digit value can occur. In this case, a value of $+16$ results if the pattern of digits $k\ 15\ l$ with $k_0 = 0$ and $l_4 = 0$ are entered into a P-recoder. Again, this is not problematic since this recoded format is only within the LDD circuits and the position of the leading non-zero digit will be correctly detected as described below. However, this is why it was important to note above that in P-recoding, this split of the zero in s_{i-1_4} is to occur only if s_{i-1} is not exactly equal to zero in order to prevent the new s_{i-1} from becoming -16. Otherwise, there would have been a difficulty to distinguish between the out of bound $+16$ and the -16 resulting from subtracting 16 from a digit that is already zero.

The condition $u_i = s_{i_0} \bar{s}_{i-1_4} \bar{z}_{i-1}$ is not a tight condition. The P-recoding causing insignificant digits deletion occurs when $s_{i-1} = 15 (= 01111_{bin})$ and $s_i = -1$ $(= 11111_{bin})$ or $s_i = 15$ (assuming that it is preceded by some $s = -1$). So, the strictest condition is $u_i = s_{i_3} s_{i_2} s_{i_1} s_{i_0} \bar{s}_{i-1_4} s_{i-1_3} s_{i-1_2} s_{i-1_1} s_{i-1_0}$. Other conditions between those two extremes can also be used like for example $u_i = s_{i_3} s_{i_2} s_{i_1} s_{i_0} \bar{s}_{i-1_4} s_{i-1_3}$.

For each digit after the recodings, an indicator n_i is used to specify the sign of the digit. Another indicator $\bar{z}_i$ is kept to indicate if the digit is not zero. The final decision of the LDD is based on those n_i's and $\bar{z}_i$'s. The first digit that has $\bar{z}_i = 1$ is the leading digit and its sign is the corresponding n_i.

Either the N-recoding or the P-recoding can be done first, there is no strict order. For example, using $u_i = s_{i_3} s_{i_2} s_{i_1} s_{i_0} \bar{s}_{i-1_4} s_{i-1_3}$. The case of $P(N(s))$ yields:

$$\begin{aligned} u_i^{pn} &= u_i \\ n_i^{pn} &= (\bar{s}_{i+1_0} s_{i_4} + u_{i+1}) \bar{u}_i \\ \bar{z}_i^{pn} &= (\bar{s}_{i+1_0} s_{i_4} + s_{i_3} + s_{i_2} + s_{i_1} + s_{i_0} \bar{s}_{i-1_4}) \bar{u}_i \\ &\quad + u_i \bar{u}_{i+1} (s_{i+1_0} + \bar{s}_{i_4}) \end{aligned}$$

And the case of $N(P(s))$ yields:

$$\begin{aligned} u_i^p &= u_i \\ n_i^{np} &= (\bar{s}_{i+1_0} s_{i_4} + u_{i+1}) \bar{u}_i \\ \bar{z}_i^{np} &= (\bar{s}_{i+1_0} s_{i_4} + s_{i_3} + s_{i_2} + s_{i_1} \\ &\quad + s_{i_0}(s_{i-1_4} \oplus \bar{u}_{i-1})) \bar{u}_i + u_i \bar{s}_{i+1_0} \bar{s}_{i_4} \end{aligned}$$

The leading non-zero digit is then determined by using the $\bar{z}$ bits out of the recodings. A tree network can be used

Table 1. Rounding value for the four IEEE modes and different fractional ranges

range	RNE	RZ	RP		RM	
			+ve	-ve	+ve	-ve
$-1 < f < -0.5$	-1	-1	0	-1	-1	0
-0.5	$-L$	-1	0	-1	-1	0
$-0.5 < f < 0$	0	-1	0	-1	-1	0
0	0	0	0	0	0	0
$0 < f < 0.5$	0	0	1	0	0	1
0.5	L	0	1	0	0	1
$0.5 < f < 1$	1	0	1	0	0	1

to encode the position of the first non-zero digit and this amount is forwarded to the left shifters to normalize the result. Obviously, the n bits out of the recodings should be shifted as well. The final sign of the number is that of the leading digit as determined by its n bit. Based on this either the result or its negation is chosen and the sign of the whole floating point result is affected.

As mentioned earlier, the fine adjustment is performed in the rounding stage. This is the other piece of challenging logic in the design at hand. In the proposed format the MSD has four bits. The rounding stage must determine the leading one among those four bits in order to decide on the approximate bit location for the rounding. The fine adjustment is then when another circuit determines if the remaining part of the number below the leading bit of the MSD is positive or negative. Those two indicators allow for the decision on the correct bit location to apply the IEEE rounding. A fractional value f_i at bit location i of a signed digit binary number $\cdots x_{i+1} x_i x_{i-1} \cdots x_0$ can be defined as $f_i = (\Sigma_{j=0}^{j=i-1} 2^j \times x_j)/2^i$. The decision of the digit added for rounding is then determined by the fractional value at the rounding position. However, the value to add in order to achieve the correct rounding does not depend only on the fractional range but also on the IEEE rounding mode. In RP and RM modes, the sign of the floating point number affects the decision as well. Any such rounding of the proposed format does not propagate a carry through the whole number as the rounding in conventional adders do. SD addition is used instead and the addition of a ± 1 digit representing the rounding decision is easily handled. The decision is according to Table 1 where L is the bit at the rounding location. It is important to note that in this format the rounding to zero mode is not a simple truncation. If the fractional value is negative a -1 must be added to perform the correct rounding to zero.

In our design the fractional range is estimated and the rounding value decided speculatively for each bit location in the Least Significant Digit (LSD). The resulting potential new LSDs after adding each rounding value are also calculated. Then, based on the circuits indicating the leading bit of the MSD and the fine adjustment the final rounded LSD

Table 2. Time delay of various components in terms of number of *FO4* delays.

Part	Delay
Adder	$5 + 2 \times \lceil \log_{f-1}(\lceil \frac{n}{f} \rceil - 1) \rceil$
[4 : 2] compressors	3
(3, 2) counters	2
Mux, in to out	1
Mux, select to out	$\lceil \log_4(n) \rceil + 1$
Signed digit adder	$8 + 2 \times \lceil \log_{f-1}(\lceil \frac{r+1}{f} \rceil - 1) \rceil + \lceil \log_4(r+1) \rceil$
Shifter	$\lceil \log_2(n) \rceil$
Other (no design details)	$\lceil \log_f(n) \rceil$

is chosen.

4. Time delay modeling and rationale for postponed rounding

The current authors had previously proposed [6] a parametric time delay model to compare floating point unit implementations. In that analytical model, the operand width n, the fan-in f of the logic gates and the radix r of the redundant format are used as parameters. The model gives an estimate of the number of equivalent elementary delay units in the critical path of the floating point hardware. The floating point unit delay is presented in "fanout of 4"(*FO4*) delays, or the delay of an inverter driving a load that is four times its own size. The model is validated through transistor simulation of different circuits. The different parts of the model are summarized in Table 2. Using units of *FO4* delays makes the model independent of the technology scaling to a large degree since this elementary gate scales almost linearly with the technology [11]. However, the model does not include any assumptions about long wires across the chip and the time delay associated with them.

This simple model allows us to have a sense of the complexity of some parts of the floating point addition algorithms. We see that shifting and adding are operations whose time delay is an $\mathcal{O}(\log n)$. In conventional designs, rounding adds a small value to the result and could cause a carry propagation through the whole number and it is also an $\mathcal{O}(\log n)$ operation. It is usually combined with the addition step [17] so that both time delays overlap. Another part that is of $\mathcal{O}(\log n)$ is the leading one prediction (if fine adjustment is performed) as mentioned in section 3. Simply using a redundant format that makes the significand addition independent of n will not enhance the speed by much if nothing is done to all those other parts. That is why postponed rounding and the other features in the proposed two-path adder are integral to the design and are as important as the redundant format. To quantify this argument let us use the assumptions of the analytical model to estimate the time delay of the adder design described above.

The critical path of the design starts with the exponent difference. This is a 15 bit adder and not an 11 bit one as in conventional adders using double precision because of the special format used in this design. In fact, the exponent width in this format $expWF$ is equal to the conventional exponent width $expW$ (which is dependent on n as specified by the IEEE standard) expanded to allow for the normalization of denormalized numbers minus $\log_2 r$ when the radix is 2^r. The significand in this format is also larger than the corresponding significand for the conventional designs because of the redundancy. The significand width is $\lceil \frac{n}{r} \rceil \times (r+1) - 1$. The swapping multiplexers must be as wide as the significand and the output of the exponent difference is used to drive the select lines. Up to this point, the delay is estimated to be $5 + 2 \times \lceil \log_{f-1}(\lceil \frac{expWF}{f} \rceil - 1) \rceil + \lceil \log_4(\lceil \frac{n}{r} \rceil \times (r+1) - 1) \rceil + 1$ *FO4* delays. The operand then passes through a $\lceil \frac{n}{r} \rceil$-way shifter which adds $\lceil \log_2(\lceil \frac{n}{r} \rceil) \rceil$ *FO4* delays. The following multiplexer adds one more *FO4* delay. The signed digit adder takes $8 + 2 \times \lceil \log_{f-1}(\lceil \frac{r+1}{f} \rceil - 1) \rceil + \lceil \log_4(r+1) \rceil$ *FO4* delays. The select lines of the last multiplexer partially depend on the output of the adder in order to determine if there is a need to adjust to the right by one bit. Hence, there is a delay from the select lines to the output equal to $\lceil \log_4(\lceil \frac{n}{r} \rceil \times (r+1) - 1) \rceil + 1$ *FO4* delays. The total delay for this design is thus:

$$\begin{aligned}\tau = \ & 16 \\ & + 2 \times \lceil \log_{f-1}(\lceil \frac{expWF}{f} \rceil - 1) \rceil \\ & + 2 \times \lceil \log_4(\lceil \frac{n}{r} \rceil \times (r+1) - 1) \rceil \\ & + \lceil \log_2(\lceil \frac{n}{r} \rceil) \rceil \\ & + 2 \times \lceil \log_{f-1}(\lceil \frac{r+1}{f} \rceil - 1) \rceil + \lceil \log_4(r+1) \rceil \end{aligned}$$

From this derivation we can evaluate the benefit coming from each of the novel ideas in this design: the postponed rounding, the use of a higher radix base for the exponents and the use of an SD adder.

Once the significand addition is independent of n by using redundancy, the rounding delay must be masked by the delay of another part that is as long or longer. Otherwise, it will add to the overall delay of the adder. The exponent difference and multiplexers time delay (second line and half of the third line of the equation) are both $\mathcal{O}(\log n)$ operations that are essential and that are already on the critical path. Performing the rounding in parallel with them seems then to be the best choice. Hence, the rounding delay does not appear at all in the equation and is effectively hidden.

The higher radix base for the exponent has an effect on the alignment shifter which only shifts then to digit boundaries. The fourth line of the multi-line equation above captures this as a delay of $\lceil \log_2(\lceil \frac{n}{r} \rceil) \rceil$ rather than $\lceil \log_2(n) \rceil$ in conventional adders. Shifting is still an $\mathcal{O}(\log n)$ operation but its time delay is reduced by about $\log_2 r$ when using a higher radix.

The effect of the SD adder is shown in the fourth line (and part of the constant of the first line) where the terms $8 + 2 \times \lceil \log_{f-1}(\lceil \frac{r+1}{f} \rceil - 1) \rceil + \lceil \log_4(r+1) \rceil$ appear instead of $5 + 2 \times \lceil \log_{f-1}(\lceil \frac{n}{f} \rceil - 1) \rceil$ in a conventional adder.

It is clear that the effect of r on the shifter delay is the opposite of its effect on the SD adder delay. The amount of redundancy (reflected by r) also has an effect on the area of the circuit and the required increase in the register file as noted in section 1. Hence, depending on the choice of n and f for the implementation, the benefit from using a different radix for the exponent may be more than the benefit from the redundancy or vice versa. To compare this design to other designs some assumptions regarding those parameters are needed. For practical CMOS designs, the fan-in is usually limited to 3 or 4. The majority of the floating point adders are currently designed to handle double precision numbers ($n = 53$) or larger. For this range, the design proposed with r set to 4 or 8 provides the best performance as presented in our previous work [6].

5. The multiplication, division and elementary functions

The design of the proposed multiplier is shown in Fig. 3. The least significant digit and the rounding bits of each operand are used to determine the rounding values r_x and r_y. Simultaneously, the multiplier Y operand is forwarded to a modified Booth recoding block and the multiplicand X is used to generate the partial products. The multiplication proceeds as $(X + r_x)(Y + r_y) = XY + r_xY + r_yX + r_xr_y$. The rounding correction block produces the last three terms of this equation and delivers them to the reduction tree with the rest of the partial products.

As noted in section 1, the extra bits in the format are negatively valued. Hence, for the multiplier operand Y, the Booth 2 recoding scheme is modified to take into consideration those extra negative bits. The extra negative bits of the multiplicand, X, are dealt with in a slightly different way. The significand of X is taken as having two components: P the positively valued bits and E the negatively valued extra bits. The output of the Booth recoders are used as select lines in multiplexers to generate the required partial products. The positive vectors are then summed by a tree of $[4:2]$ compressors while the negative vectors are summed by a separate tree of $[4:2]$ compressors. The output of each tree is in carry save format. The positive and negative vectors are then added using a $[4:2]$ compressor followed by a signed digit adder to form the final result.

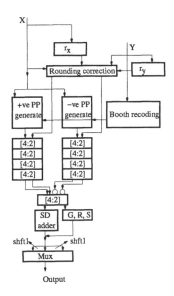

Figure 3. General block diagram of the multiplier.

To perform division and other elementary functions, a design from the literature is adapted [8, 9]. This arithmetic unit provides rapid convergence based on higher-order Newton-Raphson and series expansion techniques. To adapt the original design to the format proposed here, a short adder is used to eliminate the redundancy from the most significant part of the divisor operand by subtracting the extra bits. This non-redundant part is used to access the lookup table while the rest of the operand is fully transformed into a non-redundant form. In parallel, another adder is used to convert the dividend into a non-redundant form as well. The unit then works on those two operands as in the original design and at the end a signed digit adder is used instead of the regular carry propagate adder. The delay of the proposed unit is not much different from the original design since the extra delay of the short adder at the start is compensated by the reduced delay in the final addition.

6. Simulation results

A scalable CMOS technology was used to design the adder and the multiplier at the transistor level with $n = 53$, $f = 3$ and $r = 4$ implementing all the IEEE rounding modes.

Both circuits mostly use static CMOS technology gates with only few parts using NMOS pass transistors (namely the shifters). The designs were simulated for functionality at the logic level using *verilog* and for speed at the transistor level using the switch level simulator *irsim*.

Exhaustive testing of the functionality is obviously not practical for such designs. However, with the multiplier

Table 3. Circuit statistics and simulation results for the adder and multiplier.

	Adder	Multiplier
number of nodes	46845	76523
NMOS transistors	63589	104695
PMOS transistors	61649	105037
Test vectors	5000	10000
Model delay	$34FO4$	$35FO4$
Sim delay($0.6\mu m$)	$14ns$ ($33.35FO4$)	$14.8ns$ ($35.25FO4$)
Sim delay($0.3\mu m$)	$6ns$ ($32.40FO4$)	$6.4ns$ ($34.60FO4$)

for example successfully passing more than 2.4 million random test vectors (with a random rounding mode selected for each test), we are confident enough in the implementation. Our focus then changed to the speed simulation and sizing the different gates and transistors. Technology files ranging from $0.6\mu m$ down to $0.3\mu m$ were used. On that range of scaling factors, the adder and multiplier perform as predicted by the analytical model when compared to the delay of $FO4$ inverters at the same scaling factor. At that level fewer test vectors are used due to the longer time the simulation takes. So, only 5 000 and 10 000 random test vectors are used for the adder and the multiplier respectively. The multiplier being a larger circuit it is tested more thoroughly. The results are given in Table 3.

7. Conclusions

The proposed internal format with the proposed algorithms and arithmetic units provide a complete arithmetic system allowing all the IEEE rounding modes. The elimination of carry propagation from the arithmetic operations enhances the performance of the functional units. The proposed arithmetic unit architecture includes further enhancements that increase the floating point performance such as a hexadecimal based number format and postponed rounding techniques. The proposed system pushes the performance boundary of the design space and provides a means to achieve the computational demands of numerically intensive applications. For the double (and specially larger) precision units using standard CMOS technologies, the designs presented here are predicted to yield the highest performance.

References

[1] IEEE standard for binary floating-point arithmetic, Aug. 1985. (ANSI/IEEE Std 754-1985).

[2] J. D. Bruguera and T. Lang. Leading-one prediction with concurrent position correction. *IEEE Transactions on Computers*, 48(10):1083–1097, Oct. 1999.

[3] M. Daumas and D. Matula. Recoders for partial compression and rounding. Research Report No. RR97-01, Laboratoire de l'Informatique du Parallélisme, Ecole Normale Supérieure de Lyon, Jan. 1997. Available at http://www.ens-lyon.fr/LIP/Pub/rr1997.html.

[4] M. Daumas and D. Matula. Further reducing the redundancy of a notation over a minimally redundant digit set. Research Report No. RR2000-09, Laboratoire de l'Informatique du Parallélisme, Ecole Normale Supérieure de Lyon, Mar. 2000. Available at http://www.ens-lyon.fr/LIP/Pub/rr2000.html.

[5] H. A. H. Fahmy, A. A. Liddicoat, and M. J. Flynn. Improving the effectiveness of floating point arithmetic. In *Thirty-Fifth Asilomar Conference on Signals, Systems, and Computers, Asilomar, California, USA*, volume 1, pages 875–879, Nov. 2001.

[6] H. A. H. Fahmy, A. A. Liddicoat, and M. J. Flynn. Parametric time delay modeling for floating point units. In *The International Symposium on Optical Science and Technology, SPIE's 47th annual meeting (Arithmetic session), Seattle, Washington, USA*, July 2002.

[7] P. M. Farmwald. *On the Design of High Performance Digital Arithmetic Units*. PhD thesis, Stanford University, Aug. 1981.

[8] A. A. Liddicoat. *High-Performance Arithmetic for Division and The Elementary Functions*. PhD thesis, Stanford University, Feb. 2002.

[9] A. A. Liddicoat and M. J. Flynn. High-performance floating point divide. In *Proceedings of the Euromicro Symposium on Digital System Design*, pages 354–361, Sept. 2001.

[10] D. W. Matula and A. M. Nielsen. Pipelined packet-forwarding floating point: I. foundations and a rounder. In *Proceedings of the 13th IEEE Sympsoium on Computer Arithmetic, Asilomar, California, USA*, pages 140–147, July 1997.

[11] G. W. McFarland. *CMOS Technology Scaling and Its Impact on Cache Delay*. PhD thesis, Stanford University, June 1997.

[12] A. M. Nielsen, D. W. Matula, C. N. Lyu, and G. Even. An IEEE compliant floating-point adder that conforms with the pipelined packet-forwarding paradigm. *IEEE Transactions on Computers*, 49(1):33–47, Jan. 2000.

[13] B. Parhami. Generalized signed-digit number systems: A unifying framework for redundant number representations. *IEEE Transactions on Computers*, 39(1):89–98, Jan. 1990.

[14] B. Parhami. On the implementation of arithmetic support functions for generalized signed-digit number systems. *IEEE Transactions on Computers*, 42(3):379–384, Mar. 1993.

[15] D. S. Phatak and I. Koren. Hybrid signed-digit number systems: A unified framework for redundant number representations with bounded carry propagation chains. *IEEE Transactions on Computers*, 43(8):880–891, Aug. 1994.

[16] N. Quach and M. J. Flynn. Leading one prediction—implementation, generalization, and application. Technical Report No. CSL-TR-91-463, Computer Systems Laboratory, Stanford University, Mar. 1991.

[17] N. T. Quach. *Reducing the latency of floating-point arithmetic operations*. PhD thesis, Stanford University, Dec. 1993.

Session 5:
Decimal Arithmetic and Revisions to the IEEE 754 Standard

Chair: Eric Schwarz

Decimal Floating-Point: Algorism for Computers

Michael F. Cowlishaw

IBM UK Ltd., P.O. Box 31, Birmingham Road, Warwick CV34 5JL, UK
or Department of Computer Science, University of Warwick, Coventry CV4 7AL, UK
mfc@uk.ibm.com

Abstract

Decimal arithmetic is the norm in human calculations, and human-centric applications must use a decimal floating-point arithmetic to achieve the same results.

Initial benchmarks indicate that some applications spend 50% to 90% of their time in decimal processing, because software decimal arithmetic suffers a 100× to 1000× performance penalty over hardware. The need for decimal floating-point in hardware is urgent.

Existing designs, however, either fail to conform to modern standards or are incompatible with the established rules of decimal arithmetic. This paper introduces a new approach to decimal floating-point which not only provides the strict results which are necessary for commercial applications but also meets the constraints and requirements of the IEEE 854 standard.

A hardware implementation of this arithmetic is in development, and it is expected that this will significantly accelerate a wide variety of applications.

1. Introduction

Algorism, the decimal system of numeration, has been used in machines since the earliest days of computing. Mechanical computers mirrored the manual calculations of commerce and science, and were almost all decimal. Many early electronic computers, such as the ENIAC[1], also used decimal arithmetic (and sometimes even decimal addressing). Nevertheless, by 1961 the majority were binary, as shown by a survey of computer systems in the USA[2] which reported that "131 utilize a straight binary system internally, whereas 53 utilize the decimal system (primarily binary coded decimal)...". Today, few computing systems include decimal hardware.

The use of binary arithmetic in computing became ascendant after Burks, Goldstine, and von Neumann[3] highlighted simplicity as the primary advantage of binary hardware (and by implication, greater reliability due to the reduced number of components). They concluded that for a general-purpose computer, used as a scientific research tool, the use of binary was optimal. However, Bucholtz[4] pointed out later that they

"did not consider the equally important data processing applications in which but few arithmetic steps are taken on large volumes of input-output data. If these data are expressed in a form different from that used in the arithmetic unit, the conversion time can be a major burden."

and suggested that the combination of binary addressing with decimal data arithmetic was more powerful, a conclusion echoed by many other authors (see, for example, Schmid[5]). Inevitably, this implied that computers needed at least two arithmetic units (one for binary address calculations and the other for decimal computations) and so, in general, there was a natural tendency to economize and simplify by providing only binary arithmetic units.

The remainder of this section explains why decimal arithmetic and hardware are still essential. In section 2, the variety of decimal datatypes in use is introduced, together with a description of the arithmetic used on these types, which increasingly needs floating-point. In section 3, earlier designs are summarized, with a discussion of why they have proved inadequate. Section 4 introduces the new design, which allows integer, fixed-exponent, and floating-point numbers to be manipulated efficiently in a single decimal arithmetic unit.

1.1. The need for decimal arithmetic

Despite the widespread use of binary arithmetic, decimal computation remains essential for many applications. Not only is it required whenever numbers are presented for human inspection, but it is also often a necessity when fractions are involved.

Decimal fractions (rational numbers whose denominator is a power of ten) are pervasive in human endeavours, yet most cannot be represented by binary fractions;

the value 0.1, for example, requires an infinitely recurring binary number. If a binary approximation is used instead of an exact decimal fraction, results can be incorrect even if subsequent arithmetic is exact.

For example, consider a calculation involving a 5% sales tax on an item (such as a $0.70 telephone call), rounded to the nearest cent. Using double-precision binary floating-point, the result of multiplying 0.70 by 1.05 is a little under 0.73499999999999999 whereas a calculation using decimal fractions would yield exactly 0.735. The latter would be rounded up to $0.74, but using the binary fraction the result returned would be the incorrect $0.73.

For this reason, financial calculations (or, indeed, any calculations where the results achieved are required to match those which might be calculated by hand), are carried out using decimal arithmetic.

Further, numbers in commercial databases are predominately decimal. During a survey of commercial databases (the survey reported by Tsang[6]) the column datatypes of databases owned by 51 major organizations were analyzed. These databases covered a wide range of applications, including airline systems, banking, financial analysis, insurance, inventory control, management reporting, marketing services, order entry and processing, pharmaceuticals, and retail sales. In these databases, over 456,420 columns contained identifiably numeric data, and of these 55% were decimal (the SQL NUMERIC type[7]). A further 43.7% were integer types which could have been stored as decimals[8].

This extensive use of decimal data suggested that it would be worthwhile to study how the data are used and how decimal arithmetic should be defined. These investigations showed that the nature of commercial computation has changed so that decimal floating-point arithmetic is now an advantage for many applications.

It also became apparent that the increasing use of decimal floating-point, both in programming languages and in application libraries, brought into question any assumption that decimal arithmetic is an insignificant part of commercial workloads.

Simple changes to existing benchmarks (which used incorrect binary approximations for financial computations) indicated that many applications, such as a typical Internet-based 'warehouse' application, may be spending 50% or more of their processing time in decimal arithmetic. Further, a new benchmark, designed to model an extreme case (a telephone company's daily billing application), shows that the decimal processing overhead could reach over 90%[9].

These applications are severely compute-bound rather than I/O-bound, and would clearly benefit from decimal floating-point hardware. Such hardware could be two to three orders of magnitude faster than software.

The rest of this paper discusses the requirements for decimal arithmetic and then introduces a proposed design for floating-point decimal arithmetic. This design is unique in that it is based on the strongly typed decimal representation and arithmetic required for commercial and financial applications, yet also meets the constraints and requirements of the IEEE 854 standard[10]. The design has been implemented as a C library in software, and a hardware implementation is in development.

2. Decimal arithmetic in practice

In early computers decimal floating-point in hardware was unstandardized and relatively rare. As a result, programming languages with decimal types almost invariably describe a decimal number as an integer which is scaled (divided) by a power of ten (in other words, effectively encoding decimal values as rational numbers). The number 2.50, for example, is held as the integer 250 with a scale of 2; the scale is therefore simply a negative exponent.

Depending on the language, the scale might be fixed (as in Ada fixed point[11], PL/I fixed decimal[12], or SQL NUMERIC) or it might be variable, hence providing a simple floating-point type (as in COBOL numerics[13], Rexx strings[14], Java BigDecimal[15], Visual Basic currency[16], and C# decimal[17]). Whether fixed or floating, this scaled approach reflects common ways of working with numbers on paper, especially in school, in commerce, and in engineering, and is both effective and convenient.

For many applications, a floating scale is especially advantageous. For example, European regulations[18,19], dictate that exchange rates must be quoted to 6 digits instead of to a particular scale. All the digits must be present, even if some trailing fractional digits are zero.

Preserving the scale associated with a number is also important in engineering and other applications. Here, the original units of a measurement are often indicated by means of the number of digits recorded, and the scale is therefore part of the datatype of a number.

If the scale is not preserved, measurements and contracts may appear to be more vague (less precise) than intended, and information is lost. For example, the length of a beam might be specified as 1.200 meters; if this value is altered to 1.2 meters then a contractor could be entitled to provide a beam that is within 5 centimeters of that length, rather than measured to the nearest millimeter. Similarly, if the scale is altered by computation, it must be possible to detect that change (even if the value of the result is exact) so that incorrect changes to algorithms can be readily detected.

For these and other reasons, the scaled integer representation of decimal data is pervasive. It is used in all commercial databases (where the scale is often an attribute of a column) as well as in programming. The integer coefficient is encoded in various forms (including binary, binary coded decimal (BCD), and base 100), and the scale is usually encoded in binary.

2.1. Arithmetic on decimal numbers

Traditionally, calculation with decimal numbers has used exact arithmetic, where the addition of two numbers uses the largest scale necessary, and multiplication results in a number whose scale is the sum of the scales of the operands (1.25 × 3.42 gives 4.2750, for example).

However, as applications and commercial software products have become increasingly complex, simple rational arithmetic of this kind has become inadequate. Repeated multiplications require increasingly long scaled integers, often dramatically slowing calculations as they soon exceed the limits of any available binary or decimal integer hardware.

Further, even financial calculations need to deal with an increasingly wide range of values. For example, telephone calls are now often costed in seconds rather than minutes, with rates and taxes specified to six or more fractional digits and applied to prices quoted in cents. Interest rates are now commonly compounded daily, rather than quarterly, with a similar requirement for values which are both small and exact. And, at the other end of the range, the Gross National Product of a country such as the USA (in cents) or Japan (in Yen) needs 15 digits to the left of the decimal point.

The manual tracking of scale over such wide ranges is difficult, tedious, and very error-prone. The obvious solution to this is to use a floating-point arithmetic.

The use of floating-point may seem to contradict the requirements for exact results and preservation of scales in commercial arithmetic; floating-point is perceived as being approximate, and normalization loses scale information.

However, if unnormalized floating-point is used with sufficient precision to ensure that rounding does not occur during simple calculations, then exact scale-preserving (type-preserving) arithmetic is possible, and the performance and other overheads of normalization are avoided. Rounding occurs in the usual manner when divisions or other complex operations are carried out, or when a specific rounding operation (for example, rounding to a given scale or precision) is applied.

This eclectic approach especially benefits the very common operations of addition or subtraction of numbers which have the same scale. In this case, no alignment is necessary, so these operations continue to be simple decimal integer addition or subtraction, with consequent performance advantages. For example, 2.50 can be stored as 250×10^{-2}, allowing immediate addition to 12.25 (stored as 1225×10^{-2}) without requiring shifting. Similarly, after adding 1.23 to 1.27, no normalization shift to remove the 'extra' 0 is needed.

The many advantages of scaled-integer decimal floating-point arithmetic have led to its being widely adopted in programming languages (including COBOL, Basic, Rexx, Java, and C#) and in many other software libraries. However, these implementations have not in the past provided the full floating-point arithmetic facilities which are now the norm in binary floating-point libraries and hardware.

To see how these can be supported too, it is helpful to consider some earlier decimal floating-point designs.

3. Previous decimal floating-point designs

Decimal floating-point (DFP) arithmetics have been proposed, and often implemented, for both hardware and software. A particular dichotomy of these designs is the manner by which the coefficient (sometimes called the mantissa or significand) is represented. Designs derived from scaled arithmetic use an integer for this, whereas those built for mathematical or scientific use generally use a normalized fraction.

Both of these approaches appear in the following designs. (Here, the notation {p,e} gives the maximum precision and exponent range where known.)

- In 1955, Perkins[20] described the EASIAC, a wholly DFP virtual computer implemented on the University of Michigan's MIDAC machine {7+, ±20}. The form of the coefficient seems to have been a fraction.
- The Gamma 60 computer[21], first shipped by Bull in 1960, had a DFP calculation unit with fractional coefficients {11–19, ±40}.
- In 1962, Jones and Wymore[22] gave details of the normalized variable-precision DFP Feature on the IBM Type 1620 computer {100, ±99}.
- The Burroughs B5500 computer[23], shipped in 1964, used an integer or fixed-point coefficient {21–22, ±63}. Neely[24] noted later that this "permits mixed integer and real arithmetic without type conversion".
- Mazor, in 1966[25], designed the Fairchild Symbol II DFP unit. This used normalized variable-precision and most-significant-digit–first processing {99, ±99?}.
- In 1969, Duke[26] disclosed a hardware design using unnormalized binary integers of unspecified length for both coefficient and exponent.
- Also in 1969, Taub, Owen, and Day[27] built the IBM

Schools computer; this experimental teaching computer used a scaled integer format {6, −6 to 0}.

- Fenwick, in 1972[28], described a representation similar to Duke's, and highlighted the advantages of unnormalized DFP.
- By the early 1970s, the mathematical requirements for decimal floating-point were becoming understood. Ris, working with Gustavson, Kahan, and others, proposed a unified DFP architecture[29] which had many of the features of the later binary floating-point standard. It was a normalized DFP with a fractional coefficient, three directed rounding modes, and a trap mechanism. Three precisions were defined, up to {31, ±9999}. Ris's representations were the first to use the encoding invented by Chen and Ho[30], which allowed the 31-digit DFP numbers, with 4-digit exponent, to be represented in 128 bits.
- In 1975, Keir[31] described the advantages of an unnormalized integer coefficient, noting that it is "exactly as accurate as normalized arithmetic".
- Hull, in 1978[32], proposed a DFP with controlled (variable) precision and a fractional coefficient. This was later refined and implemented in the controlled-precision CADAC arithmetic unit by Cohen, Hull, and Hamacher[33,34].
- In 1979, Johannes, Pegden, and Petry[35] discussed the problems of efficient decimal shifts in a two-binary-integer DFP representation (problems which were revisited by Bohlender in 1991[36]).
- In 1981, Cowlishaw added arbitrary-precision DFP {10^9, ±10^9;} to the Rexx programming language[37]; this was unnormalized. Later implementations of the Rexx family of languages have used a variety of representations, all with integer coefficients.
- In 1982, Sacks-Davis[38] showed that the advantage of redundant number representations (addition time is independent of operand length) can be applied to DFP, though no implementation of this is known.
- Also in the early 1980s, the standardization of binary floating-point arithmetic was completed, with the publication of the IEEE 754 standard[39]. A later generalization of that standard, IEEE 854, extended the principles to DFP, as explained by Cody *et al*[40].

These standards formalized earlier work, recorded by Kahan[41], and, unlike earlier designs using integer coefficients, prescribe *gradual underflow* as well as infinities and NaNs.

Compliance with IEEE 854 does not require either normalization or fractional coefficients. Values may be encoded redundantly, as in the Burroughs B5500, and hence may use integer coefficients.

- IEEE 854 was first implemented in the HP 71B calculator[42,43]; this held numbers as {12, ±499} fractions, expanded to {15, ±49999} for calculations.
- In 1987, Bohlender and Teufel[44] described a bit-slice DFP unit built for PASCAL-SC (a version of Pascal designed for scientific computation). This used a BCD fractional coefficient {13, −98 to +100}.
- *Circa* 1990, Visual Basic added a floating-point currency class with a 64-bit binary integer coefficient. This was later extended to 96 bits and formed the basis of the C# decimal type {28–29, −28 to 0}.
- In 1996, Java added the exact arithmetic BigDecimal class. This is arbitrary-precision, with a binary integer coefficient.
- Finally, hand calculators almost all use decimal floating-point arithmetic. For example, the Texas Instruments TI-89[45,46] uses a BCD fractional coefficient {14, ±999}, Hewlett Packard calculators continue to use a 12-digit decimal format, and Casio calculators have a 15-digit decimal internal format.

Looking back at these designs, it would seem that those which assumed a fractional coefficient were designed with mathematical rather than commercial uses in mind. These did not meet the strong type requirements of scaled decimal arithmetic in commercial applications, and for other applications they were eclipsed by binary floating-point, which provides better performance and accuracy for a given investment in hardware. (The notable exception to this generalization is the hand-held calculator, where performance is rarely an issue and decimal floating-point is common.)

Those designs which use integer coefficients, however, have survived, and are widely used in software, probably due to their affinity with decimal data storage and exact rational arithmetic. These designs tend to follow the traditional rules of decimal arithmetic, and although this is in effect floating-point, little attempt was made to incorporate the improvements and advantages of the floating-point system defined in IEEE 854 until recently.

At first reading it may seem that the rules of IEEE 854, which appear to assume a fractional coefficient, must be incompatible with unnormalized scaled-integer arithmetic. However, a fractional coefficient is not necessary to meet the requirements of IEEE 854. All the mathematical constraints of the standard can be met when using an integer coefficient; indeed, it turns out that subnormal values are simpler to handle in this form.

Normalization, too, is not required by IEEE 854. It does offer an advantage in binary floating-point, where in effect it is a compression scheme that increases the length of the coefficient by one bit. However, in decimal arithmetic (where in any case only one value in ten could be implied in the same way) normalized arithmetic is harmful, because it precludes the exact and strongly typed arithmetic with scale preservation which is essential for many applications.

With these observations, it becomes possible to extend the relatively simple software DFP in common use today to form a richer arithmetic which not only meets the commercial and financial arithmetic requirements but also has the advantages of the IEEE 854 design: a formally closed arithmetic with gradual underflow and defined exception handling.

4. The decimal arithmetic design

There is insufficient space here to include every detail of the decimal floating-point design; this is available at http://www2.hursley.ibm.com/decimal.
Also, the arithmetic is independent of specific concrete representations, so these are not discussed here.

In summary, the core of the design is the abstract model of finite numbers. In order to support the required exact arithmetic on decimal fractions, these comprise an integer coefficient together with a conventional sign and signed integer exponent (the exponent is the negative of the scale used in scaled-integer designs). The numerical *value* of a number is given by $(-1)^{sign} \times coefficient \times 10^{exponent}$.

It is important to note that even though the concept of scale is preserved in numbers (the numbers are essentially two-dimensional), they are not the *significant numbers* deprecated by Delury[47]. Each number has an exact value and, in addition, an exact exponent which indicates its type; the number may be thought of as the sum of an integral number of discrete values, each of magnitude $10^{exponent}$. The arithmetic on these numbers is exact (unless rounding to a given precision is necessary) and is in no sense a 'significance' arithmetic.

Given the parameters just described, much of the arithmetic is obvious from either the rules of mathematics or from the requirements of IEEE 854 (the latter, for example, define the processing of NaNs, infinities, and subnormal values). The remainder of this section explains some less obvious areas.

4.1. Context

In this design, the concept of a *context* for operations is explicit. This corresponds to the concept of a 'floating-point control register' in hardware or a context object instance in software.

This context includes the flags and trap-enablers from IEEE 854 §7 and the rounding modes from §4. There is also an extra rounding mode and a precision setting.

4.1.1. Commercial rounding.
The extra rounding mode is called *round-half-up*, which is a requirement for many financial calculations (especially for tax purposes and in Europe). In this mode, if the digits discarded during rounding represent greater than or equal to half (0.5) of the value of a one in the next left position then the result should be rounded up. Otherwise the discarded digits are ignored. This is in contrast to *round-half-even*, the default IEEE 854 rounding mode, where if the discarded digits are exactly half of the next digit then the least significant digit of the result will be even.

It is also recommended that implementations offer two further rounding modes: *round-half-down* (where a 0.5 case is rounded down) and *round-up* (round away from zero). The rounding modes in IEEE 854 together with these three are the same set as those available in Java.

4.1.2. Precision.
The *working precision* setting in the context is a positive integer which sets the maximum number of significant digits that can result from an arithmetic operation. It can be set to any value up to the maximum length of the coefficient, and lets the programmer choose the appropriate working precision.

In the case of software (which may well support unlimited precision), this lets the programmer set the precision and hence limit computation costs. For example, if a daily interest rate multiplier, R, is 1.000171 (0.0171%, or roughly 6.4% per annum), then the exact calculation of the yearly rate in a non-leap year is R^{365}. To calculate this to give an exact result needs 2191 digits, whereas a much shorter result which is correct to within one unit in the last place (ulp) will almost always be sufficient and could be calculated very much faster.

In the case of hardware, precision control has little effect on performance, but allows the hardware to be used for calculations of a different precision from the available 'natural' register size. For example, one proposal[48] for a concrete representation suggests a maximum coefficient length of 33 digits; this would be unsuitable for implementing the new COBOL standard (which specifies 32-digit intermediate results) if precision control in some form were not available.

Note that to conform to IEEE 854 §3.1 the working precision should not be set to less than 6.

4.2. Arithmetic rules

The rules for arithmetic are the traditional exact decimal rational arithmetic implemented in the languages and databases described earlier, subject to the context used for the operation. These rules can all be described in terms of primitive integer operations, and are defined in such a way that integer arithmetic is itself a subset of the full floating-point arithmetic. The lack of automatic normalization is essential for this to be the case.

It is the latter aspect of the design which permits both integer and floating-point arithmetic to be carried out in the same processing unit, with obvious economies in either a hardware or a software implementation.

The ability to handle integers as easily as fractions avoids conversions (such as when multiplying a cost by a number of units) and permits the scale (type) of numbers to be preserved when necessary. Also, since the coefficient is a 'right-aligned' integer, conversions to and from other integer representations (such as BCD or binary) are simplified.

To achieve the necessary results, every operation is carried out as though an infinitely precise mathematical result is first computed, using integer arithmetic on the coefficient where possible. This intermediate result is then coerced to the precision specified in the context, if necessary, using the rounding algorithm also specified in the context. Rounding, the processing of overflow and underflow conditions, and the production of subnormal results are defined in IEEE 854.

The following subsections describe the required operators (including some not defined in IEEE 854), and detail the rules by which their initial result (before any rounding) is calculated.

The notation {*sign, coefficient, exponent*} is used here for the numbers in examples. All three parameters are integers, with the third being a signed integer.

4.2.1. Addition and subtraction.
If the exponents of the operands differ, then their coefficients are first aligned; the operand with the larger exponent has its original coefficient multiplied by 10^n, where n is the absolute difference between the exponents.

Integer addition or subtraction of the coefficients, taking signs into account, then gives the exact result coefficient. The result exponent is the minimum of the exponents of the operands.

For example, {0, 123, −1} + {0, 127, −1} gives {0, 250, −1}, as does {0, 50, −1} + {0, 2, +1}.

Note that in the common case where no alignment or rounding of the result is necessary, the calculations of coefficient and exponent are independent.

4.2.2. Multiplication.
Multiplication is the simplest operation to describe; the coefficients of the operands are multiplied together to give the exact result coefficient, and the exponents are added together to give the result exponent.

For example, {0, 25, 3} × {0, 2, 1} gives {0, 50, 4}.

Again, the calculations of coefficient and exponent are independent unless rounding is necessary.

4.2.3. Division.
The rules for division are more complex, and some languages normalize all division results. This design, however, uses exact integer division where possible, as in C#, Visual Basic, and Java. Here, a number such as {0, 240, −2} when divided by two becomes {0, 120, −2} (not {0, 12, −1}).

The precision of the result will be no more than that necessary for the exact result of division of the integer coefficient. For example, if the working precision is 9 then {0, 241, −2} ÷ 2 gives {0, 1205, −3} and {0, 241, −2} ÷ 3 gives, after rounding, {0, 803333333, −9}.

This approach gives integer or same-scale results where possible, while allowing post-operation normalization for languages or applications which require it.

4.2.4. Comparison.
A comparison compares the numerical values of the operands, and therefore does not distinguish between redundant encodings. For example, {1, 1200, −2} compares equal to {1, 12, 0}. The actual values of the coefficient or exponent can be determined by conversion to a string (or by some unspecified operation). For type checking, it is useful to provide a means for extracting the exponent.

4.2.5. Conversions.
Conversions between the abstract form of decimal numbers and strings are more straightforward than with binary floating-point, as conversions can be exact in both directions.

In particular, a conversion from a number to a string and back to a number can be guaranteed to reproduce the original sign, coefficient, and exponent.

Note that (unless deliberately rounded) the length of the coefficient, and hence the exponent, of a number is preserved on conversion from a string to a number and vice versa. For example, the five-character string "1.200" will be converted to the number {0, 1200, −3}, not {0, 12, −1}.

One consequence of this is that when a number is displayed using the defined conversion there is no hidden information; "what you see is exactly what you have". Further, the defined conversion string is in fact a valid and complete concrete representation for decimal numbers in the arithmetic; it could be used directly in an interpreted scripting language, for example.

4.2.6. Other operations.
The arithmetic defines a number of operations in addition to those already described. **abs**, **max**, **min**, **remainder-near**, **round-to-integer**, and **square-root** are the usual operations as defined in IEEE 854. Similarly, **minus** and **plus** are defined in order to simplify the mapping of the prefix − and prefix + operators present in most languages.

divide-integer and **remainder** are operators which

provide the truncating remainder used for integers (and for floating-point in Java, Rexx, and other languages). If the operands *x* and *y* are given to the divide-integer and remainder operations, resulting in *i* and *r* respectively, then the identity $x = (i \times y) + r$ holds.

An important operator, **rescale**, sets the exponent of a number and adjusts its coefficient (with rounding, if necessary) to maintain its value. For example, rescaling the number {0, 1234567, −4} so its exponent is −2 gives {0, 12346, −2}. This example is the familiar, and very heavily used, 'round to cents' operation, although rescale has many other uses (**round-to-integer** is a special case of rescale, for example).

Finally, the **normalize** operator is provided for reducing a number to its most succinct form. Unlike the binary equivalent, this normalization removes trailing rather than leading zeros from the coefficient; this means that it generalizes to arbitrary-precision implementations.

5. Conclusion

The new data type described here combines the advantages of algorism and modern floating-point arithmetic. The integer coefficient means that conversions to and from fixed-point data and character representations are fast and efficient. The lack of normalization allows strongly typed decimal numbers and improves the performance of the most common operations and conversions. The addition of the IEEE 854 subnormal and special values and other features means that full floating-point facilities are available on decimal numbers without costly and difficult conversions to and from binary floating-point. These performance and functional advantages are complemented by easier programming and the reduced risk of error due to the automation of scaling and other operations.

6. Acknowledgements

The author is indebted to Kittredge Cowlishaw, Roger Golliver, Michael Schulte, Eric Schwarz, and the anonymous referees for their many helpful suggestions for improvements to this paper.

References

[1] H. H. Goldstine and Adele Goldstine, "The Electronic Numerical Integrator and Computer (ENIAC)", *IEEE Annals of the History of Computing, Vol. 18 #1*, pp10–16, IEEE, 1996.

[2] Martin H. Weik, "A Third Survey of Domestic Electronic Digital Computing Systems, Report No. 1115", 1131pp, Ballistic Research Laboratories, Aberdeen Proving Ground, Maryland, March 1961.

[3] Arthur W. Burks, Herman H. Goldstine, and John von Neumann, "Preliminary discussion of the logical design of an electronic computing instrument", 42pp, Inst. for Advanced Study, Princeton, N. J., June 28, 1946.

[4] Werner Buchholz, "Fingers or Fists? (The Choice of Decimal or Binary representation)", *Communications of the ACM, Vol 2 #12*, pp3–11, ACM, December 1959.

[5] Hermann Schmid, "Decimal Computation", ISBN 047176180X, 266pp, Wiley, 1974.

[6] Annie Tsang and Manfred Olschanowsky, "A Study of DataBase 2 Customer Queries", *IBM Technical Report TR 03.413*, 25pp, IBM Santa Teresa Laboratory, San Jose, CA, April 1991.

[7] Jim Melton *et al*, "ISO/IEC 9075:1992: Information Technology – Database Languages – SQL", 626pp, ISO, 1992.

[8] Akira Shibamiya, "Decimal arithmetic in applications and hardware", 2pp, *pers. comm.*, 14 June 2000.

[9] M. F. Cowlishaw, "The 'telco' benchmark", URL: `http://www2.hursley.ibm.com/decimal`, 3pp, IBM Hursley Laboratory, May 2002.

[10] W. J. Cody *et al*, "IEEE 854-1987 IEEE Standard for Radix-Independent Floating-Point Arithmetic", 14pp, IEEE, March 1987.

[11] S. Tucker Taft and Robert A. Duff, "ISO/IEC 8652:1995: Information Technology – Programming Languages – Ada (Ada 95 Reference Manual: Language and Standard Libraries)", ISBN 3-540-63144-5, 552pp, Springer-Verlag, July 1997.

[12] J. F. Auwaerter, "ANSI X3.53-1976: American National Standard – Programming Language PL/I", 421pp, ANSI, 1976.

[13] JTC-1/SC22/WG4, "Proposed Revision of ISO 1989:1985 Information technology – Programming languages, their environments and system software interfaces – Programming language COBOL", 905pp, INCITS, December 2001.

[14] Brian Marks and Neil Milsted, "ANSI X3.274-1996: American National Standard for Information Technology – Programming Language REXX", 167pp, ANSI, February 1996.

[15] Sun Microsystems, "BigDecimal (Java 2 Platform SE v1.4.0)", URL: `http://java.sun/com/products`, 17pp, Sun Microsystems Inc., 2002.

[16] Microsoft Corporation, "MSDN Library Visual Basic 6.0 Reference", URL: `msdn.microsoft.com/library`, Microsoft Corporation, 2002.

[17] Rex Jaeschke, "C# Language Specification", *ECMA-TC39-TG2-2001*, 520pp, ECMA, September 2001.

[18] European Commission, "The Introduction of the Euro and the Rounding of Currency Amounts", 29pp, European Commission Directorate General II Economic and Financial Affairs, 1997.

[19] European Commission Directorate General II, "The Introduction of the Euro and the Rounding of Currency Amounts", *II/28/99-EN Euro Papers No. 22.*, 32pp, DGII/C-4-SP(99) European Commission, March 1998, February 1999.

[20] Robert Perkins, "EASIAC, A Pseudo-Computer", *Communications of the ACM, Vol. 3 #2*, pp65–72, ACM, April 1956.

[21] M. Bataille, "The Gamma 60: The computer that was ahead of its time", *Honeywell Computer Journal Vol 5 #3*, pp99–105, Honeywell, 1971.

[22] F. B. Jones and A. W. Wymore, "Floating Point Feature On The IBM Type 1620", *IBM Technical Disclosure Bulletin, 05-62*, pp43–46, IBM, May 1962.

[23] Burroughs Corporation, "Burroughs B5500 Information Processing Systems Reference Manual", 224pp, Burroughs Corporation, Detroit, Michigan, 1964.

[24] Peter M. Neely, "On conventions for systems of numerical representation", *Proceedings of the ACM annual conference, Boston, Massachusetts*, pp644–651, ACM, 1972.

[25] Stan Mazor, "Fairchild decimal arithmetic unit", 9pp, *pers. comm.*, July–September 2002.

[26] K. A. Duke, "Decimal Floating Point Processor", *IBM Technical Disclosure Bulletin, 11-69*, pp862–862, IBM, November 1969.

[27] D. M. Taub, C. E. Owen, and B. P.. Day, "Experimental Computer for Schools", *Proceedings of the IEE, Vol 117 #2*, pp303–312, IEE, February 1970.

[28] Peter M. Fenwick, "A Binary Representation for Decimal Numbers", *Australian Computer Journal, Vol 4 #4 (now Journal of Research and Practice in Information Technology)*, pp146–149, Australian Computer Society Inc., November 1972.

[29] Frederic N. Ris, "A Unified Decimal Floating-Point Architecture for the Support of High-Level Languages", *ACM SIGNUM Newsletter, Vol. 11 #3*, pp18–23, ACM, October 1976.

[30] Tien Chi Chen and Irving T. Ho, "Storage-Efficient Representation of Decimal Data", *CACM Vol 18 #2*, pp49–52, ACM, January 1975.

[31] R. A. Keir, "Compatible number representations", *Conf. Rec. 3rd Symp. Comp. Arithmetic CH1017-3C*, pp82–87, IEEE Computer Society, 1975.

[32] T. E. Hull, "Desirable Floating-Point Arithmetic and Elementary Functions for Numerical Computation", *ACM Signum Newsletter, Vol. 14 #1 (Proceedings of the SIGNUM Conference on the Programming Environment for Development of Numerical Software)*, pp96–99, ACM, 1978.

[33] Marty S. Cohen, T. E. Hull, and V. Carl Hamacher, "CADAC: A Controlled-Precision Decimal Arithmetic Unit", *IEEE Transactions on Computers, Vol. 32 #4*, pp370–377, IEEE, April 1983.

[34] T. E. Hull and M. S. Cohen, "Toward an Ideal Computer Arithmetic", *Proceedings of the 8th Symposium on Computer Arithmetic*, pp131–138, IEEE, May 1987.

[35] J. D. Johannes, C. Dennis Pegden, and F. E. Petry, "Decimal Shifting for an Exact Floating Point Representation", *Computers and Electrical Engineering, Vol. 7 #3*, pp149–155, Elsevier, September 1980.

[36] Gerd Bohlender, "Decimal Floating-Point Arithmetic in Binary Representation", *Computer arithmetic: Scientific Computation and Mathematical Modelling (Proceedings of the Second International Conference, Albena, Bulgaria, 24-28 September 1990)*, pp13–27, J. C. Baltzer AG, 1991.

[37] M. F. Cowlishaw, "The Design of the REXX Language", *IBM Systems Journal, Vol 23 #4*, pp326–335, IBM (Offprint # G321-5228), 1984.

[38] R. Sacks-Davis, "Applications of Redundant Number Representations to Decimal Arithmetic", *The Computer Journal, Vol 25 #4*, pp471–477, November 1982.

[39] David Stevenson *et al*, "IEEE 754-1985 IEEE Standard for Binary Floating-Point Arithmetic", 20pp, IEEE, July 1985.

[40] W. J. Cody *et al*, "A Proposed Radix- and Word-length-independent Standard for Floating-point Arithmetic", *IEEE Micro magazine*, pp86–100, IEEE, August 1984.

[41] W. Kahan, "Mathematics Written in Sand", *Proc. Joint Statistical Mtg. of the American Statistical Association*, pp12–26, American Statistical Association, 1983.

[42] Hewlett Packard Company, "Math Reference", *HP-71 Reference Manual, Mfg. #0071-90110, Reorder #0071-90010*, pp317–318, Hewlett Packard Company, October 1987.

[43] Hewlett Packard Company, "Chapter 13 – Internal Data Representations", *Software Internal Design Specification for the HP-71, Vol. 1 Part #00071-90068*, pp13.1–13.17, Hewlett Packard Company, December 1983.

[44] G. Bohlender and T. Teufel, "A Decimal Floating-Point Processor for Optimal Arithmetic", *Computer arithmetic: Scientific Computation and Programming Languages*, ISBN 3-519-02448-9, pp31–58, B. G. Teubner Stuttgart, 1987.

[45] Texas Instruments, "TI-89/TI-92 Plus Developers Guide, Beta Version .02", 1356pp, Texas Instruments, 2001.

[46] Texas Instruments, "TI-89/TI-92 Plus Sierra C Assembler Reference Manual, Beta Version .02", 322pp, Texas Instruments, 2001.

[47] Daniel B. Delury, "Computation with Approximate Numbers", *The Mathematics Teacher 51*, pp521–530, November 1958.

[48] Michael F. Cowlishaw, Eric M. Schwarz, Ronald M. Smith, and Charles F. Webb, "A Decimal Floating-Point Specification", *Proceedings of the 15th IEEE Symposium on Computer Arithmetic*, ISBN 0-7695-1150-3, pp147–154, IEEE, June 2001.

Panel:

Revisions to the IEEE 754 Standard for Floating-Point Arithmetic

Organized by:
Dr. Eric Schwarz, *IBM Corporation*

Almost twenty years ago the IEEE 754 binary floating-point standard was adopted. Since then almost every microprocessor as well as many programming languages have defined the floating-point arithmetic to be IEEE 754 compliant. From the many years experience in implementing the standard in hardware and writing floating-point programs, there have been numerous suggestions for revisions. All IEEE standards must undergo a review process every 5 years or be dropped as an active standard. For past reviews this standard was extended without much discussion. But finally in January 2001 an in-depth review was started. A committee was formed and over the past two years many revisions have been evaluated. The most extensive change to the standard is to adopt formats for decimal floating-point arithmetic. This proposal creates decimal floating-point data formats for 32, 64, and 128 bits. Decimal floating-point arithmetic provides an exact representation of displayed numbers and provides a precise round at decimal radix point. This type of arithmetic is required in financial calculations. Some experts argue that decimal will replace binary due to its ability to represent decimal numbers exactly, while others think that binary will remain the key floating-point format due to its speed of execution and its more regular spacing of intervals. Another once controversial proposal is the addition of fused multiply-add. This operation only causes one rounding error, while in most implementations, provides twice the performance of separate operations. Other additions to the standard include a quadword format and many predicate functions such as comparison operators like greater than. Also operators for maximum and minimum have been accepted that after hours of arguing now favor a numeric result over a NaN. There are also deletions such as the single extended and double extended formats. And there are some items that are deleted in one meeting and resurrected in the following meeting such as signaling NaNs. Over the past two years of committee review there has been many proposals discussed. This panel discussion will enlighten the audience to the additions, deletions, and some of the current controversial proposals. The panel will consist of :

- David Hough, Sun Microsystems, Editor of the Standard – Overview
- Mike Cowlishaw, IBM Corp., Decimal Floating-Point Software Advocate
- David Bailey, Lawrence Berkeley Lab., Quadword Precision Advocate
- David Matula, Southern Methodist University, Academics / Industry Consultant
- Eric Schwarz, IBM Corp., Decimal Floating-Point Hardware – Panel Chair

Session 6:
Elementary Functions

Chair: Renato Stefanelli

"Partially rounded" Small-Order Approximations for Accurate, Hardware-Oriented, Table-Based Methods

Jean-Michel Muller
CNRS - Laboratoire LIP
ENSL/CNRS/INRIA Arenaire Project
Ecole Normale Supérieure de Lyon
46 Allee d'Italie
69364 Lyon Cedex 07
FRANCE
Jean-Michel.Muller@ens-lyon.fr

Keywords Computer arithmetic, elementary and special functions, table-based methods, polynomial approximations.

Abstract

We aim at evaluating elementary and special functions using small tables and small, rectangular, multipliers. To do that, we show how accurate polynomial approximations whose order-1 coefficients are small in size (a few bits only) can be computed. We compare the obtained results with similar work in the recent literature.

1 Introduction

This paper deals with hardware-oriented methods for implementing elementary (sine, cosine, exponential, etc.), special (gamma, erf, Bessel, etc.), or special-purpose functions. We assume that a rather low precision (say, from 10 to 32 bits) is required.

Various methods have been suggested for tackling with this problem. CORDIC-like algorithms [15, 17, 2, 6, 16, 1] may be attractive for some functions, but they are made up using some algebraic relations (such as $\cos(x + y) = \cos(x)\cos(y) - \sin(x)\sin(y)$) that are satisfied by the elementary functions only. The other methods are almost all built from at least one of the following two ideas:

- Since we can easily implement additions, multiplications (and possibly, divisions), the first idea that springs in mind is to approximate a function by combinations of these basic operations, that is, by polynomial (and possibly, rational) functions;

- the continuing progress of VLSI technology allows the implementation of larger and larger tables, it therefore makes sense to directly tabulate a function (when very low precision is at stake), or to combine tabulation and a few arithmetic operations.

Let us now quickly present some recent methods, that make use of the above ideas in quite different ways.

1.1 The bipartite method

The bipartite method was originally introduced by Das Sarma and Matula [10], with the aim of getting accurate reciprocals. Later on, generalizations to "symmetric" and "multipartite" tables and/or improvements have been suggested by Schulte and Stine [11, 12, 13, 14], Muller [7], and De Dinechin and Tisserand [3].

Assume an n-bit, binary fixed-point system, and – to simplify the presentation – assume that n is a multiple of 3, $n = 3k$. We wish to design a table-based implementation of function f. The straightforward method would consist in tabulating all possible 2^n values of $f(x)$. This would lead to a table of size $n \times 2^n$ bits. Instead of that, let us split the binary representation of the input value into 3 k-bit words x_0, x_1 and x_2, that is,

$$x = x_0 + x_1 2^{-k} + x_2 2^{-2k}$$

where x_0, x_1 and x_2 are multiples of 2^{-k} that are less than 1. The original bipartite method consists in approximating the order-1 Taylor expansion

$$\begin{aligned}f(x) &= f\left(x_0 + x_1 2^{-k}\right) \\ &+ x_2 2^{-2k} f'\left(x_0 + x_1 2^{-k}\right) \\ &+ x_2^2 2^{-4k} f''(\xi), \\ \xi &\in [x_0 + x_1 2^{-k}, x]\end{aligned}$$

by
$$f(x) = f\left(x_0 + x_1 2^{-k}\right) + x_2 2^{-2k} f'(x_0).$$

That is, $f(x)$ is approximated by the sum of two functions $\alpha(x_0, x_1)$ and $\beta(x_0, x_2)$, where

$$\begin{cases} \alpha(x_0, x_1) &= f\left(x_0 + x_1 2^{-k}\right) \\ \beta(x_0, x_2) &= x_2 2^{-2k} f'(x_0) \end{cases}$$

The error of this approximation is roughly proportional to 2^{-3k} (see references [10, 11, 12, 13, 14, 7] for a more detailed error analysis). Instead of directly tabulating function f, we tabulate functions α and β. Since they are functions of $2k$ bits only, each of these tables has $2^{2n/3}$ entries. This results in a total table size of $2n \times 2^{2n/3}$ bits, which is a very significant improvement.

1.2 Methods using tabulation and a few multiplications.

Another solution is to split the input interval into some number of small sub-intervals, and store in a table, for each sub-interval, the coefficients of a low-degree polynomial approximation. The rationale behind that choice is that, for a given degree, the accuracy of an approximation is drastically improved if the size of the interval of approximation decreases. This is illustrated by Fig. 1.

Many variants to this general idea have been suggested. For instance, Piñeiro et al. [8] divide the input interval into around 2^8 subintervals. They store, for each subinterval, a degree-2 minimax approximation, and accumulate the partial terms in a fused accumulation tree. Cao et al. [5] store function values instead of coefficients and perform interpolation, using fewer look-up table memory entries, at the expense of additional hardware and extra time for calculating the coefficients on-the-fly.

Before introducing our method, we need to recall some classical results on "minimax" polynomial approximation, that will be used in the sequel of this paper.

1.3 Some reminders on minimax approximation

We denote by $\mathcal{P}_n$ the set of the polynomials of degree less than or equal to n. In the following, $||f - p||_\infty$ denotes the *distance*:

$$||f - p||_\infty = \max_{a \leq x \leq b} |f(x) - p(x)|.$$

We look for a polynomial p^* that satisfies:

$$||f - p^*||_\infty = \min_{p \in \mathcal{P}_n} ||f - p||_\infty.$$

The polynomial p^* is called the *minimax* degree-n polynomial approximation to f on $[a, b]$. The following result, due to Chebyshev, gives a characterization of the minimax approximations to a function[1].

Theorem 1 (Chebyshev) *p^* is the minimax degree-n approximation to f on $[a, b]$ if and only if there exist at least $n + 2$ values*

$$a \leq x_0 < x_1 < x_2 < \ldots < x_{n+1} \leq b$$

such that:

$$p^*(x_i) - f(x_i) = (-1)^i [p^*(x_0) - f(x_0)] \\ = \pm ||f - p^*||_\infty.$$

An algorithm, due to Remez [4, 9], computes the minimax degree-n approximation to a continuous function iteratively. That algorithm is implemented on many packages such as Maple or Mathematica. For instance, to compute some of the approximations used in this paper, we have used the `minimax` function provided in the Maple computer algebra package.

1.4 Our goals

The methods – such as the bipartite method – that do not use multipliers are very helpful for small precision implementation (say, up to 16 bits), but larger precisions cannot be reached without requiring huge tables. Even the best improvements to the bipartite method cannot dismiss the fact that that method is an order-1 method: with tables with p address bits, it seems difficult to get more than around $2p$ bits of accuracy.

Hence, we focus on methods that require a few multiplications. Our main interest will be on *order-2 methods*. When using such methods, the coefficients of the polynomial approximations are, in general, not exactly representable with a small number of bits. Thus, they are truncated or rounded to the nearest, say k-bit, number. Choosing k results from a compromise between accuracy of approximation, and multiplier size and delay.

With an order-2 approximation, the truncation of the order-2 coefficient has a small effect only (unless k is very small), on the final accuracy. And yet, the truncation of the order-1 coefficient may have a strong influence on the accuracy of the approximation. The question that immediately springs in mind is *how much is the truncated best polynomial approximation to f close to the best approximation among the "truncated polynomials"?* This is the very question that we try to address in this paper. We start from the minimax approximations to some functions, round their order-1 coefficient, and try to get better approximations than

---

[1] Although Chebyshev worked on both kinds of approximation, the minimax approximation should not be confused with the polynomial approximation that uses orthogonal Chebyshev polynomials. In practice, the minimax approximation is slightly better than the other one.

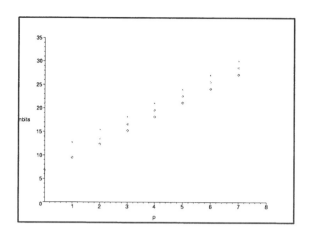

Figure 1. *Number of bits of accuracy of the degree-2 minimax piecewise approximations to* $\sin(x)$ *(circles),* $\exp(x)$ *(diamonds) and* $\sqrt{1+x}$ *(crosses), for* $x \in [0,1]$. *The interval is split into* 2^p *subintervals of equal size. For each subinterval a minimax approximation (called "subapproximation") is computed. We give here, as a function of p, the number of bits of accuracy of the less accurate subapproximation. Roughly speaking, this number of bits of accuracy grows linearly with p.*

the "truncated best one" by partially compensating (with the other coefficients) for the modification of the order-1 coefficient. Examples of such new approximations are given in Table 4. Similar "compensations" have already been done by Piñeiro, Bruguera and Muller [8]. From an existing polynomial approximation $a_0 + a_1 x + a_2 x^2$ to some function f, they round a_1 to the nearest k-bit number a_1^* and recompute a new polynomial approximation $a_0^* + a_1^* x + a_2^* x^2$ by noticing that

$$a_0^* + a_2^* X \approx f(\sqrt{X}) - a_1^* \sqrt{X}$$

where $X = x^2$, and computing a minimax computation of an approximation to $f(\sqrt{X}) - a_1^* \sqrt{X}$.

Here, we will show that there is no need to compute again a minimax approximation (a_0^* and a_2^* are easily deducible from a_0, a_2 and $a_1 - a_1^*$), and we will be able to predict how much accuracy is saved by such a compensation: around 3 bits.

2 Accurate "truncated" order-2 approximations

We aim at building degree-2 polynomial approximations to some regular enough function f, for which the coefficients of degree 1 are representable with a very small number of bits only. Let x be the input value. We assume that x is represented with n bits in fixed point, and is between 0 and 1. Let us denote $0.x_1 x_2 \ldots x_n$ this representation.

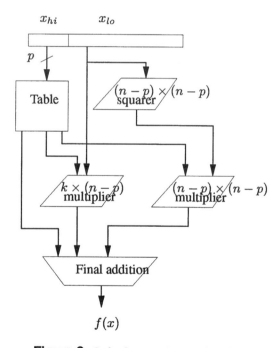

Figure 2. *Order-2 approximation (see for instance [8].*

Fig. 2 shows the main blocks of an architecture implementing an order-2 approximation. The most p significant bits of x are used as address bits, to look up in a table a degree-2 approximation to $f(x)$ in the interval $[h, h+2^{-p}]$, where $h = 0.x_1 x_2 \ldots x_p$. Define $\ell = x - h = 0.000 \cdots 0 x_{p+1} x_{p+2} \cdots x_n$. To design a suitable approximation, we will start from the degree-2 minimax approximation to f in $[h, h+2^{-p}]$, expressed as a function of ℓ:

$$P(\ell) = a_0 + a_1 \ell + a_2 \ell^2 \approx f(x).$$

Define $\epsilon_{\text{minimax}}$ as the error of this approximation, that is:

$$\epsilon_{\text{minimax}} = \max_{\ell \in [0, 2^{-p}]} |P(\ell) - f(h + \ell)|.$$

We wish to compute (and then to store in a table) a slightly different polynomial approximation to f, for which the degree-1 coefficient has a binary representation with a small number, say k, of bits. Let us denote

$$P^*(\ell) = a_0^* + a_1^* \ell + a_2^* \ell^2$$

that new approximation. We wish $P^*(\ell)$ to be as close as possible to $P(\ell)$ for $\ell \in [0, 2^{-p}]$. This means

$$(a_1 - a_1^*)\ell \approx (a_0^* - a_0) + (a_2^* - a_2)\ell^2$$
$$\text{for } \ell \in [0, 2^{-p}].$$

Our method consists in first choosing a_1^* as the k-bit number that is closest to a_1. By doing that, we now have to find an approximation

$$\delta_0 + \delta_2 \ell^2$$

to $(a_1 - a_1^*)\ell$. The coefficients a_0^* and a_2^* will be obtained by adding δ_0 and δ_2 to a_0 and a_2, respectively. Define $L = \ell^2$. Our problem reduces to finding in the interval $[0, 2^{-2p}]$ (i.e., the interval where L lies) an order-1 approximation to $(a_1 - a_1^*)\sqrt{L}$. Such an approximation is obtained by multiplying by $(a_1 - a_1^*)$ an approximation to $\sqrt{L}$. Hence, in the next section, we get minimax approximations to the square root function.

2.1 Order-1 minimax approximations to the square-root function

Concerning degree-1 approximations to the square root, there is no need to run Remez' algorithm. Theorem 1 makes it possible to directly get minimax approximations.

Theorem 2 *The degree-1 minimax approximation to $\sqrt{L}$ in the interval $[0, 2^{-2p}]$ is*

$$2^{-p-3} + 2^p L,$$

and the error of this approximation is 2^{-p-3}.

Proof. From Theorem 1, the maximum distance between the linear approximation and the square root is reached at 3

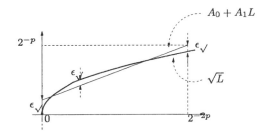

Figure 3. *Minimax order-1 approximation to the square-root in the interval $[0, 2^{-2p}]$.*

points. The concavity of the square root function implies that two of these points are 0 and 2^{-2p} (see Figure 3). Let us call α the third point. Let us denote $A_0 + A_1 L$ the linear approximation, and $\epsilon_\sqrt{}$ the error of approximation. We have

$$\begin{cases} A_0 &= \epsilon_\sqrt{} \\ A_0 + A_1 2^{-2p} - \sqrt{2^{-2p}} &= \epsilon_\sqrt{} \\ \sqrt{\alpha} - A_0 - A_1 \alpha &= \epsilon_\sqrt{} \end{cases} \quad (1)$$

Moreover, since function $\sqrt{L} - A_0 - A_1 L$ reaches its maximum value at $L = \alpha$, the derivative of this function is zero at this point. Therefore

$$\frac{1}{2\sqrt{\alpha}} - A_1 = 0 \quad (2)$$

Elementary calculation from (1) and (2) gives the result.

2.2 Coefficients and error bounds

The previous two subsections allow us to get the coefficients of P^*. These coefficients a_0^*, a_1^* and a_2^* are

$$\begin{cases} a_1^* &= a_1 \text{ rounded to } k \text{ bits} \\ a_0^* &= a_0 + (a_1 - a_1^*) 2^{-p-3} \\ a_2^* &= a_2 + (a_1 - a_1^*) 2^p \end{cases} \quad (3)$$

and the approximation error is upper-bounded by[2]

$$\begin{aligned} \epsilon_{\text{method}} &= \epsilon_{\text{minimax}} + |a_1 - a_1^*| \epsilon_\sqrt{} \\ &= \epsilon_{\text{minimax}} + |a_1 - a_1^*| 2^{-p-3} \end{aligned} \quad (4)$$

Now, we can easily get a bound on the error committed without using our method if we just round a_1 in the initial

---

[2]Of course (5) is an upper bound, and to get a tighter error bound, it is much preferable to directly calculate

$$\max_{\ell \in [0, 2^{-p}]} |P^*(\ell) - f(h + \ell)|.$$

approximation. The error will be $\epsilon_{\text{minimax}}$ plus the maximum value of $|a_1 - a_1^*|\ell$, that is

$$\epsilon_{\text{round}} = \epsilon_{\text{minimax}} + |a_1 - a_1^*|2^{-p} \quad (5)$$

This shows that when $\epsilon_{\text{minimax}}$ is much smaller than $|a_1 - a_1^*|2^{-p}$ (which happens in all practical cases), our method is 8 times more accurate than the naive rounding of coefficient a_1. Our strategy saves three bits of accuracy. This clearly appears in Figures 1, 2 and 3. In these figures, we have compared, for some very common functions (sine, exponential and $\log(1+x)$), the errors of the standard degree-2 minimax approximation ("exact", i.e., not truncated coefficients), the "rounded" minimax approximation (the order-1 coefficient is rounded to k bits), and our method. We also put the error of the order-1 minimax approximation (with "exact" coefficients). Table 4 gives the obtained coefficients for the exponential function with $p = k = 4$.

3 Using these results

We now give some examples that show how can our approximation method be used. We also compare the obtained results with some examples presented in the literature.

3.1 Exponential function with $p = 6$ and $k = 8$

Consider the case of the exponential function with $p = 6$ and $k = 8$ (that is, the table will contain 64 elements, and a_1^* will be an 8-bit number). Table 2 shows that the accuracy of approximation of the polynomials generated by our method is 18 bits. Hence, the final accuracy of an implementation, due to the rounding of a_0^*, a_2^* (and possibly ℓ in the squaring) cannot be better than or equal to 18 bits. Let us try to achieve 17 bits. To do that, let us try to make the error on the computation of $a_0^* + a_2^*\ell^2$ less than 2^{-18}, as follows:

- a_0^* will be rounded to the nearest number exactly representable with 18 fractional bits, say $\hat{a_0^*}$. This will give

$$\left|\hat{a_0^*} - a_0^*\right| \leq 2^{-19}$$

- we have to make sure that the computed value of $a_2^*\ell^2$ is at a distance from the exact value that is less than 2^{-19}.

The first question that should be addressed is how many bits of a_2^* do we keep, and what is the required accuracy when computing ℓ^2. Define $\hat{a_2^*}$ as a_2 rounded to some k' fractional bits. The number ℓ is less than 2^{-6}. Let ϵ be the error on ℓ^2 (either due to the fact that we truncate ℓ before computing ℓ^2, or to the fact that we truncate ℓ^2 before multiplying it by $\hat{a_2^*}$). From (3), the largest value of a_2^* is less than 2.

We have

$$\left|a_2^*\ell^2 - \hat{a_2^*}(\ell+\epsilon)^2\right| \approx \left|a_2^* - \hat{a_2^*}\right|\ell^2 + 2\hat{a_2^*}\epsilon\ell$$
$$\leq 2^{-12-k'-1} + 4\epsilon\ell.$$

To make this value less than 2^{-19} it suffices to choose $k' = 7$ and $\epsilon \leq 2^{-16}$. Again, to get $\epsilon \leq 2^{-16}$, it suffices to keep 8 bits of ℓ. Therefore, for each of the $2^8 = 64$ subintervals, the number of bits that must be stored is:

- 18 for a_0^*;
- 8 for a_1^*;
- 8 for a_2^* (7 for the fractional part, and 1 for the integer part).

Hence, to get a final accuracy of 17 bits, our method will require a table of $(18+8+8) \times 2^8$ bits = 1088 bytes. To get a similar accuracy, the bipartite method would require around 12 Kbytes of table.

3.2 Sine function with $p = 8$ and $k = 10$

A very similar calculation, with $p = 8$, $k = 10$, the sine function in $[0, 1]$ and the figures given by Table 1 shows that we can achieve 21 bits of accuracy with 7 stored bits for a_2^* and 22 stored bits for a_0^*. This leads to a table of 1.184 Kbytes. We can compare this figure with the best known multipartite decomposition, suggested by De Dinechin and Tisserand [3], who achieve 16 bits of accuracy with a table of similar size. And yet, our design requires the additional delay[3] of a 10×17 bit multiplication. By the way, that delay can be reduced to the delay of 5 additions if a_1^* is stored booth-recoded (to do that, we need 15 bits instead of 10 to store a_1^*).

3.3 Getting seed-values for Newton-Raphson division

The original bipartite method was designed in order to generate seed values (initial approximations to the reciprocal of a number) for Newton-Raphson division. We can as well use our method to generate accurate reciprocal approximations at low cost.

For instance, for reciprocals of mantissas of floating-point numbers (this reduces to $f = 1/(1+x)$ for $x \in [0, 1)$), the choice $p = 3$ and $k = 4$ makes it possible to get an accuracy of more than 10 bits with an extremely small table (40 bytes) and very small multiplications (4 bits of a_1^* and 4 bits of a_2^* do suffice).

---
[3]The multiplication that computes ℓ^2 is done in parallel with the table lookup.

Table 1. *Accuracy of various degree-2 approximations (expressed in number of bits) to the sine function in $[0, 1]$, assuming various values of p (number of subintervals) and k (size of a_1^*). We compare the errors of the standard degree-2 minimax approximation, called here "best possible" (no limitation to the size of a_1), the "rounded" minimax approximation (a_1 is rounded to k bits and the other coefficients remain unchanged), our method and the degree-1 minimax approximation.*

p	k	best possible degree 2	rounded	our method	best possible degree 1
4	3	19.58	8.00	11.00	12.28
	4		9.00	11.99	
	5		10.05	13.04	
	6		11.06	14.03	
	7		12.43	15.36	
6	6	25.58	13.00	16.00	16.26
	7		14.00	17.00	
	8		15.01	18.00	
	10		17.01	19.99	
	12		19.06	21.93	
8	8	31.58	17.00	20.00	20.25
	10		19.00	22.00	
	12		21.00	23.99	
	14		23.01	25.99	
10					24.25

Table 2. *Accuracy of various degree-2 approximations (expressed in number of bits) to the exponential function in $[0, 1]$, assuming various values of p and k.*

p	k	best possible degree 2	rounded	our method	best possible degree 1
4	4	18.18	7.10	10.10	10.60
	5		8.24	11.23	
	6		9.44	12.41	
5	4	21.16	8.09	11.09	14.57
	5		9.08	12.08	
	6		10.31	13.30	
8	8	30.14	15.00	18.00	18.56
	10		17.04	20.04	
	12		19.06	22.06	
10					22.55

Table 3. *Accuracy of various degree-2 approximations (expressed in number of bits) to $\log(1+x)$ in $[0,1]$, assuming various values of p and k.*

p	k	best possible degree 2	rounded	our method	best possible degree 1
4	4	18.71	9.06	12.05	12.08
	5		10.03	13.03	
	6		11.02	14.00	
6	6	24.61	13.02	16.02	16.02
	7		14.00	17.00	
	8		15.02	18.01	
8	8	30.59	17.00	20.00	20.00
	10		19.00	22.00	

Table 4. *Coefficients of the degree-2 approximation to the exponential function in $[0,1]$ that corresponds to $p=4$ (i.e., 16 subintervals) and $k=4$.*

interval	degree 0	degree 1	degree 2
$[0, \frac{1}{16}]$	0.111111111111111111110	1.000	0.100000101000
$[\frac{1}{16}, \frac{1}{8}]$	1.000100000011000111010	1.001	-0.011011001110
$[\frac{1}{8}, \frac{3}{16}]$	1.001000100000110100001	1.001	0.101101010101
$[\frac{3}{16}, \frac{1}{4}]$	1.001101001011010011111	1.010	-0.000101011101
$[\frac{1}{4}, \frac{5}{16}]$	1.010010001100011100010	1.010	1.001100110000
$[\frac{5}{16}, \frac{3}{8}]$	1.010111011110010011100	1.011	0.100100010000
$[\frac{3}{8}, \frac{7}{16}]$	1.011101000110001100010	1.100	0.000001011001
$[\frac{7}{16}, \frac{1}{2}]$	1.100011001001100100000	1.100	1.100100100011
$[\frac{1}{2}, \frac{9}{16}]$	1.101001100001111010001	1.101	1.001110000110
$[\frac{9}{16}, \frac{5}{8}]$	1.110000010100110110111	1.110	0.111110011101
$[\frac{5}{8}, \frac{11}{16}]$	1.110111100100000111010	1.111	0.110110000011
$[\frac{11}{16}, \frac{3}{4}]$	1.111111010001011111111	10.00	0.110101011000
$[\frac{3}{4}, \frac{13}{16}]$	10.000111100010111111011	10.00	10.111100111001
$[\frac{13}{16}, \frac{7}{8}]$	10.010000001110100101010	10.01	1.0011010010101
$[\frac{7}{8}, \frac{15}{16}]$	10.011001011110100101001	10.10	-0.0110010100010
$[\frac{15}{16}, 1]$	10.100011011101001011010	10.10	10.0010100011011

Conclusion

We have suggested a way of partially compensating for the loss of accuracy due to truncation or rounding of the order-1 coefficient of a polynomial approximation to some function. Our method can be used for designing hardware implementation of functions that require much smaller tables than the bipartite (and, more generally, than the order-1) methods, and that only need small arithmetic operators.

References

[1] Elisardo Antelo, Tomas Lang, and Javier D. Bruguera. Very-high radix circular CORDIC: Vectoring and unified Rotation/Vectoring. *IEEE Transactions on Computers*, 49(7):727–739, July 2000.

[2] P. W. Baker. Suggestion for a fast binary sine/cosine generator. *IEEE Transactions on Computers*, C-25(11), November 1976.

[3] F. de Dinechin and A. Tisserand. Some improvements on multipartite table methods. In Burgess and Ciminiera, editors, *Proc. of the 15th IEEE Symposium on Computer Arithmetic (Arith-15)*. IEEE Computer Society Press, 2001.

[4] J. F. Hart, E. W. Cheney, C. L. Lawson, H. J. Maehly, C. K. Mesztenyi, J. R. Rice, H. G. Thacher, and C. Witzgall. *Computer Approximations*. Wiley, New York, 1968.

[5] B. Wei J. Cao and J. Cheng. High-performance architectures for elementary function generation. In Burgess and Ciminiera, editors, *Proc. of the 15th IEEE Symposium on Computer Arithmetic (Arith-15)*. IEEE Computer Society Press, 2001.

[6] J.M. Muller. *Elementary Functions, Algorithms and Implementation*. Birkhauser, Boston, 1997.

[7] J.M. Muller. A few results on table-based methods. *Reliable Computing*, 5(3), 1999.

[8] J.A. Pineiro, J.D. Bruguera, and J.M. Muller. Faithful powering computation using table look-up and a fused accumulation tree. In Burgess and Ciminiera, editors, *Proc. of the 15th IEEE Symposium on Computer Arithmetic (Arith-15)*. IEEE Computer Society Press, 2001.

[9] E. Remez. Sur un procédé convergent d'approximations successives pour déterminer les polynômes d'approximation. *C.R. Académie des Sciences, Paris*, 198, 1934.

[10] D. Das Sarma and D. W. Matula. Faithful bipartite rom reciprocal tables. In S. Knowles and W. McAllister, editors, *Proceedings of the 12th IEEE Symposium on Computer Arithmetic*, Bath, UK, July 1995. IEEE Computer Society Press, Los Alamitos, CA.

[11] M. Schulte and J. Stine. Symmetric bipartite tables for accurate function approximation. In In T. Lang, J.M. Muller, and N. Takagi, editors, *Proceedings of the 13th IEEE Symposium on Computer Arithmetic*. IEEE Computer Society Press, Los Alamitos, CA, 1997.

[12] Michael J. Schulte and James E. Stine. Accurate function evaluation by symmetric table lookup and addition. In Thiele, Fortes, Vissers, Taylor, Noll, and Teich, editors, *Proceedings of the IEEE International Conference on Application-Specific Systems, Architectures and Processors (Zurich, Switzerland)*, pages 144–153. IEEE Computer Society Press, 1997.

[13] M.J. Schulte and J.E. Stine. Approximating elementary functions with symmetric bipartite tables. *IEEE Transactions on Computers*, 48(8):842–847, Aug. 1999.

[14] James E. Stine and Michael J. Schulte. The symmetric table addition method for accurate function approximation. *Journal of VLSI Signal Processing*, 21:167–177, 1999.

[15] J. Volder. The CORDIC computing technique. *IRE Transactions on Electronic Computers*, EC-8(3):330–334, 1959. Reprinted in E. E. Swartzlander, *Computer Arithmetic*, Vol. 1, IEEE Computer Society Press Tutorial, Los Alamitos, CA, 1990.

[16] Jack E. Volder. The birth of CORDIC. *Journal of VLSI Signal Processing Systems*, 25(2):101–105, June 2000.

[17] J. Walther. A unified algorithm for elementary functions. In *Joint Computer Conference Proceedings*, 1971. Reprinted in E. E. Swartzlander, *Computer Arithmetic*, Vol. 1, IEEE Computer Society Press Tutorial, Los Alamitos, CA, 1990.

An Overview of Floating-Point Support and Math Library on the Intel® XScale^{TM} Architecture

Cristina Iordache and Ping Tak Peter Tang
Intel Corporation
cristina.s.iordache@intel.com, peter.tang@intel.com

Abstract

New microprocessor architectures often require software support for basic arithmetic operations such as divide, or square root. The Intel® XScale^{TM} processor, designed for low power mobile devices, provides no hardware support for floating-point. We show that an efficient software implementation of the basic operations and math library routines can achieve competitive performance, and effectively hide the lack of hardware floating-point for most applications.

1. Introduction

The Intel® XScale^{TM} processor is a 32-bit RISC microarchitecture based on the architecture by Advanced RISC Machines (ARM*). Unlike processors to be used in general purpose computing such as PC or enterprise servers, the target of the Intel® XScale^{TM} processors are embedded platforms. These include high-end mobile telephones, PDAs, communicators, and wireless Web browsers.

Because of the large number of possible platforms as well as the need for software environment for product development on these processors, floating-point support for the XScale^{TM} processor (which does not have an FP unit) is indispensable. Although floating-point support on integer based processors is not new, the much enhanced performance and instruction set of this new generation of integer processors do bring the whole effort of floating-point support to a new level.

The purpose of this paper is to illustrate that through careful utilization of the XScale^{TM} microarchitecture and customized algorithms, floating-point performance on these low-power integer-based processors can rival that of PCs of just a few years ago. The outline of the paper is as follows. Section 2 discusses some of the important instructions we used in support of floating point. Section 3 discusses the emulation of IEEE single and double precision basic operations. Section 4 discusses the support of a floating-point run-time library with emphasis on the algorithmic methodology as well as software architecture. Section 5 provides timing results of these floating-point components and some overall observations.

2. The XScale^{TM} Integer Instruction Set

The Intel®XScale^{TM} processor complies with the ARM* V5TE architecture specifications. It implements the integer instruction set of ARM* Version 5, the Thumb instruction set (ARM* V5T), and the ARM* V5E DSP extensions.

The ARM* integer instruction set includes 32-bit logical, arithmetic, and test instructions (see table 1 for a list of the most important instructions, and [1] for more complete information). The destination register is specified explicitly in standard ARM mode, and thus does not need to be one of the operand registers. All of these data processing instructions have an option to update the status flags in the CPSR (Current Program Status Register) according to the result of the operation. For most instructions, execution can be conditional based on status flag values, by adding an appropriate suffix. For example:

// $r0 = r1 - r2$, and update status flags:
sub$s$ $r0, r1, r2$
// Set $r3 = 2$ if Carry flag=1, Zero flag =0 (i.e. $r0 > 0$):
movhi $r3, \#2$
// Set $r3 = 1$ if Zero flag = 1 (i.e. $r0 == 0$):
moveq $r3, \#1$
// Set $r3 = 0$ if Carry flag =0 (i.e. $r0 < 0$):
movcc $r3, \#0$

Conditional execution helps improve performance by eliminating branches and branch penalties (where appropriate), and at the same time reduces code size.

Most register-to-register instructions take 1 cycle on XScale^{TM}. The most notable exceptions are the multiply and multiply-add instructions, which have latencies between 2 and 6 cycles, depending on the range of the multiplier, and a resource latency (throughput) only one cycle shorter than the instruction latency. Long multiply and

---
*Other brands and names are property of their respective owner

multiply-add instructions that provide the full 64-bit product of 32-bit operands are helpful in the development of floating-point simulation code.

One distinctive feature of many ARM* V5 instructions is the ability to use a shift amount for the second operand, e.g.:

// $r0 = r1 + (r2 \gg 6)$:
add $r0, r1, r2$, LSR #6
// $r3 = r0 + (r2 \gg r4)$:
sub $r3, r0, r2$, LSL $r4$

This eliminates the need for separate shifts and can save code space, used register space (if the second operand value is reused), and can improve overall performance.

A related feature is the ability to rotate the carry flag into the destination register, or one of the source register bits into the carry flag.

Saving and restoring registers to/from memory is relatively expensive and should be avoided, if possible. It takes at least $4 + 2 * N$ cycles to save and restore N registers. The software conventions allow 5 scratch registers ($r0$, $r1$, $r2$, $r3$, $r12$); the values of these registers can be modified by the called routine, without saving them first. 15 general purpose registers are visible at any one time; one of them ($r13$) is defined as the stack pointer.

Table 1. Main Types of ARM* Arithmetic and Logical Instructions

Move	$Rd := Op_2$
Move NOT	$Rd := $ 0xffffffff XOR Op_2
Add	$Rd := Rn + Op_2$
Add with carry	$Rd := Rn + Op_2 + Carry$
Subtract	$Rd := Rn - Op_2$
Subtract with carry	$Rd := Rn - Op_2 - NOT(Carry)$
Reverse SUB	$Rd := Op_2 - Rn$
Reverse SUB w. carry	$Rd := Op_2 - Rn - NOT(Carry)$
$32 \times 32 \to 32$ multiply	$Rd := (Rm * Rs)$ [31:0]
$32 \times 32 \to 32$ mul.-add	$Rd := ((Rm * Rs) + Rn)$[31:0]
Long signed multiply	$RdHi, RdLo := (Rm * Rs)$
Long unsigned multiply	$RdHi, RdLo := (Rm * Rs)$
Long signed mul.-add	$RdHi, RdLo := (RdHi, RdLo) + Rm * Rs$
Long unsigned mul.-add	$RdHi, RdLo := (RdHi, RdLo) + Rm * Rs$
Count leading zeroes	$Rd := $ count of leading zeroes in Rm
Logical AND	$Rd := Rn$ AND Op_2
Logical XOR	$Rd := Rn$ EOR Op_2
Logical OR	$Rd := Rn$ EOR Op_2
Logical Bit Clear	$Rd := Rn$ AND NOT Op_2
Test logical	Update flags on Rn AND Op_2
Test equivalence	Update flags on Rn XOR Op_2
Arithmetic compare	Update flags on $Rn - Op_2$
Negative compare	Update flags on $Rn + Op_2$

3. Basic FP Operations

The compiler generates a routine call for each basic floating-point operation (add, subtract, multiply, divide). The arguments and the result are passed in integer registers, as specified by software conventions. For single precision, the operands are passed in $r0$ and $r1$, and the result is returned in $r0$. For double precision, the operands are passed in $(r0, r1)$, $(r2, r3)$, and the result is returned in $(r0, r1)$. These registers hold the memory representation format of the floating-point values used.

Typically, the argument values are unpacked, the significands and exponents are used in separate computations, and then combined to form the result at the end. Special cases (infinities, NaNs, denormals, overflow, underflow) are eliminated from the main path as quickly as possible, and processed separately. Our main goal was to speed up normal cases.

3.1. The Add/Subtract Routines

The implementations of these simple operations take advantage of the special characteristics of the ARM* architecture.

In all normal cases, the sign is determined as the sign of the argument larger in absolute value. The arguments can be swapped if necessary, so that the first one (A) determines the sign of the result.

The single precision routine does not need to completely separate the argument exponents (which also helps by reducing the number of used registers). Given that only the lower 8 bits of a register specify the shift amount, the following value, obtained by rotating the initial arguments, can be used to correctly add the significands:

// $r1 = xx...x$ s_B exp_B ($r1$ was exp_B $xx...x$ s_B):
mov $r1, r1$, ROR #24
// $r1 = xx...x$ $(s_A\text{-}s_B)$ $(exp_A\text{-}exp_B)$, $exp_A \geq exp_B$:
rsb $r1, r1, r0$, ROR #24
// test bit 8 of $r1$ $(s_A - s_B)$, to see if signs are different:
tst $r1$, #0x100
// change sign of significand B, if bit 8 is set in $r1$:
rsbne $r3, r3$, #0
// $significand = signif_A + (signif_B \gg (exp_A - exp_B))$:
add $r2, r2, r3$, ASR $r1$

For rounding, the last significand bit (l) and the round bit (r) are tested by rotation to the carry flag, and the $r = 0$ and $l = r = 1$ cases are eliminated early. The sticky bit is computed only for $l = 0, r = 1$.

The double precision routine uses similar techniques. Given that the significands cannot fit in just one 32-bit register, further speedup is obtained by selecting different code paths based on the operation (add or subtract), and the exponent difference (shift amount for the second significand).

3.2. Multiply

The single precision multiply routine is based on the long $32 \times 32 \to 64$ multiply instruction (UMULL) and is relatively straightforward. The input significands are scaled by

2^8, so that their product is in the range $[2^{62}, 2^{64})$. The output significand and the round bit are then obtained by shifting right the upper half (upper register) of the product. The sticky bit is the OR sum of the remaining bits.

The double precision multiply routine uses the long multiply and multiply-add instructions (UMULL, UMLAL) to get the full 106-bit product.

3.3. Divide

3.3.1. Single Precision Divide.
The single precision divide algorithm uses an 8-bits-in, 8-bits-out reciprocal lookup table, and computes the leading 25 bits of the quotient in 5 iterations, followed by one simple correction.

As usual, the sign and exponent are computed separately, and the significand of the result is obtained by rounding the quotient of the scaled argument significands A/B, where $A = 1a_1a_2\ldots a_{23}$, and $B = 1b_1b_2\ldots b_{23}$.

The reciprocal approximation $B_r = rcp\_table[b_1b_2\ldots b_8]$ is such that $B * B_r = 2^{31} \cdot (1 - \epsilon)$, where $\epsilon \in [0, \frac{1}{2^8+1})$.

Given that $A/B < 2$, we have $A \cdot B_r = (A/B) \cdot (B \cdot B_r) < 2^{32} \cdot (1 - \epsilon)$. The quotient will be computed as $(A \cdot B_r)/(B \cdot B_r)$.

Let $A_{r,0} = A \cdot B_r$, and let us assume that after each iteration, we have $A_{r,j} < 2^{32-5 \cdot j} \cdot (1 - \epsilon)$. (This is true for $A_{r,0}$.) Then we also have
$$A_{r,j} = 2^{31-5 \cdot j - 5} \cdot (d_j + \alpha_j),$$
where $d_j < 2^6$ is an integer, and $\alpha_j \in [0, 1)$.

Then $A_{r,j+1} = A_{r,j} - (d_j \cdot 2^{-5}) \cdot (B \cdot B_r) = 2^{31-5 \cdot (j+1)} \cdot (d_j \cdot \epsilon + \alpha_j)$.

Since $\epsilon \in [0, \frac{1}{2^6+1})$, we have
$$A_{r,j+1} < 2^{32-5 \cdot (j+1)} \cdot (1 - \epsilon).$$

Now let $R = 2^{31} - B \cdot B_r$, $R < 2^{31-6}$. The following iteration, repeated 5 times, computes the quotient to about 25 bits of accuracy:
$$A_{tmp} = (A_{r,j} \ll 5) \text{ AND } Mask,$$
where $Mask = 2^{31} - 1$
$$d_j = A_{r,j} \gg (31 - k)$$
$$A_{r,j+1} = A_{tmp} + R \cdot d_j$$
$$Q_{j+1} = (Q_j \ll 5) + d_j$$

Before rounding, one more step is needed to ensure that the leading 25 bits of the quotient are all correct:

If $(A_{r,5} >= B \cdot B_r)$
then $Q_5 = Q_5 + 1$

The exact round bit is bit 25 of the quotient. For normal cases, round-to-nearest can be performed correctly based on the round bit value only, given that the quotient cannot fall at a midpoint between two floating-point values of the same precision as the argument.

3.3.2. Double Precision Divide.
The double precision divide algorithm computes $1/b$ to over 12 bits of accuracy with a bipartite scheme using a 7-bits-in, 16-bits-out table, and a 7-bits-in, 8-bits-out table. 5 iterations followed by a simple correction are used to compute the first 55 significant bits of the quotient. Bipartite table reciprocal computation schemes were first proposed in [2], [3], [4].

Let $B = 1b_1b_2\ldots b_{52}$ be the significand of the divisor scaled by 2^{52}, and let $f = b_8b_9\ldots b_{14}$.

A scaled reciprocal approximation is computed as follows:
$B_r = T_H + 2^{15} \cdot (1 - 2^{-7} - f \cdot 2^{-7}) \cdot T_L$, where
$T_H = \lfloor \frac{2^{16}}{1.b_1\ldots b_7+2^{-7}} \rfloor \cdot 2^7$, and
$T_L = \lfloor \frac{2^8}{(1.b_1\ldots b_7+2^{-7})^2} \rfloor$.

We then have $B \cdot B_r = 2^{75} \cdot (1 - \epsilon)$, and $\epsilon \in [0, \frac{1}{2^{12}+1})$.

Let $R = 2^{75} - B \cdot B_r = 2^{75} \cdot \epsilon$. Note that $R < 2^{75-12} = 2^{63}$, so it can be represented using 2 32-bit registers.

The quotient significand is computed as $A \cdot B_r / B \cdot B_r$, in a manner similar to our single precision algorithm:
$A_{r,0} = A \cdot B_r < 2^{76} \cdot (1 - \epsilon)$, and we can show that after each iteration, the remainder is $A_{r,j} < 2^{76-11 \cdot j} \cdot (1 - \epsilon)$:
$A_{r,j} = 2^{75-11 \cdot (j+1)} \cdot (d_j + \alpha_j)$, $d_j < 2^{12}$, and $\alpha_j \in [0, 1)$.

Then $A_{r,j+1} = A_{r,j} - (d_j \cdot 2^{-11}) \cdot (B \cdot B_r) = 2^{75-11 \cdot (j+1)} \cdot (d_j \cdot \epsilon + \alpha_j)$, so $A_{r,j+1} < 2^{76-11 \cdot (j+1)} \cdot (1 - \epsilon)$.

Note that $d_j = \lfloor \frac{A_{r,j} \cdot 2^{11 \cdot j}}{2^{64}} \rfloor$. $A_{r,j}$ is 76 bits long and requires 3 32-bit registers for representation, but d_j can be obtained directly (with no additional shifting) from the register that holds the leading 12 bits of $A_{r,j}$. $A_{r,j+1} \cdot 2^{11 \cdot (j+1)}$ is obtained by shifting the remaining 2 registers 11 positions to the left, and adding $R \cdot d_j$:
$$A'_{j+1} = (A'_{j,low} \ll 11) + R \cdot d_j$$
$$Q_{j+1} = (Q_j \ll 11) + d_j$$

After the fifth iteration, a final correction ensures that the leading 55 bits of the quotient are all correct:

If $A'_5 > B \cdot B_r$
then $Q_5 = Q_5 + 1$

Rounding to nearest is performed based on the value of the round bit (bit 54). The sticky bit is not needed, since it is always 1, with the exception of exact cases.

3.3.3. 32-bit Integer Divide.
Integer divide is implemented using a 6-bits-in, 8-bits-out table that provides a short reciprocal approximation for the divisor. The quotient is computed iteratively (about 6 bits per iteration). A special path is used for short quotients (below 2^7), which are computed 1 bit at a time.

In this case as well, we compute the quotient as $Q = (A \cdot B_r)/(B \cdot B_r)$, where $B \cdot B_r$ is close to 2^{32}. B_r is obtained by scaling a table value that approximates $2^{14}/1b_1b_2\ldots b_6$ (the leading 7 bits of B). Given that the table is accurate to at least 5.75 bits, we have
$$R = 2^{32} - B \cdot B_r \in [0, 2^{32-5.75}).$$

Let $A \cdot B_r = A_{0H} \cdot 2^{32} + A_{0L}$, with A_{0L} an integer in $[0, 2^{32})$.

We have $Q = \frac{A_{0H} \cdot 2^{32} + A_{0L}}{2^{32} - R}$, so clearly $Q \geq A_{0H}$. If we set $Q_0 = A_{0H}$ in a first iteration, the remainder is
$A \cdot B_r - A_{0H} \cdot (2^{32} - R) = A_{0L} + R \cdot A_{0H} = A_{1H} \cdot 2^{32} + A_{1L}$

For any $B \in [5, 17)$, the table entries are such that $R \leq 2^{32-6}$. Given that $A_{0L} < 2^{32}$, we get $A_{1H} \leq \lceil A_{0H} \cdot 2^{-6} \rceil$.

In a similar manner we can show that after iteration j, the remainder is $A_{jH} \cdot 2^{32} + A_{jL}$, with $A_{jH} \leq \lceil A_{0H} \cdot 2^{-6 \cdot j} \rceil$.

Since $B \geq 5$ and $A_{0H} = \lfloor \frac{A \cdot B_r}{2^{32}} \rfloor < B_r < \frac{2^{32}}{B}$, A_{0H} does not exceed $51 \cdot 2^{24}$, so $A_{4H} \leq 51$. This ensures that the next remainder is $A_{5H} \cdot 2^{32} + A_{5L} < 51 \cdot R + 2^{32} < 2^{32} + (2^{32} - 2 \cdot R)$. The computation can stop after the fifth iteration.

For $B \geq 17$, $A_{0H} \leq \lfloor \frac{2^{32}}{17} \rfloor < 61 \cdot 2^{22}$.
$R < 2^{32 - 5.75}$, and we can show that after iteration j, the remainder is $A_{jH} \cdot 2^{32} + A_{jL}$, with $A_{jH} \leq \lceil A_{0H} \cdot 2^{-5.75 \cdot j} \rceil$.

Then $A_{4H} \leq 31$, and after the fifth iteration, the remainder is $A_{5H} \cdot 2^{32} + A_{5L} < 31 \cdot R + 2^{32} < 2^{32} + (2^{32} - 2 \cdot R)$ (since $R < \frac{2^{32}}{2^{5.75}} < \frac{2^{32}}{33}$).

In this case as well, 5 iterations are sufficient to perform 32-bit integer divide.

A long multiply-add is used to get the next remainder in each iteration:
$A_{jH} = (R \cdot A_{(j-1)H} + A_{(j-1)L})_{high}$
$A_{jL} = (R \cdot A_{(j-1)H} + A_{(j-1)L})_{low}$
The quotient is updated as
$Q_j = Q_{j-1} + A_{jH}$
After the fifth iteration, a final correction is applied:
if $A_{5L} + R > 2^{32}$
then $Q_5 = Q_5 + 1$

4. Transcendental Functions

In theory, once the basic floating-point operations are emulated, any high-level language implementation of floating-point run-time library based on floating-point type can be used to provide the functionality required. However, such an approach has several major drawbacks. First, the performance will be low. Consider for example a floating-point exponential function calculation based on floating-point type. Every basic operation will be translated to a function call. Moreover, a good portion of the work done in each basic floating-point operation emulation is redundant or unnecessary. The floating-point encoding is unpacked and packed at each operations, and most of the effort to maintain IEEE compliant rounding in the intermediate calculations are unnecessary. Second, and less obvious, is that the machine's capability is not fully utilized. The native 32-bit signed integer can carry 31 significant bits of accuracy while IEEE single precision format only offers 24. Similarly, a 64-bit signed integer carries 63 significant bits of accuracy while IEEE double precision only carries 53. Because of the high potential in performance advantage, a set of integer-based implementation of floating-point transcendental function was developed.

4.1. Methodology

The natural accuracy characteristic of floating-point arithmetic is relative accuracy, while that of integer arithmetic is absolute accuracy. A fundamental question about using integer arithmetic to implement a floating-point function is whether absolute accuracy suffices for a core part of the calculation. When absolute accuracy suffices, integer computation can be applied in a most natural and efficient manner. In the computation of an elementary transcendental function, one important step is often the computation of the function near a special point such as a root. That is, we need to approximate the underlying transcendental function by computing a function f of the form

$$f(x) = x^p(c_0 + c_1 x + c_2 x^2 + \cdots + c_n x^n),$$

where $|x| \ll 1$. The right hand side can be expressed as

$$g(x) + x^n \frac{f(x) - g(x)}{x^n} = g(x) + x^n h(x).$$

Here $g(x)$ is generally a leading portion of $f(x)$ such as the first term or two of $f(x)$, and n is no bigger than the leading exponent of the function $f(x) - g(x)$. With appropriate choices of $g(x)$ and n, $h(x)$ needs only to be calculated to a prescribed absolute accuracy in order that the final expression carries enough relative accuracy. This is because the magnitude of f is usually comparable to that of h. x^n appears as a scale factor on h, scaling any absolute error there into a relative error with respect to f. This kind of decomposition is made in each transcendental function we implemented; and we describe the $h(x)$ functions as our absolute accuracy core. The decomposition is by no means unique, and can be traded off with performance and accuracy. Some specific forms of $h(x)$ are

$$\frac{e^x - (1+x)}{x^2}, \frac{\sin(x) - x}{x^2}, \frac{\cos(x) - 1}{x^2}.$$

Closely related to this absolute accuracy core is a fixed-point computation of a polynomial. This is best illustrated by the simple case of computing via Horner's recurrence:

$$c_0 + x \times (c_1 + x \times (c_2 + \cdots x \times c_n))$$

Suppose the variable x and coefficients are stored in 32-bit integers

$$X \approx 2^{32+d} x, \quad \text{and} \quad C_j \approx 2^s c_j$$

for all j and some $d \geq 0$. Then the following integer computation yields the desired expression with an absolute accuracy comparable to 2^{-32}:

$P := C_N$
for $j = N-1, N-2, \ldots, 0$ do
$\quad P := $ high 32-bit$(P * X)$
$\quad P := C_j + (P \gg d)$
end do

While the computation of the absolute accuracy core in this native integer computation sequence is an important part, there are other general computations where relative accuracy has to be maintained. In general, data items are represented as scaled integers. That is, an integer value I corresponding to the significand is kept. A scaled factor Q is carried either explicitly or implicitly. The convention we adopt is $(I, Q) = I/2^Q$. When two values (I_1, Q_1), (I_2, Q_2) are operated on, the resulting value (I, Q) is obtained. For example in the case of multiplication of 32-bit I's and Q's, the high part of the product is retained and the resulting value is $(I, Q_1 + Q_2 - 32)$. This method more or less preserves 30 bits of relative accuracy.

Finally, the algorithms employed are the well known table driven methods [5]. This class of methods typically separates the function computation process into three stages: argument reduction, core approximation, and final reconstruction.

Argument reduction is performed mostly by integer arithmetic operated on the significand of the input argument. For example, in the calculation of $\exp(x)$, we perform the following computation.

$$\begin{aligned} W &= X \times P \\ &= N + R \end{aligned}$$

where X is the significand of the input value, and P is a 31-bit approximation to $1/\log(2)$ suitably scaled up by a power of 2. The resulting quantities N and R satisfy

$$x/\log(2) \approx N/8 + R/2^{34}.$$

Hence $\exp(x) \approx 2^{N/8} \times 2^r$ where $r = R/2^{34}$. For single precision calculation, the table size is usually chosen so that the reduced argument r is small enough so that $r \times 2^{32+d}$ fits in a 32-bit signed integer for some $d \geq 0$. This will allow easy integer computation of the absolute accuracy core.

The core approximation consists of a simple sequence of integer computations followed by general scaled integer computations (that is computation of the significands together with explicit computation of scale factors). The final reconstruction usually requires general scaled integer computations.

4.2. Implementation

The performance advantage of an integer-based implementation of transcendental function library over one that is based on software emulated basic floating-point operations is quite obvious. However, the implementation effort required can vary depending on the strategy employed.

We decided not to use assembly coding because of the labor intensity in both the initial library creation and its subsequent maintenance. Moreover, a hierarchical method is used in that most functions share some basic routines. For example, natural, base-2, and base-10 logarithms use the same basic underlying routine; inverse sine and inverse cosine make use of the inverse tangent routine. However, to maximize performance, we do not use common routines in the standard sense. Our common "routines" are really macros expanded into common code sequences. Thus, the library is modular and call overhead free within each function. The tradeoff is an increased code size, which is deemed acceptable as the code size of a transcendental function library is not big to begin with.

Here is an example of the single-precision natural logarithm function:

```
sgl_cvt_flt2fix(y, sc_y, x);
sgl_log(n, z, sc_z, y, sc_y);
... ... ... ...
sgl_cvt_fix2flt(y, w, sc_w);
```

The first macro unpacks an IEEE input x into a scaled integer representation. The integers y and sc_y are such that $y/2^{sc\_y}$ is the input value. This macro is used in almost every function in the very beginning. The macro `sgl_log` computes the natural logarithm of an input value represented in scaled integer form and return the result in three integers n, z, and sc_z where

$$\log(y/2^{sc\text{-}y}) \approx n\log(2) + z/2^{sc\text{-}z}.$$

Some native integer code then computes the value on the right hand side in scaled integer format. Finally, a packing macro `sgl_cvt_fix2flt` is used to convert the scaled integer value into an IEEE encoding. Clearly, this packing macro is used extensively.

The macro `sgl_log` is reused for base-2 and base-10 logarithms. We only need to change the calculation following it to yield the base-2 or base-10 logarithm, respectively. Our experience finds that modularity and reuse greatly enhance our efficiency in the overall software development process.

4.3. Example: The Single Precision Exponential Function

The computation of the single precision exponential function is based on the mathematical identity $e^x = 2^{x/\log(2)} =$

$2^{n+j/8+\gamma}$ where $-4 \leq j < 3$ and $|\gamma| \leq 1/16$. Thus

$$e^x = 2^n(2^{j/8} + 2^{j/8}(2^\gamma - 1))$$
$$\approx 2^n(2^{j/8} + 2^{j/8}\gamma p(\gamma))$$

where $p(\gamma)$ is a polynomial that approximates $(2^\gamma - 1)/\gamma$. The 8 possible values of $2^{j/8}$ are computed beforehand and stored in a table. We now outline the key implementation details involved.

The input value x is unpacked and represented by $x = (X, Q_x)$. Because the exponential function has a rather limited range of input (as it overflows and underflows easily), Q_x is quite limited in range.

The constant $1/\log(2)$ is stored as a 32-bit integer with another 8-bit extension thus $1/\log(2) \approx (\text{InvL}, 30) + (\text{dInvL}, 38)$. Using the 32-bit long integer multiplication instruction we compute $X \times \text{InvL}$ and obtain n, j and R such that

$$x \cdot (\text{InvL}/2^{30}) = n + j/8 + (R, 34).$$

Moreover, we multiply just a few leading bits of X to dInvL to obtain a correction dR to R so that $(R, 34) + (dR, 34)$ approximates γ accurately. We modify R to $R + dR$.

Next, a 5-coefficient polynomial is evaluated at R using fixed-point computation: Set $P := C_5$, and for $j = 4, 3, 2, 1$, compute $P := C_j + \langle P \times R \rangle \gg 2$, where $\langle \cdot \rangle$ denotes the high 32 bits of a product. Each of the C_j has a scale factor of 31, and thus in the end

$$(2^r - 1)/r \approx (P, 31), \qquad r = (R, 34).$$

Finally, T_j is fetched from a table where $(T_j, 30) \approx 2^{j/8}$. The value $Y := T_j + \langle T \times \langle R \times P \rangle \gg 2 \rangle$ is computed. Hence $2^n(Y, 30) \approx e^x$. n and Y are (rounded) and packed into IEEE single precision format.

4.4. Example: The Double Precision Natural Logarithm

The computation of the double precision log is based on the mathematical identity

$$log(2^k \cdot x) = k \cdot log(2) + log(x \cdot r) - log(r).$$

In the following, we denote the unbiased exponent of the double precision input as k, and the mantissa by $x = 1.x_1 x_2 \ldots x_{52}$.

For arguments sufficiently far from 1.0, the identity above is applied twice in our computation:

$$log(2^k \cdot x) = k \cdot log(2) - log(r_1) - log(r_2) + log(x \cdot r_1 \cdot r_2).$$

r_1 and r_2 are selected such that $x \cdot r_1 \cdot r_2$ is very close to 1, and so $log(x \cdot r_1 \cdot r_2)$ can be approximated by a short polynomial of argument $t = x \cdot r_1 \cdot r_2 - 1$.

In particular, for $x = (X, 52)$, we select $r_1 = (\lceil \frac{2^{11}}{X \gg 47} \rceil, 6)$. r_1 is a 6-bit value, and can be accessed from a table indexed by $x_1 x_2 \ldots x_5$, the first 5 mantissa bits after the leading 1. We choose to store $-log(r_1)$ as $(T1, 75)$, accurate to at least 70 bits, in a table using the same index.

Clearly, $x \cdot r_1 - 1 \in [0, 2^{-4})$.

Let $x \cdot r_1 = (X1, 58)$, where the binary representation of $X1$ is $10000 b_5 b_6 \ldots b_{58}$. We select $r_2 = (R2, 13)$, where $R2$ is $\frac{2^{13}}{(1.0000 b_5 \ldots b_9 1)_2}$, rounded to the nearest integer. r_2, as well as $-log(r_2)$ can be read from a table indexed by $b_5 \ldots b_9$. $-log(r_2)$ is stored as $(T2, 75)$, to an accuracy of at least 66 bits.

For $t = x \cdot r_1 \cdot r_2 - 1$, we how have $|t| < 2^{-10} + 2^{-13}$, so $log(1+t)$ can be estimated to at least 61 bits of accuracy by a degree 6 polynomial. This is sufficient to ensure that the maximum error of the final result will be less than 0.55 ulp. The high order terms of the polynomial can be safely estimated with 32×32-bit multiplies, while 64×64 multiplies are used where accuracy is critical.

The sum of polynomial terms is scaled to $(P, 75)$, and added to the two table values and the input exponent (also scaled to 75). The result is then converted from scaled integer to floating point format.

For arguments near 1 ($|x-1| < 2^{-10}$), only a polynomial evaluation as described above is used to get the final result.

5. Timings and Conclusion

Our basic single precision operation implementations (add, subtract, multiply, divide) are about 10 times faster than the corresponding GNU implementations, and comparable to the ARM* v2.5 implementations, which are also highly optimized. Our double precision basic operations are 2 to 3 times faster than the ARM* implementations, however, and 10 to 20 times faster than the GNU implementations.

Our square root assembly implementation is about 15 times faster than the GNU implementation. In double precision, our sqrt is about 3 times faster than the ARM* sqrt. In single precision, our routine is more than 7 times faster (also due to the fact that ARM* does not provide specialized single precision math functions, so the single precision result is based on a call to the double precision function).

Since all other math functions are implemented in C, their performance is compiler dependent. The timings provided in this section were obtained for a library built with the Intel Saturn compiler. The compiler generated code is not optimal, but the Saturn generated code for these functions is generally better than ARM* or GNU generated code. An assembly implementation of expf runs in 108 cycles (worst case), while the Saturn generated code for the same algorithm takes about 167 cycles (relatively close). In double precision, the current Saturn generated code for exp takes 525 cycles, while less than 250 are sufficient for an assembly implementation of the same algorithm. Most single precision routines run in 100 to 250 cycles, when the Intel

Saturn compiler is used to build the library. Double precision routines are typically 3 to 4 times slower than their single precision counterparts. We are still working with the Saturn compiler team to improve the performance of their code. All of our math library routines have maximum ulp errors below 0.55.

We found that when used with the Saturn compiler basic numerics library, the performance of the GNU libm (which calls the basic FP op routines), improved 8 to 10 times. Our integer-based libm implementation is still at least 4 times faster.

Table 2. Table of latencies

Function	Latency [cycles] Single Precision	Latency [cycles] Double Precision
Add	34	59
Multiply	35	46
Divide	57	118
SQRT	84	183
ASIN	250	520
EXP	167	525
LOG	215	433
SIN	181	607

In conclusion, given the relatively low latencies provided by software floating-point support, non floating-point extensive applications can achieve performance as if floating-point is natively supported in hardware.

References

[1] ARM* Developer Suite, *Assembler Guide*, ARM Limited, 2000.

[2] D. DasSarma, D.W. Matula, "Finite Precision Reciprocal Computation: I. Bipartite Tables, expanded version of Faithful Bipartite ROM Reciprocal Tables", *Proc. 12th IEEE Symp. Comput. Arithmetic*, 1995, pp. 17-28.

[3] H. Hassler, N. Takagi, "Function Evaluation By Table Look-up and Addition", *Proc. 12th IEEE Symp. Comput. Arithmetic*, 1995, pp. 10-16.

[4] N. Takagi, "Generating a Power of an Operand by a Table Look-up and a Multiplication", *Proc. 13th IEEE Symp. Comput. Arithmetic*, 1997, pp. 126-131.

[5] Ping Tak Peter Tang, "Table-driven implementation of the logarithm function in IEEE floating-point arithmetic", *ACM Transactions on Mathematical Software*, vol. 16, no. 4, December 1990, pp. 378–400.

Theorems on Efficient Argument Reductions

Ren-Cang Li*
Department of Mathematics
University of Kentucky
Lexington, KY 40506
Email: rcli@ms.uky.edu

Sylvie Boldo, Marc Daumas
Laboratoire de l'Informatique du Parallélisme
UMR 5668 CNRS - ENS de Lyon - INRIA
Email: Sylvie.Boldo@ens-lyon.fr
Marc.Daumas@ens-lyon.fr

Abstract

A commonly used argument reduction technique in elementary function computations begins with two positive floating point numbers α and γ that approximate (usually irrational but not necessarily) numbers $1/C$ and C, e.g., $C = 2\pi$ for trigonometric functions and $\ln 2$ for e^x. Given an argument to the function of interest it extracts z as defined by $x\alpha = z + \varsigma$ with $z = k2^{-N}$ and $|\varsigma| \leq 2^{-N-1}$, where k, N are integers and $N \geq 0$ is preselected, and then computes $u = x - z\gamma$. Usually $z\gamma$ takes more bits than the working precision provides for storing its significand, and thus exact $x - z\gamma$ may not be represented exactly by a floating point number of the same precision. This will cause performance penalty when the working precision is the highest available on the underlying hardware and thus considerable extra work is needed to get all the bits of $x - z\gamma$ right. This paper presents theorems that show under mild conditions that can be easily met on today's computer hardware and still allow $\alpha \approx 1/C$ and $\gamma \approx C$ to almost the full working precision, $x - z\gamma$ is a floating point number of the same precision. An algorithmic procedure based on the theorems is obtained. The results will enhance performance, in particular on machines that has hardware support for fused-multiply-add (fma) instruction(s).

1 Introduction

Table-based methods to compute elementary functions rely on efficient argument reduction techniques. The idea is to reduce an argument x to u that falls into a tiny interval to allow efficient polynomial approximations (see [5, 10, 11, 12, 13, 15, 17, 18, 19, 20] and references therein).

---

*This work was supported in part by the National Science Foundation under Grant No. ACI-9721388 and by the National Science Foundation CAREER award under Grant No. CCR-9875201. Part of this work was done while this author was on leave at Hewlett-Packard Company.

By default in this paper, all floating point numbers (FPNs), unless otherwise explicitly stated, are binary and of the same type and with p bits in the significand, hidden bits (if any) included, and thus the machine roundoff is

$$\epsilon_m = 2^{-p}.$$

Also we shall assume the default rounding mode is *round-to-nearest* or to even in the case of a tie [1, 6, 14] unless otherwise explicitly stated differently. The underlying machine hardware conforms to the IEEE floating point standard [1].

A commonly used argument reduction technique begins with two positive FPNs α and γ that approximate (usually irrational but not necessarily) numbers $1/C$ and $C > 0$, and thus $\alpha\gamma \approx 1$. Examples include $C = \pi/2$ or π or 2π for trigonometric functions $\sin x$ and $\cos x$, and $C = \ln 2$ for exponential function e^x. Let x be a given argument, a FPN of course. The argument reduction starts by extracting z as defined by

$$x \cdot \frac{1}{C} \approx x\alpha = z + \varsigma = \boxed{k2^{-N}} \boxed{\varsigma}, \quad (1.1)$$

where k is an integer, and $|\varsigma| \leq 2^{-N-1}$, where $N \geq 0$ is an integer. Then it computes a reduced argument

$$u = x - z\gamma. \quad (1.2)$$

For IEEE single precision elementary functions, this u is often good enough, provided $\alpha \approx 1/C$ and $\gamma \approx C$ are IEEE double precision approximations carefully chosen and IEEE double precision arithmetic is used. But sometimes better approximations to C than γ may be necessary for accuracy considerations, e.g., when creating IEEE double precision elementary functions on any of today's RISC (Reduced Instruction Set Computers) machines. If this is the case, often another FPN γ_L, roughly containing the next p bits in the significand of C so that the unevaluated $\gamma + \gamma_L \approx C$ to about $2p$ bits in the significand, is made available to overwrite the u in (1.2) by

$$u - z\gamma_L \quad (1.3)$$

Whether the u by (1.2) or this updated one is accurate enough for computing the elementary function in question is subject to further error analysis on function-by-function basis. But this is out of the scope of this paper.

On machines that have hardware support for the fused-multiply-add (fma) instructions, such as machines with HP/Intel Itanium Microprocessors [11] and IBM PowerPC Microprocessors. The computation of z can be done efficiently as

$$\{x\alpha + \sigma\}_{\mathsf{fma}} - \sigma,$$

where σ is a pre-chosen constant[1]. Given the trend of getting fma as a callable function (inlinable by compilers at certain optimization level) to the language standards such as the C99 standard [2] and the new FORTRAN standard currently under development, this technique is available to users who program only in high level languages.

Notice that if k is an ℓ bit integer, it takes up to $p + \ell$ bits to store the significand of $z\gamma$. It is conceivable that some bits of x and $z\gamma$ will cancel each other, but it is not clear how many of them will and under what condition(s), and consequently if accuracy calls for $x - z\gamma$ to be calculated exactly (or to more than p bits in the significand), how do we get these bits efficiently? This question is especially critical if the working precision is the highest available on the underlying computing platform. In this paper, we will show with mild conditions that can be easily met, $x - z\gamma$ can be represented exactly by a FPN, and thus it can be computed by an instruction of the fma type without error. While this does not exclude the possibility of any further updating as in (1.3) if deemed necessary, it does eliminate any expensive procedure[2] to compute correctly all the bits of $x - z\gamma$ had we not known that it were a FPN. Our results will enhance performance, in particular on machines that have hardware support for fma instructions.

The phenomena of $x - z\gamma$ being a FPN was mentioned in [5], but no further detail was provided there.

Throughout this paper, all FPNs in question are normalized. This is not as restrictive an assumption as it seems. Because those α and γ from elementary function computations are far from subnormal FPNs, and when x is subnormal, it is so tiny that no argument reduction is ever needed. Even if subnormal x is, say, passed to $\{x\alpha + \sigma\}_{\mathsf{fma}} - \sigma$, z will be computed to 0 and thus $u = x - z\gamma = x$ as we would like it to.

The rest of this paper is organized as follows. Section 2 presents a theorem on the number of cancelled bits of two close FPNs that will be used repeatedly in the next section. The theorem, which is of interest in its own right, is an extension of the well-known theorem due to Sterbenz [16]. Our main result is given in Section 3. In Section 4, we analyze how to satisfy the conditions of the theorem in Section 3. Combining the results of Sections 3 and 4, Section 5 presents an algorithm for α and γ, given C and $p(\geq 3)$, and its applications to $C = \ln 2$ for the exponential function $\exp(x)$ and $C = 2\pi$ for the trigonometric functions. Section 6 concludes the work of this paper.

Notation. Throughout, $\oplus, \ominus, \otimes, \oslash$ denote the floating point addition, substraction, multiplication, and division, respectively. $\{\mathcal{X}\}_{\mathsf{fma}}$ denotes the result by an instruction of the fused-multiply-add type, i.e., the exact $\mathcal{X}$ after only one rounding, where $\mathcal{X}$ is one of $\pm a \pm bc$ and $\pm a \mp bc$. ":=" defines the left-hand side to be the right-hand side. $\lfloor a \rfloor$ is the biggest integer that is no greater than a. round_to_nearest(a) is the FPN obtained from rounding a in the round-to-nearest mode, and round_up(a) is the smallest FPN that is no smaller than a, and ulp(b) := 2^{m-p+1} is the unit in the last place of a FPN $b = \pm 2^m \times 1.b_1 b_2 \cdots b_{p-1}$.

2 Exact Subtraction Theorems

Throughout this section a and b are assumed nonnegative. But minor modifications can make all results valid for non-positive a and b, too.

A well-known property [4, 6, 16] of the floating point subtraction is the following.

Theorem 2.1 (Sterbenz) *Let a and b be two FPNs. If $b/2 \leq a \leq 2b$, then $a \ominus b = a - b$, i.e., $a \ominus b$ is exact.*

We now extend this theorem to

Theorem 2.2 *Let a and b be two FPNs with $p + \ell$ bits in the significand, where integer $\ell \geq 0$. If*

$$b/2 \leq (1 + 2^{-\ell})^{-1} b \leq a \leq (1 + 2^{-\ell})b \leq 2b, \quad (2.1)$$

then $a - b$ is a (default) FPN, i.e., $a - b$ can be represented exactly by a FPN with p bits in the significand.

Remark 2.1 Like Theorem 2.1 of Sterbenz, Theorem 2.2 can be shown to hold for radix other than 2. An automatic machine proof in Coq for this is available from the second author upon request or by visiting

---

[1] This idea appeared in Markstein [11, Chap. 10] who told the first author that he got it from Clemens Roothaan.

[2] Without knowing $x - z\gamma$ is a FPN, a typical procedure to compute it exactly may look like this piece of code:

$$a_{\mathsf{H}} = z \otimes \gamma;$$
$$v = x \ominus a_{\mathsf{H}}; \quad a_{\mathsf{L}} = \{z\gamma - a_{\mathsf{H}}\}_{\mathsf{fma}};$$
$$u_{\mathsf{H}} = v \ominus a_{\mathsf{L}};$$
$$b = v \ominus u_{\mathsf{H}};$$
$$u_{\mathsf{L}} = b \ominus a_{\mathsf{L}};$$

This is at least five times as slow as by $\{x - z\gamma\}_{\mathsf{fma}}$ had we known that it were a FPN. Here we assume that there must be some cancellations in the leading bits of x and a_{H} and thus $|v| \geq |a_{\mathsf{L}}|$. The return value is $u_{\mathsf{H}} + u_{\mathsf{L}}$ unevaluated, and $u_{\mathsf{H}} + u_{\mathsf{L}} = x - z\gamma$ exactly.

http://www.ens-lyon.fr/~sboldo/coq/FArgReduct.html. The interested reader is referred to [3, 7] for more detail about Coq and machine proving.

Proof of Theorem 2.2: $(1 + 2^{-\ell})^{-1}b \leq a \leq (1 + 2^{-\ell})b$ implies that $a = 0 \Leftrightarrow b = 0$. Without loss of generality we may assume $a, b > 0$. Write

$$a = 2^{m_a}1.a_1 \cdots a_{p+\ell-1}, \quad b = 2^{m_b}1.b_1 \cdots b_{p+\ell-1},$$

where $a_i, b_i \in \{0, 1\}$. We claim that $|m_a - m_b| \leq 1$; Otherwise if $m_a - m_b \geq 2$, then

$$\frac{a}{b} = 2^{m_a - m_b}\frac{1.a_1 \cdots a_{p+\ell-1}}{1.b_1 \cdots b_{p+\ell-1}} > 2^2 \frac{1}{2} = 2,$$

a contradiction; Similarly if $m_a - m_b \leq -2$, we have $b/a > 2$, a contradiction as well. Also without loss of generality (scale by $2^{-\max\{m_a, m_b\}}$), we may assume $\max\{m_a, m_b\} = 0$, and thus $a - b$ takes this form

$$a - b = \pm d_0.d_1 d_2 \cdots d_{p+\ell}$$

and $d_{p+\ell} = 0$ if $m_a = m_b = 0$.

- If $b \leq a$, then $b \leq a \leq (1 + 2^{-\ell})b$ which implies $0 \leq a - b \leq 2^{-\ell}b$. Now if also $m_a = m_b = 0$, then $d_0 = d_1 = \cdots = d_{\ell-1} = 0 = d_{p+\ell}$, and thus $a - b = 2^{-\ell}d_\ell.d_{\ell+1} \cdots d_{p+\ell-1}$, representable exactly by the default FPN system; On the other hand if $m_a = 0 > m_b = -1$, then $d_0 = d_1 = \cdots = d_\ell = 0$, and thus $a - b = 2^{-(\ell+1)}d_{\ell+1}.d_{\ell+2} \cdots d_{p+\ell}$, also exactly representable.

- If $b > a$, then $b > a \geq (1 + 2^{-\ell})^{-1}b$ which implies $0 < b - a \leq 2^{-\ell}a$. The rest of the proof is the same as for the case $b \leq a$.

This completes the proof. ■

Equivalently, (2.1) can be restated as

$$\frac{1}{1 + 2^{-\ell}} \leq \frac{a}{b} \leq 1 + 2^{-\ell} \qquad (2.2)$$

unless $a = b = 0$.

3 Main Result

We now present the conditions under which $x - z\gamma$ can be represented exactly by a FPN, and thus it can be computed by $\{x - z\gamma\}_{\text{fma}}$ without error. As in Section 1, $\alpha \approx 1/C$ and $\gamma \approx C > 0$. For the rest of this paper set

$$\delta := \alpha\gamma - 1, \qquad (3.1)$$

and suppose

the last q consecutive significant bits of γ are zeros. (3.2)

q is allowed to be zero in which case the last significant bit of γ is 1. Let z be as defined by (1.1) with the conditions on z and ς given there. Assume for the moment that $k \neq 0$ and thus $z \neq 0$. We have

$$\begin{aligned}
\frac{x}{z\gamma} &= \frac{x\alpha}{z\alpha\gamma} \\
&= \frac{z + \varsigma}{z\alpha\gamma} \\
&= \left(1 + \frac{\varsigma}{z}\right)\frac{1}{1 + \delta}.
\end{aligned}$$

Noticing that $|\varsigma/z| \leq 1/(2k)$, we get

$$\left(1 - \frac{1}{2k}\right)\frac{1}{1+\delta} \leq \frac{x}{z\gamma} \leq \left(1 + \frac{1}{2k}\right)\frac{1}{1+\delta}. \qquad (3.3)$$

Theorem 3.1 $x - z\gamma$ *is a FPN if the following conditions are met:*

$$-1/4 \leq \delta \leq 1/2, \qquad (3.4)$$

$$\gamma \leq \mathsf{round\_up}(1/\alpha), \qquad (3.5)$$

$$2^{\lfloor \log_2 |k| \rfloor + 1} \leq \begin{cases} \frac{(2^q - 1) + (2 + 2^q)\delta + \sqrt{\Delta_-}}{-2\delta}, & \text{if } \delta < 0, \\ \frac{2^q - 1 - 2\delta + \sqrt{\Delta_+}}{2\delta}, & \text{if } \delta > 0. \end{cases} \qquad (3.6)$$

where Δ_- and Δ_+ are defined by

$$\Delta_- := (2^q - 2)^2\delta^2 + 2[(2^q)^2 - 3 \cdot 2^q - 2]\delta + (2^q - 1)^2, \qquad (3.7)$$

$$\Delta_+ := 4\delta^2 + 4\delta + (2^q - 1)^2. \qquad (3.8)$$

One implication of this theorem is that the fma makes explicitly storing the extra bits in $z\gamma$ and then subtracting it carefully from x unnecessary.

Remark 3.1 Conditions (3.4) — (3.6) essentially restrict the selection of α and γ, as approximations to $1/C$ and C. It is easy for (3.4) to hold, and the range of feasible k is tied up with δ. Therefore the major hurdle is to satisfy (3.5), while making $\alpha \approx 1/C$ and $\gamma \approx C$ as accurately as possible. Section 4 will presents a detailed analysis in this regard.

Remark 3.2 Notice that for $k > 0$

$$k = 2^{\log_2 k} \leq 2^{\lfloor \log_2 k \rfloor + 1} \leq 2^{\log_2 k + 1} = 2k.$$

Thus (3.6) holds if

$$|k| \leq \begin{cases} \frac{(2^q - 1) + (2 + 2^q)\delta + \sqrt{\Delta_-}}{-4\delta}, & \text{if } \delta < 0, \\ \frac{2^q - 1 - 2\delta + \sqrt{\Delta_+}}{4\delta}, & \text{if } \delta > 0. \end{cases} \qquad (3.9)$$

(3.6) and (3.9) leave a bound on $|k|$ undefined if $\delta = 0$ for which case, there is no constraint on k. The case $\delta = 0$ happens only when both α and γ are powers of two, a case that is not very interesting for elementary function computations.

Remark 3.3 Let us examine asymptotically in δ how big the bound by (3.9) can be because δ is very tiny in the interesting cases. To do so, what we essentially need is to expand $\sqrt{\Delta_-}$ and $\sqrt{\Delta_+}$ at $\delta = 0$. Both Δ_- and Δ_+ are quadratic in δ with the constant terms vanish at $q = 0$. Therefore the expansions should be done depending on whether $q = 0$ or not. We have

- When $q = 0$,

$$|k| \leq \begin{cases} \frac{1}{\sqrt{-2\delta}} - \frac{3}{4} + \mathcal{O}(\sqrt{-\delta}), & \text{if } \delta < 0, \\ \frac{1}{2\sqrt{\delta}} - \frac{1}{2} + \mathcal{O}(\sqrt{\delta}), & \text{if } \delta > 0. \end{cases} \quad (3.10)$$

- When $q \geq 1$,

$$|k| \leq \begin{cases} \frac{2^q-1}{-2\delta} - \frac{[2^q]^2 - 2^q - 2}{2(2^q-1)} + \mathcal{O}(\delta), & \text{if } \delta < 0, \\ \frac{2^q-1}{2\delta} - \frac{2^q - 2}{2(2^q-1)} + \mathcal{O}(\delta), & \text{if } \delta > 0. \end{cases} \quad (3.11)$$

If $\delta = \mathcal{O}(\epsilon_m) = \mathcal{O}(2^{-p})$, these bounds say that $|k|$ can grow as big as of $\mathcal{O}(2^{p/2})$ if $q = 0$, and of $\mathcal{O}(2^p)$ if $q \geq 1$. Later in Algorithm 5.1 and Remark 5.1 of Section 5, we shall show that with a slight modification to the last significant bit of γ such that $\gamma \approx C$ with error less than 1.5 ulp, q can be made $q \geq 1$.

Proof of Theorem 3.1: No proof is needed if $k = 0$ and thus $z = 0$. Assume that $|k| \geq 1$. Then x and k have the same sign, and

$$|k|2^{-N} - 2^{-N-1} \leq |x\alpha| = |k2^{-N} + \varsigma| \leq |k|2^{-N} + 2^{-N-1}.$$

From now on we consider the case $x > 0$ only, and the other case $x < 0$ can be handled in a similar way.

Suppose $k = 1$. Then $z\gamma$ is a FPN, and

$$2^{-N-1} \leq x\alpha \leq 3 \cdot 2^{-N-1}$$

which implies

$$2^{-N-1}/\alpha \leq x \leq 3 \cdot 2^{-N-1}/\alpha.$$

Let $x_{\min}$ be the smallest FPN such that $x_{\min} \geq 2^{-N-1}/\alpha$, i.e.,

$$x_{\min} = 2^{-N-1} \text{round\_up}(1/\alpha).$$

Thus $\gamma \leq 2^{N+1} x_{\min}$ by (3.5). Now $z = 2^{-N}$, and thus

$$\frac{x}{z\gamma} \geq \frac{x_{\min}}{z\gamma} = \frac{x_{\min}}{2^{-N}\gamma} \geq \frac{1}{2}.$$

On the other hand,

$$\frac{x}{z\gamma} \leq \frac{3 \cdot 2^{-N-1}/\alpha}{2^{-N}\gamma} = \frac{3}{2}\frac{1}{1+\delta} \leq 2,$$

if $\delta \geq -1/4$ which holds by (3.4). Thus $x - z\gamma$ is a FPN by Theorem 2.1.

Suppose $k \geq 2$. Inequality (3.3) yields

$$\frac{3}{4}\frac{1}{1+\delta} \leq \frac{x}{z\gamma} \leq \frac{5}{4}\frac{1}{1+\delta}.$$

Therefore

$$\frac{1}{2} \leq \frac{x}{z\gamma} \leq 2 \quad \text{if } -3/8 \leq \delta \leq 1/2 \quad (3.12)$$

which is guaranteed by (3.4). Let

$$\ell := \lfloor \log_2 k \rfloor + 1$$

which is the number of bits to store k exactly. Then

$$2^{\ell-1} \leq k < 2^\ell - 1.$$

Now if $k = 2^{\ell-1}$, then $z\gamma$ is a FPN. So $x - z\gamma$ is a FPN by (3.12) and Theorem 2.1. Notice that $z\gamma$ is a FPN with at most $p - q + \ell$ significant bits. If $q \geq \ell$, $z\gamma$ is also a FPN (in the default format). By (3.12) and Theorem 2.1, $x - z\gamma$ is a FPN. Assume now that $q < \ell$ and $k \geq 2^{\ell-1} + 1$. Then (3.3) implies

$$\left(1 - \frac{1}{2^\ell + 2}\right)\frac{1}{1+\delta} \leq \frac{x}{z\gamma} \leq \left(1 + \frac{1}{2^\ell + 2}\right)\frac{1}{1+\delta}.$$

We claim that under the condition of the theorem

$$\left(1 + \frac{1}{2^\ell + 2}\right)\frac{1}{1+\delta} \leq 1 + \frac{2^q}{2^\ell}, \quad (3.13)$$

$$\left(1 - \frac{1}{2^\ell + 2}\right)\frac{1}{1+\delta} \geq \left(1 + \frac{2^q}{2^\ell}\right)^{-1}. \quad (3.14)$$

Therefore we have

$$\left(1 + 2^{-\ell+q}\right)^{-1}(z\gamma) \leq x \leq \left(1 + 2^{-\ell+q}\right)(z\gamma).$$

Since $z\gamma$ is a FPN with no more than $p - q + \ell$ bits in the significand, by Theorem 2.2, $x - z\gamma$ is a FPN (in the default format), as expected.

We have to prove (3.13) and (3.14).

Inequality (3.13) is equivalent to

$$-[2^\ell]^2\delta - [(2^q - 1) + (2 + 2^q)\delta]2^\ell - 2 \cdot 2^q(1 + \delta) \leq 0. \quad (3.15)$$

This inequality holds if $\delta \geq 0$. Assume $\delta < 0$, and notice that Δ_- is the resultant of the quadratic polynomial in 2^ℓ on the left-hand side of (3.15)

$$\Delta_- = [(2^q - 1) + (2 + 2^q)\delta]^2 - 4 \times \delta \times 2 \cdot 2^q(1 + \delta).$$

$\Delta_- \geq 0$ for all $q \geq 0$ and all $\delta < 0$. This can be checked directly for $q = 0$ and $q = 1$ for which Δ_- is $\delta^2 - 8\delta$ and

$1 - 8\delta$ respectively. For $q \geq 2$, Δ_- as a polynomial in δ never vanishes because its resultant

$$4[(2^q)^2 - 32^q - 2]^2 - 4 \times (2^q - 2)^2 \times (2^q - 1)^2$$
$$= 32 \cdot 2^q(3 - 2^q) < 0.$$

This combining with the fact that $\Delta_- > 0$ at $\delta = 0$ implies that $\Delta_- > 0$ always. It can now be seen that (3.15) is guaranteed by (3.6).

Inequality (3.14) is equivalent to

$$-[2^\ell]^2 \delta + [(2^q - 1) - 2\delta]2^\ell + 2^q \geq 0. \quad (3.16)$$

This inequality holds if $\delta \leq 0$. Assume $\delta > 0$, and notice that Δ_+ is the resultant of the quadratic polynomial in 2^ℓ on the left-hand side of (3.16)

$$\Delta_+ = [(2^q - 1) - 2\delta]^2 - 4 \times (-\delta) \times 2^q.$$

$\Delta_+ \geq 0$ for all $q \geq 0$ and all $\delta > 0$. It can now be seen that (3.16) is guaranteed by (3.6).

The proof is now completed. ∎

4 Analysis of Constraints Between α and γ

In using Theorem 3.1 to come up with α and γ for argument reductions, we essentially need to consider making α and γ to satisfy (3.5) and, if necessary, forcing the last bit of γ to be 0 because for modest p, $|\delta|$ is easily made to be much less than $1/4$ and because the constraints on k are results of the two. In fact, it is easy to make δ as tiny as ϵ_m. For functions like exponentials, k cannot be much bigger before overflow or underflow takes over and thus the range imposed on k by Theorem 3.1 is sufficient, even for $q = 0$; While for others, the range imposed on k by Theorem 3.1 for $q \geq 1$ is also sufficient for reasons as follows. It is conceivable that (1.1) simulates extracting in exact arithmetic the certain number of leading significant bits of x/C, while (1.2) simulates $x - zC$. However $x\alpha$ if represented exactly has up to $2p$ significant bits with only about p leading bits trustworthy as an approximation to x/C because in general $\alpha \approx 1/C$ with relative error about ϵ_m. Therefore in order for (1.1) and (1.2) to mimic what they are intended, the number of extracted bits in z in (1.1) should be made no bigger than p, or equivalently $|k| \leq 2^{p-1} = 2^{-1}\epsilon_m^{-1}$.

In what follows, we shall concentrate on how to make

$$\alpha \approx 1/C \text{ and } \gamma \approx C$$

as accurate as possible while not violating the assumptions of Theorem 3.1. Naturally best possible α and γ are

$$\alpha = \text{round\_to\_nearest}(1/C) \text{ and } \gamma = \text{round\_to\_nearest}(C), \quad (4.1)$$

but that cannot always be done as will be shown by Example 4.1 below. Our best choice is to make γ approximates its target as accurately as possible while α approximates $1/\gamma$ in the best way, i.e.,

$$\alpha = 1 \oslash \gamma. \quad (4.2)$$

Lemma 4.1 *Let $a > 0$ be a FPN. Then*

$$\text{round\_up}\left(\frac{1}{1 \oslash a}\right) \geq a.$$

Proof: Without loss of generality, we scale a such that $1 \leq a < 2$. Write $1 \oslash a = 1/a + \eta$ with $|\eta| \leq 2^{-p-1}$. No proof is needed if $a = 1$. For $1 < a < 2$, it suffices to show that

$$\frac{1}{1 \oslash a} > a - 2^{-p+1}. \quad (4.3)$$

If $\eta \leq 0$, then $1 \oslash a \leq 1/a$, and thus (4.3) holds. Assume $\eta > 0$. We have

$$\frac{1}{1 \oslash a} = \frac{a}{1 + a\eta} = a[1 - (a\eta) + (a\eta)^2 - \cdots]$$
$$= a - a[(a\eta) - (a\eta)^2 + \cdots].$$

Notice that $(a\eta) - (a\eta)^2 + \cdots$ is an alternating series and thus it is bounded strictly by its first term, i.e.,

$$0 < a[(a\eta) - (a\eta)^2 + \cdots] < a(a\eta) < 4 \times 2^{-p-1} = 2^{-p+1}$$

which yields (4.3). ∎

A related result but only for those a which can be scaled by a power of two to fall between 1 and $\sqrt{2}$

$$1 \oslash (1 \oslash a) = a$$

is due to W. Kahan [9, Exercise 27 of §4.22 and its solution].

Theorem 4.1 *(3.5) holds if $\alpha\gamma \leq 1$ which is true if*

either $\alpha = \text{round\_down}(1/\gamma)$ or $\gamma = \text{round\_down}(1/\alpha)$. (4.4)

Proof: $\alpha\gamma \leq 1$ implies $\gamma \leq 1/\alpha$ and thus (3.5). Note that (4.4) implies $\alpha \leq (1/\gamma)(1 - \epsilon)$ for some $0 \leq \epsilon \leq 2\epsilon_m$, and thus $\alpha\gamma \leq 1$. ∎

Theorem 4.2 *(3.5) holds if (4.1) such that either $\alpha \leq 1/C$ or $\gamma \leq C$ or C can be scaled by a power of two to fall in $[1, \sqrt{2})$.*

The restriction that the scaled C to fall in $[1, \sqrt{2})$ is quite unpleasant but necessary as the following example shows. It is invoked in the later proof when both $\alpha \leq 1/C$ and $\gamma \leq C$ are violated.

Example 4.1 $C = 2\pi$ for which it can be verified that with (4.1), (3.4) and (3.5) hold for $3 \leq p \leq 197$ but (3.5) fails for $p = 198$.

Proof of Theorem 4.2: Without loss of generality, we may scale C by a power of 2 such that $1 \leq C < 2$. No proof is needed if $C = 1$. Assume $1 < C < 2$. Then

$$\alpha = 1/C + \delta_1, \quad \gamma = C + \delta_2$$

and $|\delta_1| \leq 2^{-p-1}, |\delta_2| \leq 2^{-p}$. First if $\delta_1 \leq 0$, then

$$\alpha \leq 1/C \Rightarrow C \leq 1/\alpha,$$

and thus

$$\begin{aligned}\gamma &= \text{round\_to\_nearest}(C) \\ &\leq \text{round\_up}(C) \\ &\leq \text{round\_up}(1/\alpha),\end{aligned}$$

as expected. Next if $\delta_2 \leq 0$, then

$$\begin{aligned}\gamma \leq C &\Rightarrow 1/C \leq 1/\gamma \\ &\Rightarrow \alpha \leq 1 \oslash \gamma \\ &\Rightarrow \frac{1}{\alpha} \geq \frac{1}{1 \oslash \gamma},\end{aligned}$$

and thus by Lemma 4.1

$$\text{round\_up}\left(\frac{1}{\alpha}\right) \geq \text{round\_up}\left(\frac{1}{1 \oslash \gamma}\right) \geq \gamma.$$

It is remained to prove the claim for the case $\delta_1 > 0$ and $\delta_2 > 0$. This is the situation where we need $C < \sqrt{2}$. Notice that $\gamma = C + \delta_2 > C > 1$. Let γ' be the biggest FPN that is smaller than γ, i.e., $\gamma' = \gamma - 2^{-p+1}$. $\delta_2 > 0$ implies that C is above or at the middle point between γ' and γ as show below.

It suffices to prove $1/\alpha > \gamma'$ for (3.5) to hold. Notice

$$\begin{aligned}\frac{1}{\alpha} = \frac{C}{1 + C\delta_1} &= C[1 - (C\delta_1) + (C\delta_1)^2 - \cdots] \\ &= C - C[(C\delta_1) - (C\delta_1)^2 + \cdots].\end{aligned}$$

$C < \sqrt{2}$ implies

$$0 < C[(C\delta_1) - (C\delta_1)^2 + \cdots] < C^2 \delta_1 < 2 \times 2^{-p-1} = 2^{-p}$$

and consequently

$$\frac{1}{\alpha} > C - 2^{-p} \geq \gamma',$$

as expected. ∎

Theorem 4.3 *(3.5) holds if (4.2).*

Proof: It is a consequence of Theorem 4.2 by taking $C = \gamma$, and thus $\gamma \leq C$ holds, and α and γ are defined as in (4.1). ∎

5 An Algorithm for α and γ

Thanks to the results of Sections 3 and 4, we suggest the following algorithm for picking $\alpha \approx 1/C$ and $\gamma \approx C$ for all p that is not too small, say $p \geq 3$. (This is to make (3.4) always satisfied.)

Algorithm 5.1 *Given C, the following steps produce α and γ such that $x - z\gamma$ is a FPN.*

1. *Compute α and γ as in (4.1);*

2. *Verify (3.5). This is automatically satisfied if we know beforehand that C can be scaled by a power of two to fall in $[1, \sqrt{2})$. Otherwise either verify directly (3.5), or check if one of the following conditions is satisfied:*

   $$\alpha\gamma \leq 1, \text{ or } \alpha \leq 1/C, \text{ or } \gamma \leq C$$

   according to Theorems 4.1 and 4.2.

3. *Compute an upper bound on all applicable k as given by (3.9). If, however, the bound by (3.9) is too small for the computation of the elementary function in question (This may happen when $q = 0$), we can either add 1 ulp to or subtract 1 ulp from γ, and then take*

   $$\alpha = 1 \oslash \gamma$$

   as in (4.2). Doing so makes (3.5) automatically satisfied by Theorem 4.3.

Remark 5.1 In the 3rd step of Algorithm 5.1, adding 1 ulp or subtracting 1 ulp, if done carefully, can make $q \geq 2$. In fact, this can be accomplished by adding 1 ulp if the last two bits of γ are 11_2 or subtracting 1 ulp if the last two bits of γ are 01_2.

Next we shall present two examples: $C = \ln 2$ from the computation of $\exp(x)$, and $C = 2\pi$ from the computation of radian trigonometric functions. When it comes to write a library of the elementary mathematical functions, we often use the floating point arithmetic of the highest precision available on any given hardware. This means to use the IEEE double precision arithmetic on the existing RISC machines, and Intel double-extended precision arithmetic (64 bits in the significand) on machines equipped with Intel processors. Therefore the parameters α and γ are either of IEEE double precision or of Intel double-extended precision. In what follows, however, we do give IEEE single precision α and γ just to show the applicability of our theorems and algorithm.

Example 5.1 Consider the computation of $\exp(x)$ based on [11]

$$\exp(x) = 2^{x \log_2 e} = 2^{x/\ln 2},$$

where $e = \exp(1)$ the natural number. Here $C = \ln 2$. Because $2\ln 2 \approx 1.386$, C can be scaled by a power of two to fall in $[1, \sqrt{2})$. Thus (3.5) holds with (4.1). For $\exp(x)$, $k2^{-N}$ is extracted to serve two purposes: the exponent M of $\exp(x)$ and a table lookup index m

$$k2^{-N} = M + m2^{-N}$$

where M and m are integers, where $0 \leq m \leq 2^N - 1$. Then

$$\exp(x) \approx 2^M \times 2^{m2^{-N}} \times \exp(t),$$

where $t = x - z\gamma$ (or $t \approx x - z\ln 2$ if some γ_L is used so that unevaluated $\gamma + \gamma_L \approx \ln 2$ to about $2p$ bits). Typically $2^{m2^{-N}}$ is obtained through table lookup. In order not to use too big a table, N may be chosen, say, no bigger than 10. On the other hand, 2^M quickly overflows or underflows as $|M|$ gets bigger, e.g., for IEEE double precision, it overflows if $M \geq 1023$ and underflows to zero if $M < -1022 - 53$. For this reason, interesting $|k|$ is no bigger than about 2^{11+N}. Therefore the constraints imposed on k by Theorem 3.1 for $q = 0$ or $q \geq 1$ are acceptable. In what follows, a string with the subscript 16 denotes a hexadecimal number, and that without the subscript is a decimal number of the usual radix 10.

a) IEEE single precision:

$$\begin{aligned}
\alpha &= 2^{-3} \times B.8AA3B_{16}, \\
\gamma &= 2^{-4} \times B.17218_{16}, \\
\delta &\approx -1.06 \times 10^{-08}, \\
|k| &\leq 13AD5D94_{16}.
\end{aligned}$$

Here $q = 3$.

b) IEEE double precision:

$$\begin{aligned}
\alpha &= 2^0 \times 1.71547652B82FE_{16}, \\
\gamma &= 2^{-1} \times 1.62E42FEFA39EF_{16}, \\
\delta &\approx -4.76 \times 10^{-17}, \\
|k| &\leq 61C6EC2_{16}.
\end{aligned}$$

Here $q = 0$. Thus the largest possible $|k|$ is about $\epsilon_m^{-1/2}$. If we add 1 ulp to γ and take $\alpha = 1 \oslash \gamma$ as suggested by Step 3) of Algorithm 5.1, it will make $q = 4$. Doing so, we get

$$\begin{aligned}
\alpha &= 2^0 \times 1.71547652B82FD_{16}, \\
\gamma &= 2^{-1} \times 1.62E42FEFA39F0_{16}, \\
\delta &\approx -4.13 \times 10^{-17}, \\
|k| &\leq 2851984E2E90048_{16}.
\end{aligned}$$

c) Intel double-extended precision: (64 bits in the significand)

$$\begin{aligned}
\alpha &= 2^{-3} \times B.8AA3B295C17F0BC_{16}, \\
\gamma &= 2^{-4} \times B.17217F7D1CF79AC_{16}, \\
\delta &\approx 3.57 \times 10^{-20}, \\
|k| &\leq 2464972759AF9B334_{16}.
\end{aligned}$$

Example 5.2 Consider the computation of $\sin(x)$ and $\cos(x)$. They are two of the most difficult elementary functions to compute. Here $C = 2\pi$ which is their period. Therefore only the fractional part of $x/(2\pi)$ is interesting. But getting enough correct fractional bits of $x/(2\pi)$ can be tricky and costly for x of huge magnitude or extremely close to an integral multiple of $\pi/2$. Presumably arguments x as such are rare in any given application. In order not to slow down the speed for most common arguments, often a fast reduction just as outlined at the beginning of this paper is performed and then some quick checking is done to see if a more careful reduction procedure is needed and if that is the case, the code is branched to perform a careful argument reduction which is rare and slow. The interested reader is referred to [8, 11, 12, 13] and references therein for more detail. In this paper, however, we are only interested in speeding up the fast reduction part.

a) IEEE single precision:

$$\begin{aligned}
\alpha &= 2^{-6} \times A.2F983_{16}, \\
\gamma &= 2^{-1} \times C.90FDB_{16}, \\
\delta &\approx -1.25 \times 10^{-08}, \\
|k| &\leq 18B0_{16}.
\end{aligned}$$

Here $q = 0$. Subtracting 1 ulp from γ and taking $\alpha = 1 \oslash \gamma$ yield $q = 2$ and

$$\begin{aligned}
\alpha &= 2^{-6} \times A.2F982_{16}, \\
\gamma &= 2^{-1} \times C.90FDC_{16}, \\
\delta &\approx -3.03 \times 10^{-08}, \\
|k| &\leq 2F4A062_{16}.
\end{aligned}$$

b) IEEE double precision:

$$\begin{aligned}
\alpha &= 2^{-3} \times 1.45F306DC9C883_{16}, \\
\gamma &= 2^2 \times 1.921FB54442D18_{16}, \\
\delta &\approx 2.28 \times 10^{-17}, \\
|k| &\leq 22066D471BD6D2D_{16}.
\end{aligned}$$

Here $q = 3$.

c) Intel double-extended precision:

$$\begin{aligned}
\alpha &= 2^{-6} \times A.2F9836E4E44152A_{16}, \\
\gamma &= 2^{-1} \times C.90FDAA22168C235_{16}, \\
\delta &\approx 1.72 \times 10^{-20}, \\
|k| &\leq E2ED4431_{16}.
\end{aligned}$$

Here $q = 0$. Subtracting 1 ulp from γ and taking $\alpha = 1 \oslash \gamma$ yield $q = 2$ and

$$\begin{aligned}
\alpha &= 2^{-6} \times A.2F9836E4E44152B_{16}, \\
\gamma &= 2^{-1} \times C.90FDAA22168C234_{16}, \\
\delta &\approx 3.34 \times 10^{-20}, \\
|k| &\leq 26FA94EFA25DF2177_{16}.
\end{aligned}$$

6 Conclusions

We have presented theorems that prove the correctness and effectiveness of the commonly used argument reduction technique in elementary function computations, especially on machines that have hardware support for fused-multiply-add instructions. The conditions of these theorems are easily met as our analysis indicates. While we showed it is not always possible to use the best possible parameters as defined by (4.1) under all circumstances, an almost best possible selection as in (4.2) can be used at all times. On case-by-case basis, however, it is possible to use (4.1) by verifying individually the conditions in our main theorem in Section 3 while none of the theorems of Section 4 apply, e.g., $C = 2\pi$ and $3 \leq p \leq 197$. Based on our results in Sections 3 and 4, a 3-step algorithm is presented to derive argument reduction parameters α and γ.

While Theorem 3.1 as of now is sufficient in the sense that effective parameters for efficient argument reductions can be obtained without any difficulty, it would be interesting to know if some of the conditions of Theorem 3.1 are necessary, i.e., $x - z\gamma$ is not a FPN if one or more of the conditions fails. But we could not either prove it or find a counterexample at this point. We shall work on this in the future.

Finally we comment that the results in this paper are extensible to floating point number systems with radix other than 2 (see Remark 2.1). But we omit details here.

Acknowledgment

The authors are indebted to referees' constructive suggestions which improve the presentation considerably. One of the referees read the early draft so carefully that he compiled a long list of inappropriate English used, along with many valuable technical comments. They are especially grateful for his/her diligent readership.

Part of this work was done while the first author was on leave at Hewlett-Packard Company. He is grateful for help received from Jim Thomas, John Okada, and Peter Markstein of HP Itanium floating point and elementary math library team at Cupertino, California. Working with these people has been a pleasure.

References

[1] American National Standards Institute and Institute of Electrical and Electronic Engineers. IEEE standard for binary floating-point arithmetic. *ANSI/IEEE Standard, Std 754-1985*, New York, 1985.

[2] ANSI/ISO/IEC 9899:1999. *Programming Languages – C.* 1999.

[3] M. Daumas, L. Rideau, and L. Théry. A generic library of floating-point numbers and its application to exact computing. In *14th International Conference on Theorem Proving in Higher Order Logics*, pages 169–184, Edinburgh, Scotland, 2001.

[4] D. Goldberg. What every computer scientist should know about floating-point arithmetic. *ACM Computing Surveys*, 23(1):5–47, Mar. 1991.

[5] J. Harrison, T. Kubaska, S. Story, and P. T. P. Tang. The computation of transcendental functions on the IA-64 architecture. *Intel Technology Journal*, (Q4):1–7, November 1999.

[6] N. J. Higham. *Accuracy and Stability of Numerical Algorithms*. SIAM, Philadephia, 1996.

[7] G. Huet, G. Kahn, and C. Paulin-Mohring. The Coq proof assistant: a tutorial: version 7.2. Technical Report 256, Institut National de Recherche en Informatique et en Automatique, Le Chesnay, France, 2002. Available at ftp://ftp.inria.fr/INRIA/publication/publi-pdf/RT/RT-0256.pdf.

[8] W. Kahan. Minimizing q*m-n. At the beginning of the file "nearpi.c", available electronically at http://http.cs.berkeley.edu/~wkahan/testpi, 1983.

[9] D. Knuth. *The Art of Computer Programming*, volume 2. Addison Wesley, Reading, MA, 1981.

[10] R.-C. Li. Always Chebyshev interpolation in elementary function computations. submitted for publication, June 2001.

[11] P. Markstein. *IA-64 and Elementary Functions: Speed and Precision*. Prentice Hall, New Jersey, 2000.

[12] J.-M. Muller. *Elementary Functions: Algorithms and Implementation*. Birkhäuser, Boston•Basel•Berlin, 1997.

[13] K. C. Ng. Argument reduction for huge arguments: Good to the last bit. Technical report, SunPro, 1992. available electronically at http://www.validgh.com/.

[14] M. L. Overton. *Numerical Computing with IEEE Floating Point Arithmetic*. SIAM, Philadelphia, 2001.

[15] M. J. D. Powell. On the maximum errors of polynomial approximations defined by interpolation and by least squares criteria. *The Computer Journal*, 9(4):404 – 407, February 1967.

[16] P. H. Sterbenz. *Floating-Point Computation*. Prentice-Hall series in automatic computation. Prentice-Hall, Englewood Cliffs, NJ, USA, 1974.

[17] P. T. P. Tang. Table-driven implementation of the exponential function in IEEE floating-point arithmetic. *ACM Transactions on Mathematical Software*, 15(2):144–157, June 1989.

[18] P. T. P. Tang. Table-driven implementation of the logarithm function in IEEE floating-point arithmetic. *ACM Transactions on Mathematical Software*, 16(4):378–400, Dec. 1990.

[19] P. T. P. Tang. Table lookup algorithms for elementary functions and their error analysis. In P. Kornerup and D. W. Matula, editors, *Proceedings of the 10th IEEE Symposium on Computer Arithmetic*, pages 232–236, Grenoble, France, June 1991. IEEE Computer Society Press, Los Alamitos, CA.

[20] P. T. P. Tang. Table-driven implementation of the `expm1` function in IEEE floating-point arithmetic. *ACM Transactions on Mathematical Software*, 18(2):211–222, June 1992.

Accelerating Sine and Cosine Evaluation with Compiler Assistance

Peter Markstein
Hewlett-Packard Laboratories
1501 Page Mill Road
Palo Alto, CA 94062, U.S.A.
peter.markstein@hp.com

Abstract

Some software libraries add special entry points to enable both the sine and cosine to be evaluated with one call for performance purposes. This paper proposes another method which does not involve new function names. By having the compiler front end recognize trigonometric function invocations, and replace them with a call to a common function which executes the code common to all the functions, followed by a short routine to produce the desired computation, it is possible to compute both the sine and cosine, when needed in about the same time as to compute only one of them.

1 Introduction

The routines to evaluate the sine and cosine functions have much code in common. If a program invokes both the sine and cosine routines for the same argument, most of the work in evaluating the first function is repeated when invoking the second. Some program libraries exploit this fact by providing alternative routines that return both $\sin x$ and $\cos x$ with one function call. A variant returns the complex quantity $\operatorname{cis} x = \cos x + i \sin x$. The drawback to these new functions is that they are not part of the standard programming interface, and codes that exploit them thereby become non-portable.

Some compilation systems recognize calls to the sine and cosine routines with the same argument, and replace them with a call to a "sincos" function that returns both values. This represents a new style of optimization, and it requires that the programmer know exactly the scope over which the transformation can apply.

In addition, the routines that return $\sin x$ and $\cos x$ concurrently may compute them in a different manner than the standard routines, altering their rounding behavior, accuracy or edge-case behavior, in the interest of speed. The search for means to accelerate the sine and cosine function arises because in practice, both functions are frequently called with the same argument. This is frequently the case in applications involving complex variables.

This paper discusses how a compiler can decompose invocations of sines and cosines into a pair of subroutine invocations, and then use existing compiler transformations and optimizations to achieve the same economies as the special purpose routines. The advantage of this approach is that it does not require an extended application program interface (API) to achieve performance improvements and it always uses the same algorithm. Therefore the characteristics of the sine and cosine routines as proposed here are not affected by whether they are evaluated one at a time or together. In addition, the compiler is not required to explicitly find instances where calls to sine and cosine are in near proximity. The technique can also cope with the tangent function, and it is readily adoptable to other function families which have much code in common, such as the hyperbolic functions, or the square root and reciprocal square root functions.

2 Method

Evaluation of the trigonometric functions involves several steps[1]:

1. reducing the argument to the first octant or smaller

2. extracting from a table, the sine and cosine of a value based on the argument reduction

3. computing the sine and cosine of the reduced argument by a Taylor, Chebychev, or Remez approximation

4. combining the table-lookups and approximation results to compute the desired function

The first three steps are common to the evaluation of sine and cosine. It is only how the results of these steps are combined that differentiates the sine and cosine functions. If the tangent is computed by dividing sine by cosine, then the tangent also fits into this paradigm.

To be specific, given an argument x, the first three steps generally lead to a reduced argument w, where $w = x - 2n\pi - t$, with n an integer and $|t| \leq \pi$. The variable t is chosen to make w smaller than a predetermined value, and t holds a value for which $\sin t$ and $\cos t$ can be obtained by table-lookup. Then $\sin x = \sin(w+t) = \sin w \cos t + \cos w \sin t$. Similarly, $\cos x = \cos w \cos t - \sin w \sin t$. Only $\sin w$ and $\cos w$ must be approximated by code, and both are needed for the evaluation of the sine and cosine functions. By constraining w to be sufficiently small, polynomials of low degree will suffice to compute $\sin w$ and $\cos w$ to the required precision. From the above discussion, the sine and cosine functions differ only in how these four values are combined in the last step.

The common steps can be exploited by introducing new functions into the repertoire of functions known by the compiler: trigstart, sinfinish, cosfinish, and perhaps tanfinish. The trigstart function performs the first 3 steps outlined above, and returns a structure, in the floating point registers, with the multiple computed values. The sinfinish, cosfinish and tanfinish routines take the structure produced by trigstart, and from it compute sin, cos, and tan respectively of the argument that had been given to trigstart. (On some platforms, the linkage conventions may not permit return of a structure in floating point registers. In such cases the values must be returned via storage, imposing a small overhead to retrieve these values. But since the return values would most likely be in a nearby cache, the penalty for streaming the structure back to registers should be no worse than one cache access).

The function trigstart could return a structure containing four values: $\sin t$, $\cos t$, $\sin w$, and $\cos w$. In practice, it has been useful to also return w, to allow the results of $\sin(\pm 0)$ and $\tan(\pm 0)$ to retain the sign of the argument. It is also possible for trigstart to return $(\sin w)/w$ instead of $\sin w$, and $(\cos w - 1)/w$ instead of $\cos w$, in which case the code for the three finishing routines would be slightly different, but slightly more precise[3]. Computing $\sin w / w$ does not involve a division; it merely computes $(1 - w^2/3! + w^4/5! - ...)$, and similarly for the expression involving $\cos w$. So trigstart becomes two operations shorter to compensate for the extra instruction introduced into each finish routine. On a machine with fused multiply-add (FMA), the finishing routines each require 3 operations. For example, the sine function would be computed

$$\sin x = \sin(w+t) = \sin t \cos w + \cos t \sin w$$
$$= \sin t + w(\sin t \frac{\cos w - 1}{w} + \cos t \frac{\sin w}{w})$$

The alternate form allows the result to be expressed as a small adjustment to the table lookup value. For the parenthesized expression, it is better to compute the smaller product directly with a multiplication operation, and then the full expression with an FMA. Here, either product could be the smaller. But the first is bounded by w, and the second by 1.0, so our code computes the first product directly.

The front end of the compiler is modified to decompose the trigonometric functions as follows:

```
sin(x) = sinfinish(trigstart(x))
cos(x) = cosfinish(trigstart(x))
tan(x) = tanfinish(trigstart(x))
```

In the compiler's intermediate language, trigstart is treated like an ordinary arithmetic operation (a pure function), which always computes the same result for a given argument, without any side effects. During early stages of optimization, trigstart can be subjected to all the usual optimizations, such as common sub-expression elimination and code motion[4]. If a program contains the code: a = sin(x); b = cos(x); the compiler would generate:

```
R100 = trigstart(x)
A    = sinfinish(R100)
R100 = trigstart(x)
B    = cosfinish(R100)
```

and an optimization would remove the second computation of trigstart [1]. Even constructions such as:

```
if (x > y) z = sin(x);
else z = cos(x);
```

where the invocations of sine and cosine are in different control flow paths would result in compiled code which behaves as though it had been written as:

```
tmp = trigstart(x);
if (x > y) z = sinfinish(tmp);
else z = cosfinish(tmp);
```

The compiler achieves this transformation by code motion or code hoisting, to move the computation of trigstart(x) in front of the conditional branch.

The names trigstart, sinfinish, cosfinish, and tanfinish need never be made part of the API. In actual practice, the name given to trigstart would be outside the user's name space. The names of the finishing routines are used here only for expository purposes as the compiler is expected to expand these 3-operation functions inline (see the example in section 3).

2.1 Hyperbolic functions

As an example of another function family amenable to this treatment, the hyperbolic functions can be expressed in

---

[1] Some compilers may assign a different symbolic register to the second computation of trigstart, but would then know how to remove the second computation and alias its result to that of the first computation.

terms of $\text{expm1}(x) = e^x - 1$ as follows:

$$\sinh x = \frac{1}{2}\text{expm1}(x) + \frac{\text{expm1}(x)}{2(\text{expm1}(x)+1)}$$
$$\cosh x = \frac{[\text{expm1}(x)]^2}{2(\text{expm1}(x)+1)} + 1$$

A hyperbolic start function could return two values: $u = \text{expm1}(x)$ and $v = \frac{1}{2}\cdot\text{expm1}(x)/(\text{expm1}(x)+1)$. Then one can compute $\sinh x = \frac{1}{2}u + v$, and $\cosh x = uv + 1$.

An alternate method of computing the hyperbolic functions [3] allows a treatment similar to that shown for the trigonometric functions. Decompose the argument x as $x = n\ln 2 + t$, with n an integer and $|t| < \frac{1}{2}\ln 2$. The hyperbolic start function returns a structure with 4 or 5 values similar to those returned by `trigstart`. These values are combined according to the hyperbolic function addition formulas in the finishing routines to obtain the desired results. Within the start routine, $\sinh(n\ln 2)$ is computed simply as $(2^n - 2^{-n})/2$, and $\cosh(n\ln 2)$ as $(2^n + 2^{-n})/2$.

3 Results

The HP-UX compiler for Itanium has implemented the methodology described above for trigonometric functions by modification of the front ends of the C/C++ and Fortran compilers to recognize invocations of the sine, cosine, and tangent functions. The optimizations to remove and/or reposition redundant `trigstart` operations occur without any change to the compiler's optimizer, since `trigstart` is regarded as a primitive operation during early rounds of optimization. In C, the user must include a reference to the header `math.h` to enable the compiler to decompose the trigonometric functions as described above. The code

```
s = sin(x);
c = cos(x);
```

results in the compiler-generated assembly code shown in Figure 1 (comments inserted by the author).

Instructions between double semicolons execute concurrently in Itanium systems with two floating point units, allowing the sine and cosine to finish in the same time together as either one would alone. Instructions with the ".s1" completer do not modify the user-visible floating point status register, and they are executed in double-extended precision. Floating point instructions with the ".d.s0" completer may affect the floating point status register, and their results are rounded to double precision.

The HP-UX Itanium compiler at high optimization (+O3) also allows complete inlining of the trig functions, in which case the `trigstart` function would be expanded in the calling routine. When this occurs in a loop, code motion will move the invariant instructions to the point just before

```
//argument x in fp reg. 8
//the instruction below is Itanium's call instruction
br.call.sptk.few rp = trigstart;;
//on return, f8 = cos t; f9 = (cos w - 1)/w, f10 = sin t;
// f11 = sin w/w; f12 = w.
//Note: fma a=b,c,d computes a=d+b*c;
// fnma a=b,c,d computes a=d-b*c
fmpy.s1 f6 = f10, f9      // f6 = sin t * (cos w - 1.0) / w
fmpy.s1 f7 = f8, f9 ;;    // f7 = cos t * (cos w - 1.0) / w
fma.s1 f6 = f8, f11, f6   // f6 = f6 + cos t * (sin w) / w
fnma.s1 f7 = f10, f11, f7;;  // f7 = f7 - sin t * (sin w) / w
fma.d.s0 f6 = f12, f6, f10 // s = sin(x) = sin t + w * f6
fma.d.s0 f8 = f12, f7, f8;; // c = cos(x) = cos t + w * f7
```

Figure 1. Compiler-Generated Assembly Language Code for Computing Sin(x) and Cos(x)

loop entry. Thus all the constants used by circular functions would be loaded into registers before the start of the loop, and only accessed from registers during execution of the loop. In a loop such as:

```
for (i = 0; i < n; i++) {
    s[i] = sin(x[i]);
    c[i] = cos(x[i]);
}
```

the invocation of both sine and cosine requires only three floating point operations beyond only invoking either one. The loop will be software pipelined, giving performance associated with vector machines, as shown in Table 1.

Of course, there is only one way to compute these trig functions; no shortcuts are taken during `trigstart`, whether or not it is inlined. HP-UX's `trigstart` implements the careful Payne-Hanek [5] argument reduction, but it is clever enough to only perform it when needed[3]. On the average, it is invoked once in 16000 evaluations. It adds about 40 cycles, when needed, to the evaluation of trigstart, or to a standard closed sine or cosine function.

Using C99 [6] the function $\text{cis}\, x = \cos x + i \sin x$ can be realized efficiently with a tiny program.

```
#include <math.h>
#include <complex.h>
//In C99 complex arithmetic
//I is the imaginary unit.
double complex cis(double x) {
    return cos(x) + I * sin(x);
}
```

With compiler assistance as proposed here, optimal assembly code is produced, similar to the code shown in Figure 1.

Table 1 shows the timing of the double precision sine and cosine functions, when invoked in isolation or together. For

	Closed Routine	Software Pipelined
$\sin x$	53	7.37
$\cos x$	53	7.37
$\sin x$ and $\cos x$ (compiler assist)	53	9.16
$\sin x$ and $\cos x$ (separate calls)	106	15.37

Table 1. Timings for Itanium II (in machine cycles)

computing both functions, timings are given for the method described in this paper, as well as for computation by two function calls. The software pipelined loops were not unrolled.

4 Conclusion

By decomposing each trigonometric function into two routines:

a routine with operations common to all the functions, and

a short finishing routine that combines the results of the common routine,

it is now possible to invoke several trigonometric functions of the same argument without introducing a new API for special functions that compute them with one invocation. The advantage is that the standard function names are the only names that need be used. Standard compiler optimization removes redundant computations of the code common to trigonometric functions of the same argument.

Whether the trigonometric functions are evaluated by closed routines or inlined into what will become a vectorized loop, only the standard function names need be referenced. If more than one trigonometric function is evaluated for the same argument, only one instance of the trigstart evaluation is performed, and in no case are the precision or corner case behavior of the routines compromised for performance. The desired effect of computing both $\sin x$ and $\cos x$ together in the same time as it takes to compute one of them is accomplished without introducing new special functions into the library.

Two similar approaches pertain to the hyperbolic functions. Square root and reciprocal square root also respond to a similar treatment. This approach is particularly advantageous for hardware which lacks a square root instruction; on such machines the square root is typically computed in terms of the reciprocal square root.

5 Acknowledgements

Realizing the method described here in HP-UX was the result of close collaboration of many people. Jim Thomas led the effort, and was supported by Ren-Cang Li and Jon Okada from his Math Library project[2]. On the compiler side, Kevin Crozier, Theresa Johnson and David Gross designed and implemented the required compiler modifications.

References

[1] S. Gal, B. Bachelis, *An Accurate Elementary Mathematical Library for the IEEE Floating Point Standard*, ACM Transactions on Mathematical Software 17, 1 (March 1991), 26-45

[2] Ren-Cang Li, Peter Markstein, Jon P. Okada, James W. Thomas, *The Libm Library and Floating-Point Arithmetic in HP-UX for Itanium 2*, 2002, http://h21007.www2.hp.com/dspp/files/unprotected/Itanium/FP_White_Paper_v2.pdf

[3] Peter Markstein, *IA-64 and Elementary Functions: Speed and Precision*, Prentice-Hall, New Jersey, 2000.

[4] Steven S. Muchnick, *Advanced Compiler Design and Implementation*, Morgan Kaufman, San Francisco, CA 1997.

[5] Jean-Michel Muller, *Elementary Functions - Algorithms and Implementation*, Birkhauser, Boston·Basel·Berlin, 1997.

[6] *Programming Languages – C*, International Standard ISO/IEC9899:1999(E), 1999.

Session 7:
Testing and Error Analysis

Chair: Tanya Vladimirova

Worst Cases and Lattice Reduction

Damien Stehlé
ENS Paris
45 rue d'Ulm
F-75005 Paris
damien.stehle@ens.fr

Vincent Lefèvre
LORIA/INRIA Lorraine
Technopôle de Nancy-Brabois
615 rue du Jardin Botanique
F-54602 Villers-lès-Nancy Cedex
lefevre@loria.fr

Paul Zimmermann
LORIA/INRIA Lorraine
Technopôle de Nancy-Brabois
615 rue du Jardin Botanique
F-54602 Villers-lès-Nancy Cedex
zimmerma@loria.fr

Abstract

We propose a new algorithm to find worst cases for correct rounding of an analytic function. We first reduce this problem to the real *small value problem — i.e. for polynomials with real coefficients. Then we show that this second problem can be solved efficiently, by extending Coppersmith's work on the* integer *small value problem — for polynomials with integer coefficients — using lattice reduction [4, 5, 6].*

For floating-point numbers with a mantissa less than N, and a polynomial approximation of degree d, our algorithm finds all worst cases at distance $< N^{\frac{-d^2}{2d+1}}$ from a machine number in time $O(N^{\frac{d+1}{2d+1}+\varepsilon})$. For $d = 2$, this improves on the $O(N^{2/3+\varepsilon})$ complexity from Lefèvre's algorithm [15, 16] to $O(N^{3/5+\varepsilon})$. We exhibit some new worst cases found using our algorithm, for double-extended and quadruple precision. For larger d, our algorithm can be used to check that there exist no worst cases at distance $< N^{-k}$ in time $O(N^{\frac{1}{2}+O(\frac{1}{k})})$.

1 Introduction

The IEEE-754 standard for binary floating-point arithmetic [11], approved in 1985 by the IEEE Standards Board and the American National Standards Institute, requires that all four basic arithmetic operations ($+$, $-$, $\times$, $\div$) and the square root are correctly rounded. For a given function, floating-point inputs for which it is difficult to guarantee correct rounding, called *worst cases*, are numbers for which the exact result — as computed in infinite precision — is near a machine number, or near the middle of two consecutive machine numbers. This is the famous "Table Maker's Dilemma" problem (TMD for short). Several authors [13, 21, 12, 14, 19] have shown that for the class of algebraic functions, such worst cases cannot be too near from a machine number or the middle of two consecutive machine numbers. Such bounds enable one to design some efficient algorithms that guarantee correct rounding for division and square root, and less efficient algorithms for other algebraic functions.

However, for non-algebraic functions, number theory bounds are not sharp enough, which makes correct rounding harder to implement. This is probably the reason why the IEEE-754 standard does not require correct rounding for those functions. Muller and other authors proposed in [20] to introduce different levels of quality for transcendental functions. This proposal was presented by Markstein at the May 2002 meeting of the IEEE-754 revision group, but the conclusion was that *"we're not yet ready to standardize"*.

Systematic work on the Table Maker's Dilemma was done by Lefèvre and Muller [16], who published worst cases for many elementary functions in double precision ($N = 2^{53}$), over the full range for some functions. Alas, their approach is too expensive to deal with the quadruple precision, which is included in the current revision of the IEEE-754 standard. Thus currently the only possible approaches for higher precisions are either to guess a reasonable bound on the precision required for the hardest to round cases and to write a library computing up to that precision, or to write a generic multiple-precision library. For instance, Ziv's MathLib library does the former, where the guessed bound is 768 bits for double precision [22].

Having an efficient algorithm to find the hardest to round cases, for a given function and a given floating-point format, would help to replace guessed bounds — which are usually overestimated — by sharper and rigorous bounds. It would thus enable one to design very efficient libraries with correct rounding [7, 8]. Then there would be no good reason any more to exclude those functions from the correct rounding requirements of the IEEE-754 standard.

Exhaustive search methods consist in finding the hardest to round cases of the given function in the given range. They give the best possible bound, but are very time-consuming. Moreover, a search for a given precision gives little knowledge for another precision. We propose here a new algorithm belonging to that class. It naturally extends the first algorithm proposed by Lefèvre [15], and is based on Coppersmith's ideas.

Previous related work was done by Elkies, who gives in [9] a new algorithm using lattice reduction to find all rational points of small height near a plane curve; for example, his record:

$$5853886516781223^3 - 447884928428402042307918^2 = 1641843$$

corresponds to a worst case of the function $x^{3/2}$ for a 53-bit input and a 79-bit output; his other example

$$2220422932^3 - 283059965^3 - 2218888517^3 = 30$$

corresponds to a worst case of the function $(x^3 + y^3)^{1/3}$ in 32-bit arithmetic. More recently Gonnet [10] also used lattice reduction to find worst cases, however his approach seems equivalent to Lefèvre's algorithm.

Our paper is organized as follows: Section 2 explains in mathematical terms the problem we want to solve, recalls Lefèvre's algorithm and analyzes its complexity. Section 3 describes our new algorithm, after a short survey on lattice reduction and Coppersmith's work, which we heavily use. Section 4 presents some new worst cases found with our algorithm for the 2^x function, in double-extended precision and quadruple precision. Section 5 discusses some ideas for possible improvements and open questions.

2 Preliminaries

2.1 Definitions and Notations

We assume we work here with floating-point numbers with a mantissa of n bits. Let $N = 2^n$; for instance, $N = 2^{53}$ corresponds to double precision, $N = 2^{64}$ corresponds to double-extended precision, and $N = 2^{113}$ corresponds to quadruple precision. A worst case for a function f is a floating-point number x such that $f(x)$ has m identical bits after the round bit. If all those m bits equal (resp. differ from) the round bit, x is a worst case for directed rounding (resp. rounding to nearest).

For sake of simplicity, we consider here directed rounding only (towards $-\infty$, towards $+\infty$, towards zero), since worst cases at precision n for all rounding modes are worst cases at precision $n + 1$ for directed rounding. (In general, we are interested in the worst cases for the inverse function too, in which case inputs are also chosen at precision $n + 1$ [16].) To find worst cases for directed rounding, we throw away the first n significant bits of the result

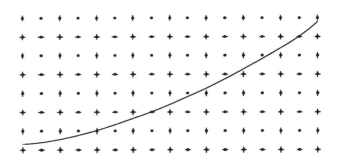

Figure 1. A function graph and the grid of machine numbers. Worst cases correspond to grid points with a small vertical distance to the curve.

mantissa. Then a worst case of length m corresponds to $|Nf(x) \bmod 1| < 2^{-m}$, where $x \bmod 1 := x - \lfloor x \rceil$ denotes the "centered" fractional part (see Fig. 1).

We also consider that both argument x and result $y = f(x)$ are normalized, i.e. $\frac{1}{2} \leq x, f(x) < 1$. This is easy to achieve by multiplying x or $f(x)$ by some fixed powers of 2, unless the exponent of $f(x)$ varies a lot in the considered range. This excludes the case of numerically irregular functions like $\sin x$ for large x. Given a polynomial approximation $P(t)$ to $Nf(\frac{t}{N})$ (for example, a Taylor expansion), the Table Maker's Dilemma can be reduced to the following problem.

REAL SMALL VALUE PROBLEM (REAL SVALP): Given positive integers M and T, and a polynomial P with real coefficients, find all integers $|t| < T$ such that

$$|P(t) \bmod 1| < \frac{1}{M}. \tag{1}$$

REMARK 1: The mantissa bound N does not appear explicitly in the real SValP, however the polynomial $P(t)$ depends on N, and so does the error made in the polynomial approximation.

REMARK 2: If the fractional bits of the function behave randomly, we can expect $\approx \frac{T}{M}$ worst cases. Therefore we may assume $T \ll M$ if we want only few worst cases.[1]

2.2 Lefèvre's Algorithm

Lefèvre's algorithm [15, 16] works as follows. One considers a linear approximation to the function f on small intervals. Those approximations are computed from higher order polynomial approximations on larger intervals, using an efficient scheme based on the "table of differences" method. On each small interval, worst cases are found using a modified version of the Euclidean algorithm, which gives a lower bound for $|Nf(\frac{t}{N}) \bmod 1|$ on that interval.

---
[1]The notation $x \ll y$ is equivalent to $x = O(y)$.

Assume $f(x) = a_0 + a_1(x-x_0) + a_2(x-x_0)^2 + O((x-x_0)^3)$ around x_0. Since we neglect the terms of order two or more in $Nf(\frac{t}{N})$, we must have $|a_2\frac{T^2}{N}| \ll \frac{1}{M}$ so that the error coming from the polynomial approximation does not exceed the distance $\frac{1}{M}$. Together with $T \ll M$, it follows $T \ll N^{1/3}$. Therefore the complexity of Lefèvre's algorithm is $O(N^{2/3+\varepsilon})$, since we have to consider $\frac{N}{T} \approx N^{2/3}$ small intervals to check a complete mantissa range.

In practice, Lefèvre's algorithm is expensive but feasible for the double precision ($N^{2/3} \approx 4 \cdot 10^{10}$), near from the limits of the current processors for double-extended precision ($N^{2/3} \approx 7 \cdot 10^{12}$), and out of reach for quadruple precision ($N^{2/3} \approx 5 \cdot 10^{22}$).

3 A New Algorithm Using Lattice Reduction

In this section, we first state some basic facts about lattices — we refer to [18] for an introduction to that subject — and we explain Coppersmith's technique, on which our algorithm is based. Then we introduce the algorithm, we prove its correctness and we analyze its complexity.

3.1 Some Basic Facts in Lattice Reduction Theory

A *lattice* L is a discrete subgroup of $\mathbb{R}^n$, or equivalently the set of all linear integral combinations of $\ell \leq n$ linearly independent vectors over $\mathbb{R}$, that is:

$$L = \{\sum_{i=1}^{\ell} x_i \mathbf{b}_i | x_i \in \mathbb{Z}\}.$$

We define the *determinant*, also called the *volume*, of the lattice L as: $\det(L) = \prod_{i=1}^{\ell} ||\mathbf{b}_i^*||$, where $||.||$ is the Euclidean norm and $[\mathbf{b}_1^*, ..., \mathbf{b}_\ell^*]$ is the Gram-Schmidt orthogonalization of $[\mathbf{b}_1, ..., \mathbf{b}_\ell]$. The *basis* $[\mathbf{b}_1, ..., \mathbf{b}_\ell]$ of L is not unique and on an algorithmic point of view, only bases which consist of short linearly independent vectors of L are of interest. Those so-called *reduced bases* always exist and can be computed in polynomial time with the well-known LLL algorithm [17].

Theorem 1 *Given a basis $[\mathbf{b}_1, ..., \mathbf{b}_\ell]$ of a lattice $L \subset \mathbb{Z}^n$, the LLL algorithm provides in polynomial time in the bit length of the input, a basis $\{v_1, ..., v_\ell\}$ satisfying*[2]:

1. $||v_1|| \leq 2^{\frac{\ell}{2}} \det(L)^{\frac{1}{\ell}}$;

2. $||v_2|| \leq 2^{\frac{\ell}{2}} \det(L)^{\frac{1}{\ell-1}}$.

Coppersmith (see [4, 5], or [6] for a better description) recently found an important consequence of this theorem: one can compute in a time polynomial in $\log A$ the small roots of a modular multivariate polynomial modulo an integer A. His method proved very powerful to factorize integers when some bits of the factors are known and to forge cryptographic schemes (see [2, 3, 4] for example). Our new algorithm intensely uses that technique.

3.2 The Integer Small Value Problem

The problem that will prove interesting in our case is the following: given a univariate polynomial $P \in \mathbb{Z}[x]$ of degree d, find on which small integer entries it has small values modulo a large integer A. Equivalently, we are looking for the small integer roots of the bivariate polynomial:

$$Q(x, y) = P(x) + y \quad (\text{mod } A).$$

We now explain how Coppersmith's technique helps solving it. First let α be a positive integer (that will grow later to infinity), and assume (x_0, y_0) is a root of Q modulo A. We consider the family of polynomials $Q_{i,j}(x,y) = x^i Q^j(x,y) A^{\alpha-j}$ with $0 \leq i + dj \leq d\alpha$. Then (x_0, y_0) is a root modulo A^α of each $Q_{i,j}$, whence of each linear combination of them.

Our goal is to build two integer combinations of those polynomials, $v_1(x, y)$ and $v_2(x, y)$, which take small values — i.e. less than A^α — for small x and y, more precisely $|x| \leq X$ and $|y| \leq Y$ for fixed bounds X and Y. Thus, if (x_0, y_0) is a small root of v_1 and v_2 modulo A^α, (x_0, y_0) is also a root of v_1 and v_2 over $\mathbb{Z}$. Finally, (x_0, y_0) will be found by looking at the integer roots of the resultant $\text{Res}_y(v_1, v_2) \in \mathbb{Z}[x]$.

It remains to explain how to find those two polynomials. For this we consider the lattice of dimension $\frac{(\alpha+1)(d\alpha+2)}{2}$ generated by the vectors associated with the polynomials $Q_{i,j}(X\tau, Yv)$: the vector linked to a bivariate polynomial $\sum_{i,j} a_{i,j} \tau^i v^j$ has its $\tau^i v^j$ coordinate equal to $a_{i,j}$. We give here the shape of the matrix we get in the case $d = 3$ and $\alpha = 2$.

$$\begin{pmatrix}
A^2 & & & & & & & & & & \\
 & A^2X & & & & & & & & & \\
 & & A^2X^2 & & & & & & & & \\
 & & & A^2X^3 & & & & & & & \\
 & & & & A^2X^4 & & & & & & \\
 & & & & & A^2X^5 & & & & & \\
 & & & & & & A^2X^6 & & & & \\
- & - & - & - & - & - & - & AY & & & \\
- & - & - & - & - & - & - & & AXY & & \\
- & - & - & - & - & - & - & & & AX^2Y & \\
- & - & - & - & - & - & - & & & & AX^3Y \\
- & - & - & - & - & - & - & - & - & - & - & Y^2
\end{pmatrix}$$

(row labels i/j: 0/0, 1/0, 2/0, 3/0, 4/0, 5/0, 6/0, 0/1, 1/1, 2/1, 3/1, 0/2)

Since we get a triangular matrix, the calculation of the determinant is obvious [3]:

$$\det(L) = A^{\frac{d}{3}\alpha^3 + o(\alpha^3)} \cdot X^{\frac{d^2}{6}\alpha^3 + o(\alpha^3)} \cdot Y^{\frac{d}{6}\alpha^3 + o(\alpha^3)}.$$

Therefore, by Theorem 1, where here the lattice dimension satisfies $\ell \sim \frac{d}{2}\alpha^2$, the LLL algorithm gives us two

---
[2] This is not the strongest result, but is sufficient for our needs.

[3] The $o()$ makes sense when α grows to infinity

vectors $\mathbf{v}_1$ and $\mathbf{v}_2$ of norms that are asymptotically less than $A^{\frac{2}{3}\alpha+o(\alpha)} \cdot X^{\frac{d}{3}\alpha+o(\alpha)} \cdot Y^{\frac{1}{3}\alpha+o(\alpha)}$. Those vectors $\mathbf{v}_1$ and $\mathbf{v}_2$ correspond to two polynomials $v_1(X\tau, Yv)$ and $v_2(X\tau, Yv)$. Moreover if $|x| \leq X$ and $|y| \leq Y$, then $|v_k(x,y)| \leq \sum_{i,j} |v_{i,j}^{(k)}| X^i Y^j \frac{|x|^i |y|^j}{X^i Y^j} \leq c \cdot \max |v_{i,j}^{(k)}| X^i Y^j| \leq c \cdot ||\mathbf{v}_k||$ for a certain constant c. Thus, to get $|v_k(x,y)| < A^\alpha$, it is sufficient that:

$$c \cdot A^{\frac{2}{3}\alpha+o(\alpha)} \cdot X^{\frac{d}{3}\alpha+o(\alpha)} \cdot Y^{\frac{1}{3}\alpha+o(\alpha)} < A^\alpha,$$

which asymptotically gives the bound $X^d Y \ll A$.

Using Coppersmith's technique, one can thus solve the integer SValP in polynomial time as long as $X^d Y < A^{1-\epsilon}$. In fact, this is not completely true because we used an argument we cannot prove: we assumed that $\mathrm{Res}_y(v_1, v_2) \neq 0$. This heuristic has been made very often in cryptography (see [2, 3, 4]).

3.3 The SLZ Algorithm

The real SValP is the following problem: Given a polynomial P, find for which integers t, $P(t)$ is near an integer. We solve this problem by reducing it to the integer SValP. The difficulty is that $P(t)$ has real coefficients, and the LLL algorithm does not work well with real input. The following algorithm overcomes that difficulty (we present here a complete algorithm to solve the Table Maker's Dilemma, in which case $P(t) = Nf(t/N)$, but the sub-algorithm consisting of steps 3 to 11 may be of interest to solve the real SValP itself).

Input: a function f, positive integers N, T, M, d, α
Output: all worst cases at distance $< 1/M$ for $f(\frac{t}{N})$ for $|t| \leq T$

1. Let $P(t)$ be the Taylor expansion of $Nf(\frac{t}{N})$ up to order d, and $n = \frac{(\alpha+1)(d\alpha+2)}{2}$
2. Compute a bound ϵ such that $|P(t) - Nf(\frac{t}{N})| < \epsilon$ for $|t| \leq T$
3. Let $M' = \lfloor \frac{1/2}{1/M+\epsilon} \rfloor$, $C = (d+1)M'$, and[4] $P'(\tau) = \lfloor CP(T\tau) \rceil$, $Q(t) = T^d P'(\frac{t}{T})$
4. Let $\{e_1, \ldots, e_n\} \leftarrow \{\tau^i v^j\}$ for $0 \leq i + dj \leq d\alpha$
5. Let $Y = (d+1)T^d$, $A = CT^d$ and $\{g_1, \ldots, g_n\} \leftarrow \{(T\tau)^i (Q(T\tau) + Yv)^j A^{\alpha-j}\}$ for $0 \leq i + dj \leq d\alpha$
6. Form the $n \times n$ integral matrix L where $L_{k,l}$ is the coefficient of the monomial e_k in g_l
7. $V \leftarrow \mathrm{LatticeReduce}(L)$
8. Let $\mathbf{v}_1, \mathbf{v}_2$ be the two smallest vectors from V, and $Q_1(T\tau, Yv)$ and $Q_2(T\tau, Yv)$ the corresponding polynomials

9. If there exists $(t, y) \in [-T, T] \times [-Y, Y]$ with $|Q_1(t, y)| \geq A^\alpha$ or $|Q_2(t, y)| \geq A^\alpha$, return(FAIL)
10. $p(t) \leftarrow \mathrm{Res}_y(Q_1(t, y), Q_2(t, y))$; if $p(t) = 0$ then return(FAIL)
11. **for** each t_0 in IntegerRoots($p(t), [-T, T]$) **do**
    if $|Nf(\frac{t_0}{N}) \bmod 1| < 1/M$ **then** output t_0.

3.4 Correctness of the Algorithm

Theorem 2 *In case algorithm SLZ does not return FAIL, it behaves correctly, i.e. it prints exactly all integers $t \in [-T, T]$ such that $|Nf(\frac{t}{N}) \bmod 1| < 1/M$.*

PROOF. Because of the final check in step 11, we only have to verify that no worst case is missed. Suppose there is $t_0 \in [-T, T]$ with $|Nf(\frac{t_0}{N}) \bmod 1| < 1/M$. We first prove that $|Q(t_0) \bmod CT^d| \leq (d+1)T^d$. From the definition of P, $|P(t_0) \bmod 1| < 1/M + \epsilon \leq \frac{1}{2M'}$. Since $|CP(T\tau) - P'(\tau)| \leq \frac{d+1}{2}$ for $|\tau| \leq 1$, by choosing $\tau = t_0/T$ we get $|Q(t_0) \bmod CT^d| < \frac{(d+1)T^d}{2} + \frac{(d+1)T^d}{2}$.

Whence $Q(t) + y = 0 \bmod CT^d$ has a root (t_0, y_0) with $|t_0| \leq T$ and $|y_0| < (d+1)T^d$. Since $Q_1(t,y)$ and $Q_2(t,y)$ are linear combinations of $Q(t) + y$ and its powers, then (t_0, y_0) is a common root of $Q_1(t, y)$ and $Q_2(t, y)$ modulo CT^d, and even over the reals since $|Q_1|, |Q_2| < CT^d$. Thus t_0 is an integer root of $\mathrm{Res}_y(Q_1(t,y), Q_2(t,y))$, and will be found at step 11. □

3.5 Choice of Parameters and Complexity Analysis

3.5.1 Coppersmith's Bound

Because of the use of Coppersmith's technique in our algorithm, to insure the algorithm does not return FAIL at step 9, the bound "$X^d Y \ll A$" has to be verified. In our case, X corresponds to T, Y to $(d+1)T^d$ and A to CT^d, so we get:

$$T \ll M^{\frac{1}{d}}.$$

3.5.2 Choice of the Degree d With Respect to T

Let $(a_i)_i$ the Taylor coefficients of f. Since we neglect Taylor coefficients of degree $d+1$ and greater, the error made in the approximation to $Nf(\frac{t}{N})$ by $P(t)$ is $\approx a_{d+1} T^{d+1} N^{-d}$. Since we are looking for worst cases with $|P(t) \bmod 1| < 1/M$, we want $T^{d+1} N^{-d} \ll 1/M$, i.e. $T^{d+1} \ll N^d/M$.

3.5.3 Complexity Analysis

Thus we have two bounds for T: the first one $T \ll M^{1/d}$ comes from Coppersmith's method, the second one

---
[4] The notation $\lfloor CP(T\tau) \rceil$ means that we round to the nearest integer each coefficient of $CP(T\tau)$. This provides an element of $\mathbb{Z}[\tau]$.

$T^{d+1} \ll N^d/M$ comes from the accuracy of the Taylor expansion. Therefore for $M \ll N^{\frac{d^2}{2d+1}}$, Coppersmith's bound wins and implies $T \ll M^{1/d}$, whereas for $M \gg N^{\frac{d^2}{2d+1}}$, Lagrange's bound gives $T^{d+1} \ll N^d/M$. The largest bound for T is obtained for $M \sim N^{\frac{d^2}{2d+1}}$, with $T \ll N^{\frac{d}{2d+1}}$. For $d = 1$, we find the constraint $T \ll N^{1/3}$ from Lefèvre's method; for $d = 2$, this gives $T \ll N^{2/5}$ with $M \sim N^{4/5}$; for $d = 3$, this gives $T \ll N^{3/7}$ with $M \sim N^{9/7}$. With $M \sim N^k$, we get a best possible interval length $T \sim N^{\frac{1}{2} - \frac{1}{8k} + o(\frac{1}{k})}$.

3.5.4 Working Precision

In step 1, we can use floating-point coefficients in the Taylor expansion $P(t)$ instead of symbolic coefficients, as long as it introduces no error in step 3 while computing $P'(\tau)$. Let a_i be the ith Taylor coefficient of f. Then to get $P'(\tau)$ correct at step 3, the error on $CN(T/N)^i a_i$ must be less than $1/2$, thus the error on a_i must be less than $1/(2CN)(N/T)^i$. Since $N \geq T$, it thus suffices to compute a_i with $\log_2(2CN)$ bits after the binary point.

REMARK 3: When searching worst cases with $M \ll N$, degree 2 is enough. Indeed, $N^{1-d}T^d \ll N^{1-d}M$ since $T^d \ll M$ (Coppersmith's bound), and for $d \geq 3$, $N^{1-d}T^d \ll N^{2-d} \ll 1/N \ll 1/M$. Thus all Taylor terms of degree ≥ 3 give a negligible contribution to $Nf(\frac{t}{N})$, and the largest value of T is $N^{2/5}$, giving a complexity of $N^{3/5}$ to search a whole range of $N/2$ values. More generally, for $M \ll N^k$, degree $2k$ is enough, giving a complexity of $N^{\frac{2k}{4k+1}}$.

4 Experimental Results

We have implemented algorithm SLZ in the Pari/GP system (version 2.2.4-alpha) [1] and experimented it on a Athlon XP 1600+ under Linux. We have chosen the 2^x function since it is the easiest one, with only one exponent range to study. Fig. 2 shows for each target precision (double, double-extended, quadruple), and for $M \approx N$ and $M \approx N^2$, the best parameters (T, d, and α) for our method, together with the estimated time to check the whole exponent range, i.e. $N/2$ floating-point numbers. For each precision, the first row gives the best parameters for the $d = \alpha = 1$ case, which is what Gonnet considers in [10]; comparing that first row to the following ones shows the speedup obtained. For $M \approx N$, the speedup increases from 3 to 17, whence is not dramatic. However for $M \approx N^2$, we get a speedup of about 1000 in quadruple precision with respect to the naive method ($d = \alpha = 1$), with $(d, \alpha) = (4, 2)$.

Fig. 3 shows a few worst cases found using algorithm SLZ for double-extended and quadruple precision. These experiments tend to show that with a carefully tuned implementation, and several computers running a few months, solving the Table Maker's Dilemma for the double-extended precision is nowadays feasible for simple elementary functions.

	N	M	T	d	α	est. time
double	2^{53}	2^{28}	2^{15}	1	1	560 days
	2^{53}	2^{53}	2^{20}	2	2	120 days
precision	2^{53}	2^{106}	2^{25}	4	2	45 days
double	2^{64}	2^{32}	2^{19}	1	1	140 years
extended	2^{64}	2^{64}	2^{24}	2	2	43 years
precision	2^{64}	2^{128}	2^{30}	4	2	9 years
quadruple	2^{113}	2^{70}	2^{35}	1	1	1600 Gyears
	2^{113}	2^{113}	2^{43}	2	2	94 Gyears
precision	2^{113}	2^{226}	2^{53}	4	2	1.6 Gyears

Figure 2. Best experimental parameters for double, double-extended and quadruple precision, and estimated time for an exponent range of $N/2$ values.

N	t_0	$N2^{-1/2+t_0/N} \mod 1$
2^{64}	586071771766963	$0.1 1^{47} 001111...$
2^{64}	594068190588573	$0.0 0^{48} 100010...$
2^{64}	891586182147388	$0.1 1^{50} 001000...$
2^{64}	9014384889202147	$0.0 1^{53} 010011...$
2^{64}	9602866023852631	$0.0 0^{54} 111001...$
2^{113}	1119374922072865495	$0.0 1^{63} 000000...$
2^{113}	8923960372306650064	$0.0 0^{64} 101011...$
2^{113}	43616445401128570224	$0.0 1^{65} 011110...$
2^{113}	53608038600996804036	$0.0 1^{67} 000001...$

Figure 3. Some worst cases found for the 2^x function in double-extended and quadruple precision.

5 Possible Improvements, Open questions

We have presented a new algorithm, based on lattice reduction, to search for worst cases for correct rounding of analytic functions. The first experimental results show that algorithm SLZ is quite efficient, especially to detect worst cases at distance much less than 2^{-n}, where n is the target precision. However the efficiency largely depends on the function considered, like in Lefèvre's algorithm.

Several open questions remain. Does this approach extend like in the modular case ([4]) to functions of two variables like x^y or $\arctan \frac{x}{y}$?

Our algorithm is complementary to that of Elkies [9],

which works well when $M \ll N$ (in our notation), i.e. when we expect many worst cases, whereas our algorithm is more efficient when $M \gg N$, i.e. when we expect only few worst cases, or none. However, in the case of $f(x) = x^{3/2}$, related to Hall's conjecture, Elkies proposes a special-purpose algorithm to find all worst cases at distance $< 1/N$ in $O(N^{1/2+\varepsilon})$. Does this algorithm generalize to other algebraic functions?

References

[1] C. Batut, K. Belabas, D. Bernardi, H. Cohen, and M. Olivier. *User's Guide to PARI/GP*, 2000. ftp://megrez.math.u-bordeaux.fr/pub/pari/manuals/users.pdf.

[2] D. Boneh and G. Durfee. Cryptanalysis of RSA with private key d less than $n^{0.292}$. In *Proceedings of Eurocrypt'99*, volume 1592 of *Lecture Notes in Computer Science*, pages 1–11. Springer-Verlag, 1999.

[3] D. Boneh, G. Durfee, and N. Howgrave-Graham. Factoring $n = p^r q$ for large r. In *Proceedings of Eurocrypt'99*, volume 1592 of *Lecture Notes in Computer Science*, pages 326–337. Springer-Verlag, 1999.

[4] D. Coppersmith. Finding a small root of a bivariate integer equation; factoring with high bits known. In *Proceedings of Eurocrypt'96*, volume 1070 of *Lecture Notes in Computer Science*, pages 178–189. Springer-Verlag, 1996.

[5] D. Coppersmith. Finding a small root of a univariate modular equation. In *Proceedings of Eurocrypt'96*, volume 1070 of *Lecture Notes in Computer Science*, pages 155–165. Springer-Verlag, 1996.

[6] D. Coppersmith. Finding small solutions to small degree polynomials. In *Proceedings of CALC'01*, volume 2146 of *Lecture Notes in Computer Science*, pages 20–31. Springer-Verlag, 2001.

[7] D. Defour and F. de Dinechin. Software carry-save for fast multiple-precision algorithms. Research Report 2002-08, Laboratoire de l'Informatique du Parallélisme, 2002. 10 pages.

[8] D. Defour, F. de Dinechin, and J.-M. Muller. Correctly rounded exponential function in double precision arithmetic. Research Report 2001-26, Laboratoire de l'Informatique du Parallélisme, 2001. 21 pages.

[9] N. Elkies. Rational points near curves and small nonzero $|x^3 - y^2|$ via lattice reduction. In *Proceedings of ANTS-IV*, volume 1838 of *Lecture Notes in Computer Science*, pages 33–63. Springer-Verlag, 2000.

[10] G. Gonnet. A note on finding difficult values to evaluate numerically. http://www.inf.ethz.ch/personal/gonnet/FPAccuracy/NastyValues.ps, Sept. 2002. 3 pages.

[11] IEEE standard for binary floating-point arithmetic. Technical Report ANSI-IEEE Standard 754-1985, New York, 1985. approved March 21, 1985: IEEE Standards Board, approved July 26, 1985: American National Standards Institute, 18 pages.

[12] C. S. Iordache and D. W. Matula. Infinitely precise rounding for division, square root, and square root reciprocal. In *Proceedings of the 14th IEEE Symposium on Computer Arithmetic*, pages 233–240. IEEE Computer Society, 1999.

[13] W. Kahan. A test for correctly rounded SQRT. Lecture note, University of California at Berkeley, 1996. http://www.cs.berkeley.edu/~wkahan/SQRTest.ps, 4 pages.

[14] T. Lang and J.-M. Muller. Bounds on runs of zeros and ones for algebraic functions. In *Proceedings of the 15th IEEE Symposium on Computer Arithmetic*, pages 13–20. IEEE Computer Society, 2001.

[15] V. Lefèvre. *Moyens arithmétiques pour un calcul fiable*. Thèse de doctorat, École Normale Supérieure de Lyon, Jan. 2000.

[16] V. Lefèvre and J.-M. Muller. Worst cases for correct rounding of the elementary functions in double precision. In N. Burgess and L. Ciminiera, editors, *Proceedings of the 15th IEEE Symposium on Computer Arithmetic (ARITH'15)*, pages 111–118. IEEE Computer Society, 2001.

[17] A. K. Lenstra, H. W. Lenstra, and L. Lovász. Factoring polynomials with rational coefficients. *Mathematische Annalen*, 261:515–534, 1982.

[18] L. Lovász. An algorithmic theory of numbers, graphs and convexity. *SIAM lecture series*, 50, 1986.

[19] L. D. McFearin and D. W. Matula. Generation and analysis of hard to round cases for binary floating point division. In *Proceedings of the 15th IEEE Symposium on Computer Arithmetic*, pages 119–127. IEEE Computer Society, 2001.

[20] J.-M. Muller. Proposals for a specification of the elementary functions. In *Abstracts of SCAN'2002*, pages 54–55. Laboratory LIP6, Paris, France, 2002.

[21] M. Parks. Number-theoretic test generation for directed rounding. In *Proceedings of the 14th IEEE Symposium on Computer Arithmetic*, pages 241–248. IEEE Computer Society, 1999.

[22] A. Ziv. Fast evaluation of elementary mathematical functions with correctly rounded last bit. *ACM Trans. Math. Softw.*, 17(3):410–423, 1991.

Isolating critical cases for reciprocals using integer factorization

John Harrison
Intel Corporation, JF1-13
2111 NE 25th Avenue
Hillsboro OR, USA
johnh@ichips.intel.com

Abstract

One approach to testing and/or proving correctness of a floating-point algorithm computing a function f is based on finding input floating-point numbers a such that the exact result f(a) is very close to a "rounding boundary", i.e. a floating-point number or a midpoint between them. In the present paper we show how to do this for the reciprocal function by utilizing prime factorizations. We present the method and show examples, as well as making a fairly detailed study of its expected and worst-case behavior. We point out how this analysis of reciprocals can be useful in analyzing certain reciprocal algorithms, and also show how the approach can be trivially adapted to the reciprocal square root function.

1 Background

Suppose we have a floating-point algorithm computing a function that approximates a true mathematical function $f : \mathbb{R} \to \mathbb{R}$. For example, consider the following algorithm for the Intel® Itanium® architecture designed to compute a floating-point square root $\sqrt{a}$ using an initial reciprocal square root approximation followed by a sequence of fused multiply-adds. (In the actual implementation, the initial approximation instruction deals with special cases including $a = 0$.)

1. $y_0 = \texttt{frsqrta}(a)$
2. $H_0 = \frac{1}{2} y_0$ $S_0 = a y_0$
3. $d_0 = \frac{1}{2} - S_0 H_0$
4. $H_1 = H_0 + d_0 H_0$ $S_1 = S_0 + d_0 S_0$
5. $d_1 = \frac{1}{2} - S_1 H_1$
6. $H_2 = H_1 + d_1 H_1$ $S_2 = S_1 + d_1 S_1$
7. $d_2 = \frac{1}{2} - S_2 H_2$ $e_2 = a - S_2 S_2$
8. $H_3 = H_2 + d_2 H_2$ $S_3 = S_2 + e_2 H_2$
9. $e_3 = a - S_3 S_3$
10. $S = S_3 + e_3 H_3$

If an algorithm is, like this one, implemented by composing basic floating-point operations (rather than, say, some more complicated analysis of bit-patterns), then the value computed can usually be represented as the result of rounding some approximation $f^*(x) \approx f(x)$, the value before the final rounding. In this case, the final S results from rounding the exact value $S_3 + e_3 H_3$.

The algorithm will therefore round correctly for all inputs x such that $f^*(x)$ and $f(x)$ round to the same number (for all the rounding modes under consideration). In the concrete square root example, this means that $\sqrt{a}$ and $S_3 + e_3 H_3$ should always round the same way.

A sufficient condition for equivalent rounding behavior is that the two values $f^*(x)$ and $f(x)$ should never be separated by a rounding boundary, i.e. a floating-point number (for directed rounding) or a midpoint (for round-to-nearest). That is, there is never a rounding boundary m with $f(x) \leq m \leq f^*(x)$ or $f^*(x) \leq m \leq f(x)$, unless $f^*(x) = f(x)$. (Not quite a necessary condition in the round-to-nearest mode since if one is exactly equal to the rounding boundary and the other on the "right" side, the correct result will be obtained.) This is usually hard to establish by analytic reasoning. However, it is usually easy to establish some sort of relative error bound ϵ such that:

$$|f^*(x) - f(x)| \leq \epsilon |f(x)|$$

Therefore, misrounding can occur only when

$$|f(x) - m| \leq \epsilon |f(x)|$$

It is therefore interesting for purposes of both testing and proving correctness to deliberately concoct test points x to make the relative distance from a rounding boundary $|f(x) - m|/|f(x)|$ as small as possible. Indeed, irrespective of the details of the algorithms we are concerned with, these test points might be expected to display greatest sensitivity to the accuracy of $f^*(x)$ and so show up errors most easily.

For some basic algebraic functions, such special x can be found analytically using number-theoretic techniques [14, 11], in such a way that the very worst examples (having the smallest relative distance from a rounding boundary) are isolated. For transcendental functions, this is more difficult, but one can still generate good cases by exploiting local linearity and solving congruences. For double-precision it is feasible, though costly, to isolate the very worst examples [6].

One use of the points so obtained is to test floating-point functions. Indeed, Parks [11] reports that such testing exposed a bug in a commercial microprocessor. A more ambitious goal, realized for square root algorithms by Cornea [1], is to isolate a sufficiently large set of points that the correct behavior of the algorithm on these, in conjunction with an analytical proof that covers all other cases, gives a complete correctness proof of the algorithm in all cases. For example, if we can prove analytically that for all floating-point numbers x we have:

$$|f^*(x) - f(x)| \leq \epsilon |f(x)|$$

and that some set S_ϵ contains all points x where $|m - f(x)| \leq \epsilon |f(x)|$ for some rounding boundary m, the correctness of the algorithm in all cases is equivalent to the correctness just for the points in S_ϵ. If such sets can be found easily and they are not too large, this gives a very effective methodology for proofs of algorithms. The goal of this paper is to show how to isolate such special cases for the reciprocal (and reciprocal square root) function and demonstrate their applicability in such correctness proofs of algorithms.

2 Critical cases for quotient and reciprocal

We will in what follows consider a single floating-point format with precision p, which contains all the floating-point numbers concerned and is also the destination format for the result. We also ignore the possibility of overflow and underflow in computation sequences. This keeps the presentation simpler and accords well with the intended applications where all input numbers are double-extended and additional exponent range (but not precision) is available for intermediate computations. The results that follow can straightforwardly be refined for mixed-precision applications.

It's instructive to examine the problem for the general case of quotients, and then contrast the restriction to the reciprocal. In general, we seek floating-point numbers x and y such that x/y lies close to some w that is either itself a floating-point number or a midpoint between two floating-point numbers. Without loss of generality, we can assume:

$$x = 2^{e_x} a \quad 2^{p-1} \leq a < 2^p$$
$$y = 2^{e_y} b \quad 2^{p-1} \leq b < 2^p$$
$$w = 2^{e_w} m \quad 2^p \leq m < 2^{p+1}$$

where p is the floating-point precision and a, b and m, as well as the various e_i, are integers. Note that even values of m correspond to floating-point numbers and odd values correspond to midpoints. We are interested in how small the relative difference $|x/y - w|/|x/y|$ can become. This relative difference can be rewritten as:

$$\frac{|x/y - w|}{|x/y|} = |1 - wy/x| = |1 - 2^{-q} mb/a|$$

where $q = e_x - (e_w + e_y)$, and so

$$\frac{|mb - 2^q a|}{2^q a}$$

Given the ranges of the values a, b and m, we have

$$2^{2p-1} \leq mb < 2^{2p+1}$$

and

$$2^{q+p-1} \leq 2^q a < 2^{q+p}$$

It turns out that the only interesting cases are when $q = p$ or $q = p + 1$. For if $q \leq p - 1$ then $q + p \leq 2p - 1$ so we have

$$2^q a \leq 2^q (2^p - 1) < 2^{2p-1} \leq mb$$

(remember that the values a, b and m are integers so when $< 2^r$ they are actually $\leq 2^r - 1$) and so

$$\frac{|mb - 2^q a|}{2^q a} \geq 2^q / (2^q a) = 1/a > 2^{-p}$$

Similarly if $q = p + 2$ we have:

$$mb \leq (2^p - 1)(2^{p+1} - 1) < 2^{2p+1} \leq 2^{p+2} a < 2^{2p+2}$$

and therefore

$$\frac{|mb - 2^q a|}{2^q a} \geq (2^{p+1} + 2^p - 1)/(2^q a) > 2^{p+1}/2^{2p+2} = 2^{-(p+1)}$$

Finally, if $q \geq p + 3$ then $2^q a > 2mb$ and so

$$\frac{|mb - 2^q a|}{2^q a} > 1/2$$

In all these cases, the distance is at least $2^{-(p+1)}$. Therefore, when seeking cases where the distance is of order 2^{-2p} (for realistic p) we need only consider $q \in \{p, p+1\}$. This being the case, the denominator $2^q a$ is constrained to within a factor of 4, so the essential problem is to find how small

$$|mb - 2^q a|$$

can become for $q \in \{p, p+1\}$. Since the value is an integer, we can try to find small values by explicit consideration of the various possibilities in succession:

$$\begin{aligned}
mb &= 2^p a + 1 \\
mb &= 2^p a - 1 \\
mb &= 2^{p+1} a + 1 \\
mb &= 2^{p+1} a - 1 \\
mb &= 2^p a + 2 \\
mb &= 2^p a - 2 \\
mb &= 2^{p+1} a + 2 \\
mb &= 2^{p+1} a - 2 \\
mb &= 2^p a + 3 \\
&\ldots
\end{aligned}$$

It seems that the number of possible solutions of these equations is too large for this to be a practical approach. On the other hand, if we fix any one of the values a, b and m, the problem becomes tractable. If we fix either m or b then the problem becomes a set of linear congruences (with additional range restrictions filtering the possible solution set), which are easy to solve. If we consider the special case of the reciprocal, then we fix $a = 2^{p-1}$. This problem is also tractable, as we shall see, but has a somewhat different character. We just need to consider

$$\begin{aligned}
mb &= 2^{2p-1} + \delta \\
mb &= 2^{2p} + \delta
\end{aligned}$$

for successive small integers δ. In fact, the situation is even better, because once again no small values can arise in the former case because of the range limitation, except for the trivial $mb = 2^{2p-1}$; the next case must be $(2^p + 1)2^{p-1} = 2^{2p-1} + 2^{p-1}$. So we need only be concerned with solutions to

$$mb = 2^{2p} + \delta$$

for integers $2^{p-1} \leq b < 2^p$ and $2^p \leq m < 2^{p+1}$. Indeed, for small δ, it is easy to see that the two upper bounds imply the lower ones.

3 Factorization distribution

Our approach to the problem of finding all solutions to $mb = 2^{2p} + \delta$ (with p and δ fixed) is quite straightforward. We find the prime factorization of $2^{2p} + \delta$, and consider all possible ways of distributing these prime factors into two parts m and b subject to the appropriate range limitation $m < 2^{p+1}$ and $b < 2^p$. In general, we will refer to a factorization $n = ab$ of n with $a < A$ and $b < B$ as an (A, B)-*balanced* factorization.

Consider, for illustration, the case $p = 6$ and $\delta \in \{\pm 1, \pm 2, \pm 3\}$. In each case we find the prime factorization of $2^{2p} + \delta$:

$$\begin{aligned}
2^{12} + 1 &= 17 \cdot 241 \\
2^{12} - 1 &= 3^2 \cdot 5 \cdot 7 \cdot 13 \\
2^{12} + 2 &= 2 \cdot 3 \cdot 683 \\
2^{12} - 2 &= 2 \cdot 23 \cdot 89 \\
2^{12} + 3 &= 4099 \\
2^{12} - 3 &= 4093
\end{aligned}$$

In the cases $2^{12}+1$, $2^{12}+2$, $2^{12}+3$ and $2^{12}-3$, the largest factor is already $> 2^{p+1} = 128$, so there is no possible distribution obeying the range restrictions. For $2^{12} - 2$ there is exactly one such distribution:

$$m \cdot b = 89 \cdot (2 \cdot 23) = 89 \cdot 46$$

Note that the 'symmetrical' distribution is not admissible because $89 > 2^p$. For $2^{12} - 1$, there are four possible distributions:

$$\begin{aligned}
m \cdot b &= (3^2 \cdot 13) \cdot (5 \cdot 7) = 117 \cdot 35 \\
m \cdot b &= (3 \cdot 5 \cdot 7) \cdot (3 \cdot 13) = 105 \cdot 39 \\
m \cdot b &= (7 \cdot 13) \cdot (3^2 \cdot 5) = 91 \cdot 45 \\
m \cdot b &= (5 \cdot 13) \cdot (3^2 \cdot 7) = 65 \cdot 63
\end{aligned}$$

Note that the corresponding m are all odd, and therefore represent midpoints. Thus, we can say that $|1/y - w| \geq 4/2^{12}|1/y|$ for any midpoint w except in the cases where y's significand b is in the set $\{35, 39, 45, 46, 63\}$; for $b = 46$ we get a $2/2^{12}$ relative distance and for 35, 39, 45 and 63 we get $1/2^{12}$. Since the above lists exhausts all m, even or odd, we see that $|1/y - w| \geq 4/2^{12}|1/y|$ for any floating-point number w, except for the special cases when y is a power of 2 and so its reciprocal is exactly representable (i.e. $1/y = w$).

4 Implementation

The implementation of the above idea is straightforward, given any reasonable programming language. We have used Objective CAML, a very high-level functional language that we have previously used extensively for implementation of theorem proving code:

> http://www.ocaml.org/

This already has a multiprecision integer and rational function datatype available. It does not, however, have a built-in library for factoring numbers, and we did not want to write our own code for this operation — since the numbers can be as large as 2^{226} (for quad precision reciprocals), factorization is a non-trivial problem. We used the factoring code included in the PARI / GP system:

> http://www.parigp-home.de/

The documentation says:

> **factorint**(n, $\{flag = 0\}$): factors the integer n using a combination of the Shanks SQUFOF and Pollard Rho method (with modifications due to Brent), Lenstra's ECM (with modifications by Montgomery), and MPQS (the latter adapted from the LiDIA code with the kind permission of the LiDIA maintainers), as well as a search for pure powers with exponents ≤ 10.

We are not experts in the topic of factorization, but have been quite impressed with how fast it usually factors numbers. Only for quad precision, when the numbers are of the order 2^{226}, does it start to slow down noticeably. Rather than a strict primality test, the factors are only subjected to a strong probabilistic primality test. Therefore, out of paranoia, we have developed our own code to certify primality, by constructing prime certificates in the style of Pratt [12], appealing to Lucas's theorem. That is, to certify that each p occurring in PARI/GP's factorization is prime, we show that there is a primitive root a modulo p such that $a^{p-1} \equiv 1 \pmod{p}$ but $a^{\frac{p-1}{q}} \not\equiv 1 \pmod{p}$ for any prime factor q of $p-1$. (The primitive root a is found randomly, and the factors q of $p-1$ are found by using PARI/GP's factorization recursively, certifying those factors as primes too.) This certification slows down the factorization process by a moderate amount, so we sometimes switch it off when experimenting.

Once we have the prime factors, we need to test all ways of distributing them over two numbers subject to range restrictions. As noted, we need only apply the upper range restrictions $m < 2^{p+1}$ and $b < 2^p$. Roughly, we just naively enumerate all possibilities. In order to cut off choice points as soon as possible, we start distributing from the largest prime factors, i.e. consider the prime factors $p_1^{\alpha_1} \cdot p_2^{\alpha_2} \cdots p_k^{\alpha_k}$ in decreasing order $p_1 > p_2 > \cdots > p_k$. We first consider all $\alpha_1 + 1$ ways of distributing $p_1^{\alpha_1}$ into two parts. If any of these distributions already violate the range restriction, they are abandoned. Otherwise, for each one, we consider the $\alpha_2 + 1$ ways of distributing $p_2^{\alpha_2}$, and so on. The algorithm is very straightforward to program recursively in OCaml.

It might be doubted whether such a naive distribution algorithm is acceptably efficient. At least it has been adequate to obtain some results quite quickly for the main precisions that interest us, $p \in \{24, 53, 63, 113\}$. We first look at some of these results and then turn to a detailed performance analysis.

5 Results

Table 1 presents a small sample of the results obtained using the methods outlined above. For each of the four major precisions $p = 24, 53, 64, 113$, we list the 66 floating-point significands whose reciprocals are closest either to floating-point numbers or midpoints. This distance, as a multiple of the corresponding 2^{-2p}, is given in the 'd' columns. When, as often happens, several reciprocals have the same 'd' value we order them in decreasing order, and cut the table off on that basis. The asterisk means that the distance is from a floating-point number (and hence may be unimportant if we are concerned only with round-to-nearest).

Larger lists for d up to a few thousand can be generated for all these precisions without requiring more than a few days of runtime on a modern machine. And of course, it is trivial to parallelize the task since it consists of a separate subtask for each d considered.

6 Applications

We can use the techniques set out above in the design and verification of algorithms for correctly rounded reciprocals. These might be substituted by the programmer, or by the compiler if it can recognize that in an expression a/b, the constant a is guaranteed to be 1. (This could be generalized to any power of 2.) For example, the following algorithm is normally used for double-extended precision division (precision $p = 64$) on Intel® Itanium® processors.

1. $y_0 = \text{frcpa}(b)$
2. $d = 1 - by_0$ $q_0 = ay_0$
3. $d_2 = dd$ $d_3 = dd + d$
4. $y_1 = y_0 + y_0 d_3$ $d_5 = d_2 d_2 + d$
5. $y_2 = y_0 + y_1 d_5$ $r_0 = a - bq_0$
6. $e = 1 - by_2$ $q_1 = q_0 + r_0 y_2$
7. $y_3 = y_2 + ey_2$ $r = a - bq_1$
8. $q = q_1 + ry_3$

Single precision Mantissa	d	Double precision Mantissa	d	Extended precision Mantissa	d	Quad precision Mantissa	d
0x800000	0*	0x10000000000000	0*	0x8000000000000000	0*	0x10000000000000000000000000000000	0*
0xFFFFFF	1	0x1FFFFFFFFFFFFF	1	0xFFFFFFFFFFFFFFFF	1	0x1FFFFFFFFFFFFFFFFFFFFFFFFFFFFFFF	1
0xFE01FF	1	0x1FFFFFF8000001	1	0xD6329033D6329033	1	0x1FFFFFFFFFFFFFFE00000000000001	1
0xFC3237	1	0x1FD8CD299E8D79	2*	0xB7938C6947D97303	1	0x1B52F1BB6F8DC3F0D920E2F3D449B	1
0xF0FF0F	1	0x1FC94266515BC9	2*	0x99D0C486A0FAD481	1	0x19C1ECF3420D27F8729BA7E1AB31D	1
0xF02A3B	1	0x1F739BD459BEA2	2	0x989E556CADAC2D7F	1	0x17ABDE305BAC595488190B4AD7657	1
0xF00FF1	1	0x1F65FAD23B0D86	2	0x8E05E117D9E786D5	1	0x14367E6C7D1CD9E2833D2900EE8D5	1
0xEE4BC5	1	0x1EF7930608393E	2	0xFFFFFFFFFFFFFFFE	4*	0x1FFBA28E4810FB56A9FDD85058227	2*
0xEC7EC7	1	0x1EDA43AEE3120F	2*	0xFFFFFFFE00000002	4*	0x1FF0231E35F73DFF14F89AADF10C2	2
0xE25473	1	0x1ED31F284BA183	2	0xFFFFF000007FFFFE	4*	0x1FE5A1913A4EF66DEF762D8053282	2
0xE1368B	1	0x1E9A473949BF6	2	0xFF801FFA00FFE002	4*	0x1FE1696E4EFFB6A84655C0D432D92	2
0xE05475	1	0x1E8D517D09C5C2	2	0xFF007FC01FF007FE	4*	0x1FDA070427995BB524AB4B13DC457	2
0xDE86A9	1	0x1E756F08DF1792	2	0xFE421D63446A3B34	4	0x1FAA42B2AE532A32F819FE18EDEAF	2*
0xDC23DD	1	0x1E4599DD926B71	2*	0xFDC1EAD583108905	4*	0x1F97117BE0A19F4B8279CEBB8A682	2
0xD43D43	1	0x1E20ADBC4078A2	2	0xFC41DF1077C41DF3	4*	0x1F76B18346B7182CE92732C773FB2	2
0xD25D25	1	0x1DE4A0D00FA9B2	2	0xFC07FFE03FFF01FE	4*	0x1F742DB89E4A0B81D5A2FE647E4EB	2*
0xD0DD0D	1	0x1DE441D5331432	2	0xFC07F01FC07F01FE	4*	0x1F490212CC8000000003E92042599	2*
0xD0AC19	1	0x1DA83EEDD80267	2*	0xFBFC17DFE0BEFF04	4	0x1F40436566B31BF75C99DF44F291F	2*
0xC23DC3	1	0x1DA210DAEB138E	2	0xFB20F95555168D17	4*	0x1F361A9D498B669732A60AFCF9461	2*
0xC100C1	1	0x1D7F8AC20F7A3F	2	0xFB0089D7241D10FC	4	0x1F25136B121FE2DD08B9975B8DBD2	2
0xB84A93	1	0x1D7B72B82BAE23	2	0xFA0BF7D05FBE82FC	4	0x1F1DFB37ABDE94B6800C5550CD152	2
0xAD1367	1	0x1D5B9032F086BE	2	0xF98AF433A85E62BF	4*	0x1F182E16A52503DCEDEBC24CA2B4E	2
0xAB8BE1	1	0x1D5616F7BA44B9	2*	0xF96BA24DC930852A	4*	0x1F140333F6B5946A06272DBD508B7	2*
0xA6449F	1	0x1C8ECFA282734B	2*	0xF93AB02081C9D1D6	4*	0x1F0DD51725F05CC5C752AA05A4311	2*
0xA24CF7	1	0x1C69BF28EBA166	2	0xF912590F016D6D04	4	0x1F001BEA0DE009CE597A0A8CE4B02	2
0xA0DDD1	1	0x1C67CF42F20D11	2*	0xF858A9FE5A20550D	4*	0x1EC36516E1240A243EF66232D4BA7	2*
0x9BEAAF	1	0x1C4D3AABD478F6	2	0xF84CEE8E701FC266	4*	0x1EBC4C4507F9CE8304761C8F703D2	2
0x986799	1	0x1C2693DCF34742	2	0xF774DD7F912E1F54	4	0x1E9EDBD047D1D813FB315AB469B2E	2
0x909909	1	0x1BEA3278B789D2	2	0xF7444DFBF7B20EAC	4	0x1E89A9332E4A8C84E2AA22A6DF7F1	2*
0x8EFA43	1	0x1BB2278C9B2F97	2*	0xF6F0243D8121FB7A	4*	0x1E8119576512C73436A03607DCB9B	2
0x87CC45	1	0x1B962F9EBB9659	2*	0xF6640F754B4E709A	4*	0x1E765D90D920CEEEBD7F5E0E0BBA9	2
0x869913	1	0x1B227794E85702	2	0xF39E8B657E24734AC	4	0x1E5DF4F7C4BB8E29C00588956CF0009	2*
0xCA6691	2*	0x1B0FD7099EB189	2*	0xF36EE790DE069D54	4	0x1E587973506590E472C4A72A35CF1	2*
0xA1E58F	2*	0x1B0942AAAE0BD3	2*	0xF363A464E2DCD8EB	4*	0x1E40DECFCF36257C367CACDAD3F77	2*
0xFFFFFE	4*	0x1AE6849E786AD2	2	0xF286AD7943D79434	4	0x1E2BE9D384CE2D85FD8013E21ECF2	2
0xFFE002	4*	0x1ABBEB8E009CE1	2*	0xEF9DA1D868469215	4*	0x1E15BB4DD7AA987E15487C533C649	2*
0xAAAAAC	4	0x1AA7C88EE59082	2	0xEDF09CCC53942014	4	0x1E0C9181ECD8418355A1A49887852	2
0x8C1284	4	0x1A6F41DAB98CB2	2	0xEDE957FFFC485AA	4*	0x1E0697C8651B43A9309DE9E6F021E	2
0x801001	4*	0x1A2CE4D7478A06	2	0xEDE95090B57B7A56	4*	0x1DEE59C5D9CC8CB8613C8C6AD453F	2*
0x800001	4*	0x1A149BAD85DE72	2	0xEDBAA0922AFB6EAA	4*	0x1DBA39CE33CA8DEF599F5DA2A534F	2
0x94F105	5	0x1A0E795098FF63	2*	0xEC4B058D0F7155BC	4	0x1DAC85098ABA5E144E44187FB5467	2*
0x92ABAB	5	0x1A0B8FFFFCBE8E	2	0xEC1CA6DB6D7BD444	4	0x1D99C15392893EA6B5200AB3E8819	2*
0xE401C8	8	0x19F142D24E1352	2	0xEB443F5A21FAD10E	4*	0x1D983DBB99EAC626064F81D7BF4D2	2
0xE071F9	8*	0x19BD2D9FD24AD7	2*	0xEA6EE2D972746ED1	4*	0x1D86CC4938A03D4525C152AB8505E	2
0xD6D764	8*	0x19B8D7C084EE43	2*	0xEA40E197842DA6AF	4*	0x1D573D7B5CFE2D277AD5E05BAC65E	2
0xD443F2	8*	0x199E1B447E99C2	2	0xE934A8E070ACB65D	4*	0x1D4D79E1F152354E10F583D4A65C9	2*
0xCFBA38	8	0x19939800033273	2*	0xE84BDA12F684BDA3	4*	0x1D4CF86C34F75247D8FA16202FA29	2*
0xB5C2F1	8*	0x1975E059B82E49	2*	0xE775FF856986AE74	4	0x1D4562A76F879A38EEE86F526D231	2*
0xB447BC	8*	0x1960A45D1A71E6	2	0xE6944AE6502F8A22	4*	0x1D3168C71EC1068F69A433D0DF9B7	2
0xAE4F88	8	0x19385F4F83B2B1	2	0xE5CB972E5CB972E4	4	0x1D26EDCA8F70B604EC3E7797A93A1	2
0x9A5F6E	8*	0x190759A7F39561	2*	0xE597116BD81B26A3	4*	0x1D16B15A15C76FF2477078355A9AE	2
0x988597	8*	0x18F187FFFCE1CF	2*	0xE58C38D1342FBE3A	4*	0x1D12CF093CC27703DA21DC7D68CE7	2*
0x91FEDC	8*	0x18DFA37A569E47	2*	0xE58469F0234F72C4	4	0x1CEC5ADBF01E9685CD487AD8F3327	2*
0xFFFFFD	9	0x1879574AF5FBB1	2*	0xE511C4648E2332C4	4	0x1CE72221273FE0035FEC64CBB3DBF	2*
0xA013D1	11	0x184A12EFEF626E	2	0xE3FC771FE3B8FF1C	4	0x1CE4C2D686D170738B75E2AFECF3E	2
0xF56DA7	12*	0x18401CBCDB5596	2	0xE3C845B18BD25EC6	4*	0x1CD0C2468D84F6ACF871D5E1FCBA9	2
0x858376	12*	0x181EFE51EAD722	2	0xE318DE3C8E6370E4	4	0x1CCB50FE42CD1B95A59CA8AD6EB99	2*
0xCF7D05	13	0x1806C89FCB9452	2	0xE301201062C997DE	4*	0x1CABCCB01B54CE7E2A63A99B9D9C2	2
0xEE5223	14*	0x17F52093014F0E	2	0xE23B9711DCB88EE4	4	0x1C9ED60CFD93F55F117571C3FDA0E	2
0xE528AB	14*	0x17D93736C115FF	2*	0xE231188C46231187	4	0x1C820E19C1A66CE04C62A562E9111	2*
0xBF3621	14*	0x17A6B0778D60C1	2	0xE1F00785C1FF0F82	4	0x1C673D52FCD6E005D2A3D3D40EDCF	2
0xAB5ED9	14*	0x178CB7D5D6E322	2	0xE159BE4A8763011C	4	0x1C670773DF1678DF0A4336D3FE21E	2
0x8EFE15	14*	0x17641C46F799EE	2	0xE0A72F05397829CA	4*	0x1C4D4290F01337DEE39B3A7862BDE	2
0x897ECD	14*	0x175D929C3C2FC9	2*	0xE0A720FAC6F829CA	4	0x1C4BD3136A6DB6DB6DB351F3546E2	2
0xFFFFFC	16*	0x1733B8284238F1	2*	0xE073C0EE938231F9	4*	0x1C49777FF62E0B0DA1A4CDB58587F	2*
0xF83F04	16*	0x17255CA25B68E1	2*	0xDF738B7CF7F482E4	4	0x1C4814DFE06ECB4EB88D00BA934F2	2

Table 1. Some numbers with reciprocals closest to numbers(*) and midpoints

As usual in algorithms of this kind, each operation uses a fused multiply-add (*not* a separate multiplication and addition), all steps but the last are performed in round-to-nearest mode with additional exponent range precluding the possibility of intermediate overflow or underflow, and the last operation is done in the intended rounding mode and target precision.

Embedded in this algorithm is the computation of a very accurate reciprocal approximation y_3. Originally, in the design of algorithms of this kind, the correctness of the final rounding of q was justified by a theorem whose precondition requires perfect rounding of y_3 [9], and only later was it noted by the present author that a relative error $y_3 = \frac{1}{b}(1+\epsilon)$ for $|\epsilon| < 2^{-p}$ suffices, which can be satisfied by a relatively weak error condition on y_2 and the analysis of a few special cases [3, 8]. However, if we are in a situation where $a = 1$ we might consider, instead of using the entire sequence, unpicking the algorithm for reciprocation to be used directly, since its latency is shorter by 1 operation, and it uses only 9 floating-point operations instead of 14:

1. $y_0 = \mathtt{frcpa}(b)$
2. $d = 1 - by_0$
3. $d_2 = dd$ $\quad\quad d_3 = dd + d$
4. $y_1 = y_0 + y_0 d_3$ $\quad d_5 = d_2 d_2 + d$
5. $y_2 = y_0 + y_1 d_5$
6. $e = 1 - by_2$
7. $y = y_2 + ey_2$

Now the question of whether y is always correctly rounded becomes critical. First we will consider round-to-nearest. The initial approximation returned by $\mathtt{frcpa}$ will satisfy $y_0 = \frac{1}{b}(1+\epsilon_0)$ for some $|\epsilon_0| \le 2^{-8.886}$. A routine relative error analysis, assuming each rounding $rn(x)$ yields $x(1+\epsilon)$ for some $|\epsilon| \le 2^{-64}$, shows that y^*, the value of y before the last rounding, satisfies

$$y^* = \frac{1}{b}(1+\epsilon)$$

where $|\epsilon| \le 2^{-123.37}$. Therefore, the only cases where incorrect rounding can occur are those closer than this relative distance to a midpoint. The potentially failing significands b can be isolated by finding all $(2^{65}, 2^{64})$-balanced factorizations $mb = 2^{128} + d$ for integers $|d| \le 24$ (since $24 + 1 > 2^{-123.37}/2^{-128}$) and m odd. The set of b values that we need to consider are the following 134 (ordered in decreasing size, not according to their closeness to a midpoint):

```
0xFFFFFFFFFFFFFFFF 0xFFFFFFFFFFFFFFFD 0xFE421D63446A3B34
0xFBFC17DFE0BEFF04 0xFB940B119826E598 0xFB0089D7241D10FC
0xFA0BF7D05FBE82FC 0xF912590F016D6D04 0xF774DD7F912E1F54
0xF7444DFBF7B20EAC 0xF39B8657E24734AC 0xF36EE790DE069D54
0xF286AD7943D79434 0xEDF09CCC53942014 0xEC4B058D0F7155BC
0xEC1CA6DB6D7BD444 0xE775FF856986AE74 0xE5CB972E5CB972E4
0xE58469F0234F72C4 0xE511C4648E2332C4 0xE3FC771FE3B8FF1C
0xE318DE3C8E6370E4 0xE23B9711DCB88EE4 0xE159BE4A8763011C
0xDF738B7CF7F482E4 0xDEE256F712B7B894 0xDEE24908EDB7B894
0xDE86505A77F81B25 0xDE03D5F96C8A976C 0xDDFF059997C451E5
0xDB73060F0C3B6170 0xDB6DB6DB6DB6DB6C 0xDB6DA92492B6DB6C
0xDA92B6A4ADA92B6C 0xD9986492DD18DB7C 0xD72F32D1C0CC4094
0xD6329033D6329033 0xD5A004AE261AB3DC 0xD4D43A30F2645D7C
0xD33131D2408C6084 0xD23F53B88EADABB4 0xCCCE6669999CCCD0
0xCCCE666646633330 0xCCCCCCCCCCCCCCCD 0xCBC489A1DBB2F124
0xCB21076817350724 0xCAF92AC7A6F19EDC 0xC9A8364D41B26A0C
0xC687D6343EB1A1F4 0xC54EDD8E76EC6764 0xC4EC4EC362762764
0xC3FCF61FE7B0FF3C 0xC3FCE9E018B0FF3C 0xC344F8A627C53D74
0xC27B1613D8B09EC4 0xC27B09EC27B09EC4 0xC07756F170EAFBEC
0xBDF3CD1B9E68E8D4 0xBD5EAF57ABD5EAF4 0xBCA1AF286BCA1AF4
0xB9B501C68DD6D90C 0xB880B72F050B57FC 0xB85C824924643204
0xB7C8928A28749804 0xB7A481C71C43DDFC 0xB7938C6947D97303
0xB38A7755BB835F24 0xB152958A94AC54A4 0xAFF5757FABABFD5C
0xAF4D99ADFEFCAAFC 0xAF2B32F270835F04 0xAE235074CF5BAE64
0xAE0866F90799F954 0xADCC548E46756E64 0xAD5AB56AD5AB56AC
0xAD5AAA952AAB56AC 0xAB55AAD56AB55AAC 0xAAAAAB55555AAAAAC
0xAAAAAAAAAAAAAAAC 0xAAAAA00000555554 0xA93CFF3E629F347D
0xA80555402AAA0154 0xA8054ABFD5AA0154 0xA7F94913CA4893D4
0xA62E84F95819C3BC 0xA5889F09A0152C44 0xA4E75446CA6A1A44
0xA4428F8DCDEF5BC 0xA27E096B503396EE 0x9E9B8FFFFFD8591C
0x9E9B8B0B23A7A6E4 0x9E7C6B0C1CA79F1C 0x9DFC78A4EEEE4DCB
0x9C15954988E121AB 0x9A585968B4F4D2C4 0x99D0C486A0FAD481
0x99B831EEE01FB16C 0x990C8B8926172254 0x990825E0CD75297C
0x989E556CADAC2D7F 0x97DAD92107E19484 0x9756156041DBBA94
0x95C4C0A72F501BDC 0x94E1AE991B4B4EB4 0x949DE0B0664FD224
0x942755353AA9A094 0x9349AE0703CB65B4 0x92B6A4ADA92B6A4C
0x9101187A01C04E4C 0x907056B6E018E1B4 0x8F808E79E77A99C4
0x8F64655555317C3C 0x8E988B8B3BA3A624 0x8E05E117D9E786D5
0x8BEB067D130382A4 0x8B679E2B7FB0532C 0x887C8B2B1F1081C4
0x8858CCDCA9E0F6C4 0x881BB1CAB40AE884 0x87715550DCDE29E4
0x875BDE4FE977C1EC 0x86F71861FDF38714 0x85DBEE9FB93EA864
0x8542A9A4D2ADB5EC 0x8542A150A8542A14 0x84BDA12F684BDA14
0x83AB6A090756D410 0x83AB6A06F8A92BF0 0x83A7B5D13DAE81B4
0x8365F2672F9341B4 0x8331C0CFE9341614 0x82A5F5692FAB4154
0x8140A05028140A04 0x8042251A9D6EF7FC
```

One can show by explicit computation that the algorithm works correctly on these values. It therefore rounds correctly on all values in round-to-nearest.

For directed rounding modes, the situation is less good. Once again the relative error condition gives rise to a set of test points, this time 227 of them. The algorithm works correctly on 220 of them, but not on floating-point numbers with one of the following 7 significands, the last of these representing exact powers of 2, for which the true result is exactly representable. Cognoscenti who perform a back-of-envelope calculation will not be surprised by the failure on exactly representable results, since correctness here would require y_2 already to be the correct result, which our relative error cannot quite guarantee.

```
0x8c82da588adc6416 0x84fdf027ef813f7b 0x827b9b8059090ab2
0x8080402010080401 0x8000080000400001 0x8000000000000001
0x8000000000000000
```

This analysis indicates that the algorithm will produce correctly rounded results if the ambient rounding mode is known to be round-to-nearest, but will not always guarantee correct rounding in other rounding modes. Moreover, note that for the same reason, the 'inexact' flag will be incorrectly set in round-to-nearest mode in the special cases where b is a power of 2. (As noted, the penultimate approximation y_2 cannot be the exact reciprocal in such cases, for otherwise we would obtain $e = 0$ and correct rounding in all modes.) However, if this is considered important, it would be easy to detect and fix the problem with special case code without affecting overall latency.

7 Feasibility study

Although the previous sections show that the method is usefully applicable to some real problems, it's worth analyzing how practical the approach is likely to be in general. In attempting to use the method, three potential practical problems might arise

- Too many special points are isolated for further analysis to be feasible
- The factorization of some of the numbers is not feasible
- The distribution of prime factors is not feasible

We will not analyze the feasibility of factorization, since we do not understand the details of its implementation. We will however make the empirical observation that all factorizations for precisions up to $p = 64$ seem to be very straightforward for PARI / GP, taking a fraction of a second, while those for $p = 113$ usually take several seconds and, exceptionally, minutes.

Average density of balanced factorizations

It is not difficult to see that "on average" we obtain a fairly modest number of balanced factorizations per value examined. First note that the number of (A, B)-balanced products of numbers $\leq n$ is the number of lattice points contained both within the rectangle $0 \leq x \leq A, 0 \leq y \leq B$ and under the curve $xy = n$. We can get a good estimate by ignoring "edge effects" and just considering the plane area, integrating to obtain:

$$C(n) = n(1 + \ln(\frac{AB}{n}))$$

Differentiating with respect to n yields the expected density, i.e. the average number of (A, B)-balanced product representations of a number close to n:

$$D(n) = \ln(AB/n)$$

Of course, these gross averages do not reflect small-scale fluctuations. Nevertheless, the agreement is fairly good with some empirical results obtained by sampling. In the following table, we examine the density of $(2^p, 2^p)$-balanced products for several p, looking in each case at 31 regions close to $\frac{k+1/2}{32}2^{2p}$ for $0 \leq k \leq 31$ and sampling 1024 successive points in each. The final figures at the bottom give the mean value. This indicates how accurate the sampling process is on average; perfectly representative sampling would give exactly 1 here. (We avoid sampling at $\frac{k}{32}2^p$ because that would lead to strong correlations between the sets of numbers at different k.)

$\ln(2^{2p}/n)$	$p = 24$	$p = 53$	$p = 64$
4.1588	4.4785	4.6835	3.3300
3.0602	2.8496	5.6621	3.2734
2.5494	2.4570	2.7070	2.2753
2.2129	2.0332	2.2421	2.2089
1.9616	2.0000	1.6953	2.3417
1.7609	1.9101	1.5664	1.5585
1.5939	1.5742	1.9140	1.2128
1.4508	1.3632	1.4765	1.5625
1.3256	1.3144	1.0839	1.2558
1.2144	1.2050	1.2187	1.2890
1.1143	1.0175	1.0996	1.4296
1.0233	1.0273	0.9335	0.9687
0.9400	0.7539	0.9062	0.8828
0.8630	0.7636	0.8613	0.8789
0.7915	0.6875	0.7187	0.6875
0.7248	0.6933	0.6621	0.7832
0.6623	0.6621	0.5976	0.7656
0.6035	0.5878	0.5468	0.6445
0.5479	0.5546	0.6210	0.5683
0.4953	0.4941	0.5136	0.6289
0.4453	0.4394	0.3847	0.3652
0.3976	0.3984	0.4453	0.4277
0.3522	0.3417	0.3476	0.3242
0.3087	0.3203	0.2890	0.3593
0.2670	0.2382	0.2285	0.2773
0.2270	0.2480	0.2070	0.3007
0.1885	0.1347	0.2207	0.2148
0.1515	0.1347	0.1640	0.1562
0.1158	0.0839	0.0976	0.1015
0.0813	0.0917	0.1054	0.0761
0.0480	0.0449	0.0371	0.0527
0.0157	0.0078	0.0078	0.0156
1.0000	0.9660	1.0701	0.9755

So much for the average case. What about the worst case? This seems a more difficult question to address theoretically, but in the next section we will show how to obtain a pessimistic upper bound.

Feasibility of distribution algorithm

Although the final number of values produced depends on the number of balanced factorizations, the process by which the balanced factorizations are enumerated involves examination of many dead-end paths, so the runtime of the distribution process may be very large relative to the final number of possibilities produced. A reasonable, though pessimistic, bound on the runtime of the distribution algorithm for a value n is $d(n)$, the *total* number of divisors of n, regardless of balance. For even without early cutoffs owing to range limitations, the algorithm cannot examine, given

$$n = \Pi_{i=1}^{i=k} p_i^{\alpha_i}$$

more than

$$d(n) = \Pi_{i=1}^{i=k}(1+\alpha_i)^n$$

possibilities, since each factor $p_i^{\alpha_i}$ can, without range cutoffs, be distributed in $1 + \alpha_i$ ways.

It is well known that the average number of divisors $d(n)$ of a number near n is approximately $d(n) = \ln(n)$. This can easily be derived using the same sort of argument as we used above for balanced products [2]. This suggests that on average, the distribution process will not have many cases to examine; even for quad precision, we have $n \leq 2^{230}$ and so $\ln(n) \leq 160$.

What about the worst case? The number of divisors of a number can be much larger than $\ln(n)$. In fact [2], *almost all* numbers (in a precise sense) have about $\ln(n)^{\ln(2)}$ divisors, with the larger overall average of $\ln(n)$ resulting from a small proportion of numbers with many more divisors. Asymptotically, it is known [2] that $d(n)$ has an upper limit of exactly $2^{\ln(n)/\ln(\ln(n))}$, or more precisely, that if $\epsilon > 0$ then $d(n) < 2^{(1+\epsilon)\ln(n)/\ln(\ln(n))}$ for all sufficiently large n, while $d(n) > 2^{(1-\epsilon)\ln(n)/\ln(\ln(n))}$ for infinitely many n.

This asymptotic limit needs refinement to be useful to us for the concrete ranges we are interested in. We can obtain a more refined estimate of the maximum $d(n)$ for all n below some limit N we are interested in as follows. The key to efficient search is to seek the *minimal* n with the *maximal* number of divisors possible for $n \leq N$. The minimality constraint forces strong patterns onto the prime factorization. Suppose that n has the following prime factorization:

$$n = \Pi_{i=1}^{i=k} p_i^{\alpha_i}$$

Let $p_i < p_j$ be two primes (not necessarily appearing with nonzero index in the above factorization) such that $p_i^\beta < p_j < p_i^{\beta+1}$ for some nonnegative integer β. Then it is easy to see that if n has the minimality property, the following relationships hold between the α's:

$$\beta \alpha_j \leq \alpha_i \leq (\beta + 1)\alpha_j + 2\beta$$

For if the first inequality failed we could get a smaller number with at least as many divisors by replacing $p_i^{\alpha_i} p_j^{\alpha_j}$ with $p_i^{\alpha_i + \beta} p_j^{\alpha_j - 1}$, while if the second inequality failed we could likewise replace it with $p_i^{\alpha_i - (\beta+1)} p_j^{\alpha_j + 1}$.

This observation includes the case where p_j is the first prime beyond those appearing in the factorization, and in this case $\alpha_i \leq 2\beta$. For example, if 17^α appears in the factorization, so must $3^{2\alpha}$ and $2^{4\alpha}$, while if no power of 17 appears in the factorization then the highest possible power of 2 appearing is 2^8, and the highest power of 3 is 3^6. Note in particular that the factorization of the minimal n must contain the first k consecutive primes without gaps, for some k.

These observations cut down the search space dramatically enough that we can easily perform an exhaustive search for the precise worst numbers up to quite large values, say 2^{3000}. The following table shows, for various values of p up to 230, the minimal $n \leq 2^p$ with the largest number of divisors possible in that range. For each such n, we show $\log_2(n)$ and $\log_2(d(n))$ (where $d(n)$ is the number of divisors of n), as well as the ratio with the expected limit superior $r(n) = \log_2(d(n))/(\ln(n)/\ln(\ln(n)))$ and the actual factorization of n.

p	$\log_2(n)$	$\log_2(d(n))$	$r(n)$	Factorization of that worst n
10	9.71	5.00	1.416	$2^3\ 3\ 5\ 7$
20	19.45	7.90	1.525	$2^4\ 3^2\ 5\cdots 13$
30	29.45	10.39	1.535	$2^6\ 3^3\ 5^2\ 7\cdots 17$
40	39.80	12.71	1.528	$2^6\ 3^4\ 5^2\ 7\cdots 23$
50	49.84	14.75	1.512	$2^5\ 3^3\ 5^2\ 7^2\ 11\cdots 31$
60	59.96	16.71	1.498	$2^6\ 3^4\ 5^3\ 7^2\ 11\cdots 37$
70	69.42	18.49	1.488	$2^7\ 3^4\ 5^2\ 7^2\ 11\cdots 43$
80	79.88	20.33	1.474	$2^8\ 3^5\ 5^3\ 7^2\ 11\cdots 47$
90	89.90	22.07	1.463	$2^8\ 3^4\ 5^3\ 7^2\ 11\cdots 59$
100	99.88	23.75	1.453	$2^7\ 3^5\ 5^3\ 7^2\ 11^2\ 13\cdots 61$
110	109.64	25.33	1.443	$2^8\ 3^5\ 5^3\ 7^2\ 11\cdots 71$
120	119.87	26.97	1.435	$2^7\ 3^6\ 5^3\ 7^2\ 11^2\ 13\cdots 73$
130	129.87	28.56	1.427	$2^7\ 3^6\ 5^3\ 7^2\ 11^2\ 13^2\ 17\cdots 79$
140	139.99	30.12	1.420	$2^{10}\ 3^5\ 5^4\ 7^2\ 11^2\ 13^2\ 17\cdots 83$
150	149.74	31.66	1.416	$2^9\ 3^5\ 5^3\ 7^2\ 11^2\ 13^2\ 17\cdots 97$
160	159.79	33.14	1.408	$2^8\ 3^6\ 5^3\ 7^3\ 11^2\ 13^2\ 17\cdots 101$
170	169.83	34.66	1.404	$2^9\ 3^5\ 5^3\ 7^2\ 11^2\ 13^2\ 17\cdots 107$
180	179.99	36.14	1.398	$2^8\ 3^6\ 5^3\ 7^3\ 11^2\ 13^2\ 17\cdots 109$
190	189.82	37.56	1.393	$2^9\ 3^5\ 5^4\ 7^2\ 11^2\ 13^2\ 17^2\ 19\cdots 113$
200	199.88	39.02	1.388	$2^{10}\ 3^6\ 5^3\ 7^3\ 11^2\ 13^2\ 17\cdots 127$
210	209.93	40.43	1.383	$2^{10}\ 3^6\ 5^3\ 7^3\ 11^2\ 13^2\ 17\cdots 137$
220	219.87	41.83	1.379	$2^8\ 3^5\ 5^4\ 7^3\ 11^2\ 13^2\ 17^2\ 19\cdots 139$
230	229.92	43.21	1.375	$2^{10}\ 3^5\ 5^3\ 7^3\ 11^2\ 13^2\ 17\cdots 151$

We can see that even for double-extended precision, the number of factorizations that could possibly need to be examined is about 2^{28}. Although a fairly large number, this is definitely feasible. (And of course in practice such cases are exceptional and not all factorizations would be examined.) For quad precision, on the other hand, it is entirely possible for the search to be infeasible. We have not yet encountered this phenomenon in practice, however.

Note that $d(n)$ also gives an upper bound to the number of balanced factorizations. It is, of course, pessimistic, but testing on some of the values above suggests that the the number of balanced factorizations is a reasonable proportion (say 10%) of the total number of divisors. Naturally, it would be better to refine all these estimates to consider only numbers very close to the powers of 2, which is what we are really interested in.

The special numbers that we searched for above are particular cases of *highly composite numbers* [13]. For a detailed survey of the subject see [10], while Achim Flammenkamp's Web page seems to give a more efficient algorithm for generating HCNs:

    http://wwwhomes.uni-bielefeld.de/achim/highly.html

The sequence of highly composite numbers is A002182 in Sloane's Encyclopedia of Integer Sequences.

8 Extension to reciprocal square root

It is interesting to note that a similar factor distribution technique can be used to attempt to find exceptional cases for the reciprocal square root. In this case, we seek floating-point numbers or midpoints w and floating-point numbers y such that

$$\frac{|w - \frac{1}{\sqrt{y}}|}{|\frac{1}{\sqrt{y}}|}$$

is small. We can rewrite this as:

$$|\sqrt{y}(w - \frac{1}{\sqrt{y}})| = |w\sqrt{y} - 1|$$

In the critical cases where $w\sqrt{y} - 1$ is very small, then $w\sqrt{y} + 1$ is almost exactly 2 and so:

$$|w\sqrt{y} - 1| = \frac{|w^2 y - 1|}{|w\sqrt{y} + 1|} \approx \frac{|w^2 y - 1|}{2}$$

Once again, let us scale the values w and y to integers m and b:

$$y = 2^{e_y} b \quad 2^{p-1} \le b < 2^p$$
$$w = 2^{e_w} m \quad 2^p \le m < 2^{p+1}$$

and then the distance we are interested in is then:

$$\frac{|m^2 b - 2^q|}{2^{q+1}}$$

where $q = -(2e_w + e_y)$. So we seek cases where $d = m^2 b - 2^q$ is as small as possible. Keeping in mind the range restrictions, we see that $2^{3p-1} \le m^2 b < 2^{3p+2}$. As with simple reciprocals, it is impossible to come very close to the extremal powers of 2, but we do now need to consider two cases, $q = 3p$ and $q = 3p + 1$.

The reciprocal square root function is of some theoretical interest because it seems *prima facie* possible that $d = m^2 b - 2^q$ could be very small, perhaps even ± 1, yet no precisions where it is much smaller than 2^p have ever been found, and one might expect on naive statistical grounds that it is unlikely. (We only have 2^{2p} different choices of pairs m and b, and are scattering the resulting $m^2 b$'s somehow over an interval of size about 2^{3p}.) Li [7] proves that *assuming* the ABC conjecture from number theory holds, the distance is indeed of order 2^p for all sufficiently large p. Even if the ABC conjecture were proven, however, it's not clear whether it would be possible to constructivize the proof in order to obtain useful bounds for specific precisions. Iordache and Matula [4] observe that $d = 1$ is impossible in general, allowing the accuracy required to be lowered slightly, but add that 'trying to lower it is not an easy problem, even for a fixed p'. Although the present work does not touch the general case, and nor can it fully bridge the gap between expected and provable bounds, it *does* allow us quite easily to improve the provable bound for the typical p we are interested in by a reasonable factor.

We can take over the prime distribution function with little change. The only difference is that we now need to distribute the prime factors among $m^2 b$. This has the immediate consequence that only even powers of primes can be allocated to the m^2 part, and so any prime appearing to an odd power in the prime factorization of $2^q + d$ must be allocated at least once to b. This is almost always enough to render the distribution immediately impossible. We have made some searches for double-extended precision ($p = 64$) and quad precision ($p = 113$). For double-extended, we have shown that $d \le 1024$ is impossible, and it would be easy to continue the search much further. For quad precision, the cost of factoring numbers is now a serious bottleneck, with a single number sometimes taking a day of CPU time and one of the factorizations for the $d = 6$ case apparently defeating factorization in a reasonable time. Nevertheless we have at least shown that $d < 6$ is impossible, which represents some improvement. For smaller precisions, it seems likely that other algorithms based on an (intelligent) exhaustive analysis of the whole space of significands would be more efficient. For example Lang and Muller [5] have performed a complete analysis of the double-precision case $p = 53$ (and found that the minimal distance is about 2^{-110}).

9 Conclusion

The methods described here allow reasonably effective isolation of the 'worst cases' for the reciprocal function. This opens the way to correctness proofs of reciprocal algorithms using the same kind of two-part approach used by Cornea [1] for square roots. In the absence of new theoretical advances, the method described may also be the best available means of improving the difficulty bounds on the reciprocal square root functions for larger precisions. Although our method has feasibility problems for the extreme case of quad-precision reciprocal square roots, it would be possible to explore alternative factoring algorithms. The numbers we are interested in factoring are very close (in relative terms) to powers of 2, so it is possible that algorithms such as the Special Number Field Sieve (SNFS) would give much better results.

Acknowledgements

The author is grateful to the anonymous referees, who made a number of excellent suggestions, and pointed out connections of which the author had been unaware.

References

[1] M. Cornea-Hasegan. Proving the IEEE correctness of iterative floating-point square root, divide

and remainder algorithms. *Intel Technology Journal*, 1998-Q2:1–11, 1998. Available on the Web as http://developer.intel.com/technology/itj/q21998/articles/art_3.htm.

[2] G. H. Hardy and E. M. Wright. *An Introduction to the Theory of Numbers*. Clarendon Press, 5th edition, 1979.

[3] J. Harrison. Formal verification of IA-64 division algorithms. In M. Aagaard and J. Harrison, editors, *Theorem Proving in Higher Order Logics: 13th International Conference, TPHOLs 2000*, volume 1869 of *Lecture Notes in Computer Science*, pages 234–251. Springer-Verlag, 2000.

[4] C. Iordache and D. W. Matula. On infinitely precise rounding for division, square root, reciprocal and square root reciprocal. In I. Koren and P. Kornerup, editors, *Proceedings, 14th IEEE symposium on on computer arithmetic*, pages 233–240, Adelaide, Australia, 1999. IEEE Computer Society. See also Technical Report 99-CSE-1, Southern Methodist University.

[5] T. Lang and J.-M. Muller. Bounds on runs of zeros and ones for algebraic functions. Research Report 4045, INRIA, 2000.

[6] V. Lefèvre and J.-M. Muller. Worst cases for correct rounding of the elementary functions in double precision. Research Report 4044, INRIA, 2000.

[7] R.-C. Li. The ABC conjecture and correctly rounded reciprocal square root. Preprint, 2002.

[8] P. Markstein. *IA-64 and Elementary Functions: Speed and Precision*. Prentice-Hall, 2000.

[9] P. W. Markstein. Computation of elementary functions on the IBM RISC System/6000 processor. *IBM Journal of Research and Development*, 34:111–119, 1990.

[10] J.-L. Nicholas. On highly composite numbers. In *Ramanujan Revisited: Proceedings of the Centenery Conference*, pages 215–244. Academic Press, 1988.

[11] M. Parks. Number-theoretic test generation for directed rounding. *IEEE Transactions on Computers*, 49:651–658, 2000.

[12] V. Pratt. Every prime has a succinct certificate. *SIAM Journal of Computing*, 4:214–220, 1975.

[13] S. Ramanujan. Highly composite numbers. *Proceedings of the London Mathematical Society*, 14:347–409, 1915.

[14] P. T. P. Tang. Testing computer arithmetic by elementary number theory. Preprint MCS-P84-0889, Mathematics and Computer Science Division, Argonne National Labs, 1989.

Solving Range Constraints for Binary Floating-Point Instructions

Abraham Ziv, Merav Aharoni, Sigal Asaf
IBM Research Labs
Haifa University, Mount Carmel
Haifa 31905, ISRAEL
E-mail address: ziv@il.ibm.com

Abstract

We present algorithms that solve the following problem: given three ranges of floating-point numbers R_x, R_y, R_z, a floating-point operation (op), and a rounding-mode (round), generate three floating-point numbers $\bar{x}$, $\bar{y}$, $\bar{z}$ such that $\bar{x} \in R_x$, $\bar{y} \in R_y$, $\bar{z} \in R_z$, and $\bar{z} = round(\bar{x}$ op $\bar{y})$. This problem, although quite simple when dealing with intervals of real numbers, is much more complex when considering ranges of machine numbers. We provide full solutions for add and subtract, and partial solutions for multiply and divide. We use range constraints on the input operands and on the result operand of floating-point instructions to target corner cases when generating test cases for use in verification of floating-point hardware. The algorithms have been implemented in a floating-point test-generator and are currently being used to verify floating-point units of several processors.

1. Introduction

Floating-point unit verification presents a unique challenge in the field of processor verification. The particular complexity of this area stems from the vast test-space, which includes many corner cases that should each be targeted, and from the intricacies of the implementation of floating-point operations. Verification by simulation involves executing a subset of tests which is assumed to be a representative sample of the entire test-space [2]. In doing so, we would like to be able to define a particular subspace, which we consider as "interesting" in terms of verification, and then generate tests selected at random out of the subspace.

Consider the following cases as examples of this approach. It is interesting to check the instruction FP-SUB, where the two inputs are normalized numbers and the output is denormal, since this means the inputs had values that were close to each other, resulting in a massive cancellation of the most significant bits of the inputs and underflow in the result.

As a more general example, consider the following 10 basic ranges of floating-point numbers: ±normalized numbers, ±denormal numbers, ±zero, ±infinity, and ±NaNs. For a binary instruction, it is interesting to generate test cases (when possible) for all the combinations of selecting two inputs and the output out of these ranges (1000 combinations).

Subsets of floating-point numbers may be represented in various ways. Some examples of common representations are:

1. Ranges of floating-point numbers.
2. Masks. In this representation each bit may take on one of the values 0, 1, X, where X specifies that both 0 and 1 are possible. For example $01X1X$ represents the set $\{01010, 01011, 01110, 01111\}$.
3. Number of 1s (0s). This defines the set of all floating-point numbers that have a given number of 1s (0s).

Probably the most natural of these, in the context of test-generation, is the representation as ranges. The primary advantage in representing sets of numbers as ranges, is its simplicity. All the basic types of floating-point numbers are readily defined in terms of ranges. Many interesting sets of numbers can be efficiently represented as a union of ranges. Test generation for the FP-ADD instruction, where the input sets are described as masks, is discussed in [3]. Examples of other algorithms that may be used for test generation may be found in [6], [8], and [9].

In this paper, we describe algorithms that provide random solutions given constraints on the instructions add, subtract, multiply, and divide when the two inputs and the output sets are all represented as ranges. In other words, given a floating-point operation $op \in \{add, subtract, multiply, divide\}$, a rounding mode $round \in \{$ to-zero, to-positive-infinity, to-negative-infinity, to-nearest $\}$, and three ranges R_x, R_y, R_z of floating point numbers, the algorithm will produce three random floating-point numbers, $\bar{x}$, $\bar{y}$, $\bar{z}$, such that $\bar{x} \in R_x$, $\bar{y} \in R_y$, $\bar{z} \in R_z$, and $\bar{z} = round(\bar{x}\ op\ \bar{y})$.

The difficulty in solving range constraints stems from the fact that although a floating-point range appears to be a continuous set of numbers, each range is in fact a set of discrete machine numbers. To illustrate this, consider two floating-point numbers $\bar{x} \in R_x$ and $\bar{z} \in R_z$. It is quite possible that there exists a real number y, in the real interval defined by the endpoints of R_y, for which $round(\bar{x} \times y) = \bar{z}$, but that there does not exist such a machine number.

The algorithms presented have been implemented in a floating-point test-generator, FPgen, in the IBM Haifa Research Labs [4]. FPgen is an automatic floating-point test-generator, which receives as input the description of a floating-point coverage task [2] and outputs a random test that covers this task. A coverage task is defined by specifying a floating-point instruction and a set of constraints on the inputs, on the intermediate result(s), and on the final result. For each task, the generator produces a random test that satisfies all the constraints. FPgen employs various algorithms, both analytic and heuristic, to solve the various constraint types. Solving the general constraint problem is NP-Complete. When the constraints are given as ranges, FPgen employs the algorithms described in this paper in order to generate the test cases.

In Section, 2 we describe the problem more formally and outline our general approach. In Section 3, we present the continuous algorithm, which works efficiently for all four arithmetic operations in most cases. In Section 4, we investigate the cases in which the algorithm does not work efficiently for multiplication. In Section 5, we present a second algorithm for addition, which solves the cases that are too difficult for the continuous algorithm. Section 6 includes a summary of the results and some suggestions for future work in this area.

2. Problem Definition

2.1. Formalization

We investigate the following problem:
Given six machine numbers $\bar{a}_x$, $\bar{b}_x$, $\bar{a}_y$, $\bar{b}_y$, $\bar{a}_z$, $\bar{b}_z$, find an algorithm that produces two machine numbers $\bar{x}$, $\bar{y}$ such that $\bar{a}_x \leq \bar{x} \leq \bar{b}_x$, $\bar{a}_y \leq \bar{y} \leq \bar{b}_y$, and such that $\bar{z} = round(\bar{x}\ op\ \bar{y})$ satisfies $\bar{a}_z \leq \bar{z} \leq \bar{b}_z$ where $op \in \{+, -, \times, \div\}$.

round represents a rounding mode. It may be, for instance, one of the IEEE standard 754 [1] rounding modes: up, down, toward zero, to nearest/even.

Throughout the paper, we represent machine numbers using lower-case letters with an upper bar, for example $\bar{x}$. This notation is used to distinguish them from real numbers, which are represented by regular lower-case letters, for example x.

2.2. General algorithm

In the following Sections, we present two algorithms, the continuous algorithm and the discrete algorithm. The continuous algorithm is based on the similarity between the problem discussed here, and the problem of finding such a solution when the ranges are real number intervals. In the case of real numbers, the problem is relatively simple: we reduce the intervals and find a solution by using the inverse of the required operation. For machine numbers, it is not possible to use the inverse operation, therefore the algorithm that we propose differs slightly. As we will show later, the continuous algorithm is very efficient for most cases, for all four operations. Howeverer, there are certain cases for which this is not true.

The discrete algorithm is specific for addition and subtraction. It is slower than the continuous algorithm, however, if a solution exists, this algorithm will find it. Therefore, for addition, we suggest first running the continuous algorithm. If the algorithm does not terminate within a reasonable amount of time, use the discrete algorithm.

In general, our discussion focuses on a solution for IEEE 754 sets of floating-point numbers. However, the algorithms can easily be converted to non-IEEE rounding modes (such as round half down), and most of the results are valid for any floating-point format (such as decimal floating-point [5]).

Some of the discussion which follows is valid for all four arithmetic operations. However, an important part of the discussion is restricted to addition, subtraction, and multiplication. Note, the problems for addition and subtraction are actually equivalent, because $\bar{y}$ may be replaced by $-\bar{y}$. Hence subtraction is not discussed further. Throughout the paper, we indicate which parts are relevant to all operations and which parts are true only for specific operations.

2.3. Notation

- **Floating-point number representation.** Using the notation of the IEEE standard 754 [1], we represent a floating-point number v by three values: $S(v)$, $E(v)$, and $F(v)$, which denote the *sign*, the *exponent* and the *significand*, respectively. The value of v is

$$v = (-1)^{S(v)} \cdot 2^{E(v)} \cdot F(v)$$

- **p (precision).** This signifies the number of bits in the significand, F, of the floating-point number.

- **ulp(x) or unit in the last place.** This is the weight of one unit in the least significant bit of the significand of

x. The *ulp* of a floating-point number with exponent E and precision p, is 2^{E+1-p}.

3. The Continuous Algorithm

In the following Section, we describe the continuous algorithm for addition. Similar algorithms are also valid for the division and multiplication problems. For the divide continuous algorithm, the symbols $+$ and $-$ should be replaced by the symbols $\div$ and $\times$, respectively. For the multiply continuous algorithm, the symbols $+$ and $-$ should be replaced by the symbols $\times$ and $\div$, respectively.

3.1. Eliminating the *round* operation

We define the set

$$I'_z = \{z \mid \bar{a}_z \le round(z) \le \bar{b}_z\}.$$

Obviously I'_z is an interval, and $[\bar{a}_z, \bar{b}_z] \subset I'_z$. We denote the ends of this interval by a'_z, and b'_z.

Clearly $a'_z \le \bar{a}_z \le \bar{b}_z \le b'_z$, $a'_z < b'_z$, and I'_z is one of the intervals (a'_z, b'_z), $[a'_z, b'_z)$, $(a'_z, b'_z]$, $[a'_z, b'_z]$, depending on $\bar{a}_z, \bar{b}_z$ and on the rounding mode. Our problem can now be formulated as follows:

Find two machine numbers $\bar{x}, \bar{y}$ such that $\bar{a}_x \le \bar{x} \le \bar{b}_x$, $\bar{a}_y \le \bar{y} \le \bar{b}_y$, and $\bar{x} + \bar{y} \in I'_z$.

We define the set of all real number pairs that solve the above inequalities as follows

$$A_{x,y} = \{(x,y) \mid \bar{a}_x \le x \le \bar{b}_x, \ \bar{a}_y \le y \le \bar{b}_y, \ x+y \in I'_z\} \quad (1)$$

Clearly, the set of all solutions to our problem is the subset of $A_{x,y}$, which includes all the pairs in which x and y are machine numbers. In Figure 1, we depict the set of solutions for addition.

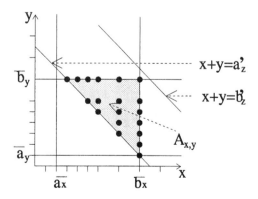

Figure 1. Solution set $A_{x,y}$ for addition

3.2. Reducing the input intervals

There are cases in which some of the intervals $[\bar{a}_x, \bar{b}_x]$, and $[\bar{a}_y, \bar{b}_y]$, I'_z can be reduced to smaller intervals, such that the problems remain equivalent. By *equivalent* we mean that the set $A_{x,y}$ is the same for both problems. We perform the reduction as follows:

For the $+$ operation and $I'_z = [a'_z, b'_z]$, and for every $(x,y) \in A_{x,y}$:

$$\bar{a}_x \le x \le \bar{b}_x, \ \bar{a}_y \le y \le \bar{b}_y, \ a'_z \le x+y \le b'_z.$$

In other words,

$$a'_z - y \le x \le b'_z - y, \ a'_z - x \le y \le b'_z - x.$$

The two relations above imply that

$$a'_z - \bar{b}_y \le x \le b'_z - \bar{a}_y, \ a'_z - \bar{b}_x \le y \le b'_z - \bar{a}_x,$$

Combining this with the original boundaries on x and y from Equation 1, we get new boundaries on x, y, and $x+y$, which we denote by $A_x, B_x, A_y, B_y, A_z, B_z$:

$$A_x = \max\{\bar{a}_x, a'_z - \bar{b}_y\} \le x \le \min\{\bar{b}_x, b'_z - \bar{a}_y\} = B_x$$

$$A_y = \max\{\bar{a}_y, a'_z - \bar{b}_x\} \le y \le \min\{\bar{b}_y, b'_z - \bar{a}_x\} = B_y$$

$$A_z = \max\{a'_z, \bar{a}_x + \bar{a}_y\} \le x+y \le \min\{b'_z, \bar{b}_x + \bar{b}_y\} = B_z.$$

We assumed that $I'_z = [a'_z, b'_z]$. However, if we have $I'_z = (a'_z, b'_z]$ or $[a'_z, b'_z)$ or (a'_z, b'_z), some of the $\le$ relations need to be replaced by $<$ relations. Apart from this, the results would be identical. We can summarize the results in the following way:

For the $+$ operation we have,

$$A_{x,y} = \{(x,y) \mid x \in I_x, \ y \in I_y, \ z = x+y \in I_z\}$$

where I_x, I_y, I_z are intervals (each end point of which is either closed or open) with the end points: A_x, B_x, A_y, B_y, A_z, and B_z, as described above.

In order to follow a similar argument for the $\times$ and the $\div$ operations, we have to first assume (with no loss of generality) that the operands are all positive.

Proposition 1 *The intervals I_x, I_y, I_z cannot be reduced any further.*

To see this, it suffices to show that for every $A_x < x < B_x$, there exists a y, such that $(x, y) \in A_{x,y}$ and likewise for every $A_y < y < B_y$ and for every $A_z < z < B_z$. We shall prove this for every $A_x < x < B_x$, with the $+$ operation. The rest of the cases have similar proofs.

Proof: Consider any x, such that $A_x < x < B_x$. From the definitions of A_x, B_x we deduce that $a'_z - \bar{b}_y < x < b'_z - \bar{a}_y$.

Consider now the expression $z - y$. The left hand side of the above double inequality is equal to $z - y$ with $z = a'_z$, $y = \bar{b}_y$. The right hand side is equal to the same expression with $z = b'_z, y = \bar{a}_y$. Therefore, by changing (y, z) continuously, from $(a'_z, \bar{b}_y)$ to $(b'_z, \bar{a}_y)$, we deduce the existence of intermediate values of y and z, such that $z - y = x$ (or $z = x + y$) exactly. For these values of y and z, we have, $\bar{a}_y < y < \bar{b}_y$, $a'_z < x + y < b'_z$ (and obviously $\bar{a}_x < x < \bar{b}_x$), therefore $(x, y) \in A_{x,y}$. ∎

3.3. Finding a random solution using the continuous algorithm

The above discussion enables us to generate random solutions in the following way:

(i) Choose a random machine number $\bar{x} \in I_x$. If no such $\bar{x}$ exists, there is no solution. Stop.

(ii) Choose a random machine number $\bar{y}$ in the interval
$$[\max\{\bar{a}_y, a'_z - \bar{x}\}, \min\{\bar{b}_y, b'_z - \bar{x}\}].$$
If no such $\bar{y}$ exists, go back to step (i).

(iii) If $\bar{a}_z \leq round(\bar{x} + \bar{y}) \leq \bar{b}_z$, return the result $(\bar{x}, \bar{y})$ and stop. Otherwise, go back to step (i).

Intuitively, we see that this method is very efficient if the projections of the machine number solutions onto the x, y and z axes give sets which are each dense in I_x, I_y and I_z respectively. At the other extreme, when the density of the machine numbers is relatively small, the algorithm may work for an unacceptably long amount of time. In the following two Sections we discuss this problem in more depth.

4. Efficiency of the Continuous Algorithm for Multiplication

For the multiply operation, the algorithm of the previous Section is adequate, except for one extreme case, when $\bar{a}_z = \bar{b}_z$.

We illustrate this using an example with binary floating-point numbers with a three bit significand:
$\bar{a}_x = (1.00)_2$, $\bar{b}_x = (1.01)_2$, $\bar{a}_y = \bar{b}_y = (1.11)_2$, $\bar{a}_z = \bar{b}_z = (10.0)_2$ with the round-up mode. Then,
$$round(\bar{x}\bar{y}) \in \{(1.11)_2, (10.1)_2\}$$
so there is no solution (step (i) cannot discover this) although $A_{x,y}$ is not empty. In this particular case, the continuous algorithm may run for an unreasonably long time.

The following theorem proves the sufficiency of the algorithm for all other cases. Note, this theorem is valid only for binary floating point systems.

Theorem 1 *For the $\times$ operation in a binary floating point system, assume that the set $A_{x,y}$ is not empty, and that $\bar{a}_z < \bar{b}_z$. Let $\bar{x}$ be a machine number in I_x and let $\bar{a}_y$ be a normalized machine number. Then there exists at least one machine number, $\bar{y}$, such that $(\bar{x}, \bar{y})$ is a solution.*

Note, the condition $\bar{a}_z < \bar{b}_z$ implies that there exist at least two machine numbers in I'_z. Therefore, the only case not covered by the theorem is the case where I'_z contains a single machine number.

The fact that $\bar{x} \in I_x$ means that there exists some $y_0 \in I_y$ (not necessarily a machine number), such that $(\bar{x}, y_0) \in A_{x,y}$.

The assumption that $\bar{a}_y$ is normalized is not really restrictive. This is because if both $\bar{x}, \bar{y}$ are denormals, their rounded product has only two possible values. Namely, 0 and the smallest positive denormalized machine number, depending on the rounding mode. This is a simple case, which can be easily treated. Therefore, in interesting cases, either $\bar{a}_x$ or $\bar{a}_y$ must be normalized.

Proof: We divide the discussion into three sub cases.
(i) $\bar{a}_y > a'_z/\bar{x}$, (ii) $\bar{b}_y < b'_z/\bar{x}$, (iii) $\bar{a}_y \leq a'_z/\bar{x}$, $\bar{b}_y \geq b'_z/\bar{x}$.

As mentioned above, there exists a real number, y_0, such that $(\bar{x}, y_0) \in A_{x,y}$. Obviously $\bar{a}_x \leq \bar{x} \leq \bar{b}_x$ in all three cases. Now, in case (i), we choose $\bar{y} = \bar{a}_y$. Consequently, $\bar{a}_y \leq \bar{y} \leq \bar{b}_y$ and
$$\bar{y} > a'_z/\bar{x} \Leftrightarrow \bar{x}\bar{y} > a'_z.$$
Because $(\bar{x}, y_0) \in A_{x,y}$ we must have $\bar{x}y_0 \leq b'_z$. However, (in case (i)) the point $(\bar{x}, \bar{a}_y) = (\bar{x}, \bar{y})$ lies on the lower boundary of $A_{x,y}$, so, $\bar{y} \leq y_0$, which means that either $\bar{y} = y_0$, hence $(\bar{x}, \bar{y}) \in A_{x,y}$, or $\bar{y} < y_0$, hence $\bar{x}\bar{y} < \bar{x}y_0 \leq b'_z$ and again $(\bar{x}, \bar{y}) \in A_{x,y}$.

In case (ii) we choose $\bar{y} = \bar{b}_y$, therefore, $\bar{a}_y \leq \bar{y} \leq \bar{b}_y$ and
$$\bar{y} < b'_z/\bar{x} \Leftrightarrow \bar{x}\bar{y} < b'_z.$$
Since $(\bar{x}, y_0) \in A_{x,y}$ we must have $\bar{x}y_0 \geq a'_z$. However, (in case (ii)), the point $(\bar{x}, \bar{b}_y) = (\bar{x}, \bar{y})$ lies on the upper boundary of $A_{x,y}$, so, $\bar{y} \geq y_0$ and either $\bar{y} = y_0$, hence $(\bar{x}, \bar{y}) \in A_{x,y}$, or $\bar{y} > y_0$, hence $\bar{x}\bar{y} > \bar{x}y_0 \geq a'_z$ and again $(\bar{x}, \bar{y}) \in A_{x,y}$.

In case (iii), all points $(\bar{x}, y)$ between the points $(\bar{x}, a'_z/\bar{x})$, $(\bar{x}, b'_z/\bar{x})$ (inclusive) must satisfy $\bar{a}_y \leq y \leq \bar{b}_y$. Hence, in order to complete the proof, it suffices to show that there exists at least one machine number, $\bar{y}$, such that $a'_z/\bar{x} < \bar{y} < b'_z/\bar{x}$:

Let $\bar{y}_1$ be the largest machine number that satisfies $\bar{y}_1 \leq a'_z/\bar{x}$. We choose $\bar{y} = \bar{y}_1 + ulp(\bar{y}_1)$. Obviously $\bar{y}$ is a machine number and $\bar{y} > a'_z/\bar{x}$. We complete the proof by showing that $\bar{y} < b'_z/\bar{x}$. We denote $z_1 = \bar{x}\bar{y}_1$, $z = \bar{x}\bar{y}$. Then,
$$z - z_1 = \bar{x}(\bar{y}_1 + ulp(\bar{y}_1)) - \bar{x}\bar{y}_1 = \bar{x}\,ulp(\bar{y}_1).$$

Now we divide $z - z_1$ by $ulp(z_1)$. Using the notation from Section 2.3,

$$\frac{z - z_1}{ulp(z_1)} = \bar{x}\frac{ulp(\bar{y}_1)}{ulp(z_1)} = \frac{z_1}{\bar{y}_1}\frac{ulp(\bar{y}_1)}{ulp(z_1)} =$$

$$\frac{2^{E(z_1)}S(z_1)}{2^{E(\bar{y}_1)}S(\bar{y}_1)}\frac{2^{E(\bar{y}_1)+1-p}}{2^{E(z_1)+1-p}} = \frac{S(z_1)}{S(\bar{y}_1)}.$$

$\bar{a}_y$ is a machine number satisfying $\bar{a}_y \leq a'_z/\bar{x}$. Therefore, $\bar{y}_1 \geq \bar{a}_y$, and $\bar{y}_1$ is normalized. Hence $S(\bar{y}_1) \geq 1$. Additionally, $0 < S(z_1) < 2$, so, $S(z_1)/S(\bar{y}_1) < 2$. We have, then, $z - z_1 < 2\,ulp(z_1)$. In addition $b'_z - a'_z \geq 2\,ulp(a'_z)$ because there exist at least two machine numbers in I'_z. This implies that $z < z_1 + 2\,ulp(z_1)$ and that $a'_z + 2\,ulp(a'_z) \leq b'_z$. Therefore,

$$z < z_1 + 2\,ulp(z_1) = \bar{x}\bar{y}_1 + 2\,ulp(\bar{x}\bar{y}_1) \leq$$

$$\bar{x}a'_z/\bar{x} + 2\,ulp(\bar{x}a'_z/\bar{x}) = a'_z + 2\,ulp(a'_z) \leq b'_z.$$

Hence,

$$\bar{x}\bar{y} < b'_z \Leftrightarrow \bar{y} < b'_z/\bar{x}$$

and the proof is complete. ∎

5. The Discrete Algorithm for the + Operation

5.1. Preliminary discussion

There are some cases in which the algorithm of Section 3.3, applied to the + operation, does not work efficiently. This occurs, for example, when there is a significant cancellation of bits; in other words, when the exponent of the result is smaller than the input exponents. We illustrate this point by an example using binary floating-point numbers with a three bit significand:
$\bar{a}_x = (1.00)_2$, $\bar{b}_x = (1.11)_2$, $\bar{a}_y = -(1.11)_2$, $\bar{b}_y = -(1.00)_2$, $\bar{a}_z = (0.00101)_2$, $\bar{b}_z = (0.00111)_2$, with round to nearest mode. Clearly the set $A_{x,y}$ in this example, is not empty. Yet, there is no solution of machine numbers. The set of solutions is sparse on the z axis, in the vicinity of the interval $[\bar{a}_z, \bar{b}_z]$. In general, the algorithm cannot detect whether such a solution exists.

Therefore, we need to find a different algorithm for addition, which solves these difficult cases as well. The focus of this algorithm will be to search for a solution within a subspace of the (x, y) plane. More specifically, the algorithm described below works only in cases where the rectangle $R = [\bar{a}_x, \bar{b}_x] \times [\bar{a}_y, \bar{b}_y]$ has the following properties:

(I) *The step size, Δy, between successive machine numbers along the y-axis, is constant throughout the rectangle.*

(II) *The step size, Δx, between successive machine numbers along the x-axis, may vary within the rectangle, but is always at least the size of Δy.*

The conditions on x and y are interchangeable.

In order to make use of this algorithm, the original rectangle R, which contains all the solution pairs, must be divided into sub-rectangles, each of which has these properties. A method for efficient partitioning of the original rectangle is described in Section 5.5.

In the most extreme case, the original rectangle may include all of the pairs of floating-point numbers (both positive and negative). In such a case, the number of sub-rectangles will be approximately $8(E_{\max} - E_{\min})$, which is close to 16000 for IEEE double precision. This algorithm is therefore slower than the one described in Section 3.3. Hence we recommend using first the continuous algorithm, and if it does not find a solution within a reasonable amount of time, to then use the discrete algorithm.

We first set up the theoretical framework underlying the algorithm, in Sections 5.2 and 5.3, by showing a necessary and sufficient condition for the existence of a solution. The algorithm itself is presented in Sections 5.4 and 5.5.

5.2. Interval lemmas

Lemma 1 *Given two intervals $[a_1, b_1]$, $[a_2, b_2]$, the condition $[a_1, b_1] \cap [a_2, b_2] \neq \emptyset$ is equivalent to $a_1 \leq b_1$, $a_2 \leq b_2$, $a_1 \leq b_2$, $a_2 \leq b_1$.*

This lemma has been formulated for closed intervals. We may similarly formulate this lemma for intervals in which one, or both, ends are open. The proof of the lemma is immediate.

Lemma 2 *Let P be a set of points on the real axis, let I be an interval on the real axis, and let I_P be a closed interval on the real axis, whose end points are both in P. Then, $I_P \cap I \cap P \neq \emptyset$ if and only if $I_P \cap I \neq \emptyset$ and $I \cap P \neq \emptyset$.*

Proof: It is easy to see that

$$I_P \cap I \cap P \neq \emptyset \Rightarrow I_P \cap I \neq \emptyset, I \cap P \neq \emptyset.$$

Now, assume that $I_P \cap I \neq \emptyset$ and that $I \cap P \neq \emptyset$. If at least one of the ends of I_P is in I, this end point is included in $I_P \cap I \cap P$. Therefore, we assume that both ends are outside of I. There are three possibilities: (i) both ends lie to the left of I, (ii) both ends lie to the right of I and (iii) the left end lies to the left of I and the right end lies to the right of I. In cases (i) and (ii) we have $I_P \cap I = \emptyset$, which contradicts our assumptions. In case (iii) we have $I \subseteq I_P$, so $I_P \cap I \cap P = I \cap P \neq \emptyset$.
∎

5.3. Condition for the existence of a solution

In what follows, we assume that the rectangle R satisfies conditions (I) and (II) of sub-Section 5.1. Let $\bar{x}$ be a machine number in $[\bar{a}_x, \bar{b}_x]$. We define the set of points

$$D(\bar{x}) = \{\bar{x} + \bar{y} \mid \bar{y} \in [\bar{a}_y, \bar{b}_y], \bar{y} \text{ is a machine number}\}.$$

Because of conditions (I) and (II), every point of $D(\bar{x})$ can be written in the form $\bar{a}_x + \bar{a}_y + k\,\Delta y$, where k is an integer. We also define $I(\bar{x})$ to be the smallest interval of real numbers that includes $D(\bar{x})$. Namely,

$$I(\bar{x}) = [\bar{x} + \bar{a}_y, \bar{x} + \bar{b}_y].$$

We will need the following lemma:

Lemma 3 *Assume that at least one point of the sequence $\bar{a}_x + \bar{a}_y + k\,\Delta y$ ($k = 0, \pm 1, \pm 2, \cdots$) is included in I'_z. Given a machine number $\bar{x} \in [\bar{a}_x, \bar{b}_x]$, a necessary and sufficient condition for the existence of a solution in $D(\bar{x})$ (i.e., $D(\bar{x}) \cap I'_z \neq \emptyset$) is the condition $I(\bar{x}) \cap I'_z \neq \emptyset$.*

Proof: We use Lemma 2 and substitute

$$P = \{\bar{a}_x + \bar{a}_y + k\,\Delta y \mid k = 0, \pm 1, \pm 2, \cdots\},$$
$$I = I'_z, \quad I_P = I(\bar{x}).$$

Obviously $D(\bar{x}) = I(\bar{x}) \cap P$. Hence, from Lemma 2 we find that

$$I_P \cap I \cap P = I(\bar{x}) \cap I'_z \cap P =$$
$$D(\bar{x}) \cap I'_z \neq \emptyset \Leftrightarrow I(\bar{x}) \cap I'_z \neq \emptyset, \; I'_z \cap P \neq \emptyset$$

Because we assumed that $I'_z \cap P \neq \emptyset$, the condition $I(\bar{x}) \cap I'_z \neq \emptyset$ is equivalent to $D(\bar{x}) \cap I'_z \neq \emptyset$. ∎

Theorem 2 *Assume that I'_z is closed on both ends and that (I) and (II) are satisfied. A necessary and sufficient condition for the existence of a solution for the + operation, is: At least one integer k, satisfies*

$$(a'_z - \bar{a}_x - \bar{a}_y)/\Delta y \leq k \leq (b'_z - \bar{a}_x - \bar{a}_y)/\Delta y$$

and at least one machine number $\bar{x}$, satisfies

$$\bar{a}_x \leq \bar{x} \leq \bar{b}_x \,, \; a'_z - \bar{b}_y \leq \bar{x} \leq b'_z - \bar{a}_y. \qquad (2)$$

If the conditions are satisfied (i.e., there exist solutions), for each $\bar{x}$ that satisfies equation 2, there exist solutions of the form $(\bar{x}, \bar{y})$, where $\bar{x} + \bar{y}$ is in $D(\bar{x})$. For each machine number $\bar{x}$, which is not of this type, there are no solutions. All solutions with $\bar{x}$ are of the form

$$(\bar{x}, \bar{a}_y + n\,\Delta y)$$

where

$$(a'_z - \bar{x} - \bar{a}_y)/\Delta y \leq n \leq (b'_z - \bar{x} - \bar{a}_y)/\Delta y$$
$$n = 0, 1, \cdots, N \,, \; N = (\bar{b}_y - \bar{a}_y)/\Delta y.$$

Remark: Theorem 2 was formulated with the assumption that I'_z is closed at both of its ends. If this is not so, the formulation is similar when using the relation $<$ to replace some of the corresponding $\leq$ relations.

Proof: A number in $D(\bar{x}) \cap I'_z$ must be of the form $\bar{x} + \bar{y} = \bar{a}_x + \bar{a}_y + k\,\Delta y$ and must be included in I'_z. Therefore, from Lemma 3 we see that necessary and sufficient conditions for the existence of a solution are $a'_z \leq \bar{a}_x + \bar{a}_y + k\,\Delta y \leq b'_z$ for some k and $I(\bar{x}) \cap I'_z \neq \emptyset$. The first of these conditions is equivalent to

$$(a'_z - \bar{a}_x - \bar{a}_y)/\Delta y \leq k \leq (b'_z - \bar{a}_x - \bar{a}_y)/\Delta y$$

and the second is equivalent to $\bar{x} + \bar{a}_y \leq b'_z, \bar{x} + \bar{b}_y \geq a'_z$ from Lemma 1, which is equivalent to

$$a'_z - \bar{b}_y \leq \bar{x} \leq b'_z - \bar{a}_y.$$

The solutions which correspond to $\bar{x}$, must clearly be of the form $(\bar{x}, \bar{a}_y + n\,\Delta y)$, with $n = 0, 1, \cdots, N$, and they must satisfy $a'_z \leq \bar{x} + \bar{a}_y + n\,\Delta y \leq b'_z$. Therefore, we must have

$$(a'_z - \bar{x} - \bar{a}_y)/\Delta y \leq n \leq (b'_z - \bar{x} - \bar{a}_y)/\Delta y,$$
$$n = 0, 1, \cdots, N.$$

∎

5.4. The discrete algorithm

Theorem 2 leads us to formulate a second algorithm for the addition. Because the theorem assumes conditions (I) and (II), one must first divide the rectangle R into sub-rectangles, each of which either satisfies these conditions, or satisfies the conditions with x and y interchanged.

The following algorithm must be applied to each of these sub-rectangles in a random order, until a solution is found.

(i) Test for the existence of an integer k, which satisfies

$$(a'_z - \bar{a}_x - \bar{a}_y)/\Delta y \leq k \leq (b'_z - \bar{a}_x - \bar{a}_y)/\Delta y.$$

If no such k exists, stop. There is no solution.

(ii) Choose, at random, a machine number $\bar{x}$ that satisfies

$$\bar{a}_x \leq \bar{x} \leq \bar{b}_x \,, \; a'_z - \bar{b}_y \leq \bar{x} \leq b'_z - \bar{a}_y$$

(i.e. $\bar{x} \in I_x$). If no such $\bar{x}$ exists, stop. There is no solution.

(iii) Choose at random, an integer n that satisfies

$$(a'_z - \bar{x} - \bar{a}_y)/\Delta y \leq n \leq (b'_z - \bar{x} - \bar{a}_y)/\Delta y$$
$$n \in \{0, 1, \cdots, N\}$$

(such an n must exist!).
Return the solution $(\bar{x}, \bar{a}_y + n\,\Delta y)$ and stop.

The remark that follows Theorem 2, applies to this algorithm as well. We need only apply this algorithm to those sub-rectangles that intersect the sub-plane $a'_z \leq x + y \leq b'_z$.

5.5. Partitioning the original ranges into sub-rectangles

Several different methods can be used to partition the original rectangle into sub-rectangles. In the simplest method, each sub-rectangle includes exactly one exponent for both x and y. Using this method, in the worst case, we get approximately $4(E_{\max} - E_{\min})^2$ rectangles, which is around 16,000,000 for double precision. Therefore, this solution is impractical.

The following method generates, in the worst case, approximately $8(E_{\max} - E_{\min})$ sub-rectangles, which is around 16,000 for double precision. For each exponent, E, we define the following two rectangles:

1. The exponent of y is E and the exponent of $x \geq E$.

2. The exponent of x is E and the exponent of $y > E$.

For each quadrant, there are about $2(E_{\max} - E_{\min})$ rectangles. Because, in the worst case, we must define these rectangles for each quadrant of the plane, in total we get about $8(E_{\max} - E_{\min})$.

Figure 2 illustrates the partition to sub-rectangles for the first quadrant.

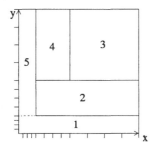

Figure 2. Dividing into rectangles

6. Summary and Future Work

We presented algorithms for addition, multiplication, and division, based on the fact the machine number ranges behave similarly to real number intervals in many cases. We pointed out that these algorithms are usually efficient and fast, and that similar algorithms may be easily formulated for other arithmetic operations.

We showed that the continuous algorithm is efficient for multiplication, whenever the result range consists of more than one value (i.e., when $\bar{a}_z \neq \bar{b}_z$). For addition, we presented the discrete algorithm, which always finds a solution, if one exists. This algorithm is slower than the continuous algorithm, and its use is recommended only in those cases in which the continuous algorithm does not work efficiently.

A natural extension of this work is to generalize the algorithm to fully support the $\div$ operation. This includes checking the effectiveness of the continuous algorithm and possibly finding an additional algorithm that solves those cases for which the continuous algorithm is insufficient. The solution for multiplication is not yet complete. We need to find an algorithm that efficiently solves the case of a single value in the result range.

It is also interesting to investigate the multiply-add instruction ($a \times b + c$) [7], which is under a standardization process, and which is known to be bug-prone. It is not difficult to formulate a continuous algorithm. But solving this instruction, obviously involves the problems that arise when solving the $+$ instruction. However, this problem is even more complex, as we have a variable step size between consecutive machine numbers in the product ($a \times b$).

References

[1] IEEE standard for binary floating point arithemetic, 1985. An American national standard, ANSI/IEEE Std 754.

[2] Raanan Grinwald and Eran Harel and Michael Orgad and Shmuel Ur and Avi Ziv. User defined coverage - a tool supported methodology for design verification. In *Proc. 38th, Design Autometation Conference (DAC38)*, pp. 158 – 163, 1998.

[3] Abraham Ziv and Laurent Fournier. Solving the generalized mask constraint for test generation of binary floating point add operation. *Theoretical Computer Science*, 291:183 – 201, 2003.

[4] S. Asaf and L. Fournier. FPgen - a deep-knowledge test-generator for floating point verification. Technical Report H-0140, IBM Israel, 2002. http://domino.watson.ibm.com/library/cyberdig.nsf/Home.

[5] M. Cowlishaw, E. Schwarz, R. Smith, and C. Webb. A decimal floating-point specification. In *Proc. IEEE 15th Symp. Computer-Arithmetic (ARITH15)*, pp. 147 – 154, 2001.

[6] W. Kahan. A test for correctly rounded sqrt. http://www.cs.berkeley.edu/~wkahan/SQRTest.ps.

[7] W. Kahan. Lectures notes on the status of IEEE standard 754 for binary floating-point arithmetic., 1996. http://www.cs.berkeley.edu/~wkahan/ieee754status/IEEE754.pdf.

[8] L. D. McFearin and D. W. Matula. Generation and analysis of hard to round cases for binary floating point division. In *IEEE Symposium on Computer Arithmetic (ARITH15)*, pp. 119 – 127, 2001.

[9] M. Parks. Number-theoretic test generation for directed rounding. In *Proc. IEEE 14th Symp. Computer-Arithmetic (ARITH14)*, pp. 241 – 248, 1999.

A Parametric Error Analysis of Goldschmidt's Division Algorithm
- Extended Abstract -

Guy Even
Dept. of Electrical Engineering Systems
Tel-Aviv University
Tel-Aviv 69978, Israel
guy@eng.tau.ac.il

Peter-Michael Seidel
Computer Science & Engineering Dept.
Southern Methodist University
Dallas, TX, 75275
seidel@engr.smu.edu

Warren E. Ferguson
warren_e_ferguson@hotmail.com

Abstract

Back in the 60's Goldschmidt presented a variation of Newton-Raphson iterations for division that is well suited for pipelining. The problem in using Goldschmidt's division algorithm is to present an error analysis that enables one to save hardware by using just the right amount of precision for intermediate calculations while still providing correct rounding. Previous implementations relied on combining formal proof methods (that span thousands of lines) with millions of test vectors. These techniques yield correct designs but the analysis is hard to follow and is not quite tight.

We present a simple parametric error analysis of Goldschmidt's division algorithm. This analysis sheds more light on the effect of the different parameters on the error. In addition, we derive closed error formulae that allow to determine optimal parameter choices in four practical settings.

We apply our analysis to show that a few bits of precision can be saved in the floating-point division (FP-DIV) microarchitecture of the AMD-K7^{TM} microprocessor. These reductions in precision apply to the initial approximation and to the lengths of the multiplicands in the multiplier. When translated to cost, the reductions reflect a savings of 10.6% in the overall cost of the FP-DIV micro-architecture.

1. Introduction and Summary

Asymptotically optimal division algorithms are based on multiplicative division methods [13, 17, 21]. Current commercial processor designs employ a parallel multiplier for performing division and square-root operations for floating-point [1, 4, 12, 15]. The parallel multiplier is used for additional operations, such as: multiplication and fused multiply-add. Since meeting the precision requirements of division operations requires more precision than other operations, the dimensions of the parallel multiplier are often determined by precision requirements of division operations. It follows that tighter analysis of the required multiplier dimensions for division operations can lead to improvements in cost, delay, and even power consumption.

The main two methods used for multiplicative division are a variation of Newton's method [7, 8] and a method introduced by Goldschmidt [9] that is based on an approximation of a series expansion. Division based on Newton's method has a quadratic convergence rate (i.e., the number of accurate bits doubles in each iteration) and is self-correcting (i.e., inaccuracies of intermediate computations do not accumulate). A rigorous error analysis of Newton's method appears in [3, 10, 14] and for various exceptional cases in [4]. The analysis in [3, 10] considers the smallest precision required per iteration. Our error analysis follows this spirit by defining separate error parameters for every intermediate computation. In addition, the analysis in [3, 10] relies on directed roundings, a method that we use as well.

Each iteration of Newton's method involves two *dependent* multiplications; namely, the product of the first multiplication is one of the operands of the second multiplication. The implication of having to compute two dependent multiplications per iteration is that these multiplications cannot be parallelized or pipelined.

Goldschmidt's division algorithm also requires two multiplications per iteration and the convergence rate is the same as for Newton's method. However, the most important feature of Goldschmidt's algorithm is that the two multiplications per iteration are *independent* and can be pipelined or computed in parallel. On the other hand, Goldschmidt's algorithm is not self-correcting; namely, inaccuracies of intermediate computations accumulate and cause the computed result to drift away from the accurate quotient. Goldschmidt's division algorithm was used in the IBM System/360 model 91 [2] and even more recently in the IBM S/390 [19] and in the AMD-K7™ microprocessor [15]. However, lack of a general and simple error analysis of Goldschmidt's division algorithm has averted most designers from considering implementing Goldschmidt's algorithm. Thus most implementations of multiplicative division methods have been based on Newton's method in spite of the longer latency due to dependent multiplications in each iteration [4, 12] (see also [22] for more references).

Goldschmidt's method is not self-correcting as explained in [11] (there is a wrong comment on this in [23]). This makes it particularly important and difficult to keep track of accumulated and propagated error terms during intermediate computations. We were not able to locate a general analysis of error bounds of Goldschmidt's algorithm in the literature. Goldschmidt's error analysis in [9] is with respect to a design that uses a serial radix-4 Booth multiplier with 61-bits. Goldschmidt's design computes the quotient of two binary numbers in the range $[1/2, 1)$, and his analysis shows that the absolute error is in the range $[-2^{56}, 0]$. Krishnamurthy [11] analyzes the error only for the case that only one multiplicand is allowed to be imprecise in intermediate computations (the second multiplicand must be precise); such an analysis is only useful for determining lower bounds for delay. Recent implementations of Goldschmidt's division algorithm still rely on an error analysis that over-estimates the accumulated error [15]. Such over-estimates lead to correct designs but waste hardware and cause unnecessary delay (since the multiplier and the initial lookup table are too large). These over-estimations were based on informal arguments that were confirmed by a mechanically verified proof that spans over 250 definitions and 3000 lemmas [18].

Agarwal *et al.* [1] presented a multiplicative division algorithm that is based on an approximate series expansion. This algorithm was implemented in IBM's Power3™. Their algorithm provides no advantages over Goldschmidt's algorithm. In double precision, their algorithm requires 8 multiplications and the longest chain of dependent multiplications consists of 4 multiplications.

We present a version of Goldschmidt's division algorithm that uses directed roundings. We develop a simple general parametric analysis of tight error bounds for our version of Goldschmidt's division algorithm. Our analysis is parametric in the sense that it allows arbitrary one-sided errors in each intermediate computation and it allows an arbitrary number of iterations. In addition, we suggest four practical simplified settings in which errors in intermediate computations are not arbitrary. For each of these four settings, we present a closed error formula. The advantage of closed formulae is in simplifying the task of finding optimal parameter combinations in implementations of Goldschmidt's division method for a given target precision.

We demonstrate the advantages of our error analysis by showing how it could lead to savings in cost and delay. For this purpose we consider Oberman's [15] floating-point micro-architecture used in the AMD-K7™ design. We present a micro-architecture that implements our version of Goldschmidt's algorithm and follows the micro-architecture described in [15]. The modules building our micro-architecture were made as similar as possible to the modules in [15]. This was done so that the issue of the precisions of the lookup table and multiplier could be isolated from other issues. Based on our analysis, we use a smaller multiplier (70×74 bits compared to 76×76 in [15]) and we allow a slightly larger initial error ($2^{-13.51}$ compared to $2^{-13.75}$ in [15]). Based on the cost models of Paul & Seidel [16] and Mueller & Paul [14], we estimate that our parameter choices for multiplier widths and initial approximation accuracy reduce the cost of the micro-architecture by 10.6% compared to the parameter choices in [15].

The paper is organized as follows. In Section 2 we present Newton's method for division and then proceed by presenting Goldschmidt's algorithm as a variation of Newton's method. In Section 3, a version of Goldschmidt's algorithm with imprecise intermediate computations is presented as well as an error analysis. In Section 4 we develop closed form error bounds for Goldschmidt's method with respect to specific settings. In Section 5 we present an alternative micro-architecture to [15] and compare costs. Due to space limitations, all proofs are omitted and can be found in the full version [6].

2. Goldschmidt's division algorithm

In this section we present Newton's method for computing the reciprocal of a given number. We then continue by describing a version of Goldschmidt's division algorithm [9] that uses precise intermediate computations. We show how Goldschmidt's algorithm is derived from Newton's method. The error analysis of Newton's method is used to analyze the errors in Goldschmidt's algorithm.

2.1. Newton's method.

Newton's method can be applied to compute the reciprocal of a given number. To compute the reciprocal of $B > 0$, apply Newton's method to the function $f(x) = B - 1/x$. Note that: (a) the root of $f(x)$ is $1/B$, which is the reciprocal we want to compute, and (b) the function $f(x)$ has a derivative $f'(x) = x^{-2}$ in the interval $(0, \infty)$. In particular, the derivative $f'(x)$ is positive.

Newton iterations are defined by the following recurrence: Let x_0 denote an initial estimate $x_0 \neq 0$ and define

x_{i+1} by

$$x_{i+1} = x_i - \frac{f(x_i)}{f'(x_i)}$$
$$= x_i - \frac{B - \frac{1}{x_i}}{x_i^{-2}} \quad (1)$$
$$= x_i \cdot (2 - B \cdot x_i).$$

Consider the *relative error* term e_i defined by

$$e_i \triangleq \frac{\frac{1}{B} - x_i}{\frac{1}{B}}$$
$$= 1 - B \cdot x_i.$$

It follows that

$$\begin{aligned} e_{i+1} &= 1 - B \cdot x_{i+1} \\ &= 1 - B \cdot x_i \cdot (2 - B \cdot x_i) \\ &= (1 - B \cdot x_i)^2 \\ &= e_i^2. \end{aligned} \quad (2)$$

Equation 2 has three implications:

1. Convergence of x_i to $1/B$ at a quadratic rate is guaranteed provided that the initial relative error is less than 1. Equivalently, convergence holds if $x_0 \in (0, \frac{2}{B})$.

2. For $i \geq 1$, the relative error e_i is non-negative, hence, $x_i \leq 1/B$. This property is referred to as *one-sided convergence*.

3. If $B \in [1, 2)$, then also the *absolute error* decreases at a quadratic rate. Hence, the number of "accurate" bits doubles every iteration, and the number of iterations required to obtain p bits of accuracy is logarithmic in p.

The disadvantage of Newton's iterations, with respect to a pipelined multiplier, is that each iteration consists of 2 *dependent* multiplications: $\alpha_i = B \cdot x_i$ and $\beta_i = x_i \cdot (2 - \alpha_i)$. Namely, the product β_i cannot be computed before the product α_i is computed.

2.2. Goldschmidt's Algorithm

In this section we describe how Goldschmidt's algorithm can be derived from Newton's method. Here, our goal is to compute the quotient A/B. Goldschmidt's algorithm uses three values N_i, D_i and F_i defined as follows:

$$N_i \triangleq A \cdot x_i$$
$$D_i \triangleq B \cdot x_i$$
$$F_i \triangleq 2 - D_i.$$

Algorithm 1 Goldschmidt-Divide(A, B) - Goldschmidt's iterative algorithm for computing A/B

Require: $|e_0| < 1$.
1: Initialize: $N_{-1} \leftarrow A, D_{-1} \leftarrow B, F_{-1} \leftarrow \frac{1-e_0}{B}$.
2: **for** $i = 0$ to k **do**
3:    $N_i \leftarrow N_{i-1} \cdot F_{i-1}$.
4:    $D_i \leftarrow D_{i-1} \cdot F_{i-1}$.
5:    $F_i \leftarrow 2 - D_i$.
6: **end for**
7: Return(N_i)

Consider Newton's iteration: $x_{i+1} = x_i \cdot (2 - B \cdot x_i)$. We may rewrite an iteration by

$$x_{i+1} = x_i \cdot F_i,$$

which when multiplied by A and B, respectively, becomes

$$N_{i+1} = N_i \cdot F_i \xrightarrow{i \to \infty} A/B$$
$$D_{i+1} = D_i \cdot F_i \xrightarrow{i \to \infty} 1.$$

Since x_i converges to $1/B$, it follows that N_i converges to A/B and D_i converges to 1.

Note that: (a) $N_i/D_i = A/B$, for every i; (b) N_i converges to A/B at the same rate that x_i converges to $1/B$; and (c) Let $A, B > 0$. Since the relative error e_i in Newton's iterations is non-negative, for $i \geq 1$, it follows that $N_i \leq A/B, D_i \leq 1$ and $F_i \geq 1$ for $i \geq 1$.

As in Newton's iterations, the algorithm converges if $|e_0| < 1$ and the relative error decreases quadratically. One could use a fixed initial approximation of the quotient. Usually a more accurate initial approximation of $1/B$ is computed by a lookup table or even a more elaborate functional unit (c.f. [5, 20]).

Algorithm 1 lists Goldschmidt's division algorithm. Given A and B the algorithm computes the quotient A/B. The listing uses the same notation used above, and iterates k times.

Observe that the two multiplications that take place in every iteration (in Lines 3-4) are independent, and therefore, Goldschmidt's division algorithm is more amenable to pipelined implementations. The initial approximation is assumed to depend on the value of B. Note that k iterations of either Newton's method or Goldschmidt's algorithm require $2k + 1$ multiplications. These $2k + 1$ multiplication must be linearly ordered in Newton's method implying a critical path of $2k + 1$ dependent multiplications. In Goldschmidt's algorithm the two multiplications that take place in every iterations are independent, hence the critical path consists only of $k + 1$ multiplications.

An error analysis of Goldschmidt's algorithm with precise arithmetic is based on the following claim.

Claim 1 *The following equalities hold for $i \geq 0$:*

$$D_i = 1 - e_0^{2^i}$$
$$F_i = 1 + e_0^{2^i}$$

The key difficulty in analyzing the error in imprecise implementations of Goldschmidt's algorithm is due to the violation of the invariant $N_i/D_i = A/B$. Consider the equality

$$\begin{aligned} N_k &= A/B \cdot D_0 \cdot F_0 \cdot F_1 \cdot \ldots \cdot F_{k-1} \\ &= A/B \cdot (1 - e_0) \cdot \prod_{i=0}^{k-1}(1 + e_0^{2^i}). \end{aligned}$$

Imprecise $D_0, F_0, \ldots, F_{k-1}$ accumulate to an imprecise approximation of A/B.

3. Imprecise Intermediate Computations

This section contains the core contribution of our paper. We present a version of Goldschmidt's algorithm with imprecise intermediate computations. In this algorithm the invariant $N_i/D_i = A/B$ of Goldschmidt's algorithm with precise arithmetic does not hold anymore. We then develop a simple parametric analysis for error bounds in this algorithm. The error analysis is based on relative errors of intermediate computations. The setting is quite general and allows for different relative errors for each computation.

We define the relative error as follows.

Definition 1 *The relative error of x with respect to y is defined by $\frac{y-x}{y}$.*

Note that one usually uses the negative definition (i.e., $(x - y)/y$). We prefer this definition since is helps clarify the direction of the directed roundings that we use.

The analysis begins by using the exact values of all the relative errors. The values of the relative errors depend on the actual values of the inputs and on the hardware used for the intermediate computations. However, one can usually easily derive upper bounds on the absolute errors of each intermediate computation. E.g., such bounds on the absolute errors are simply derived from the precision of the multipliers. Our analysis continues by translating the upper bounds on the absolute errors to upper bounds on the relative errors. Hence we are able to analyze the accuracy of an implementation of the proposed algorithm based on upper bounds on the absolute errors.

An interesting feature of our analysis is that directed roundings are used for all intermediate calculations. Surprisingly, directed roundings play a crucial role in this analysis and enable a simpler and tighter error analysis than round-to-nearest rounding (c.f. [15]).

3.1. Goldschmidt's division algorithm using approximate arithmetic

A listing of Goldschmidt's division algorithm using approximate arithmetic appears in Algorithm 2. The values corresponding to N_i, D_i, and F_i using the imprecise computations are denoted by N_i', D_i' and F_i', respectively.

Algorithm 2 Goldschmidt-Approx-Divide(A, B) - Goldschmidt's division algorithm using approximate arithmetic
---
1: Initialize: $N_{-1}' \leftarrow A$, $D_{-1}' \leftarrow B$, $F_{-1}' \leftarrow \frac{1-e_0}{B}$.
2: **for** $i = 0$ to k **do**
3: $\quad N_i' \leftarrow (1 - n_i) \cdot N_{i-1}' \cdot F_{i-1}'$.
4: $\quad D_i' \leftarrow (1 + d_i) \cdot D_{i-1}' \cdot F_{i-1}'$.
5: $\quad F_i' \leftarrow (1 - f_i) \cdot (2 - D_i')$.
6: **end for**
7: Return(N_i')

Directed roundings are used for all intermediate calculations. For example, N_i' is obtained by rounding down the product $N_{i-1}' \cdot F_{i-1}'$. We denote by n_i the relative error of N_i' with respect to $N_{i-1}' \cdot F_{i-1}'$. Since $N_{i-1}' \cdot F_{i-1}'$ is rounded down, we assume that $n_i \geq 0$. Similarly, rounding down is used for computing F_i' (with the relative error f_i) and rounding up is used for computing D_i' (with the relative error d_i).

The initial approximation of the reciprocal $1/B$ is denoted by F_{-1}'. The relative error of F_{-1}' with respect to $1/B$ is denoted by e_0. We do not make any assumption about the sign of e_0.

Our error analysis is based on the following assumptions:

1. The operands are in the range $A, B \in [1, 2)$.

2. All the relative errors incurred by directed rounding are at most $1/4$. This assumption is easily met by multipliers with more than 4 bits of precision.

3. We require that $|e_0| + 3d_0/2 + f_0 < 1/2$. Again, this assumption is easily met if the multiplications and the initial reciprocal approximation are precise enough.

4. The initial approximation F_{-1}' of $1/B$ is in the range $[1/2, 1]$. This assumption is easily met if lookup tables are used.

Definition 2 *The relative error of N_i' with respect to A/B is*

$$\rho(N_i') \triangleq \frac{A/B - N_i'}{A/B}.$$

3.2. A simplifying assumption: strict directed roundings

The following assumption about directed rounding used in Algorithm 2 helps simplify the analysis.

Definition 3 (Strict Directed (SD) rounding) *Rounding down is strict if $x \geq 1$ implies that $rnd(x) \geq 1$. Similarly, rounding up is strict if $x \leq 1$ implies that $rnd(x) \leq 1$.*

Observe that, in general, rounding down means that $rnd(x) \leq x$, for all x. Often the absolute error introduced by rounding is bounded by $\varepsilon > 0$, namely $x - \varepsilon \leq rnd(x) \leq x$. Strict rounding down requires that if $x \geq 1$, then $rnd(x) \geq 1$ no matter how close x is to 1. In non-redundant binary representation strict rounding is easily implemented as follows. Strict rounding down can be implemented by truncating. Strict rounding up can be obtained by (i) an increment by a unit in each position below the rounding position and (ii) truncation of the bit string in positions less significant than the rounding position.

Assumption 2 (SD rounding) *All directed roundings used in Algorithm 2 are strict.*

3.3. Parametric Error Analysis

Lemma 3 bounds the ranges of N'_i and F'_i in Algorithm 2 under Assumption 2. This lemma is an extension of the properties $D_i \leq 1$ and $F_i \geq 1$ (for $i \geq 1$) of Algorithm 1.

Goldschmidt already pointed out that since F_i tends to 1 from above, one could save hardware since the binary representation of F_i begins with the string $1.000\ldots$. An analogous remark holds for D_i. However, Lemma 3 refers to the inaccurate intermediate results (i.e., D'_i and F'_i) rather than the precise intermediate results (i.e., D_i and F_i). Parts 2-3 of lemma 3 show that the same hardware reduction strategy applies to Algorithm 2, even though intermediate calculations are imprecise.

Definition 4 *Define δ_i, for $i \geq 0$, as follows:*

$$\delta_i := \begin{cases} |e_0| + 3d_0/2 & \text{for } i = 0 \\ \delta_{i-1}^2 + f_{i-1} & \text{otherwise.} \end{cases}$$

Lemma 3 *(ranges of D'_i and F'_i)*
The following bounds hold:

1. $D'_0 \in [1 - \delta_0, 1 + \delta_0] \subseteq (1/2, 3/2)$.
2. $D'_i \in [1 - \delta_i, 1]$, for every $i \geq 1$.
3. $F'_i \in [1, 1 + \delta_i]$, for every $i \geq 1$.
4. $D'_i \leq D'_{i+1}$, for every $i \geq 1$.

The following claim summarizes the relative error of Goldschmidt's division algorithm using approximate arithmetic.

Theorem 4 *For every $i > 0$, the relative error $\rho(N'_i) = \frac{A/B - N'_i}{A/B}$ satisfies $\pi_i \leq \rho(N'_i) \leq \pi_i + \delta_i$, where π_i is defined by $\pi_i \triangleq 1 - (1 - n_i) \cdot \prod_{j=0}^{i-1} \frac{1-n_j}{1+d_j} \geq 0$.*

For $i = 0$ it can be verified that $|\rho(N'_0)| \leq \pi_0 + \delta_0$.

A somewhat looser (yet easier to evaluate) bound on the relative error follows from

$$\pi_i \leq \sum_{j=0}^{i} n_j + \sum_{j=0}^{i-1} d_j. \quad (3)$$

3.4. Deriving bounds on relative errors from absolute errors

In this subsection we obtain bounds on the relative errors $\rho(N'_i)$ from the absolute errors of the intermediate computations. The reason for doing so is that in an implementation one is able to easily bound the absolute errors of intermediate computations; these follow directly from the precision of the operation, the rounding used (e.g., floor or ceiling), and the representation of the results (binary, carry-save, etc.).

Consider the computation of N'_i. The relative error introduced by this computation is n_i, and N'_i equals $(1 - n_i) \cdot N'_{i-1} \cdot F'_{i-1}$. An accurate computation would produce the product $N'_{i-1} \cdot F'_{i-1}$. Hence, the absolute error is $n_i \cdot N'_{i-1} \cdot F'_{i-1}$.

Definition 5 *The absolute errors of intermediate computations are defined as follows:*

$$neps_i \triangleq n_i \cdot N'_{i-1} \cdot F'_{i-1}$$
$$deps_i \triangleq d_i \cdot D'_{i-1} \cdot F'_{i-1}$$
$$feps_i \triangleq f_i \cdot (2 - D'_i).$$

In an implementation, the exact absolute errors are unknown. Instead, we use upper bounds on the absolute errors. We denote these upper bounds as follows: $\widehat{neps}_i \geq neps_i$, $\widehat{deps}_i \geq deps_i$ and $\widehat{feps}_i \geq feps_i$.

The following claim shows how one can derive upper bounds on the relative errors from upper bounds on the absolute errors.

Claim 5 *(from absolute to relative errors) If $A, B \in [1, 2)$, then for $i \geq 2$ the relative errors are bounded by:*

$$0 \leq n_i \leq 2\widehat{neps}_i/(1 - \pi_{i-1} - \delta_{i-1})$$
$$0 \leq d_i \leq \widehat{deps}_i/(1 - \delta_{i-1})$$
$$0 \leq f_i \leq \widehat{feps}_i$$

A careful reader might be concerned by the fact that δ_{i-1} and π_{i-1} appear in the above bounds on the relative errors n_i and d_i. When analyzing the errors, one computes upper bounds for all relative errors from the first iteration to the last. These bounds are used to compute upper bounds on δ_{i-1} and π_{i-1}, which in turn are used to bound n_i and d_i. In the full version [6] bounds are given for the relative errors in the first and second iteration.

4. Closed Form Error Bounds in Specific Settings

In this section we describe specific settings of the relative errors that enable us to derive closed form error bounds. The advantage of having closed form error bounds is that such bounds simplify the task of minimizing an objective function (modeling cost or delay) subject to the required precision. Closed form error bounds also enable one to easily evaluate the effect of design choices (e.g., initial error, precision of intermediate computations, and number of iterations) on the final error. We have derived closed form error bounds in four specific settings. We are describing Setting I and Setting IV in the following. Settings II ($n_i, d_i \leq \hat{n}$ and f_i/δ_i^2 is exponential in $-k$) and Setting III ($n_i, d_i \leq \hat{n}$ and f_i/δ_i^2 is constant) are presented in the full version [6].

4.1. Setting I: $n_i, d_i \leq \hat{n}$ and $f_i = 0$.

Setting I deals with the situation that all the relative errors n_i, d_i are bounded by the same value $\hat{n}$. In addition it is assumed in this setting that $f_i = 0$, for every i. The justification for Setting I is that if all internal operands are represented by binary strings of equal length, then it is possible to bound all the relative errors n_i, d_i by the same value. The relative errors f_i can be assumed to be 0, if the computations $F_i' = (2 - D_i')$ are precise.

Using Theorem 4 and Eq. 3, the relative approximation error $\rho(N_i') = \frac{A/B - N_i'}{A/B}$ in Setting I can be bounded by:

$$0 \leq \rho(N_i') = \frac{A/B - N_i'}{A/B} \leq (2i+1)\hat{n} + (|e_0| + \frac{3}{2}\hat{n})^{2^i}.$$

4.2. Setting IV: $n_i, d_i \leq \hat{n}$ and $f_i \leq \hat{f}$ for every i.

In setting IV the assumptions are: (i) $n_i, d_i \leq \hat{n}$, for every i, and (ii) $f_i \leq \hat{f} \leq 1/8$, for every i. Hence, $\delta_i \leq \delta_{i-1}^2 + \hat{f}$ for all $i \geq 0$.

The following claim bounds the error term δ_k corresponding to the kth iteration of Algorithm 2.

Claim 6 *Let* $\alpha = (1 + \sqrt{\hat{f}})$. *For every $k > 0$ the following holds:*

$$\delta_k \leq \hat{f} + \max\{\alpha^{2^{k+1}-2} \cdot \delta_0^{2^k}, (\alpha^{2^k-2} \cdot \delta_0^{2^{k-1}} + \hat{f})^2, 9\hat{f}^2\}.$$

Note that a slightly looser bound that does not involve a max function can be written as:

$$\delta_k \leq \hat{f} + \alpha^{2^{k+1}-2} \cdot \delta_0^{2^k} + 7\hat{f}^{3/2}.$$

Based on Theorem 4 and Equation 3 the error bound in setting IV satisfies:

$$\rho(N_k') < (2k+1) \cdot \hat{n} + \hat{f}$$
$$+ \max\{\alpha^{2^{k+1}-2} \cdot \delta_0^{2^k}, (\alpha^{2^k-2} \cdot \delta_0^{2^{k-1}} + \hat{f})^2, 9\hat{f}^2\}.$$

One can easily see that, due to the first term, there is a threshold above which increasing the number of iterations (while maintaining all other parameters fixed) increases the bound on the relative error. Moreover, the contribution of the error term $\hat{f}$ to $\rho(N_k')$ does not increase with the number of iterations k (as opposed to $\hat{n}$). This implies that in a cost effective choice one would use $\hat{f} > \hat{n}$.

5. Application: An Alternative FP-DIV Micro-architecture for AMD-K7™

In this section we propose an alternative FP-DIV micro-architecture for the AMD-K7 microprocessor [15]. This alternative micro-architecture is a design that implements Algorithm 2. Our micro-architecture uses design choices that are similar to those of [15] to facilitate isolating the effect of precisions on cost. Our error analysis allows us to accurately determine the required multiplier precision and thus both save cost and reduce delay.

Overview micro-architecture. The FP-DIV micro-architecture of the AMD-K7 microprocessor is described in [15]. The micro-architecture is based on Goldschmidt's algorithm. We briefly outline this micro-architecture: (i) Round-to-nearest rounding is done in intermediate computations (as opposed to directed rounding suggested in Algorithm 2). (ii) The design contains a single 76×76-bits multiplier. This means that the absolute errors $\widehat{neps}_i$ and $\widehat{deps}_i$ are identical during all the iterations (i.e., since round-to-nearest is used, $\widehat{neps}_i = \widehat{deps}_i = 2^{-76}$). However, our alternative micro-architecture may use smaller multipliers (even multipliers in which the multiplicands do not have equal lengths) provided that the error analysis proves that the final result is accurate enough. (iii) Intermediate results are compressed and represented using non-redundant binary representation. This means that Assumption 2 on strict directed rounding is easy to implement in our alternative micro-architecture. (Recall that directed rounding is used in Algorithm 2.) (iv) The computation of F_i' is done using one's complement computation. This means that the absolute error $\widehat{feps}_i$ is identical during all the iterations, and that the error analysis of Setting IV is applicable for our alternative architecture. (v) Final rounding of the quotient is done by back multiplication. Our alternative micro-architecture uses the same final rounding simply by meeting the same error bounds needed in the final rounding of [15].

Required final precisions. The micro-architecture in [15] supports multiple precisions: single precision $(24, 8)$ in one iteration, double precision $(53, 11)$ in two iterations, an extended precision $(64, 15)$ and an internal extended precision $(68, 18)$ in three iterations. Final rounding is based on back-multiplication: namely, comparing

$N'_k \cdot B$ with A. In general, correct IEEE rounding based on back-multiplication requires that $\rho(N'_k) < 2^{-(p+1)}$, where p denotes the precision. (The description of the required precision for correct rounding in [15] is somewhat confusing since it is stated in terms of a two sided absolute error. For example, the absolute error in the 68-bit precision is bounded by 2^{-70}.)

To summarize, the upper bounds on the relative errors are as follows: (i) for single precision: $\rho(N'_1) < 2^{-25}$, (ii) for double precision: $\rho(N'_2) < 2^{-54}$, (iii) for extended double precision: $\rho(N'_3) < 2^{-65}$, and (iv) for the 68-bit precision: $\rho(N'_3) < 2^{-69}$.

Note that the bound for the 64-bit precision is weaker than the bound for the 68-bit precision. The bound for single precision is easily satisfied by the parameter choices needed to satisfy the 53-bit precision. Hence we focus below on two iterations for double precision and on three iterations for the 68-bit precision.

From relative errors to multiplier dimensions. In the full version [6], we analyze the lengths of the multiplicands in Algorithm 2. We obtain the following results. The length of the first multiplicand (used for N'_i and D'_i) should be slightly larger than $\log_2(1/\hat{n}) + 2$. The length of the second multiplicand should be greater than or equal to $log_2(1/\hat{f}) + 1 + \log_2(1/(1 - \delta_0))$.

Optimizing the error parameters. In the full version [6], we present a search for combinations of relative errors that minimize the sizes of the multiplier and lookup table. We used a cost function that is based on the cost model of Paul & Seidel [16] for Booth Radix-8 multipliers and the cost model of Paul & Mueller [14] for lookup tables, adders, etc. (Formulas for hardware costs appear in the full version [6].) The optimal parameter combination for double precision is $-\log_2(\hat{f}) = 55.67$, $-\log_2(\hat{n}) = 57.74$, and $-\log_2(e_0) = 13.92$. For the 68-bit internal precision we found the following combination: $e_0 = 2^{-13.51}$, $-\log_2 \hat{n} = 71.91$, and $-\log_2 \hat{f} = 68.9$. We conclude that multiplier dimensions 70×74 combined with a relative error bound $e_0 \leq 2^{-13.51}$ are a feasible choice of error parameters. These parameters lead to a savings in cost of 10.6% compared to the micro-architecture described in [15].

References

[1] R. Agarwal, F. Gustavson, and M. Schmookler. Series approximation methods for divide and square root in the power3 processor. In *Proceedings of the 14th IEEE Symposium on Computer Arithmetic*, pages 116–123, 1999.

[2] S. F. Anderson, J. G. Earle, R. E. Goldschmidt, and D. M. Powers. The IBM 360/370 model 91: floating-point execution unit. *IBM J. of Research and Development*, Jan. 1967.

[3] P. Beame, S. Cook, and H. Hoover. Log depth circuits for division and related problems. *SIAM Journal on Computing*, 15:994–1003, 1986.

[4] M. A. Cornea-Hasegan, R. A. Golliver, and P. Markstein. Correctness proofs outline for Newton-Raphson based floating-point divide and square root algorithms. In *Proceedings of the 14th IEEE Symposium on Computer Arithmetic*, pages 96–105, 1999.

[5] D. DasSarma and D. W. Matula. Faithful bipartite ROM reciprocal tables. In *Proc. 12th IEEE Symposium on Computer Arithmetic*, pages 17–28, 1995.

[6] G. Even, P.-M. Seidel, and W. E. Ferguson. A parametric error analysis of Goldschmidt's division algorithm. full version, submitted for Journal publication, available at request, 2003.

[7] D. Ferrari. A division method using a parallel multiplier. *IEEE Trans. on Computers*, EC-16:224–226, Apr. 1967.

[8] M. J. Flynn. On division by functional iteration. *IEEE Transactions on Computers*, C-19(8):702–706, Aug. 1970.

[9] R. Goldschmidt. Applications of division by convergence. Master's thesis, MIT, June 1964.

[10] D. Knuth. *The Art of Computer Programming*, volume 2. Addison-Wesley, 3nd edition, 1998.

[11] E. V. Krishnamurthy. On optimal iterative schemes for high-speed division. *IEEE Transactions on Computers*, C-19(3):227–231, Mar. 1970.

[12] P. Markstein. *IA-64 and Elementary Functions : Speed and Precision*. Hewlett-Packard Books. Prentice Hall, 2000.

[13] K. Mehlhorn and F. Preparata. Area-time optimal division for $t = \omega((\log n)^{1+\epsilon})$. *Information and Computation*, 72(3):270–282, 1987.

[14] S. M. Mueller and W. J. Paul. *Computer Architecture. Complexity and Correctness*. Springer, 2000.

[15] S. F. Oberman. Floating-point division and square root algorithms and implementation in the AMD-K7 microprocessor. In *Proceedings of the 14th IEEE Symposium on Computer Arithmetic*, pages 106–115, 1999.

[16] W. Paul and P.-M. Seidel. To Booth or Not to Booth? *INTEGRATION, the VLSI Journal*, 33:1–36, Jan. 2003.

[17] J. Reif and S. Tate. Optimal size integer division circuits. *SIAM Journal on Computing*, 19(5):912–924, Oct. 1990.

[18] D. Russinoff. A mechanically checked proof of IEEE compliance of a register-transfer-level specification of the amd-K7 floating-point multiplication, division, and square root instructions. *LMS Journal of Computation and Mathematics*, 1:148–200, December 1998.

[19] E. Schwarz, L. Sigal, and T. McPherson. CMOS floating point unit for the S/390 parallel enterpise server G4. *IBM J. of Research and Development*, 41(4/5):475–488, 1997.

[20] P.-M. Seidel. *On the Design of IEEE Compliant Floating-Point Units and their Quantitative Analysis*. PhD thesis, University of Saarland, Computer Science Department, Germany, 1999.

[21] N. Shankar and V. Ramachandran. Efficient parallel circuits and algorithms for division. *Information Processing Letters*, 29(6):307–313, 1988.

[22] P. Soderquist and M. Leeser. Area and performance tradeoffs in floating-point divide and square-root implementations. *ACM Computing Surveys*, 28(3):518–564, 1996.

[23] O. Spaniol. *Computer Arithmetic - Logic and Design*. Wiley, 1981.

Session 8:
Cryptography

Chair: Peter Montgomery

A Unidirectional Bit Serial Systolic Architecture for Double-Basis Division over $GF(2^m)$

Amir K. Daneshbeh and M.A. Hasan
Department of Electrical and Computer Engineering
University of Waterloo, Waterloo, Ontario, N2L 3G1 Canada
E-mail: akdanesh@engmail.uwaterloo.ca, ahasan@ece.uwaterloo.ca

Abstract

A unidirectional bit serial systolic architecture for division over Galois field $GF(2^m)$ is presented which uses both triangular and polynomial basis representations. It is suitable for hardware implementations where the dimension of the field is large and may vary. This is the typical case for cryptographic applications. This architecture is simulated in Verilog-HDL and synthesized for a clock period of 1.4 ns using Synopsys. The time and area complexities are truly linear, since no carry propagation structures are present, and the complexity measures are equivalent or excel the best designs proposed so far.

1 Introduction

Finite field arithmetic has many applications in coding theory [1], and cryptography [10]. Among the arithmetic operations over finite fields the multiplicative inversion and division are the most costly ones. Most hardware proposals to compute inversion and division over $GF(2^m)$, e.g., [2, 5, 6, 13, 14] suffer from a common problem. In applications where the dimension of the field, m, is large or may vary, their implementation becomes cumbersome or even impractical. In particular, in public key cryptographic applications, the field dimension may exceed 4000 for discrete logarithm cryptosystems or 500 for elliptic curve cryptosystems [10].

In general, three major schemes to compute multiplicative inverses exist: Fermat's little theorem, variants of extended Euclidean algorithm (EEA) or a solution of a set of linear equations. First scheme is efficient if a fast squaring or multiplication scheme is available [9]. The extended Euclidean algorithm (EEA) based schemes to compute inversion and division are the most efficient in time and area but may require variable size counter-like structures with carry propagation chain to keep track of the difference of the degree of polynomials, e.g., [5, 14, 11], in which case they are not suitable for high-performance and scalable VLSI implementations. The third method is generally inefficient for large values of field dimension. However, it is shown that a set of linear equations formed upon triangular basis representation can be solved by schemes similar to EEA [6].

Many systolic array proposals for multiplicative inversion or division over $GF(2^m)$ based on EEA or its dual (extended Stein's algorithm), i.e., [4, 8, 15, 16], all require counter-like structures with carry propagation delays. Since the carry propagation chain depends on the field dimension m, in general, it dominates the critical delay path. Only counter or comparator architectures with no carry propagation chain can be considered dimension independent. Moreover, a counter with no carry propagation chain can be easily transformed into a distributed structure.

In this paper we expand the double-basis multiplicative inversion scheme over $GF(2^m)$ proposed in [6] to present a unidirectional bit serial systolic implementation of a divider with no carry propagation structure suitable for large values of m where the divisor is received in triangular basis, and both dividend and quotient, i.e., result, are in polynomial basis. We will show that the extra cost of division in systolic implementations is minimal. Most importantly, an efficient stepwise restoring of the result polynomial is proposed as well.

2 Preliminaries

2.1 Polynomial and Triangular Basis Representations

Given an irreducible polynomial $F(x)$ of degree m over the finite field $GF(2)$ and considering one of the roots of $F(x)$ such as ω, i.e., $F(\omega) = 0$, it is well known that the set of elements $\Omega = \{1, \omega, \omega^2, \cdots, \omega^{m-1}\}$ is linearly independent and forms a basis referred to as a polynomial or canonical basis. Then, any element $A \in GF(2^m)$ can be represented as $A = \sum_{i=0}^{m-1} a_{\Omega i}\omega^i$, where $a_{\Omega i} \in \{0,1\}$ are the coordinates of A with respect to the polynomial basis Ω, or in a column vector form $\underline{a}_\Omega = [a_{\Omega 0}, a_{\Omega 1}, \cdots, a_{\Omega m-1}]^T$ over GF(2).

Given the polynomial basis Ω, the set $\Delta = \{\delta_0, \delta_1, \cdots, \delta_{m-1}\}$ is linearly independent and forms a basis if $\delta_i = \sum_{j=0}^{m-1-i} f_{i+j+1}\omega^j$, $0 \leq i \leq m-1$, where f_i's are coefficients of $F(x)$ [7]. The set Δ is referred to as a triangular basis of Ω. Next, the element A of GF(2^m) can also be represented as $A = \sum_{i=0}^{m-1} a_{\Delta i}\delta_i$, where $a_{\Delta i}$ is the ith coordinate of A w.r.t. Δ. The change of coordinates of element A between Ω and Δ bases is given by [7]

$$a_{\Omega j} = \sum_{i=0}^{m-j-1} f_{i+j+1} a_{\Delta i} \quad 0 \leq j \leq m-1 \quad (1)$$

and

$$a_{\Delta k} = \begin{cases} a_{\Omega m-1-k} & k = 0, \\ a_{\Omega m-1-k} + \sum_{i=0}^{k-1} f_{m-k+i} a_{\Delta i} & 1 \leq k \leq m-1. \end{cases} \quad (2)$$

2.2 Double-Basis Inversion Algorithm

In [6], it is said that the inversion can be seen as the solution to a set of linear equations written as

$$\underline{1}_\Delta = \mathbf{H}(\underline{a}_\Delta)\underline{b}_\Omega \quad (3)$$

where $\underline{1}_\Delta$ is a column vector representing the multiplicative identity element of GF(2^m) in triangular basis and $\mathbf{H}(\underline{a}_\Delta) = [\underline{a}_\Delta, \underline{a\omega}_\Delta, \cdots, \underline{a\omega}_\Delta^{m-1}]$ is a Hankel matrix built upon $\underline{a}_\Delta$, the column vector representation of element A with entries as its triangular basis coordinates. Then, the coordinates of the inverse element B in the polynomial basis can be computed by solving Equation (3). The Hankel matrix in (3) has $2m-1$ constant coefficients. Let these be h_i, $0 \leq i \leq 2m-2$, then the matrix entries at (i,j) is h_{i+j}. Algorithm 1 as proposed in [6] can be used to solve Equation (3). Algorithm 1 is an optimization based on a variant of one of the algorithms discussed in [12].

Two aspects of Algorithm 1 should be noted. First, the *exit* condition of Algorithm 1 is ensured by monotonically decreasing value of $R^{(i)}(x)$. Second, an invariant similar to the invariant of the basic EEA [1, 2],

$$W^{(i)}(x)x^{2m-1} + U^{(i)}(x)x^{2m-2}H(x^{-1}) = R^{(i)}(x), \quad (4)$$

is preserved at each iteration. In the following we do not discuss the $W^{(*)}(x)$ polynomials since they are not relevant to the inversion. It can be noted that the polynomial updating step follows an extended Euclidean algorithm (EEA) transformation using a long-division scheme. Basically, the polynomial updating step of Algorithm 1 performs a series

---

Algorithm 1 Double-Basis Inversion Algorithm.

input: $H(x) = h_{2m-2}x^{2m-2} + h_{2m-3}x^{2m-3} + \cdots + h_1 x + h_0$.
output: The inverse element.
Initialization:
$i \leftarrow 0$; $U^{(-1)}(x) \leftarrow 0$; $U^{(0)}(x) \leftarrow 1$;
$R^{(-1)}(x) \leftarrow x^{2m-1}$; $R^{(0)}(x) \leftarrow x^{2m-2}H(x^{-1})$;
Polynomial Updating:
while ($\deg R^{(i)}(x) > m-1$) do
$\quad i \leftarrow i+1$;
$\quad d \leftarrow \deg R^{(i-2)}(x) - \deg R^{(i-1)}(x)$;
$\quad R^{(i)}(x) \leftarrow R^{(i-2)}(x) - x^d R^{(i-1)}(x)$;
$\quad U^{(i)}(x) \leftarrow U^{(i-2)}(x) - x^d U^{(i-1)}(x)$;
$\quad$ if ($\deg R^{(i)}(x) \geq \deg R^{(i-1)}(x)$) then
$\quad\quad$ swap $(R^{(i)}(x), R^{(i-1)}(x))$;
$\quad\quad$ swap $(U^{(i)}(x), U^{(i-1)}(x))$;
return $U^{(i)}(x)$

---

of multi-stepwise long division iterations where the invariant of Equation (4) is preserved.

Algorithm 1 can be extended to perform *division* over finite fields by initializing $U^{(0)}(x)$ with the dividend whose degree is $\leq m-1$. In this case, while computing $U^{(i)}(x)$, a modulo reduction may be required, since the degree of $x^d U^{(i-1)}(x)$ may become m or more. Thus, the irreducible polynomial $F(x)$ must be input as well.

Solving Equation (3) and consequently Algorithm 1 can be considered as a two step process: the Hankel matrix entry formation in the initialization step and the polynomial updating step. In general, the formation of $2m-1$ entries of the Hankel matrix requires recursive formulae such as

$$h_i = \begin{cases} a_{\Delta i} & 0 \leq i \leq m-1, \\ \sum_{j=0}^{m-1} h_{i-m+j} f_j & m \leq i \leq 2m-2. \end{cases} \quad (5)$$

In [6], it is shown that by a suitable choice of irreducible polynomial, such as trinomials of type $x^m + x + 1$, the formation of the entries of the Hankel matrix requires no recursion and they can be easily generated using a single row of XOR gates. It can be shown that for all practical cases the Hankel matrix entry formation has a lower computational complexity compared with the polynomial updating step of Algorithm 1. In what follows a systolic implementation of the polynomial updating step of Algorithm 1 is proposed.

3 Systolization of the Polynomial Updating Step

For large values of m, the complexity of a centralized implementation of the polynomial updating step of Algorithm 1 is dominated by the presence of structures (counters or comparators) with carry propagation delay and long control interconnects. For a VLSI implementation of Algorithm 1 with large values of m a bit serial systolic architecture will be more suitable, not only because of bit-level pipeline data processing but most importantly because in this case a distributed control structure with no carry propagation can be devised.

Using a bit serial architecture of Algorithm 1, the computation of $d \leftarrow \deg R^{(i-2)}(x) - \deg R^{(i-1)}(x)$ may be implemented implicitly by a series of bitwise shifts of polynomial $R^{(i-1)}(x)$ up to an alignment condition of the leading coefficients of polynomials $R^{(i-2)}(x)$ and $R^{(i-1)}(x)$. Such an implementation has two major consequences discussed next.

3.1 Upper Bound of Number of Iterations of Algorithm 1

The *exit* condition of Algorithm 1 occurs after a variable number of iterations. However, by imposing one single degree shift of polynomial $R^{(i-1)}(x)$ per iteration, it is easy to show that in the worst case after exactly $2m$ iterations the exit condition is satisfied, that is the degree of both polynomials $R^{(i-1)}(x)$ and $R^{(i-2)}(x)$ are less than or equal $m-1$.

Hence, in a modified version of Algorithm 1, an exact number of iterations can be defined as a **for** loop of $2m$ iterations. In a bit serial systolic implementation of such an algorithm where each iteration is executed in a different PE, no explicit exit condition checking is required.

The $2m$ iterations is a worst-case. In all other cases, the exit condition may occur sooner, *i.e.*, $i < 2m$. Then a modified variant of Algorithm 1 must ensure that the returned value of $U^{(i)}(x)$ is not modified in consecutive iterations up to $2m$. This will be shown later. However, first, we must introduce the second major advantage of a single degree shift of polynomial $R^{(i-1)}(x)$ in evaluating the **swap** condition $(\deg R^{(i)}(x) \geq \deg R^{(i-1)}(x))$.

3.2 Distributed Relative Degree Tracking of Algorithm 1

In Algorithm 1, the *swap* condition ensures that in the next iteration the polynomial $R^{(i-1)}(x)$ has lower degree (relative to $R^{(i-2)}(x)$) always in order to be shifted. According to Algorithm 1, at each iteration after a new polynomial $R^{(i)}(x)$ is computed, it is necessary that the preshifted polynomial $R^{(i-1)}(x)$ and its degree to be known in the next iteration. However, in a bit serial implementation with single degree shifts, no effective restoring of the shifted polynomial is necessary in order to decide which polynomial will be shifted next if a mechanism to keep track of the relative degree of two polynomials is devised as shown below.

Let δ represent an up/down counter which counts the number of shifts of the shifting polynomial $R^{(i-1)}(x)$, and initialized to zero. Initially polynomial $R^{(-1)}(x) = x^{2m-1}$,
and if no restoring of polynomials is needed, then in all successive iterations when an alignment of two polynomials $R^{(i-1)}(x)$ and $R^{(i-2)}(x)$ occurs, the degree of both polynomials will be $2m-1$. However, δ keeps track of their relative degree.

At all iterations while the polynomial $R^{(i-1)}(x)$ is shifted, δ is increased. At an *alignment* of polynomials $R^{(*)}(x)$, if $\delta \geq 0$, an *overwriting* occurs and the sign of δ is swapped. The *overwriting* is the only iteration when both polynomials are updated. Afterwards, while δ is negative, a shift of polynomial $R^{(i-1)}(x)$ increments δ towards zero and again into positive until the next *overwriting* condition. Hence, only the sign of δ is needed to decide whether an *overwriting* is required or not. Furthermore, such an up/down counter can be easily implemented by using a distributed ring-counter without any carry propagation.

3.3 Shifted Result Problem

If Algorithm 1 is implemented by one shift per iteration as discussed in Section 3.1, then at the iteration where i becomes $2m$ (inside the while loop),

$$\deg R^{(2m)}(x) \leq m - 1 \quad \text{(restoring case)}.$$

An inherent problem to implement the polynomial updating step of Algorithm 1 without an effective restoring of polynomials $R^{(*)}(x)$ is what we may call the *shifted result* problem. In Sections 3.1, 3.2, it has been shown that by keeping track of the degree of polynomials $R^{(*)}(x)$, both the *exit condition* checking and the *swap condition* can be done while one of the two polynomials are left shifted bitwise continuously. Mathematically, this means that the degree of a non-restored polynomial $R^{(i-1)}(x)$ always equals to $2m-1$ or approaching it. Hence, in this non-restoring scheme, at the iteration where i becomes $2m$, we have

$$\deg R^{(2m)}(x) \leq 2m - 1 \quad \text{(non-restoring case)}.$$

As a consequence, in a non-restored scheme, not the correct result $U^{(2m)}(x)$ but rather a shifted result $U^{(2m)}(x)x^m$ will be returned unless a corrective action is applied. Different strategies to tackle a similar problem while computing multiplicative inverses using extended Euclidean algorithm are proposed: initializing $U^{(0)}(x) = x^{-m}$ [8]; initializing $U^{(-1)}(x) = x^m$, $U^{(0)}(x) = 0$, but at each iteration divide $U^{(i-1)}(x)$ instead of multiply such that $U^{(i)}(x) \leftarrow U^{(i-2)}(x) - U^{(i-1)}(x)/x^d$, ([1], Section 2.3), [16]; multiplying or dividing $U^{(i-1)}(x)$ alternatively [2]; or using an auxiliary polynomial [4]. In a bit serial systolic implementation, a more efficient method appears to be the use of double delay elements to perform a stepwise restoring of $U^{(*)}(x)$ as will be proposed in the following variant of Algorithm 1.

3.4 A Division Algorithm suitable for systolization

Using the optimizations suitable for an efficient systolization of Algorithm 1 introduced in Sections 3.1, 3.2, 3.3, Algorithm 2 is proposed which can perform the division as well.

Algorithm 2 Division Variant of Algorithm 1 with Restoring Result Polynomial.

input: irreducible polynomial $F(x)$,
　　dividend in polynomial basis $C(x)$, and
　　$H(x) = h_{2m-2}x^{2m-2} + h_{2m-3}x^{2m-3} + \cdots + h_1 x + h_0$
　　which is a function of $A(x)$.
output: $C(x)/A(x) \mod F(x)$.
　Initialization:
　　$\delta \leftarrow 0; U \leftarrow 0; V \leftarrow C(x);$
　　$R \leftarrow x^{2m-1}; S \leftarrow x^{2m-2}H(x^{-1});$
　Polynomial Updating:
　for $2m$ times **do**
　　if $(s_{2m-1} = 0 \ \& \ \delta \geq 0)$ **then**
　　　$[R, S] \leftarrow [R, xS];$
　　　$[U, V] \leftarrow [U/x, V];$
　　else if $(s_{2m-1} = 0 \ \& \ \delta < 0)$ **then**
　　　$[R, S] \leftarrow [R, xS];$
　　　$[U, V] \leftarrow [U, xV \mod F(x)];$
　　else if $(s_{2m-1} = 1 \ \& \ \delta < 0)$ **then**
　　　$[R, S] \leftarrow [R, x(R-S)];$
　　　$[U, V] \leftarrow [U, (x(U-V)) \mod F(x)];$
　　else if $(s_{2m-1} = 1 \ \& \ \delta \geq 0)$ **then**
　　　$\delta \leftarrow -\delta;$ 　　// overwriting case //
　　　$[R, S] \leftarrow [S, x(R-S)];$
　　　$[U, V] \leftarrow [V, (U-V) \mod F(x)];$
　　$\delta \leftarrow \delta + 1;$
　return U

Algorithm 2 receives as input the field defining irreducible polynomial $F(x)$, the dividend polynomial $C(x)$, and $2m - 1$ entries of the Hankel matrix as defined in Section 2.2. Unlike in Algorithm 1, the indices (in the superscript) of $R^{(*)}(x)$ and $U^{(*)}(x)$ polynomials are dropped in Algorithm 2 and two consecutive $R^{(*)}(x)$ polynomials (respectively $U^{(*)}(x)$) are renamed as a pair of distinct polynomials $[R, S]$ (respectively $[U, V]$).

In Algorithm 2, the *alignment* condition is shown as $(s_{2m-1} = 1)$, since $(r_{2m-1} = 1)$ always, and where s_{2m-1} is the $(2m - 1)$st degree coefficient of S. Also, the *overwriting* condition is shown as $(s_{2m-1} = 1 \ \& \ \delta \geq 0)$, where & represents a logical AND operator. The **for** loop of $2m$ represents the worst-case number of iterations to have the correct result in U, *i.e.*, at the $2m$th iteration, an overwriting occurs and $U \leftarrow V$. In all cases where U holds the final results before the $2m$th iteration, Algorithm 2 ensures that the value of U is not modified in consecutive iterations while $\delta < 0$.

The innovation of Algorithm 2 is the restoring transformation of $[U, V]$ versus the non-restoring transformation of pair $[R, S]$.

3.5 Bit Serial Systolic Architecture for Division

It is said that Algorithm 2 is optimized for systolization. A bit serial unidirectional systolic architecture for the updating step of Algorithm 2 is depicted in Figure 1. A single type PE is used and no latches external to PEs are required. Furthermore, a distributed ring-counter without any carry propagation is used.

In Figure 1, there are three sets of inputs to each PE: *datapath* signals (r, s, u, v, f); *state* signals ($dseq$, dec, $update$); and the *control* signal $start$. The *datapath* inputs are defined in Algorithm 2. The $start$ is defined as a one following by m zeroes. The $update$ is the latched value of incoming $s_{in} = s_{2m-1}$ at ($start = 1$). Signals dec and $dseq$ represent the sign and magnitude of the counter δ. The *state* signals are initialized to 1, 0, 0 respectively. The latency of this architecture is $5m - 2$ cycles. The throughput which also depends on the size of inputs is 1 division per $2m$ cycles. Hence, as a common feature of all systolic architectures two back-to-back computations can be processed.

In Figure 2, one processing element (PE) of architecture in Figure 1 is shown. Two of the *state* signals, $update$ and dec are precomputed and saved for the next PE. The signed integer δ may be represented by a sign bit, dec, and by a $(m + 1)$-bit vector $dseq$ where $dseq = 2^{|d|}$. At *initialization* $\delta = 0$, hence, $dec = 0$, and $dseq = 00 \cdots 001$. At each shift of S while $dec = 0$, $dseq$ is left shifted. At *overwriting*, dec is set, and $dseq$ is right shifted. In all consecutive iterations until $dseq$ returns back to 1, dec remains set. It will reset at $dseq = 1$. The dec in Figure 2 represents a boolean simplification of this set-reset mechanism , exactly, $dec \leftarrow \overline{dseq_{in}} \ \& \ (dec_{in} \mid s_{in})$. The overline represents a logical NOT and "|" a logical OR operator. The dec is used to check for an *overwriting* event. An internal non-latched *state* signal $ovrw$ is computed as ($update_{in} \ \& \ \overline{dec_{in}}$).

The *reduction* condition for the division algorithm is defined as $((v_{in} = v_m = 1) \ \& \ (dec_{in} = 1) \ \& \ (start_{in} = 1))$, however, mapped more efficiently by a balanced delay path for the input of three latches u_1, v_1, and $reduce_{lat}$. Accordingly, the critical delay path is $t_{cp} = max(3T_{A2}, 2T_{A2} + T_{X2}, T_{A2} + 2T_{X2}) + T_{M2}$, where T_{A2}, T_{M2} and T_{X2} are defined as the delay of a 2-input AND gate, MUX and XOR gate respectively.

4 Implementation Considerations and Comparison

In this section some implementation features and possible enhancements of the proposed bit serial systolic archi-

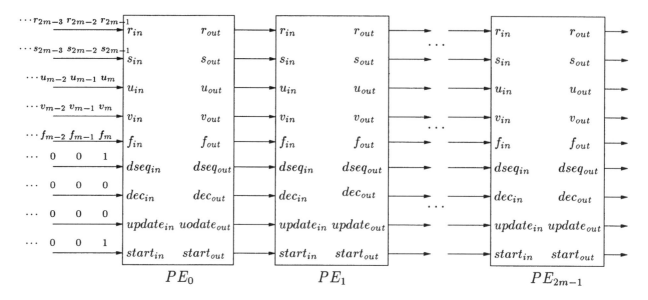

Figure 1. Bit serial unidirectional systolic architecture for Division.

tecture are discussed and a comparison with two systolic architectures are provided.

4.1 Dealing with Varying Dimension

This architecture may process any field dimension up to the size defined by the number of the PEs where no external buffering is available. Let N represent the number of PEs in this architecture. In a bit serial unidirectional systolic architecture as described, it is easy to see that if $N \geq 2m$, the coefficients of the inverse element start to appear at the output of the $2m$th PE after $4m$ cycles. Hence, using the architecture in Figure 1, it is possible to process any field dimension $m \leq N/2$ as far as the correct result can be captured at the right cycle.

Two such schemes can be considered. Either the output of the $2m$th PE should be accessible directly, or a mechanism should be in place to de-activate all remaining PEs from $2m + 1$ to N, and to capture the result at the output of Nth PE always. The former does not require any inner PE modification and provides the optimal latency [3].

4.2 Implementation Results

The divider as described in Section 3.5 is coded in Verilog-HDL, simulated and verified. The Verilog model of Figure 2 has been synthesized using Synopsys with a standard CMOS 0.18μ library for a clock period of 1.4 ns. The average setup time of the flip-flops was 0.32 ns and the clock skew set at 10%. The area reports, including the flip-flops, are equivalent to 120 2-input NAND gates. It should be pointed out that non-combinational area is the dominant part, more than 75%.

4.3 Comparison

Table 1 shows a comparison of proposed divider with two representative dividers, namely [4], and [5]. The design in [4] is chosen since it is a divider with best overall performance figures known to us and it is also in the same category as ours (a bit serial unidirectional systolic architecture). The second design [5] is chosen to show why single row parallel-in designs in particular when they use counter-like structures are not suitable for large values of m.

In Table 1, the deteriorating effect of a carry propagation counter in both area and timing complexity in the architecture of [5] w.r.t. the design in [4] and ours can be seen. Further, the single row systolic architectures with parallel inputs as in [5] are impractical for large values of m, not only due to large IO pins but mainly due to large number of IO latches for synchronization purposes. On the other hand, the architecture in [4] requires an auxiliary polynomial to restore the shifted result problem. This can be seen by an extra number of IO pin and extra MUXs required in each PE.

5 Conclusions

In this paper a unidirectional bit serial systolic architecture of a double-basis divider over $GF(2^m)$ is presented. Main contribution is the restoring mechanism introduced in Algorithm 2 and implemented by using double delay elements in Figure 2. Also, a distributed counter structure

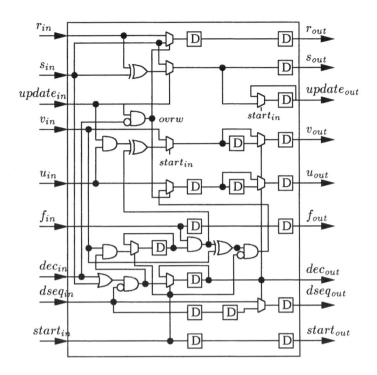

Figure 2. Processing element for the divider.

again implemented by using double delay elements is proposed. This bit serial divider is useful in applications where other field operations are performed using double-basis, and moreover, the field dimension may be large or variable such as cryptographic applications. This architecture is well suitable for high performance (high clock rates) applications. It is seen that delay and area are highly dependent on the delay elements used. Delay or area optimizations by using special latch design may be explored further.

Since the polynomial updating step of Algorithm 1 follows an extended Euclidean algorithm transformation, this architecture can be used in all applications where such algorithm is used. In particular, it can compute division over $GF(2^m)$ when the element is represented in polynomial basis as well.

References

[1] E. Berlekamp. *Algebraic Coding Theory*. McGraw-Hill Book Company, 1968.

[2] H. Brunner, A. Curiger, and M. Hofstetter. "On Computing Multiplicative Inverses in $GF(2^m)$". *IEEE Transactions on Computers*, 42(8):1010–1015, Aug. 1993.

[3] A. Daneshbeh and M. Hasan. "A Class of Scalable Unidirectional Bit Serial Systolic Architectures for Multiplicative Inversion and Division over $GF(2^m)$". http://www.cacr.math.uwaterloo.ca, Dec 2002. Technical Report CORR 2002-35.

[4] J.-H. Guo and C.-L. Wang. "Bit-serial Systolic Array Implementation of Euclid's Algorithm for Inversion and Division in $GF(2^m)$". *Proceedings of Technical Papers, International Symposium on VLSI Technology, Systems, and Applications*, pages 113–117, June 1997.

[5] J.-H. Guo and C.-L. Wang. "Systolic Array Implementation of Euclid's Algorithm for Inversion and Division in $GF(2^m)$". *IEEE Transactions on Computers*, 47(10):1161–1167, Oct. 1998.

[6] M. Hasan. "Double-Basis Multiplicative Inversion over $GF(2^m)$". *IEEE Transactions on Computers*, 47(9):960–970, Sep. 1998.

[7] M. Hasan and V. Bhargava. "Architecture for a Low Complexity Rate-Adaptive Reed-Solomon Encoder". *IEEE Transactions on Computers*, 44(7):938–942, July 1995.

[8] C.-T. Huang and C.-W. Wu. "High-Speed C-Testable Systolic Array Design for Galois-field Inversion". *Proceedings of the European Design and Test Conference*, Paris:342–346, March 1997.

[9] T. Itoh and S. Tsujii. "A Fast Algorithm for Computing Multiplicative Inverses in $GF(2^m)$ Using Normal Basis". *Information and Computation*, 78:171–177, 1988.

[10] A. Menezes, P. van Oorschot, and S. Vanstone. *Handbook of Applied Cryptography*. CRC Press, 1996.

[11] R. Schroeppel, H. Orman, S. O'Malley, and O. Spatscheck. "Fast Key Exchange with Elliptic Curve Systems". *Advances in Cryptology, EUROCRYPT '95, LNCS 921, Springer-Verlag*, pages 43–56, 1995.

[12] Y. Sugiyama. "An Algorithm for Solving Discrete-Time Wiener-Hopf Equations based on Euclid's Algorithm".

Table 1. Comparison of bit serial systolic dividers.

Circuits	Guo et al. [5]	Guo et al. [4]	Figure 2
Time Complexity	$O(m)$	$O(m)$	$O(m)$
Area Complexity	$O(m \log m)$	$O(m)$	$O(m)$
Throughput	1	$\frac{m}{m+1}$	$\frac{m}{2m}$
Latency	$8m - 1$	$5m - 4$	$5m - 4$
IO pins	$O(m)$	10	9
Unidirectional	yes	yes	yes
Critical Path	$2T_{A2} + T_{X3} + 2T_{M2}$	$T_{A2} + T_{X2} + T_{X3} + T_{M2}$	$2T_{A2} + T_{X2} + T_{M2}$
Single Cell	no	yes	yes
Gate Count:	total of different cells	per cell	per cell
IO Synch. Latch	$4m$	-	-
inner-inter Latch	$46m + 4m\lceil \log_2 m + 1 \rceil$	22	18
MUX	$35m + 2$	11	10
XOR	$11m$	5	3
AND/OR	$26m$	8	7
others	adder, zero-check	-	-

IEEE Transactions on Information Theory, 32(5):394–409, May 1986.

[13] Y. Watanabe, N. Takagi, and K. Takagi. "A VLSI Algorithm for Division in GF(2^m) Based on Extended Binary GCD Algorithm". *Transactions of The Institute of Electronics, Information and Communication Engineers, 2002. IEICE'02*, E85(5):994–999, 2002.

[14] C.-H. Wu, C.-M. Wu, M.-D. Shieh, and Y.-T. Hwang. "Systolic VLSI Realization of a Novel Iterative Division Algorithm over GF(2^m): A High Speed Low-Complexity Design". *Proceedings of the IEEE International Symposium on Circuits and Systems. ISCAS'01*, 4:33–36, 2001.

[15] C.-H. Wu, C.-M. Wu, M.-D. Shieh, and Y.-T. Hwang. "An Area-Efficient Systolic Division Circuit over GF(2^m) for Secure Communication". *Proceedings of the IEEE International Symposium on Circuits and Systems. ISCAS'02*, 5:733–736, 2002.

[16] Z. Yan and D. Sarwate. "Systolic Architectures for Finite Field Inversion and Division". *Proceedings of the IEEE International Symposium on Circuits and Systems. ISCAS'02*, V:789–792, June 2002.

Efficient Multiplication in $GF(p^k)$ for Elliptic Curve Cryptography

J.-C. Bajard, L. Imbert, C. Nègre and T. Plantard
Laboratoire d'Informatique de Robotique et de Microélectronique de Montpellier
LIRMM, 161 rue Ada, 34392 Montpellier cedex 5 – France
{bajard, imbert, negre, plantard}@lirmm.fr

Abstract

We present a new multiplication algorithm for the implementation of elliptic curve cryptography (ECC) over the finite extension fields $GF(p^k)$ where p is a prime number greater than $2k$. In the context of ECC we can assume that p is a 7-to-10-bit number, and easily find values for k which satisfy: $p > 2k$, and for security reasons $\log_2(p) \times k \simeq 160$. All the computations are performed within an alternate polynomial representation of the field elements which is directly obtained from the inputs. No conversion step is needed. We describe our algorithm in terms of matrix operations and point out some properties of the matrices that can be used to improve the design. The proposed algorithm is highly parallelizable and seems well adapted to hardware implementation of elliptic curve cryptosystems.

1. Introduction

Cryptographic applications such as elliptic or hyperelliptic curves cryptosystems (ECC, HECC) [11, 12, 13] require arithmetic operations to be performed in finite fields. This is the case, for example, for the Diffie-Hellman key exchange algorithm [6] which bases its security on the discrete logarithm problem. Efficient arithmetic in these fields is then a major issue for lots of modern cryptographic applications [14]. Many studies have been proposed for the finite field $GF(p)$, where p is a prime number [23] or the Galois field $GF(2^k)$ [4, 7, 16]. In [1], D. Bailey and C. Paar use optimal extension fields $GF(p^k)$ and they propose an efficient arithmetic solution in those fields when p is a Mersenne or pseudo-Mersenne prime [2]. Although it could result in a wider choice of cryptosystems, arithmetic over the more general finite extension fields $GF(p^k)$, with $p > 2$, has not been extensively investigated yet. Moreover it has been proved that elliptic curves defined over $GF(p^k)$ – where the curves verify the usual conditions of security – provide at least the same level of security as the curves usually defined over $GF(2^k)$ or $GF(p)$. For ECC, a good level of security can be achieved with p and k prime and about 160-bit key-length. Table 1 gives some good candidates for p and k and the corresponding key-size.

p	k	key-size	form of p
67	29	175	$64 + 3$
67	31	188	$= 1000011$
73	23	142	$64 + 8 + 1$
73	29	179	$= 1001001$
127	19	132	$128 - 1$
127	23	160	$= M_7 = 2^7 - 1$
127	29	202	$= 1111111$
257	17	136	$256 + 1$
257	19	152	$= F_4 = 2^{2^3} + 1$
257	23	184	$= 100000001$

Table 1. **Good candidates for primes p and k and the corresponding key-size in bits.**

In this paper, we introduce a Montgomery like modular multiplication algorithm in $GF(p^k)$ for $p > 2k$ (this condition comes from technical reasons that we shall explain further). Given the polynomials $A(X)$ and $B(X)$ of degree less than k, and $G(X)$ of degree k (we will give more details on $G(X)$ in section 1.2), our algorithm computes

$$A(X)\,B(X)\,G(X)^{-1} \mod N(X),$$

where $N(X)$ is a monic irreducible polynomial of degree k ; and both the operands and the result are given in an alternate representation introduced in the next section.

In the classical polynomial representation, we can consider the elements of $GF(p^k)$ as polynomials of degree less than k in $GF(p)[X]$ and we represent the field with respect to an irreducible polynomial $N(X)$ of degree k over $GF(p)$. Any element A of $GF(p^k)$ is then represented using a polynomial $A(X)$ of degree $k - 1$ or less with coefficients in $GF(p)$, i.e., $A(X) = a_0 + a_1 X + \cdots + a_{k-1} X^{k-1}$, where $a_i \in \{0, \ldots, p - 1\}$. Here, we consider an alternate solution which consists of representing the polynomials with their values at k distinct points instead of their k coefficients. As a result, if we choose k points $(e_1, e_2, \ldots, e_k)$, we represent the polynomial A with the sequence $(A(e_1), A(e_2), \ldots, A(e_k))$. Within this representation, addition, subtraction and multiplication are performed over completely independent channels which has great advantage from a chip design viewpoint.

1.1. Montgomery Multiplication in $GF(p^k)$

Montgomery's technique for modular multiplication of large integers [15] has recently been adapted to modular multiplication in $GF(2^k)$ by Koç and Acar [4]. The proposed solution is a direct translation of the original Montgomery algorithm in the field $GF(2^k)$, with X^k playing the role of the Montgomery factor ; i.e. it computes $A(X) B(X) X^{-k} \mod N(X)$, where N is a k-order irreducible polynomial with coefficients in $GF(2)$.

In turn this method easily extends to $GF(p^k)$, with $p > 2$. As in [4], we represent the field $GF(p^k)$ with respect to a monic irreducible polynomial $N(X)$ and we consider the field elements as polynomials of degree less than k in $GF(p)[X]$; i.e. we consider the elements of $GF(p)[X]/N(X)$. Thus, if we take A and B in $GF(p^k)$, we successively compute

$$Q(X) = -A(X) B(X) N(X)^{-1} \mod X^k$$
$$R(X) = [A(X) B(X) + Q(X) N(X)] X^{-k}$$

to get the result $A(X) B(X) X^{-k} \mod N(X)$. In terms of elementary operations over $GF(p)$, the complexity of this method is $k^2 + (k-1)^2$ multiplications (modulo p) and $(k-1)^2 + (k-2)^2 + k$ additions (modulo p).

1.2. Alternate polynomial representation

The general idea of our approach is the change of representation. When dealing with polynomials the idea which first comes in mind is to use a coefficient representation. However, the valued representation – where a polynomial is represented by its values at sufficiently many points – can be of use. Thanks to Lagrange's theorem we can actually represent any polynomial of degree less than k with its values at various distinct points $\{e_1, e_2, \ldots, e_k\}$. A very good discussion on polynomial evaluation and interpolation can be found in [21].

In the following of the paper, we represent a polynomial of degree at most $k-1$, say A, by the sequence $(A(e_1), A(e_2), \ldots, A(e_k))$. In the following we consider the notation $a_i = A(e_i)$. At this point, it is very important to understand that the a_is do not represent the coefficients of A, and that there is nothing to do to obtain such a representation. We directly consider the polynomial in this form. As an example, the input 100111010101 which would usually represent the polynomial $9X^2 + 13X + 5$ in the coefficient representation, is considered here as the sequence $(9, 13, 5)$. This sequence corresponds to the unique polynomial P of degree 2 which has values $P(e_1) = 9$, $P(e_2) = 13$ and $P(e_3) = 5$. We can easily compute its coefficients by means of interpolation but as we shall see further, there is no need to do so. We will use this representation during all the computational steps.

2. New algorithm

As mentioned previously, Koç and Acar used the polynomial X^k in their adaptation of Montgomery multiplication to the field $GF(2^k)$, and we have briefly shown in section 1.1 that their solution easily extends to $GF(p^k)$. In our new approach we rather consider the k-order polynomial

$$G(X) = (X - e_1)(X - e_2) \ldots (X - e_k), \quad (1)$$

where $e_i \in \{0, 1, \ldots, p-1\}$. A first remark is that this clearly implies $p > k$. As we shall see further, $2k$ distinct points are actually needed. Thus given the three polynomials $A = (a_1, a_2, \ldots, a_k)$, $B = (b_1, b_2, \ldots, b_k)$ and $N = (n_1, n_2, \ldots, n_k)$ in $GF(p^k)$; and under the condition $p > 2k$, our algorithm computes the product $A(X) B(X) G^{-1}(X) \mod N(X)$ in two stages.

Stage 1: We define the polynomial Q of degree less than k such that:

$$Q(X) = [-A(X) B(X) N^{-1}(X)] \mod G(X),$$

in other words, we compute in parallel and in $GF(p)$

$$q_i = [-a_i b_i N^{-1}(e_i)], \quad \forall i = 1 \ldots k.$$

Stage 2: Since $[A(X) B(X) + Q(X) N(X)]$ is a multiple of $G(X)$, we compute $R(X)$ of degree less than k such that

$$R(X) = \left[A(X) B(X) + Q(X) N(X)\right] G^{-1}(X)$$

In this algorithm it is important to note that it is not possible to evaluate $R(X)$ directly as mentioned in step 2. Since $[A(X) B(X) + Q(X) N(X)]$ is a multiple of $G(X)$ its representation at the points $\{e_1, e_2, \ldots, e_k\}$ is merely composed of 0. The same clearly applies for $G(X) = \prod_{i=1}^{k}(X - e_i)$. As a direct consequence the division by $G(X)$, which actually reduces to the multiplication by $G^{-1}(X)$, has neither effect nor sense. We address this problem by using k extra values $\{e'_1, e'_2, \ldots, e'_k\}$ where $e'_i \neq e_j$ for all i, j, and by computing $[A(X) B(X) + Q(X) N(X)]$ for those k extra values. In algorithm 1, the operations in step 3 are then performed for $X \in \{e'_1, e'_2, \ldots, e'_k\}$.

Steps 1 and 3 are fully parallel operations in $GF(p)$. The complexity of algorithm 1 thus mainly depends on the two polynomial interpolations (steps 2, 4).

Algorithm 1 New Multiplication in $GF(p^k)$

Step 1: For $X \in \{e_1, \ldots, e_k\}$, compute in parallel
$$Q(X) = -A(X)\,B(X)\,N^{-1}(X)$$

Step 2: Extend Q in $\{e'_1, \ldots, e'_k\}$ using Lagrange interpolation

Step 3: For $X \in \{e'_1, \ldots, e'_k\}$, compute in parallel
$$R(X) = \left[A(X)B(X) + Q(X)N(X)\right]G^{-1}(X)$$

Step 4: Extend R back in $\{e_1, \ldots, e_k\}$ using Lagrange interpolation.

2.1. Implementation

In step 1 we compute in $GF(p)$ and in parallel for all i in $\{1, \ldots, k\}$
$$q_i = (-a_i \times b_i \times \tilde{n}_i) \bmod p, \tag{2}$$

where the $\tilde{n}_i$s are precomputed constants ($\tilde{n}_i = N^{-1}(e_i)$). Then in step 2, the extension is performed via Lagrange interpolation:
$$Q(X) = \sum_{i=1}^{k} q_i \left(\prod_{j=1, j \neq i}^{k} \frac{X - e_j}{e_i - e_j} \right). \tag{3}$$

If we denote
$$\omega_{t,i} = \prod_{j=1, j \neq i}^{k} \frac{e'_t - e_j}{e_i - e_j}, \tag{4}$$

the extension of $Q(X)$ in $\{e'_1, e'_2, \ldots, e'_k\}$ becomes
$$\begin{pmatrix} q'_1 \\ q'_2 \\ \vdots \\ q'_{k-1} \\ q'_k \end{pmatrix} = \begin{pmatrix} \omega_{1,1} & \cdots & \omega_{1,k-1} & \omega_{1,k} \\ \omega_{2,1} & \cdots & \omega_{2,k-1} & \omega_{2,k} \\ \vdots & & & \\ \omega_{k-1,1} & \cdots & \omega_{k-1,k-1} & \omega_{k-1,k} \\ \omega_{k,1} & \cdots & \omega_{k,k-1} & \omega_{k,k} \end{pmatrix} \begin{pmatrix} q_1 \\ q_2 \\ \vdots \\ q_{k-1} \\ q_k \end{pmatrix}. \tag{5}$$

Operations in step 3 are performed in parallel for $i \in \{1, \ldots, k\}$. We compute
$$r'_i = (a'_i \times b'_i + q'_i \times n'_i)\,\zeta_i \bmod p, \tag{6}$$

where the ζ_is are also precomputed values.
$$\zeta_i = G(e'_i)^{-1} \bmod p = \left[\prod_{j=1}^{k}(e'_i - e_j)\right]^{-1} \bmod p.$$

It is easy to remark that $G(e'_i) \neq 0, \forall i \in \{1, \ldots, k\}$. Thus the modular inverse, $G(e'_i)^{-1} \bmod p$, always exists.

At the end of step 3, the polynomial R of degree less than k is defined by its values at $\{e'_1, e'_2, \ldots, e'_k\}$, namely $(r'_1, r'_2, \ldots, r'_k)$. If we want to reuse the obtained result as the input of other multiplications (which is frequently the case in exponentiation algorithms), we must also know the values of R at $\{e_1, e_2, \ldots, e_k\}$. This is done in step 4 again by mean of Lagrange interpolation. As in step 2, we define
$$\omega'_{t,i} = \prod_{j=1, j \neq i}^{k} \frac{e_t - e'_j}{e'_i - e'_j}, \tag{7}$$

and we compute
$$\begin{pmatrix} r_1 \\ r_2 \\ \vdots \\ r_{k-1} \\ r_k \end{pmatrix} = \begin{pmatrix} \omega'_{1,1} & \cdots & \omega'_{1,k-1} & \omega'_{1,k} \\ \omega'_{2,1} & \cdots & \omega'_{2,k-1} & \omega'_{2,k} \\ \vdots & & & \\ \omega'_{k-1,1} & \cdots & \omega'_{k-1,k-1} & \omega'_{k-1,k} \\ \omega'_{k,1} & \cdots & \omega'_{k,k-1} & \omega'_{k,k} \end{pmatrix} \begin{pmatrix} r'_1 \\ r'_2 \\ \vdots \\ r'_{k-1} \\ r'_k \end{pmatrix}. \tag{8}$$

Another implementation option would be to insert some of the multiplications by constants into the matrix operations of steps 2 and 4. We can introduce the $\tilde{n}_i$s of (2) and the n'_is of (6) in the matrix of equation (5) to gain one product in each step 1 and 3. We do not give much details about this solution because we will see further that the original matrices have some very attractive properties for the hardware implementation.

2.2. Example

In this example, we consider the finite field $GF(17^5)$ defined according to the monic irreducible polynomial $N(X) = X^5 + 4X + 1$. ($p = 17$ and $k = 5$ satisfy $p > 2k$.) The two sets of points used for Lagrange representation are $\mathsf{E} = \{2, 4, 6, 8, 10\}$ and $\mathsf{E}' = \{3, 5, 7, 9, 11\}$. For all e_i in E and e'_i in E', we have $N^{-1}(e_i) = (5, 13, 8, 15, 4)$ (for use in step 1) and $N(e'_i) = (1, 1, 6, 11, 4)$ (for use in step 3). Also used in step 3 is the vector $\zeta = (6, 3, 14, 11, 12)$. The two interpolation matrices needed in steps 2 and 4 are:

$$\omega = \begin{pmatrix} 2 & 8 & 13 & 5 & 7 \\ 7 & 1 & 10 & 11 & 6 \\ 6 & 11 & 10 & 1 & 7 \\ 7 & 5 & 13 & 8 & 2 \\ 2 & 14 & 8 & 10 & 1 \end{pmatrix}$$

and

$$\omega' = \begin{pmatrix} 1 & 10 & 8 & 14 & 2 \\ 2 & 8 & 13 & 5 & 7 \\ 7 & 1 & 10 & 11 & 6 \\ 6 & 11 & 10 & 1 & 7 \\ 7 & 5 & 13 & 8 & 2 \end{pmatrix}.$$

In Lemma 1 we will observe some symmetry between the elements of these two matrices.

Given $A(X)$ and $B(X)$ in $GF(17^5)$, known by their values at points of E and E', we compute $R(X) = A(X)B(X)G^{-1}(X) \bmod N(X)$ in the same representation. We have:

$$A(\mathsf{E}) = (3, 9, 0, 9, 4)$$
$$A(\mathsf{E'}) = (15, 0, 1, 10, 5)$$
$$B(\mathsf{E}) = (5, 3, 12, 12, 0)$$
$$B(\mathsf{E'}) = (12, 1, 8, 13, 13)$$

In step 1 of the algorithm we compute

$$Q(\mathsf{E}) = (10, 6, 0, 12, 0)$$

and we extend it in step 2 (eq. (5)) from E to E'

$$Q(\mathsf{E'}) = (9, 4, 2, 9, 3)$$

Now in step 3 (eq. (6)), we evaluate in parallel for each value of E'

$$R(\mathsf{E'}) = (12, 12, 8, 3, 6)$$

and we interpolate it back (eq. (8)) to obtain the final result in E

$$R(\mathsf{E}) = (12, 9, 7, 6, 12).$$

We can easily check that this actually is the correct result. If we consider the classical coefficient representation of A and B, we have $A(X) = 2X^4 + X + 3$, $B(X) = X^3 + 5X + 4$ and $R(X) = 4X^4 + 5X^3 + 14X^2 + 5$, which evaluated at points of E gives $R(\mathsf{E})$.

3. Arithmetic over $GF(p)$

From an hardware point of view, this method is of interest if and only if we can take advantage of an efficient arithmetic over $GF(p)$. In this section we give the idea of some algorithms for the addition and the multiplication modulo a prime p. Different solutions have been proposed but most of them only focus on large primes which are useful if one wants to implement elliptic curve cryptography over $GF(p)$ [23]. Here we only need arithmetic operations modulo small primes, say 8 to 12 bits.

3.1. Addition

When we aim at computing the modular sum $a + b \bmod m$, a classical approach consists in evaluating in parallel the quantities $a + b$ and $a + b - m$. The correct result is selected according to the sign of $a + b - m$. For a single operation, this solution gives a result less than m. However, when several additions have to be computed, we do not need to reduce the sum modulo m after each addition.

If $2^{m-1} \leqslant m < 2^m$, another solution is to keep the intermediate results less than 2^m by only performing a reduction modulo m when the partial sum becomes greater than 2^m. In other words, we perform the sum $a + b$ and we subtract m only if a carry has occurred. In [19], a redundant representation is used so that the modular addition is performed without carry propagation. The redundant addition is then used within a radix-2 and radix-4 modular multiplication algorithms.

3.2. Multiplication

Multiplication modulo special numbers have been extensively studied. For instance a multiplication modulo $2^n - 1$ is presented in [17] and modulus of the form $2^n + 1$, are used in [22] in the context of DSP applications. Other works exists for Fermat numbers $F_n = 2^{2^n} + 1$ and Mersenne primes $M_p = 2^p - 1$, with p prime.

In the general case, the product $a \times b \bmod p$ can be implemented by means of index calculus with two lookup tables and one addition. We simply use the fact that any element of the group $GF(p)$ corresponds to a power of a generator g of the group. We retrieve in a first table the values α and β such that $a = g^\alpha$ and $b = g^\beta$, we evaluate $\alpha + \beta \bmod p - 1$ and we read in the second table the result $r = ab \bmod p = g^{|\alpha+\beta|_{p-1}}$ (see figure 1). This solution has been proposed in [9] for the special case of the 4th Fermat prime $p = 257$. The advantage here is that addition modulo $p - 1$ reduces to a classical 8-bit addition.

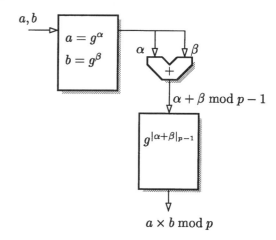

Figure 1. Index calculus modular multiplication.

An interesting suggestion for multiplication by means of look-up tables can be found in [20] under the term quarter-squarer multiplier. It is based of the following equation:

$$ab \bmod n = \left(\left(\frac{a+b}{2}\right)^2 - \left(\frac{a-b}{2}\right)^2 \right) \bmod n, \quad (9)$$

where both squares are given by a look-up table of 2^{n+1} input bits. This is illustrated in figure 2. Optimizations of

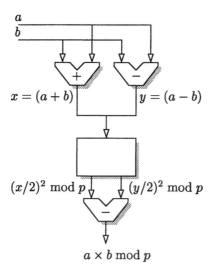

Figure 2. Look-up based modular multiplication for small operands.

this general idea are possible. One can perform the division by two (shifts) before the table look-up. This divides the size of the table by a factor two, but when $a + b$ is odd, the correcting term b must be added (modulo p) at the end. The size can be further reduced using the fact that $x^2 \equiv (-x)^2$ (mod p). In this case, the address resolution problem must be solved.

One can also consider a double-and-add method, sometimes called the Russian peasant method for multiplication [10], associated with a Booth recoding of one of the operands. A table is used to store the double (modulo p) of each value less than p. If we want to multiply two m-bit numbers, this method requires at most $m/2$ additions. For example the evaluation of $121 \times 17 \mod 29$ only requires 2 additions and 7 doublings which are just table lookups. $|121 \times 17|_{29} = 2(2(2(2(2(2(2 \times 17 \mod 29) \mod 29) \mod 29) - 17 \mod 29) \mod 29) \mod 29) + 17 \mod 29$

Modular multiplication by a constant is a lot easier. For small operands, one can simply implement the modular multiplication with some combinatorial logic implementing the function.

4. Complexity

In table 2 we count the number of additions (A), multiplications by a constant (CM) and real multiplications (M) over $GF(p)$ of algorithm 1.

The time required for a sequential implantation corresponds to the number of operations given in table 2. Since

	A	CM	M
step 1	-	k	k
step 2	$k(k-1)$	k^2	-
step 3	k	k	$2k$
step 4	$k(k-1)$	k^2	-
total	$2k^2 - k$	$2k^2 + 2k$	$3k$

Table 2. Number of additions (A), constant multiplications (CM) and real multiplications (M) over $GF(p)$ for a sequential implementation of the algorithm.

the product in $GF(p^k)$ can be totally parallelized into k streams, the time required is exactly $1/k$ times that for the sequential version. If we define T_A the time required for one addition, T_{CM} for one constant multiplication and T_M for one real multiplication respectively, we can precisely evaluate the time complexity of our algorithm on a parallel architecture. Table 3 summarizes the four steps of the algorithm.

step 1	$T_M + T_{CM}$
step 2	$T_{CM} + \sum_{j=1}^{k-1} \max(T_{CM}, T_A)$
step 3	$\max(T_M, T_{CM}) + T_A + T_{CM}$
step 4	$T(MC) + \sum_{j=1}^{k-1} \max(T_{CM}, T_A)$

Table 3. Time complexity estimation on a pipelined architecture.

5. Discussion

5.1. Simplified architecture

The major advantage of this method is that the matrices in (5) and (8) do not depend on the inputs. Thus all the operations reduce to multiplications by constants which significantly simplify the hardware implementation. Moreover, in the example presented in section 2.2 we have detected symmetries between the elements of the two matrices that can also contribute to a simplified architecture. We have the following Lemma.

Lemma 1 *As in the previous example, let us denote $e_i = 2i$ and $e'_j = 2j+1$. According to equations (4) and (7) we have*

$$\omega_{i,j} = \prod_{m=1, m \neq j}^{k} \frac{2i + 1 - 2m}{2j - 2m} \quad (10)$$

and

$$\omega'_{i,j} = \prod_{m=1, m \neq j}^{k} \frac{(2i - (2m+1))}{(2j+1 - (2m+1))}. \quad (11)$$

Then for every $i,j \in \{1,\ldots,k\}$ we have

$$\omega_{i,j} = \omega'_{k+1-i,k+1-j}. \quad (12)$$

In other words equation (8) can be implemented with the same matrix than eq. (5), by simply reversing the order of the elements of the vectors r and r':

$$\begin{pmatrix} r_k \\ r_{k-1} \\ \vdots \\ r_2 \\ r_1 \end{pmatrix} = \begin{pmatrix} \omega_{1,1} & \cdots & \omega_{1,k-1} & \omega_{1,k} \\ \omega_{2,1} & \cdots & \omega_{2,k-1} & \omega_{2,k} \\ \vdots & & & \vdots \\ \omega_{k-1,1} & \cdots & \omega_{k-1,k-1} & \omega_{k-1,k} \\ \omega_{k,1} & \cdots & \omega_{k,k-1} & \omega_{k,k} \end{pmatrix} \begin{pmatrix} r'_k \\ r'_{k-1} \\ \vdots \\ r'_2 \\ r'_1 \end{pmatrix}. \quad (13)$$

Proof: *We are going to rearrange each part of the equality to make the identity appear. Let us first focus on the right-hand part of the identity.*

$$\omega'_{k+1-i,k+1-j} =$$
$$\prod_{m \neq k+1-j} \frac{2(k+1-i) - (2m+1)}{2(k+1-j) + 1 - (2m+1)}$$
$$= \prod_{m \neq k+1-j} \frac{-2i + 2(k+1-m) - 1}{-2j + 1 + 2(k+1-m) - 1}.$$

So far we have just changed the position of $k+1$ in each term of the product. Next just by multiplying each fraction by -1, and extracting all the 2s in the denominators, we get:

$$\omega'_{k+1-i,k+1-j} =$$
$$2^{1-k} \prod_{m \neq k+1-j} \frac{2i - 2(k+1-m) + 1}{j - (k+1-m)}$$
$$= 2^{1-k} \prod_{m \neq j} \frac{2i + 1 - 2m}{j - m}.$$

Here we have reordered the indices $m \leftarrow k+1-m$. We now do the same with the left-hand expression.

$$\omega_{i,j} = \prod_{m \neq j} \frac{2i + 1 - 2m}{2j - 2m}.$$

We extract the 2s in the denominators:

$$\omega_{i,j} = 2^{1-k} \prod_{m \neq j} \frac{2i + 1 - 2m}{j - m},$$

and we conclude that the new expressions for $\omega_{i,j}$ and $\omega'_{k+1-i,k+1-j}$ are the same. □

This lemma points out symmetry properties of the matrices that mainly depend on the choice made in the example for the points of e and e'. They can be taken into account to improve the hardware architecture. Other choices of points could be more interesting and could result in very attractive chip design solutions. This is currently work in progress in our team.

5.2. Cryptographic context

In ECC, the main operation is the addition of two points of an elliptic curve defined over a finite field. Hardware implementation of elliptic curves cryptosystems thus requires efficient operators for additions, multiplications and divisions. Since division is usually a complex operation, we use projective (or homogeneous) coordinates to bypass this difficulty (only one division is needed at the very end of the algorithm).

Thus the only operations are addition and multiplication in $GF(p)$. Moreover it is worth noticing that we do not need to reduce modulo p after each addition. We only subtract p from the result of the last addition if it is greater than $2^{\lceil \log_2(p) \rceil}$ (we recall that p is odd). In other words we just have to check one bit after each addition. The exact value is only needed for the final result.

In ECC protocols, additions chains of points of an elliptic curve are needed. In homogeneous coordinates, those operations consist in additions and multiplications over $GF(p^k)$. Only one division is needed at the end and it can be performed in the Lagrange representation using the Fermat-Euler theorem which states that for all non zero value x in $GF(p^k)$, then $x^{p^k-1} = 1$. Hence we can compute the inverse of x by computing x^{p^k-2} in $GF(p^k)$.

It is also advantageous to use a polynomial equivalent to the Montgomery notation during the computations. We consider polynomials in the form

$$A'(X) = A(X) G(X) \mod N(X)$$

instead of $A(X)$. It is clear that adding two polynomials given in this notation gives the result in the same notation, and for the product, since

$$\text{Mont}(A, B, N) = A(X) B(X) G^{-1}(X) \mod N(X),$$

we have

$$\text{Mont}(A', B', N) =$$
$$A'(X) B'(X) G^{-1}(X) \mod N(X)$$
$$= A(X) B(X) G(X) \mod N(X).$$

6. Conclusion

Works from Bailey and Paar [1, 2], Smart [18] and Crandall [5] have shown that it is possible to obtain more efficient software implementation over $GF(p^k)$ than over $GF(2^k)$ or $GF(p)$ when p is carefully chosen (Mersenne, pseudo-Mersenne, generalized Mersenne primes, etc). In this article we have presented a new modular multiplication algorithm over the finite extension field $GF(p^k)$, for $p > 2k$, which is highly parallelizable and well adapted to hardware implementation. Our algorithm is particularly interesting for ECC since it seems that there exists fewer non-singular curves over $GF(p^k)$ than over $GF(2^k)$. Finding "good" curves for elliptic curve cryptography would then be easier. This could result in a wider choice of curves than in the case $p = 2$. This method can be extended to finite fields of the form $GF(2^{nm})$, where $2^n > 2m$. In this case $p = 2^n$ is no longer a prime number which forces us to choose the values of E and E$'$ in $GF(2^n)^*$. Fields of this form can also be useful for the recent tripartite Diffie-Hellamn key exchange algorithm [8] or the short signature scheme [3] which require an efficient arithmetic over $GF(p^{kl})$, where $6 < k \leqslant 15$ and l is a prime number greater than 160.

Acknowledgements

The authors would like to thank the anonymous reviewers for their very useful comments. This work has been supported by an *ACI cryptologie 2002* grant from the French ministry of education and research.

References

[1] D. Bailey and C. Paar. Optimal extension fields for fast arithmetic in public-key algorithms. In H. Krawczyk, editor, *Advances in Cryptography – CRYPTO'98*, volume 1462 of *Lecture Notes in Computer Science (LNCS)*, pages 472–485. Springer-Verlag, 1998.

[2] D. Bailey and C. Paar. Efficient arithmetic in finite field extensions with application in elliptic curve cryptography. *Journal of Cryptology*, 14(3):153–176, 2001.

[3] D. Boneh, H. Shacham, and B. Lynn. Short signatures from the Weil pairing. In *proceedings of Asiacrypt'01*, volume 2139 of *Lecture Notes in Computer Science*, pages 514–532. Springer-Verlag, 2001.

[4] Ç. K. Koç and T. Acar. Montgomery multiplication in GF(2^k). *Designs, Codes and Cryptography*, 14(1):57–69, April 1998.

[5] R. Crandall. Method and apparatus for public key exchange in a cryptographic system. U.S. Patent number 5159632, 1992.

[6] W. Diffie and M. E. Hellman. New directions in cryptography. *IEEE Transactions on Information Theory*, IT-22(6):644–654, November 1976.

[7] A. Halbutoğullari and Ç. K. Koç. Parallel multiplication in GF(2^k) using polynomial residue arithmetic. *Designs, Codes and Cryptography*, 20(2):155–173, June 2000.

[8] A. Joux. A one round protocol for tripartite Diffie-Hellman. In *4th International Algorithmic Number Theory Symposium (ANTS-IV*, volume 1838 of *Lecture Notes in Computer Science*, pages 385–393. Springer-Verlag, July 2000.

[9] G. A. Jullien, W. Luo, and N. Wigley. High Throughput VLSI DSP Using Replicated Finite Rings. *Journal of VLSI Signal Processing*, 14(2):207–220, November 1996.

[10] D. E. Knuth. *The Art of Computer Programming, Vol. 2: Seminumerical Algorithms*. Addison-Wesley, Reading, MA, third edition, 1997.

[11] N. Koblitz. Elliptic curve cryptosystems. *Mathematics of Computation*, 48(177):203–209, January 1987.

[12] N. Koblitz. *A Course in Number Theory and Cryptography*, volume 114 of *Graduate texts in mathematics*. Springer-Verlag, second edition, 1994.

[13] N. Koblitz. *Algebraic aspects of cryptography*, volume 3 of *Algorithms and computation in mathematics*. Springer-Verlag, 1998.

[14] A. J. Menezes, P. C. Van Oorschot, and S. A. Vanstone. *Handbook of applied cryptography*. CRC Press, 2000 N.W. Corporate Blvd., Boca Raton, FL 33431-9868, USA, 1997.

[15] P. L. Montgomery. Modular multiplication without trial division. *Mathematics of Computation*, 44(170):519–521, April 1985.

[16] C. Paar, P. Fleischmann, and P. Roelse. Efficient multiplier architectures for galois fields GF(2^{4n}). *IEEE Transactions on Computers*, 47(2):162–170, February 1998.

[17] A. Skavantzos and P. B. Rao. New multipliers modulo $2^N - 1$. *IEEE Transactions on Computers*, 41(8):957–961, August 1992.

[18] N. P. Smart. A comparison of different finite fields for use in elliptic curve cryptosystems. Research report CSTR-00-007, University of Bristol, June 2000.

[19] N. Takagi and S. Yajima. Modular multiplication hardware algorithms with a redundant representation and their application to RSA cryptosystem. *IEEE Transactions on Computers*, 41(7):887–891, July 1992.

[20] F. J. Taylor. Large moduli multipliers for signal processing. *IEEE Transactions on Circuits and Systems*, C-28:731–736, Jul 1981.

[21] J. Von Zur Gathen and J. Gerhard. *Modern Computer Algebra*. Cambridge University Press, 1999.

[22] Z. Wang, G. A. Jullien, and W. C. Miller. An Efficient Tree Architecture for Modulo $2n + 1$ Multiplication. *Journal of VLSI Signal Processing*, 14(3):241–248, December 1996.

[23] T. Yanik, E. Savaş, and Ç. K. Koç. Incomplete reduction in modular arithmetic. *IEE Proceedings: Computers and Digital Technique*, 149(2):46–52, March 2002.

Low Complexity Sequential Normal Basis Multipliers over $GF(2^m)$

Arash Reyhani-Masoleh
Centre for Applied Cryptographic Research
Combinatorics and Optimization Department
University of Waterloo
Waterloo, Ontario, Canada N2L 3G1
areyhani@math.uwaterloo.ca

M. Anwar Hasan
Electrical and Computer Engineering Department/
Centre for Applied Cryptographic Research
University of Waterloo
Waterloo, Ontario, Canada N2L 3G1
ahasan@ece.uwaterloo.ca

Abstract

For efficient hardware implementation of finite field arithmetic units, the use of a normal basis is advantageous. In this article, two architectures for multipliers over the finite field $GF(2^m)$ are proposed. Both of these multipliers are of sequential type – after receiving the coordinates of the two input field elements, they go through m iterations (or clock cycles) to finally yield all the coordinates of the product in parallel. These multipliers are highly area efficient and require fewer number of logic gates even when compared with the most area efficient multiplier available in the open literature. This makes the proposed multipliers suitable for applications where the value of m is large but space is of concern, e.g., resource constrained cryptographic systems. Additionally, the AND gate count for one of the multipliers is $\lfloor \frac{m}{2} \rfloor + 1$ only. This implies that if the multiplication over $GF(2^m)$ is performed using a suitable subfield $GF(2^n)$ where $n > 1$ and $n \mid m$, then the corresponding multiplier architecture will yield a highly efficient digit or word serial multiplier.

Keywords: *Finite field, Massey-Omura multiplier, optimal normal basis.*

1 Introduction

Finite field $GF(2^m)$ is a set of 2^m elements where we can add, subtract, multiply and divide (by non-zero elements) without leaving the set. Such finite field arithmetic operations widely used in error control coding and cryptography. In both of these applications, there is a need to design low complexity finite field arithmetic units. The complexity of such a unit largely depends on how the field elements are represented, and there are many ways to represent field elements. Among them, representation of field elements using a normal basis is quite attractive for efficient hardware implementation. A normal basis exists for every extended finite field. Massey and Omura are the first to propose multipliers based on the normal basis [7].

Like any finite field multiplier, a hardware implementation of a normal basis multiplier can be categorized either as a parallel or sequential type. In a typical parallel multiplier for $GF(2^m)$, once $2m$ bits of two inputs are received, m bits of the product are obtained together at the output after delays through various logic gates (if the multiplier is implemented using combinational logic gates) or after delays due to a memory access (if the multiplier is implemented using a look-up table). Such a parallel type multiplier (see for example [16, 15, 12]) requires a lot of silicon area and is considered to be impractical for cryptographic applications where finite fields with very large values of m (e.g., 4000) are used.

On the other hand, a sequential multiplier is much (about m times) more area efficient, but in general takes m iterations (or clock cycles) for one multiplication. Some sequential multipliers generate one bit of the product in each of these m cycles. In another type of sequential multipliers, all m bits of the product are kind of incrementally generated for m cycles and they are obtained together at the end of the m-th cycle. These two types multipliers are hereafter referred to as sequential multipliers with serial output (SMSO) and sequential multipliers with parallel output (SMPO), respectively. Examples of the former type includes Berlekamp's bit-serial dual basis multiplier [2] and Massey-Omura's original bit-serial normal basis multiplier [7], while those of the latter type includes the normal basis multiplier due to Agnew et al. [1] and the well known polynomial basis multiplier based on LFSR [8]. The main advantage of SMPO over SMSO is that the former can run at a much higher clock rate.

In this article, we propose two architectures for SMPO using a normal basis. Like the existing architectures of similar multiplier, the proposed ones have $O(1)$ critical path delay and take m cycles to complete the multiplication. The gate counts of the proposed architectures are given and it is shown that they are better than those of the existing architectures. Additionally, the AND gate count for one of the multipliers is $\lfloor \frac{m}{2} \rfloor + 1$ only. This implies that if the multiplication over $GF(2^m)$ is performed using a suitable subfield $GF(2^n)$ where $n > 1$ and $n \mid m$, then the corresponding multiplier architecture will yield a highly efficient digit or word serial multiplier. To the best of our knowledge, no such AND efficient bit-serial $GF(2^m)$ multiplier in other field representation such as polynomial, dual and redundant basis exists.

2 Preliminaries

2.1 Normal Basis Representation

Let a normal basis of $GF(2^m)$ over $GF(2)$ be

$$N = \{\beta, \beta^2, \cdots, \beta^{2^{m-1}}\},$$

where $\beta \in GF(2^m)$. It is well known that there exists such a normal basis for any positive integer $m > 1$ [6]. Using such a basis, any field element $A \in GF(2^m)$ can be represented as a linear combination of the elements of N, i.e., $A = \sum_{i=0}^{m-1} a_i \beta^{2^i} = (a_0, a_1, \cdots, a_{m-1})$, where $a_i \in GF(2), 0 \leq i \leq m-1$, are referred to as coordinates of A with respect to (w.r.t.) N. In hardware implementation, A^2 is almost free of cost and can be easily performed by right cyclic shifts, i.e., $A^{2^i} = (a_{m-i}, a_{m-i+1}, \cdots, a_{m-i-1})$. However, multiplication is not as easy as squaring. Below we briefly review bit-serial multipliers due to Massey-Omura [7] and Agnew et. al. [1]. The former is a sequential multiplier with serial output (SMSO) type whereas the latter is a sequential multiplier with parallel output (SMPO) type. In this paper, we present two new architectures of SMPO.

2.2 Sequential Multiplier with Serial Output

Let $A = (a_0, a_1, \cdots, a_{m-1})$ and $B = (b_0, b_1, \cdots, b_{m-1})$ be two elements of $GF(2^m)$, where a_i's and b_i's are their respective normal basis coordinates. Let $C = (c_0, c_1, \cdots, c_{m-1})$ be their product: $C = AB$. Then, any coordinates of C, say c_{m-1}, is a function u of A and B which can be obtained by a matrix multiplication, i.e., $c_{m-1} = u(A, B) = \mathbf{a} \cdot \mathbf{M} \cdot \mathbf{b}^T$, where $\mathbf{M}$ is a binary $m \times m$ matrix known as the *multiplication matrix* [5], $\mathbf{a} = [a_0, a_1, \cdots, a_{m-1}]$, $\mathbf{b} = [b_0, b_1, \cdots, b_{m-1}]$ and T denotes vector transposition. The numbers of 1's in $\mathbf{M}$ is known as the *complexity* of the normal basis N [10] and is denoted as C_N which determines the gate counts and hence time delay for a normal basis multiplier. It is well known that $C_N \geq 2m - 1$. If $C_N = 2m - 1$, then the normal basis is called an optimal normal basis.

In [7], Massey and Omura have shown that if the function $u(A, B)$ is implemented to generate c_{m-1}, then the other coordinates of C can be obtained from the same implementation with inputs appropriately shifted in cyclic fashion, more precisely $c_{m-1-i} = u(A^{2^i}, B^{2^i})$. A block diagram of such an architecture of SMSO is presented in Figure 1(a). In this figure, all coordinates of A and B are first serially loaded into the shift registers. Then in each clock cycle, one coordinate of C from c_{m-1} to c_0 is generated by the u function just by cyclic shifts of the registers. The following example is used to illustrate the complexity of the u function. The field and the normal basis presented in this example will be used in all architectures presented in this paper.

Example 1. *Consider the finite field $GF(2^5)$ generated by the irreducible polynomial $F(z) = z^5 + z^2 + 1$ and let α be its root, i.e., $F(\alpha) = 0$. If we choose $\beta = \alpha^5$, then it can be verified that $\{\beta, \beta^2, \beta^4, \beta^8, \beta^{16}\}$ is a normal basis which happens to be an optimal normal basis of type 2. Now, using Table 2 in [10], we have*

$$\mathbf{M} = \begin{bmatrix} 0 & 0 & 1 & 0 & 1 \\ 0 & 0 & 1 & 1 & 0 \\ 1 & 1 & 0 & 0 & 0 \\ 0 & 1 & 0 & 1 & 0 \\ 1 & 0 & 0 & 0 & 0 \end{bmatrix},$$

$$c_4 = a_3 b_3 + (a_0 b_2 + a_2 b_0) + (a_0 b_4 + a_4 b_0) \\ + (a_1 b_2 + a_2 b_1) + (a_1 b_3 + a_3 b_1), \quad (1)$$

and corresponding $GF(2^5)$ bit-serial multiplier is shown in Figure 1(b).

In general, the number of AND gates and XOR gates of Figure 1(a) are C_N and $C_N - 1$, respectively. Also, its critical path delay is $T_A + \lceil \log_2 C_N \rceil T_X$, where T_A and T_X are the time delays due to one AND gate and one XOR gate, respectively.

It is well known that (1) can be rearranged to reduce the AND gate count the Massey-Omura multiplier from C_N to m (see for example [4]). This increases the critical path of the multiplier from $T_A + \lceil \log_2 C_N \rceil T_X$ to $T_A + (\lceil \log_2 \rho \rceil + \lceil \log_2 m \rceil) T_X$. This differences in the critical path delays for these two variants of the Massey-Omura multipliers however disappear if an optimal normal basis is chosen where $\rho = 2$ and $C_N = 2m - 1$. For tradeoff between area and time, one can use digit serial multiplier (see for example [13]).

2.3 Sequential Multiplier with Parallel Output

In [1], Agnew et al. presented another architecture for multiplier using the normal basis. The output coordinates of this multiplier are generated in parallel after m clock cycles

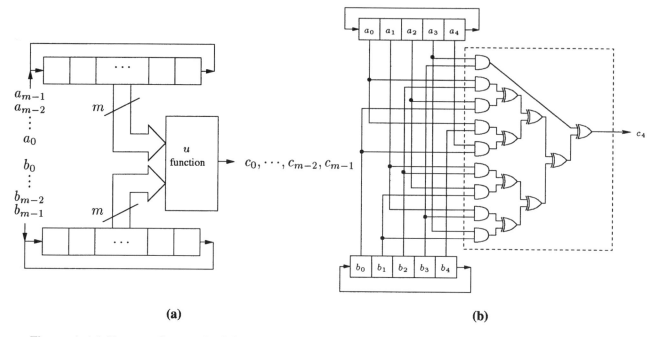

Figure 1. (a) Massey-Omura SMSO for $GF(2^m)$. (b) $GF(2^m)$ Massey-Omura multiplier of Example 1.

(i.e., it is a SMPO architecture). For the field and normal basis constructed in Example 1, the corresponding multiplier architecture is shown in Figure 2(a). In this multiplier structure, all coordinates c_i, $0 \leq i \leq 4$ are obtained using (1) as follow

$$c_i = b_i a_{i+1} + b_{i+1}(a_i + a_{i+3}) + b_{i+2}(a_{i+3} + a_{i+4}) \\ + b_{i+3}(a_{i+1} + a_{i+2}) + b_{i+4}(a_{i+2} + a_{i+4}), \quad (2)$$

where the additions in the subscript indices are reduced modulo 5. In (2), if one implements the first term, i.e., $b_0 a_1$ for c_0, the second term, i.e., $b_2(a_1 + a_4)$ for c_1 and up to the final term, i.e., $b_3(a_1 + a_3)$ for c_4, the SMPO of Figure 2(a) is obtained. The initial contents of shift registers A and B are shown in the figure. Details of the R_i cell are shown in Figure 2(b). Initially, the D_i latches of R_i's are cleared to zero and after m clock cycles the D_i's contain the coordinates of $C = AB$.

The number of AND gates and XOR gates of the SMPO can be easily obtained as m and C_N, respectively. The critical path delay of the multiplier is $T_A + (1 + \lceil \log_2 \rho \rceil)T_X$, where ρ is the maximum number of a_i terms that are XORed before being multiplied with a b_i term in (2). This parameter ρ is in fact the maximum number of 1's among all rows (or columns) of the multiplication matrix M. It is noted that $2 \leq \rho \leq m$. For optimal normal bases where $\rho = 2$, the critical path delay would be $T_A + 2T_X$ as shown in Figure 2(a) and $\leq T_A + (1 + \lceil \log_2 m \rceil)T_X$ for an arbitrary normal basis.

3 New SMPO Architectures

3.1 Formulation

As before let us consider two $GF(2^m)$ elements $A = (a_0, a_1, \cdots, a_{m-1})$ and $B = (b_0, b_1, \cdots, b_{m-1})$ and let their product be $C = (c_0, c_1, \cdots, c_{m-1})$. For $0 \leq i \leq m-1$, let

$$F_i(A, B) = a_{i-g} b_{i-g} \beta + \sum_{j=1}^{v} z_{i,j} \delta_j \quad (3)$$

where $v = \lfloor \frac{m}{2} \rfloor$, $\delta_j = \beta^{1+2^j}$, $1 \leq j \leq v$, and $g \in \{0, 1\}$ determines $z_{i,j}$ as follows. For $1 \leq j < \lceil \frac{m}{2} \rceil$,

$$z_{i,j} = \begin{cases} (a_i + a_{i+j})(b_i + b_{i+j}), & \text{if } g = 0, \\ a_i b_{i+j} + a_{i+j} b_i, & \text{if } g = 1. \end{cases} \quad (4)$$

For even values of m, we have

$$z_{i,v} = \begin{cases} b_i(a_i + a_{i+v}), & \text{if } g = 0, \\ a_i b_{i+v}, & \text{if } g = 1. \end{cases} \quad (5)$$

Note that additions and subtractions in the above subscripts are reduced modulo m. Now, we can state the following theorem which is the key equation for our new architectures.

Theorem 1. *Consider three elements A, B and $C = AB$ of $GF(2^m)$. Then,*

$$C = (((F_{m-1}^2 + F_{m-2})^2 + F_{m-3})^2 + \cdots + F_1)^2 + F_0 \quad (6)$$

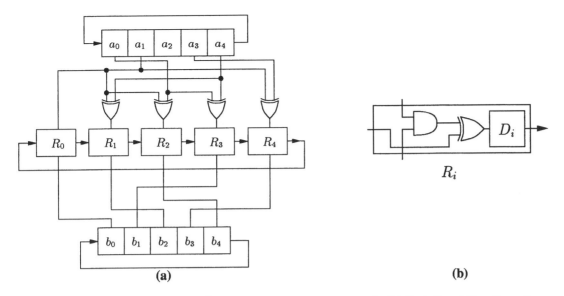

Figure 2. (a) $GF(2^5)$ SMPO due to Agnew et al. (b) Details of the R_i cell.

where $F_i \in GF(2^m)$ is a short form of $F_i(A, B)$ as defined in (3).

Proof. Using [12], we have

$$C = \begin{cases} \sum_{i=0}^{m-1} a_i b_i \beta^{2^{i+1}} + \sum_{i=0}^{m-1} \sum_{j=1}^{v} x_{i,j} \delta_j^{2^i}, & \text{for } m \text{ odd} \\ \sum_{i=0}^{m-1} a_i b_i \beta^{2^{i+1}} + \sum_{i=0}^{m-1} \sum_{j=1}^{v-1} x_{i,j} \delta_j^{2^i} + \sum_{i=0}^{v-1} x_{i,v} \delta_v^{2^i}, & \text{for } m \text{ even,} \end{cases}$$

where $x_{i,j} = a_i b_{i+j} + a_{i+j} b_i$, $0 \leq i \leq m-1$, $1 \leq j \leq v$.

Also, using [14],

$$C = \begin{cases} \sum_{i=0}^{m-1} a_i b_i \beta^{2^i} + \sum_{i=0}^{m-1} \sum_{j=1}^{v} y_{i,j} \delta_j^{2^i}, & \text{for } m \text{ odd} \\ \sum_{i=0}^{m-1} a_i b_i \beta^{2^i} + \sum_{i=0}^{m-1} \sum_{j=1}^{v-1} y_{i,j} \delta_j^{2^i} + \sum_{i=0}^{v-1} y_{i,v} \delta_v^{2^i}, & \text{for } m \text{ even,} \end{cases}$$

where $y_{i,j} = (a_i + a_{i+j})(b_i + b_{i+j})$, $0 \leq i \leq m-1$, $1 \leq j \leq v$.

Now, combining the above two equations, the following can be obtained.

$$C = \sum_{i=0}^{m-1} a_{i-g} b_{i-g} \beta^{2^i} + \sum_{i=0}^{m-1} \sum_{j=0}^{v} z_{i,j} \delta_j^{2^i},$$

where $z_{i,j}$ was previously defined in terms of g. Thus,

$$C = \sum_{i=0}^{m-1} a_{i-g} b_{i-g} \beta^{2^i} + \sum_{i=0}^{m-1} \sum_{j=0}^{v} z_{i,j} \delta_j^{2^i}$$

$$= \sum_{i=0}^{m-1} \left(a_{i-g} b_{i-g} \beta + \sum_{j=0}^{v} z_{i,j} \delta_j \right)^{2^i}$$

$$= \sum_{i=0}^{m-1} F_i^{2^i}$$

$$= (((F_{m-1}^2 + F_{m-2})^2 + F_{m-3})^2 + \cdots + F_1)^2 + F_0.$$

□

Based on the above theorem, we can now have the following algorithm for normal basis multiplication.

Algorithm 1. (Normal Basis Multiplication)
Input: $A, B \in GF(2^m)$ given w.r.t. N
Output: $C = AB$
1. Initialize $X = (a_0, a_1, \cdots, a_{m-1})$, $Y = (b_0, b_1, \cdots, b_{m-1})$, $D = (d_0, d_1, \cdots, d_{m-1}) = 0$
2. For $t = 1$ to m {
3. $\quad D = D^2 + F_{m-1}(X, Y)$ using (3)
4. $\quad X = X^2$ and $Y = Y^2$
5. }
6. $C = D$

In step 3 of this algorithm, we use the fixed function F_{m-1} with varying inputs, i.e., at the t-th iteration this function is $F_{m-1}(A^{2^{t-1}}, B^{2^{t-1}})$. It can be proven that $F_{m-t} = F_{m-1}(A^{2^{t-1}}, B^{2^{t-1}})$. Instead of using F_{m-1}, $F_{m-2}, \cdots, F_0$ with fixed inputs A and B as shown in Theorem 1, the use of only F_{m-1} with varying inputs greatly simplifies implementation of Algorithm 1. This is discussed below.

3.2 Architectures

In order to realize the above algorithm in hardware, the structure of Figure 3(a) is proposed. In the initialization step

Multiplier	#XOR	#AND
XESMPO	v	$m-1$
AESMPO	$m-1$	v

Table 1. The number of gates in the Z array.

for the multiplication operation, the coordinates of A and B are serially loaded into the corresponding shift registers as shown in the figure. This is similar to step 1 of Algorithm 1. Let $D(t) \in GF(2^m)$ denote the field element in the t-th iteration of step 3 of the algorithm whose normal basis coordinates are stored in m latches of D_i's as shown in Figure 3(a). Then, to start the multiplication operation, all D_i latches have to be cleared corresponding to step 1 of the algorithm. In this figure, the cyclic shift operation at the output of D_i's performs squaring to obtain D^2 used in the algorithm. Finally the whole operation of $D^2 + F_{m-1}$ in step 3 is realized by the Z array, the XOR array and an additional AND gate as shown in the figure. The Z array contains v number of Z blocks which generate $z_{m-1,j}$, $1 \leq j \leq v$, needed for F_{m-1} in (3) and the XOR array consists of XOR gates. Depending on the value of $g \in \{0, 1\}$, one of the two Z blocks as shown in Figures 3(b) and 3(c) is used. It is noted that for m even, the Z block for generating $z_{m-1,v}$ is different from both Figure 3(b) and 3(c). For this case, a slightly different Z block is needed which will generate $z_{m-1,v}$ corresponding to (5) and requires one AND gate for $g = 0$, and one AND gate as well as one XOR gate for $g = 1$. From now on, the multiplier architecture containing Z blocks as shown in Figure 3(b) and 3(c) are referred to as XOR efficient SMPO (XESMPO) and AND efficient SMPO (AESMPO), respectively.

3.3 Complexities

The gate complexity of the XESMPO and AESMPO depends on the number of gates in the Z array and the XOR array. The number of XOR gates and AND gates in the Z array are easy to obtain because it consists of v identical blocks[1]. These values for the proposed multipliers are shown in Table 1.

To obtain the number of XOR gates in the XOR array, first we find this number for the XESMPO/AESMPO for the field and the basis constructed in Example 1. For this example, $\delta_1 = \beta + \beta^{2^3}$ and $\delta_2 = \beta^{2^3} + \beta^{2^4}$ and substituting these into (3) for $i = m - 1 = 4$, we have $F_4 = a_{4-g}b_{4-g}\beta + z_{4,1}(\beta + \beta^{2^3}) + z_{4,2}(\beta^{2^3} + \beta^{2^4})$. Since the contents of the XOR array for both $g = 0$ and $g = 1$ are the same, here we only consider $g = 0$. Thus, $F_4 = (a_4b_4 + z_{4,1}, 0, 0, z_{4,1} + z_{4,2}, z_{4,2})$. Let $D^2 = (d_4, d_0, d_1, d_2, d_3)$

[1] For m even, one block which generates $z_{m-1,v}$ is different from the others.

then the output of the XOR array would be $D^2 + F_4 = (d_4 + a_4b_4 + z_{4,1}, d_0, d_1, d_2 + z_{4,1} + z_{4,2}, d_3 + z_{4,2})$ which can be realized using 5 XOR gates. This architecture is shown in Figure 4. The architecture of XESMPO for this example is similar to Figure 4 except that two Z blocks in the Z array and a_4b_4 in the first coordinate of F_4 should be replaced by Figure 3(b) and a_3b_3, respectively.

In general, if no terms or signals are reused, then the number of XOR gates in the XOR array of Figure 3 is upper bounded by $1 + \sum_{j=1}^{v} H(\delta_j)$, where $H(\delta_j)$ is the number of non-zero coordinates in the normal basis representation of δ_j. For m being even, this value can be reduced to $1 + \sum_{j=1}^{v-1} H(\delta_j) + 0.5H(\delta_v)$ by reusing half of the signals involved in δ_v [12]. Also, from [12]

$$C_N = \begin{cases} 1 + 2\sum_{j=1}^{v} H(\delta_j), & \text{for } m \text{ odd,} \\ 1 + 2\sum_{j=1}^{v-1} H(\delta_j) + H(\delta_v), & \text{for } m \text{ even.} \end{cases}$$

One can then conclude that the number of XOR gates in the XOR array is upper bounded by $0.5(C_N + 1)$. Therefore, based on the above discussions, one can obtain the gate counts of the proposed multipliers as stated below.

Proposition 1. *The gate complexities of the proposed XESMPO ($g = 1$) and AESMPO ($g = 0$) using normal basis are*

$$\begin{aligned} \#AND &= m, \\ \#XOR &\leq \frac{C_N + 1}{2} + \left\lfloor \frac{m}{2} \right\rfloor, \end{aligned}$$

and

$$\begin{aligned} \#AND &= \left\lfloor \frac{m}{2} \right\rfloor + 1, \\ \#XOR &\leq \frac{C_N + 2m - 1}{2}, \end{aligned}$$

respectively.

To obtain the maximum clock rate for the proposed multipliers, we have to obtain the delay of the critical path of the Z array and the XOR array in Figure 3(a). The delay of the Z array is $T_A + T_X$ for both XESMPO and AESMPO. Since the output of the XOR array is

$$D^2 + a_{i-g}b_{i-g}\beta + \sum_{j=1}^{v} z_{i,j}\delta_j, \tag{7}$$

the critical path of the XOR array depends on the normal basis representations of δ_j, for $1 \leq j \leq v$. In the worst case when all δ_j's have a common coordinate, say β^{2^k}, then the critical path is determined by the k-th coordinate of the output of the XOR array and is equal to $\lceil \log_2(v+1) \rceil T_X$ for $k \neq 0$ or $\lceil \log_2(v+2) \rceil T_X$ for $k = 0$. Since $v = \lfloor \frac{m}{2} \rfloor$, one can easily verify that the critical path delay is upper

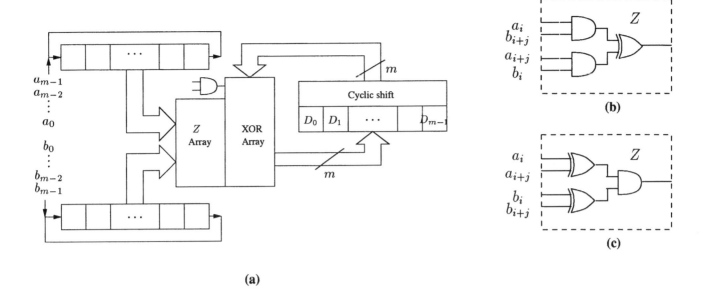

Figure 3. (a) The structure of proposed sequential multipliers of $GF(2^m)$. (b) The Z block for the XESMPO ($g = 1$). (c) The Z block for the AESMPO ($g = 0$).

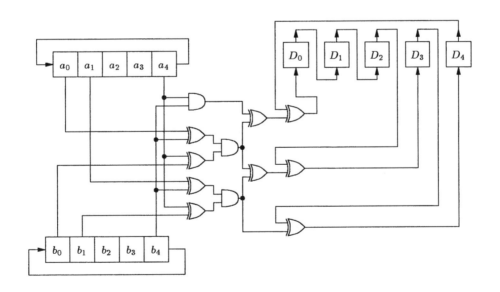

Figure 4. The proposed $GF(2^5)$ AESMPO of Example 1.

bounded by $T_A + \lceil \log_2(m+4) \rceil T_X$. Let τ be the maximum number of terms among all m coordinates in normal basis representation of (7), then the critical path of the proposed multipliers would be $T_A + (1 + \lceil \log_2 \tau \rceil)T_X$. For optimal normal bases, it can be shown that $\tau = 3$ which results in the critical path of the XOR array to be $\lceil \log_2 3 \rceil T_X = 2T_X$ as seen in Figure 4.

3.4 Comparison

Table 2 compares our proposed multipliers with the existing leading ones in terms of number of gates and registers. As seen in the table, our proposed multipliers have a fewer total number of AND and XOR gates compared to the other multipliers. Also, for the optimal normal bases our architectures has about $0.5m$ fewer number of gates compared to the best architecture in the table. This difference would increase if a non-optimal normal basis is used. For the five binary fields recommended by NIST for ECDSA (elliptic curve digital signature algorithm), i.e., $m \in \{163, 233, 283, 409, 571\}$, only $m = 233$ has an optimal normal basis.

In Table 3, multipliers are compared in terms of the critical path delay for generic and optimal normal basis. In this table, T_β denotes the time delay of a constant multiplier with inputs as $E = (0, e_1, \cdots, e_{m-1})$ and $\beta = (1, 0, 0, \cdots, 0)$ where $E, \beta \in GF(2^m)$. As seen in this table, our proposed multipliers have one more XOR gate delay than the multiplier of Agnew et al. [1] for optimal normal bases. However, compared to the other multipliers, the proposed one has a shorter critical path delay. Additionally, the proposed multipliers have a tighter upper bound for the critical path delay.

4 Conclusions

In this article, two architectures of normal basis $GF(2^m)$ multipliers namely AESMPO and XESMPO have been proposed. Their architectural level details have been presented and compared with other sequential normal basis multipliers in term of number of AND gates, XOR gates, latches and critical path delays. Compared to the existing sequential normal basis multipliers, our proposed multipliers have a fewer number of gates.

The proposed AESMPO can be extended to a word-serial multiplier for the composite field $GF(2^m)$ over $GF(2^n)$ where $m = nk$. For such an extension, one needs to simply replace the AND and XOR gates with the sub-field $GF(2^n)$ multiplications and additions, respectively. In this case, the word-serial architecture will need only $\lfloor \frac{k}{2} \rfloor + 1$ sub-field multiplications over $GF(2^n)$. This results in about half the number of gates compared to the architectures proposed in [9] for normal basis, and in [11] for polynomial basis.

Also, for software implementation of Algorithm 1, one can generate $z_{i,j}$ on the fly to obtain $F_i(A, B)$ using pre-stored δ_j's which are fixed for a specific normal basis. In this situation, the algorithm corresponding to the AESMPO will require fewer number of instructions to be executed by the processor on which the algorithm is implemented. This is because[2] $Pr\{z_{i,j} = 1 | g = 0\} = \frac{1}{4}$ and $Pr\{z_{i,j} = 1 | g = 1\} = \frac{3}{8}$ which results in a fewer number of XOR instructions on average.

Acknowledgments

This work has been supported in part by an NSERC post-doctoral fellowship awarded to A. Reyhani-Masoleh and in part by an NSERC grant awarded to M. A. Hasan.

References

[1] G. B. Agnew, R. C. Mullin, I. M. Onyszchuk, and S. A. Vanstone. "An Implementation for a Fast Public-Key Cryptosystem". *Journal of Cryptology*, 3:63–79, 1991.

[2] E. R. Berlekamp. "Bit-Serial Reed-Solomon Encoders". *IEEE Transactions on Information Theory*, 28(6):869–874, Nov. 1982.

[3] M. Feng. "A VLSI Architecture for Fast Inversion in $GF(2^m)$". *IEEE Transactions on Computers*, 38(10):1383–1386, October 1989.

[4] L. Gao and G. E. Sobelman. "Improved VLSI Designs for Multiplication and Inversion in $GF(2^M)$ over normal bases". In *Proceedings of 13th Annual IEEE International ASIC/SOC Conference*, pages 97–101, 2000.

[5] IEEE Std 1363-2000. "IEEE Standard Specifications for Public-Key Cryptography". January 2000.

[6] R. Lidl and H. Niederreiter. *Introduction to Finite Fields and Their Applications*. Cambridge University Press, 1994.

[7] J. L. Massey and J. K. Omura. "Computational Method and Apparatus for Finite Field Arithmetic". *US Patent No. 4,587,627*, 1986.

[8] E. D. Mastrovito. *VLSI Architectures for Computation in Galois Fields*. PhD thesis, Linkoping Univ., Linkoping Sweden, 1991.

[9] R. C. Mullin. "Multiple Bit Multiplier". *US Patent No. 5,787,028*, July 1998.

[10] R. C. Mullin, I. M. Onyszchuk, S. A. Vanstone, and R. M. Wilson. "Optimal Normal Bases in $GF(p^n)$". *Discrete Applied Mathematics*, 22:149–161, 1988/89.

[11] C. Paar, P. Fleishmann, and P. Soria-Rodriguez. "Fast Arithmetic for Public-Key Algorithms in Galois Fields with Composite Exponents". *IEEE Transactions on Computers*, 48(10):1025–1034, Oct. 1999.

[12] A. Reyhani-Masoleh and M. A. Hasan. "A New Construction of Massey-Omura Parallel Multiplier over $GF(2^m)$". *IEEE Transactions on Computers*, 51(5):511–520, May 2002.

---

[2] In the truth table of $z_{i,j}$ in terms of a_i, a_{i+j}, b_i and b_{i+j}, there are four and six 1s for $g = 0$, and $g = 1$, respectively.

Sequential Multiplier	#AND gates	#XOR gates	Total gate count generic N	Total gate count optimal N	#1-bit registers	Output format
Massey-Omura[7]	C_N	$C_N - 1$	$2C_N - 1$	$4m - 3$	$2m$	serial
Improved ,, [4]	m	$C_N - 1$	$m + C_N - 1$	$3m - 2$	$2m$	serial
Feng [3]	$2m - 1$	$\leq C_N + m - 1$	$\leq C_N + 3m - 2$	$\leq 5m - 3$	$3m - 2$	parallel
Agnew et al [1]	m	C_N	$m + C_N$	$3m - 1$	$3m$	parallel
XESMPO	m	$\leq \frac{C_N+1}{2} + \lfloor\frac{m}{2}\rfloor$	$\leq \frac{C_N+2m+1}{2} + \lfloor\frac{m}{2}\rfloor$	$\leq 2m + \lfloor\frac{m}{2}\rfloor$	$3m$	parallel
AESMPO	$\lfloor\frac{m}{2}\rfloor + 1$	$\leq \frac{C_N+2m-1}{2}$	$\leq \frac{C_N+2m+1}{2} + \lfloor\frac{m}{2}\rfloor$	$\leq 2m + \lfloor\frac{m}{2}\rfloor$	$3m$	parallel

Table 2. Comparison of sequential normal basis multipliers over $GF(2^m)$. Note that for an optimal normal basis $C_N = 2m - 1$, otherwise $C_N > 2m - 1$.

multiplier	generic N	optimal N	upper bound
Massey-Omura[7]	$T_A + \lceil \log_2 C_N \rceil T_X$	$T_A + (1 + \lceil \log_2 m \rceil)T_X$	$\leq T_A + 2 \lceil \log_2 m \rceil T_X$
Improved ,, [4]	$T_A + (\lceil \log_2 \rho \rceil + \lceil \log_2 m \rceil)T_X$	$T_A + (1 + \lceil \log_2 m \rceil)T_X$	$\leq T_A + 2 \lceil \log_2 m \rceil T_X$
Feng [3]	$T_A + 3T_X + T_\beta$	$T_A + 5T_X$	$\leq T_A + (3 + \lceil \log_2 m \rceil)T_X$
Agnew et al [1]	$T_A + (1 + \lceil \log_2 \rho \rceil)T_X$	$T_A + 2T_X$	$\leq T_A + (1 + \lceil \log_2 m \rceil)T_X$
XESMPO	$T_A + (1 + \lceil \log_2 \tau \rceil)T_X$	$T_A + 3T_X$	$\leq T_A + \lceil \log_2(m+4) \rceil T_X$
AESMPO	$T_A + (1 + \lceil \log_2 \tau \rceil)T_X$	$T_A + 3T_X$	$\leq T_A + \lceil \log_2(m+4) \rceil T_X$

Table 3. Comparison of critical path delays of sequential normal basis multipliers over $GF(2^m)$. Note that for an optimal normal basis $C_N = 2m - 1$, otherwise $C_N > 2m - 1$. Also, ρ and τ are defined in Subsections 2.3 and 3.3, respectively.

[13] A. Reyhani-Masoleh and M. A. Hasan. "Efficient Digit-Serial Normal Basis Multipliers over $GF(2^M)$". In *IEEE International Symposium on Circuits and Systems, ISCAS 2002*, pages 781–784, May 2002.

[14] A. Reyhani-Masoleh and M. A. Hasan. "Efficient Multiplication Beyond Optimal Normal Bases". *to appear in IEEE Transactions on Computers, Special Issue on Cryptographic Hardware and Embedded Systems*, April 2003.

[15] B. Sunar and C. K. Koc. "An Efficient Optimal Normal Basis Type II Multiplier". *IEEE Transactions on Computers*, 50(1):83–88, Jan. 2001.

[16] C. C. Wang, T. K. Truong, H. M. Shao, L. J. Deutsch, J. K. Omura, and I. S. Reed. "VLSI Architectures for Computing Multiplications and Inverses in $GF(2^m)$". *IEEE Transactions on Computers*, 34(8):709–716, Aug. 1985.

A LOW COMPLEXITY AND A LOW LATENCY BIT PARALLEL SYSTOLIC MULTIPLIER OVER GF(2^m) USING AN OPTIMAL NORMAL BASIS OF TYPE II

Soonhak Kwon
Department of Mathematics, Sungkyunkwan University
Suwon 440-746, Korea
shkwon@math.skku.ac.kr

Abstract

Using the self duality of an optimal normal basis (ONB) of type II, we present a bit parallel systolic multiplier over $GF(2^m)$ which has a low hardware complexity and a low latency. We show that our multiplier has a latency $m + 1$ and the basic cell of our circuit design needs 5 latches (flip-flops). On the other hand, most of other multipliers of the same type have latency $3m$ and the basic cell of each multiplier needs 7 latches. Comparing the gates areas in each basic cell, we find that the hardware complexity of our multiplier is 25 percent reduced from the multipliers with 7 latches.

1. Introduction

Arithmetic of finite fields, especially finite field multiplication, found various applications in many cryptographic and coding theoretical areas. Therefore an efficient design of a finite field multiplier is needed. Though one may design a finite field multiplier in a software arrangement, a hardware implementation has a strong advantage when one wants a high speed multiplier. Moreover, arithmetic of $GF(2^m)$ is easily realized in a circuit design using a few logical gates. A good multiplication algorithm depends on the choice of a basis for a given finite field. In general, there are three types of basis being used, that is, polynomial, dual and normal basis. Some popular multipliers for cryptographical and coding theoretical purposes are Berlekamp's bit serial multiplier [1,2] which use a dual basis, and a bit parallel multiplier of Massey-Omura type [4,5,17,18] which use a normal basis. Above mentioned multipliers and other traditional multipliers have some unappealing characteristics. For example, they have irregular circuit designs. In other words, their hardware structures may be quite different for varying choices of m for $GF(2^m)$, though the multiplication algorithm is basically same for each m. Moreover as m gets large, the propagation delay also increases. So deterioration of the performance is inevitable. A systolic multiplier does not suffer from above problems. It has a regular structure consisting of a number of replicated basic cells, each of which has the same circuit design. So overall structures of systolic multipliers are same and not depending on a particular choice of m for $GF(2^m)$. Furthermore since each basic cell is only connected with its neighboring cells, signals can be propagated at a high clock speed. There are systolic multipliers using a polynomial basis [7,8,10,11,12,13,15] and a dual basis [9]. A bit parallel systolic multiplier in [8] has a comparable or better longest path delay than the multipliers in [7,9,10]. However, it has bidirectional data flows, whereas multipliers in [7,9,10] have unidirectional data flows. A bit parallel systolic multiplier proposed in [10] uses an all one polynomial (AOP) basis. This AOP multiplier has a low cell complexity and a high throughput when compared with other multipliers. However, it is applicable to relatively few finite fields. To be specific, the number of $m \leq 1000$ for which an AOP multiplier in $GF(2^m)$ exists is only 68. On the other hand, most of the multipliers using a standard polynomial basis are applicable to all finite fields. In this paper, we propose a design of a bit parallel systolic multiplier using a type II optimal normal basis (ONB) in $GF(2^m)$. Our multiplier has unidirectional data flows and does not broadcast signals. We show that our bit parallel systolic multiplier has a lower hardware complexity when compared with other systolic multipliers in [7,8,9,11]. A bit parallel systolic multiplier using an AOP basis in [10] has a lower hardware complexity than ours, but our multiplier is applicable to broader class of finite fields. We also show that our multiplier has a latency $m + 1$ whereas other multipliers of the same type have latency $3m$ except for the multiplier in [10], where the latency is $m + 1$. A low latency and a low hardware complexity multiplier is introduced in [12] but it broadcasts signals and has bidirectional data flows. In practical situations such as VLSI implementation, broadcastings and bidirectional data flows should be avoided if one wants a reliable and a fault tolerant architecture. It should be mentioned that, though there exists [4,5] a bit parallel multiplier (of Massey-Omura type) using an ONB of type II, we have not yet found a systolic multiplier using the same basis. In fact, except for our multiplier and the multiplier in [10] using an AOP basis (which is not a normal basis, but roughly speaking, it may be viewed as a variant of a type I optimal normal basis), there is no other bit parallel systolic multiplier using a normal basis to our knowledge at this moment.

2. Normal basis and optimal normal basis of type II

Let $GF(2^m)$ be a finite field of 2^m elements. $GF(2^m)$ is a vector space over $GF(2)$ of dimension m. We briefly explain basic finite field arithmetic.

Definition 1. *Two bases* $\{\alpha_1, \alpha_2, \cdots, \alpha_m\}$ *and* $\{\beta_1, \beta_2, \cdots, \beta_m\}$ *of* $GF(2^m)$ *are said to be dual if the trace map,* $Tr : GF(2^m) \to GF(2)$, *with* $Tr(\alpha) = \alpha + \alpha^2 + \cdots + \alpha^{2^{m-1}}$, *satisfies* $Tr(\alpha_i \beta_j) = \delta_{ij}$ *for all* $1 \leq i, j \leq m$, *where* $\delta_{ij} = 1$ *if* $i = j$, *zero if* $i \neq j$. *A basis* $\{\alpha_1, \alpha_2, \cdots, \alpha_m\}$ *is said to be self dual if* $Tr(\alpha_i \alpha_j) = \delta_{ij}$.

Definition 2. *A basis of* $GF(2^m)$ *over* $GF(2)$ *of the form* $\{\alpha, \alpha^2, \cdots, \alpha^{2^{m-1}}\}$ *is called a normal basis for* $GF(2^m)$.

It is well known [6] that normal bases in $GF(2^m)$ exist for all m. But our main interest is the following type of normal bases which is explained in detail in [6] and also in [4].

Theorem 1. *Let* $GF(2^m)$ *be a finite field of* 2^m *elements where* $2m + 1 = p$ *is a prime. Suppose that either* (⋆) 2 *is a primitive root* (mod p) *or* (⋆⋆) -1 *is a quadratic non residue* (mod p) *and 2 generates the quadratic residues* (mod p). *Then letting* $\alpha = \beta + \beta^{-1}$ *where* β *is a primitive pth root of unity in* $GF(2^{2m})$, *we have* $\alpha \in GF(2^m)$ *and* $\{\alpha, \alpha^2, \cdots, \alpha^{2^{m-1}}\}$ *is a basis over* $GF(2)$.

Definition 3. *A normal basis in theorem 1 is called an optimal normal basis (ONB) of type II.*

Using the assumptions in the previous theorem, one finds easily (see [4].)

$$\alpha^{2^s} = (\beta + \beta^{-1})^{2^s} = \beta^{2^s} + \beta^{-2^s} = \beta^t + \beta^{-t},$$

where $0 < t < p = 2m + 1$ with $2^s \equiv t \pmod{p}$. Moreover, replacing t by $p - t$ if $m + 1 \leq t \leq 2m$, we find that $\{\alpha, \alpha^2, \cdots, \alpha^{2^{m-1}}\}$ and $\{\beta + \beta^{-1}, \beta^2 + \beta^{-2}, \cdots, \beta^m + \beta^{-m}\}$ are same sets. That is, $\alpha^{2^s}, 0 \leq s \leq m - 1$ is just a permutation of $\beta^s + \beta^{-s}, 1 \leq s \leq m$. In the design of our bit serial multiplier, we need the following self duality of a type II ONB. In fact, it is a special case of a well known fact from the theory of Gauss periods saying that a Gauss period of type (m, k) over $GF(2^m)$ is self dual if and only if k is even, which was proved in [3]. We present an elementary proof of the special case.

Lemma 1. *An optimal normal basis* $\{\alpha^{2^s} | 0 \leq s \leq m - 1\}$ *of type II in* $GF(2^m)$, *if it exists, is self dual.*

Proof. After a permutation of basis elements, it can be written as $\{\beta^s + \beta^{-s} | 1 \leq s \leq m\}$. Note that $Tr((\beta^i + \beta^{-i})(\beta^j + \beta^{-j})) = Tr(\beta^{i-j} + \beta^{-(i-j)} + \beta^{i+j} + \beta^{-(i+j)})$. If $i = j$, then we have $Tr(\beta^{2i} + \beta^{-2i}) = Tr(\alpha^{2^s})$ for some $0 \leq s \leq m - 1$. Thus the trace value is $\alpha + \alpha^2 + \cdots + \alpha^{2^{m-1}} = 1$ because of the linear independence. Therefore we may assume $i \neq j$. Then replacing $i - j$ by $|i - j|$ if $i - j < 0$ and $i + j$ by $2m + 1 - (i + j)$ if $m + 1 \leq i + j \leq 2m$ (recall $\beta^{2m+1} = 1$.), we find $Tr((\beta^i + \beta^{-i})(\beta^j + \beta^{-j})) = Tr(\beta^u + \beta^{-u} + \beta^v + \beta^{-v}) = Tr(\beta^u + \beta^{-u}) + Tr(\beta^v + \beta^{-v})$ for some $1 \leq u, v \leq m$. Since $\beta^u + \beta^{-u}, \beta^v + \beta^{-v}$ are in $\{\alpha^{2^s} | 0 \leq s \leq m - 1\}$, we have $Tr(\beta^u + \beta^{-u}) + Tr(\beta^v + \beta^{-v}) = 1 + 1 = 0$. □

3. Multiplication algorithm

The proof of lemma 1 implies that many of the notations can be simplified if we define $\alpha_s = \beta^s + \beta^{-s}, 1 \leq s \leq m$. Therefore from now on, we assume $\{\alpha^{2^s} | 0 \leq s \leq m - 1\}$ is a type II ONB in $GF(2^m)$ and $\{\alpha_s | \alpha_s = \beta^s + \beta^{-s}, 1 \leq s \leq m\}$ is a basis obtained after a permutation of the basis elements of the normal basis. For a given $x = \sum_{i=1}^{m} x_i \alpha_i$ with $x_i \in GF(2)$, by using lemma 1, we have

$$x_s = \sum_{i=1}^{m} x_i Tr(\alpha_s \alpha_i) = Tr(\alpha_s \sum_{i=1}^{m} x_i \alpha_i) = Tr(\alpha_s x),$$

for all $1 \leq s \leq m$. We extend the definition of α_s and x_s for all integers s as follows.

Definition 4. *Let* β *be a primitive pth* ($p = 2m + 1$) *root of unity in* $GF(2^{2m})$ *and let* $x \in GF(2^m)$. *For each integer* s, *define* α_s *and* x_s *as*

$$\alpha_s = \beta^s + \beta^{-s}, \qquad x_s = Tr(\alpha_s x).$$

Lemma 2. *We have* $\alpha_s = 0 = x_s$ *if* $2m+1$ *divides* s. *Also for all* s,

$$\alpha_{2m+1+s} = \alpha_s = \alpha_{2m+1-s} = \alpha_{-s}$$

and

$$x_{2m+1+s} = x_s = x_{2m+1-s} = x_{-s}.$$

Proof. We have $\alpha_s = \beta^s + \beta^{-s} = 0$ if and only if $\beta^s = \beta^{-s}$, that is, $\beta^{2s} = 1$. And this happens whenever $2m+1 = p$ divides s since β is a primitive pth root of unity. Now $\alpha_{2m+1+s} = \beta^{2m+1+s} + \beta^{-(2m+1+s)} = \beta^s + \beta^{-s} = \alpha_s$ is obvious because $\beta^{2m+1} = 1$. Also $\alpha_{2m+1-s} = \beta^{2m+1-s} + \beta^{-(2m+1-s)} = \beta^{-s} + \beta^s = \alpha_s$. The result for x_s instantly follows from the result for α_s. □

Now for any integer s and t, we have

$$\alpha_s \alpha_t = (\beta^s + \beta^{-s})(\beta^t + \beta^{-t}) = \alpha_{s-t} + \alpha_{s+t}.$$

Using above relation and lemma 2, we are ready to give the following assertion.

Theorem 2. *Let $x = \sum_{i=1}^{m} x_i \alpha_i$ and $y = \sum_{i=1}^{m} y_i \alpha_i$ be elements in $GF(2^m)$. Then we have $xy = \sum_{i=1}^{m} (xy)_i \alpha_i$, where the kth coefficient $(xy)_k$ satisfies*

$$(xy)_k = \sum_{i=1}^{2m} y_i x_{i-k} = \sum_{i=1}^{2m+1} y_i x_{i-k}.$$

Proof. By the self duality of our basis,
$$\begin{aligned}
(xy)_k &= Tr(\alpha_k xy) \\
&= Tr(\alpha_k x \sum_{i=1}^{m} y_i \alpha_i) = \sum_{i=1}^{m} y_i Tr(\alpha_k \alpha_i x) \\
&= \sum_{i=1}^{m} y_i Tr(\alpha_{i-k} x + \alpha_{i+k} x) \\
&= \sum_{i=1}^{m} y_i (Tr(\alpha_{i-k} x) + Tr(\alpha_{i+k} x)) \\
&= \sum_{i=1}^{m} y_i (x_{i-k} + x_{i+k}) = \sum_{i=1}^{m} y_i x_{i-k} + \sum_{i=1}^{m} y_i x_{i+k}.
\end{aligned}$$

On the other hand, the second summation of above expression can be written as

$$\begin{aligned}
\sum_{i=1}^{m} y_i x_{i+k} &= \sum_{i=1}^{m} y_{m+1-i} x_{m+1-i+k} \\
&= \sum_{i=1}^{m} y_{m+i} x_{m+i-k} \\
&= \sum_{i=m+1}^{2m} y_i x_{i-k},
\end{aligned}$$

where the first equality follows by rearranging the summands and the second equality follows from lemma 2. Therefore we get

$$(xy)_k = \sum_{i=1}^{m} y_i x_{i-k} + \sum_{i=1}^{m} y_i x_{i+k} = \sum_{i=1}^{2m} y_i x_{i-k}.$$

Since $y_{2m+1} = 0$ by lemma 2, $(xy)_k = \sum_{i=1}^{2m+1} y_i x_{i-k}$ is obvious from above result. □

4. Bit serial arrangement using type II optimal normal basis

Using theorem 2, we may express $(xy)_k$ as a matrix multiplication form of a row vector and a column vector,

$$(xy)_k = (x_{1-k}, x_{2-k}, \cdots, x_{2m-k}, x_{2m+1-k})$$
$$\times (y_1, y_2, \cdots, y_{2m}, y_{2m+1})^T,$$

where $(y_1, y_2, \cdots, y_{2m}, y_{2m+1})^T$ is a transposition of the row vector $(y_1, y_2, \cdots, y_{2m}, y_{2m+1})$. Then we have

$$(xy)_{k+1} = (x_{-k}, x_{1-k}, \cdots, x_{2m-1-k}, x_{2m-k})$$
$$\times (y_1, y_2, \cdots, y_{2m}, y_{2m+1})^T.$$

Since $x_{-k} = x_{2m+1-k}$ by lemma 2, we find that $(x_{-k}, x_{1-k}, \cdots, x_{2m-1-k}, x_{2m-k})$ is a right cyclic shift of $(x_{1-k}, x_{2-k}, \cdots, x_{2m-k}, x_{2m+1-k})$ by one position. From this observation, we may realize the multiplication algorithm in the shift register arrangement shown in Fig. 1. The shift register is initially loaded with $(x_0, x_1, \cdots, x_{2m})$ which is in fact $(0, x_1, \cdots, x_m, x_m, \cdots, x_1)$. After k clock cycles, we get $(xy)_k$, the kth coefficient of xy with respect to the basis $\{\alpha_1, \cdots, \alpha_m\}$.

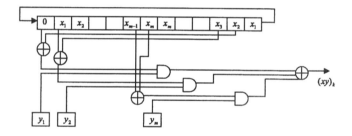

Figure 1. Bit serial multiplication using an optimal normal basis of type II.

Note that a bit parallel multiplier using an optimal normal basis of type II is discussed in [4], where it is suggested to find an efficient bit serial version of the construction in [4]. A hardware design of a bit serial multiplier using a type II ONB is proposed in a few papers [14,20,21]. The circuits in [14,20] and ours are basically the same design though the methods are slightly different. The construction in [21] is different from ours. The strong point of our approach is that our method can be applied to give a bit parallel systolic array which will be discussed in the next section and also a linear systolic array which follows directly from the bit parallel array via projection to vertical direction.

5. Bit parallel systolic architecture using an optimal normal basis of type II

A basis $\{\alpha_1, \alpha_2, \cdots, \alpha_m\}$ is used to derive an efficient bit serial multiplication of Berlekamp type in previous section. By rearranging the basis elements α_i, we may give a low latency and a low complexity bit parallel systolic multiplier. First, note that $\{\alpha_1, \alpha_2, \alpha_3, \cdots, \alpha_m\}$ and $\{\alpha_1, \alpha_3, \alpha_5, \cdots, \alpha_{2m-1}\}$ are same sets. That is, if m is odd, then $\alpha_1, \alpha_3, \cdots, \alpha_{2m-1}$ are $\alpha_1, \alpha_3, \cdots, \alpha_m, \alpha_{m+2} = \alpha_{m-1}, \alpha_{m+4} = \alpha_{m-3}, \cdots, \alpha_{2m-1} = \alpha_2$. If m is even, $\alpha_1, \alpha_3, \cdots, \alpha_{m-1}, \alpha_{m+1} = \alpha_m, \alpha_{m+3} = \alpha_{m-2}, \cdots, \alpha_{2m-1} = \alpha_2$. Thus $\{\alpha_1, \alpha_3, \alpha_5, \cdots, \alpha_{2m-1}\}$ is also a basis for $GF(2^m)$ and a basis conversion from $\{\alpha_1, \alpha_2, \alpha_3, \cdots, \alpha_m\}$ to $\{\alpha_1, \alpha_3, \alpha_5, \cdots, \alpha_{2m-1}\}$ is as obvious as shown above. Let $x = \sum_{i=1}^{m} x_i \alpha_i$ be an element of $GF(2^m)$. Recall that we defined $x_s \in GF(2)$ as $x_s = Tr(\alpha_s x)$ for any integer s in definition 4. Let $y = \sum_{i=1}^{m} y_i \alpha_i$ be another element in $GF(2^m)$. For each integer j, we define row vectors X_j and Y_j as

$$X_j = (x_j, x_{j+1}, \cdots, x_{j+2m}),$$

and

$$Y_j = (y_j, y_{j+1}, \cdots, y_{j+2m}).$$

Note that for a nonnegative integer s, X_{j+s} and Y_{j+s} are left cyclic shifts of X_j and Y_j by s positions, respectively. Also note that X_{j-s} and Y_{j-s} are right cyclic shifts of X_j and Y_j by s positions. Moreover for any j and k, by using theorem 2, we get the following expression.

$$(xy)_k$$
$$= \sum_{i=1}^{2m+1} y_i x_{i-k} = X_{1-k} Y_1^T$$
$$= (x_{1-k}, x_{2-k}, \cdots, x_{2m+1-k})(y_1, y_2, \cdots, y_{2m+1})^T$$
$$= (x_{j-k}, x_{j+1-k}, \cdots, x_{j+2m-k})(y_j, y_{j+1}, \cdots, y_{j+2m})^T$$
$$= X_{j-k} Y_j^T.$$

Theorem 3. *Let $x = \sum_{i=1}^{m} x_i \alpha_i$ and $y = \sum_{i=1}^{m} y_i \alpha_i$ be elements of $GF(2^m)$. Then we have*

$$(xy)_{2k-1} = \sum_{i=1}^{m}(y_{i+k-1}x_{i-k} + y_{i-k}x_{i+k-1}) + y_{m+1-k}x_{m+1-k}.$$

Proof. By the remark just before the statement of this theorem,

$$(xy)_{2k-1}$$
$$= X_{k-(2k-1)} Y_k^T = X_{1-k} Y_k^T$$
$$= (x_{1-k}, x_{2-k}, \cdots, x_{2m+1-k})(y_k, y_{k+1}, \cdots, y_{k+2m})^T$$
$$= \sum_{i=1}^{2m+1} y_{i+k-1} x_{i-k}$$
$$= \sum_{i=1}^{m} y_{i+k-1} x_{i-k} + y_{m+k} x_{m+1-k} + \sum_{i=m+2}^{2m+1} y_{i+k-1} x_{i-k}$$
$$= \sum_{i=1}^{m} y_{i+k-1} x_{i-k} + y_{m+1-k} x_{m+1-k} + \sum_{i=m+2}^{2m+1} y_{i+k-1} x_{i-k}.$$

On the other hand, the second summation of above expression can be written as

$$\sum_{i=m+2}^{2m+1} y_{i+k-1} x_{i-k} = \sum_{i=1}^{m} y_{2m+2-i+k-1} x_{2m+2-i-k}$$
$$= \sum_{i=1}^{m} y_{i-k} x_{i+k-1},$$

where the first equality follows by rearranging the order of summation and the second equality follows from lemma 2. Therefore we have

$$(xy)_{2k-1} = \sum_{i=1}^{m} y_{i+k-1} x_{i-k} + y_{m+1-k} x_{m+1-k}$$
$$+ \sum_{i=1}^{m} y_{i-k} x_{i+k-1},$$

which is the desired result. $\square$

Now for each $(xy)_{2k-1}$, we define a column vector

$$\mathcal{W}_k = (w_{1k}, w_{2k}, \cdots, w_{mk}, w_{(m+1)k})^T,$$

where

$$w_{ik} = y_{i+k-1} x_{i-k} + y_{i-k} x_{i+k-1}, \text{ if } 1 \leq i \leq m$$
$$w_{(m+1)k} = y_{m+1-k} x_{m+1-k}, \text{ if } i = m+1.$$

Then the sum of all entries of the column vector $\mathcal{W}_k$ is exactly $(xy)_{2k-1}$ and $\mathcal{W}_k$ appears as a kth column vector of the $m+1$ by m matrix $\mathcal{W} = (w_{ik})$ where

$$\mathcal{W} = \begin{pmatrix} w_{11} & w_{12} & w_{13} & \cdots & w_{1m} \\ w_{21} & w_{22} & w_{23} & \cdots & w_{2m} \\ w_{31} & w_{32} & w_{33} & \cdots & w_{3m} \\ \cdot & \cdot & \cdot & & \cdot \\ \cdot & \cdot & \cdot & & \cdot \\ \cdot & \cdot & \cdot & & \cdot \\ w_{m1} & w_{m2} & w_{m3} & \cdots & w_{mm} \\ w_{(m+1)1} & w_{(m+1)2} & w_{(m+1)3} & \cdots & w_{(m+1)m} \end{pmatrix}$$

For each $1 \leq i, k \leq m$, using the relation

$$w_{ik} = y_{i+k-1} x_{i-k} + y_{i-k} x_{i+k-1},$$

we have

$$w_{(i-1)(k-1)} = y_{i+k-3}x_{i-k} + y_{i-k}x_{i+k-3}.$$

That is, the signals x_{i-k} and y_{i-k} in the expression of w_{ik} come from the signals in the expression of $w_{(i-1)(k-1)}$. Also since

$$w_{(i-1)(k+1)} = y_{i+k-1}x_{i-k-2} + y_{i-k-2}x_{i+k-1},$$

we deduce that the signals x_{i+k-1} and y_{i+k-1} in the expression of w_{ik} come from the signals in the expression of $w_{(i-1)(k+1)}$. Moreover the signals in the last row come from the signals in the mth row. That is, $w_{(m+1)1} = y_m x_m$ comes from the signals y_m and x_m in the expression $w_{m1} = y_m x_{m-1} + y_{m-1} x_m$. And for each $2 \leq k \leq m$, $w_{(m+1)k} = y_{m+1-k}x_{m+1-k}$ comes from the signals y_{m+1-k} and x_{m+1-k} in the expression $w_{m(k-1)} = y_{m+k-2}x_{m+1-k} + y_{m+1-k}x_{m+k-2}$. From this observation, we may construct a bit parallel systolic multiplier with respect to the basis $\{\alpha_1, \alpha_3, \cdots, \alpha_{2m-1}\}$. The circuit of basic cell is explained in Fig. 2, where • is one bit latch (flip-flop). An output of the vertical line produces partial sum of the product.

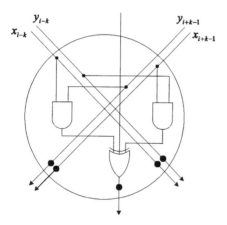

Figure 2. The circuit of (i,k) basic cell.

For convenience, assume $m = 5$ where the existence of type II ONB is well known. Then the matrix $\mathcal{W}$ is as follows;

$$\mathcal{W} = \begin{pmatrix} w_{11} & w_{12} & w_{13} & w_{14} & w_{15} \\ w_{21} & w_{22} & w_{23} & w_{24} & w_{25} \\ w_{31} & w_{32} & w_{33} & w_{34} & w_{35} \\ w_{41} & w_{42} & w_{43} & w_{44} & w_{45} \\ w_{51} & w_{52} & w_{53} & w_{54} & w_{55} \\ y_5 x_5 & y_4 x_4 & y_3 x_3 & y_2 x_2 & y_1 x_1 \end{pmatrix}$$

where $w_{ik} = y_{i+k-1}x_{i-k} + y_{i-k}x_{i+k-1}$ for $1 \leq i,k \leq m$. Letting $z = \sum_{i=1}^{m} z_i \alpha_i$ be another element in $GF(2^m)$, we may realize the product sum operation $u = xy + z$ in a bit parallel systolic arrangement shown in Fig. 3. Note that $x_0 = 0 = y_0$ in the arrangement and the output at the kth column is u_{2k-1}. We compare our multiplier with other bit parallel systolic multipliers in Table 1. None of the multipliers in the table broadcasts signals and all have unidirectional data flows except for the multipliers in [8] and [11], which have bidirectional data flows. Since the multiplier in [9] uses a dual basis, one needs a basis conversion process.

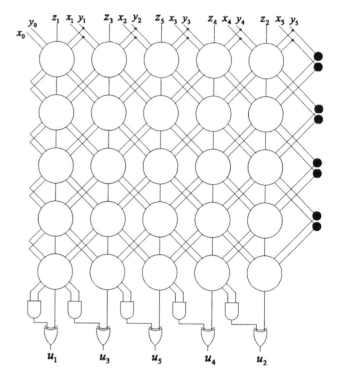

Figure 3. Systolic architecture for computing $u = xy + z$ in $GF(2^5)$.

Table 1. Comparison of our multiplier with other bit parallel systolic multipliers.

	Wang [7]	Yeh [8]	Fenn [9]	Wei [11]	Lee [10]	Fig. 3
basis	polynomial	polynomial	dual	polynomial	AOP	type II ONB
function	AB	$AB+C$	AB	AB^2+C	$AB+C$	$AB+C$
cell complexity						
AND	2	2	2	3	1	2
XOR	0	2	2	1	1	0
3XOR	1	0	2	1	0	1
Latch	7	7	7	10	3	5
number of cells	m^2	m^2	m^2	m^2	$(m+1)^2$	m^2
latency	$3m$	$3m$	$3m$	$3m$	$m+1$	$m+1$
critical path delay	$D_A+D_{3X}+D_L$	$D_A+D_X+D_L$	$D_A+D_X+D_L$	$D_A+D_{3X}+D_L$	$D_A+D_X+D_L$	$D_A+D_{3X}+D_L$

AND and XOR mean 2-input gates and 3XOR means a 3-input XOR gate. D_A, D_X, D_{3X} and D_L denote the delay time of an AND, a XOR, a 3XOR and a latch respectively. Note that the area complexity of a latch is higher than any other gate in the table.

Approximately, a latch takes 5 times more area than an AND gate and 2 times more area than a XOR gate.

It is shown in Table 1 that our multiplier has the best hardware complexity and latency except for the multiplier in [10]. The multiplier in [10] is applicable to a finite field $GF(2^m)$ when there is an all one polynomial (AOP) basis. A polynomial $1 + x + x^2 + \cdots + x^m \in GF(2)[x]$ is called an all one polynomial (AOP) of degree m. A finite field $GF(2^m)$ has an AOP basis whenever an AOP of degree m is irreducible over $GF(2)$. It is not difficult to show that an AOP basis exists in $GF(2^m)$ if and only if $m + 1 = p$ is a prime and 2 is a primitive root modulo p. A behavior of an AOP basis for moderately small values of m is well known. In fact, a table in [6, p. 100] shows that the number of $m \leq 2000$ for which an AOP basis exists is 118. For example, we have an AOP basis when $m = 2, 4, 10, 12, 18, 28, 36, 52, 58, 60, 66, 82, 100, 106, \cdots$. On the other hand, the same table says that the number of $m \leq 2000$ for which a type II optimal normal basis exists is 324. There is a type II optimal normal basis when $m = 2, 3, 5, 6, 9, 11, 14, 18, 23, 26, 29, 30, 33, 35, 39, 41, 50, 51, 53, 65, 69, 74, \cdots$. Therefore our multiplier in Fig. 3 is applicable to a broader class of m than the multiplier using an AOP basis. Moreover, since an AOP basis in [10] is a nonconventional basis having $m + 1$ basis elements in $GF(2^m)$, one needs extra logical operations to convert the basis to an ordinary basis. However, our multiplier has no such problem.

6. Conclusion

In this paper, we proposed a bit parallel systolic multiplier using an optimal normal basis (ONB) of type II. We showed, in Table 1, that our multiplier has a lower hardware complexity and a latency than corresponding multipliers in Table 1 except for the multiplier in [10], which uses an AOP basis. However, an AOP basis appears quite less frequently than an optimal normal basis of type II. A table in [6, p. 100] implies that a type II ONB is three times more likely to occur than an AOP basis. Therefore our bit parallel systolic multiplier provides an hardware efficient architecture for many finite fields, where an AOP basis does not exist and no other low complexity systolic architecture is known yet. Using the remark on gates areas in Table 1, we find that the hardware complexity of our multiplier is reduced by 25 percent from the ones in [7,8,9]. Also, our construction of Fig. 3 can be easily modified via projection to vertical direction to give a linear (one dimensional) systolic array. In this case, our linear systolic array has parallel in parallel out architecture and gives an output after m clock cycles.

References

[1] E.R. Berlekamp, "Bit-serial Reed-Solomon encoders," *IEEE Trans. Inform. Theory*, vol. 28, pp. 869–874, 1982.

[2] M. Wang and I.F. Blake, "Bit serial multiplication in finite fields," *SIAM J. Disc. Math.*, vol. 3, pp. 140–148, 1990.

[3] S. Gao, J. von zur Gathen and D. Panario, "Gauss periods and fast exponentiation in finite fields," *Lecture Notes in Computer Science*, vol. 911, pp. 311–322, 1995.

[4] B. Sunar and Ç.K. Koç, "An efficient optimal normal basis type II multiplier," *IEEE Trans. Computers*, vol 50, pp. 83–87, 2001.

[5] A. Reyhani-Masoleh and M.A. Hasan, "A new construction of Massey-Omura parallel multiplier over $GF(2^m)$," *IEEE Trans. Computers*, vol. 51, pp. 511–520, 2002.

[6] A.J. Menezes, *Applications of finite fields*, Kluwer Academic Publisher, 1993.

[7] C.L. Wang and J.L. Lin, "Systolic array implementation of multipliers for finite fields $GF(2^m)$," *IEEE Trans. Circuits Syst.*, vol. 38, pp. 796–800, 1991.

[8] C.S. Yeh, I.S. Reed and T.K. Troung, "Systolic multipliers for finite fields $GF(2^m)$," *IEEE Trans. Computers*, vol. C-33, pp. 357–360, 1984.

[9] S.T.J. Fenn, M. Benaissa and D. Taylor, "Dual basis systolic multipliers for $GF(2^m)$," *IEE Proc. Comput. Digit. Tech.*, vol. 144, pp. 43–46, 1997.

[10] C.Y. Lee, E.H. Lu and J.Y. Lee, "Bit parallel systolic multipliers for $GF(2^m)$ fields defined by all one and equally spaced polynomials," *IEEE Trans. Computers*, vol. 50, pp. 385–393, 2001.

[11] C.W. Wei, "A systolic power sum circuit for $GF(2^m)$," *IEEE Trans. Computers*, vol. 43, pp. 226–229, 1994.

[12] S.K. Jain, L. Song and K.K. Parhi, "Efficient semisystolic architectures for finite field arithmetic," *IEEE Trans. VLSI Syst.*, vol. 6, pp. 101–113, 1998.

[13] J.H. Guo and C.L. Wang, "Systolic array implementation of Euclid's algorithm for inversion and division in $GF(2^m)$," *IEEE Trans. Computers*, vol. 47, pp. 1161–1167, 1998.

[14] S. Kwon and H. Ryu "Efficient bit serial multiplication using optimal normal bases of type II in $GF(2^m)$," *Lecture Notes in Computer Science*, vol. 2433, pp. 300–308, 2002.

[15] C.Y. Lee, E.H. Lu and L.F. Sun, "Low complexity bit parallel systolic architecture for computing $AB^2 + C$ in a class of finite field $GF(2^m)$," *IEEE Trans. Circuits Syst. II*, vol. 48, pp. 519–523, 2001.

[16] W.C. Tsai, C.B Shung and S.J. Wang, "Two systolic architectures for modular multiplication," *IEEE Trans. VLSI Syst.*, vol. 8, pp. 103–107, 2000.

[17] T. Itoh and S. Tsujii, "Structure of parallel multipliers for a class of finite fields $GF(2^m)$," *Information and computation*, vol. 83, pp. 21–40, 1989.

[18] Ç.K. Koç and B. Sunar, "Low complexity bit parallel canonical and normal basis multipliers for a class of finite fields," *IEEE Trans. Computers*, vol. 47, pp. 353–356, 1998.

[19] C. Paar, P. Fleischmann and P. Roelse, "Efficient multiplier archtectures for Galois fields $GF(2^{4n})$," *IEEE Trans. Computers*, vol. 47, pp. 162–170, 1998.

[20] H. Wu, M.A. Hasan, I.F. Blake and S. Gao, "Finite field multiplier using redundant representation," *IEEE Trans. Computers*, vol. 51, pp. 1306–1316, 2002.

[21] G.B. Agnew, R.C. Mullin, I. Onyszchuk and S.A. Vanstone, "An implementation for a fast public key cryptosystem," *J. Cryptology*, vol. 3, pp. 63–79, 1991.

Session 9:
Powering, Multiplication, and Counters

Chair: Alexandre Tenca

High–Radix Iterative Algorithm for Powering Computation

J.-A. Piñeiro[§], M. D. Ercegovac[¶], J. D. Bruguera[§]

[§] Department of Electronic and Computer Engineering
Univ. Santiago de Compostela, Spain.
[¶] Computer Science Dept.
University of California, Los Angeles (UCLA), USA.
e-mail: alex,bruguera@dec.usc.es, milos@cs.ucla.edu

Abstract

A high-radix composite algorithm for the computation of the powering function (X^Y) is presented in this paper. The algorithm consists of a sequence of overlapped operations: (i) digit-recurrence logarithm, (ii) left-to-right carry-free (LRCF) multiplications, and (iii) on-line exponential. A redundant number system is used, and the selection in (i) and (iii) is done by rounding except from the first iteration, when selection by table look-up is necessary to guarantee the convergence of the recurrences. A sequential implementation of the algorithm is proposed, and the execution times and hardware requirements are estimated for single and double-precision floating-point computations, for radix $r = 128$, showing that powering can be computed with similar performance as high-radix CORDIC algorithms.

1 Introduction

Powering (X^Y) is an important operation in applications such as scientific computing, digital signal processing (DSP) and computer 3D graphics [7]. As other elementary functions, such as square root, inverse square root, logarithm and trigonometric functions, it has been traditionally computed by software routines [2, 6]. These routines provide very accurate results, but they are often too slow for numerically intensive or real-time applications. The timing constraints of these applications have led to the development of dedicated hardware for the computation of elementary functions, including the implementation of table-based algorithms [9, 15], functional iteration methods [10, 11] and digit-recurrence algorithms [1, 4].

However, accurately computing the floating-point powering function is considered difficult [9], and the prohibitive hardware requirements of a table-based implementation (note that X^Y is a 2-variable function) have led only to partial solutions, such as powering algorithms for a constant exponent [12, 16] or for very low precision [7]. A direct implementation of a digit-recurrence algorithm for powering computation is not feasible due to its high intrinsecal complexity.

In this paper we present a composite iterative algorithm for the computation of the powering function (X^Y), for a floating-point input operand $X = M_x 2^{E_x}$ and integer b-bit operand[1] Y. The final result X^Y is computed as $Z = M_z 2^{E_z} = e^{Y \ln(M_x)} 2^{Y E_x}$, through a sequence of overlapped operations. The first step consists of computing $\ln(M_x)$ by using a high-radix recurrence with selection by rounding [13]. An intermediate computation $Y_L \ln(M_x)$, with $Y_L = Y \log_2(e)$, is carried out using a high-radix left-to-right carry-free (LRCF) multiplication [3]. Another LRCF multiplication by $\ln 2$ is performed to guarantee the convergence of the algorithm and, as last step, the exponential of the resulting product is computed by an on-line high-radix ($r = 2^b$) algorithm, with on-line delay $\delta = 2$ [14].

In the stages computing the logarithm and the exponential, selection by table look-up is performed in the first iteration to guarantee the convergence of both algorithms. However, the on-line delay of 2 cycles allows the addressing of the initial tables in the exponential scheme one cycle in advance, reducing the delay of the critical path in this stage.

A sequential architecture implementing our algorithm with radix $r = 128$ is proposed in this paper, and estimates of the total area and execution time for single and double-precision floating-point computations are obtained according to an approximate model for the area and delay of the main components used in the proposed architecture.

---
[1]The computation of the powering operation with integer exponent in the range $[1, 128]$ is useful for instance in computer graphics computations (the lighting specular component in the OpenGL graphics pipeline [7]).

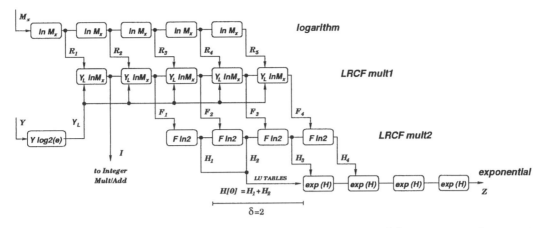

Figure 1. Operation flow of the powering algorithm (single-precision computation, $r = 128$)

2 Algorithm

In this section we present a new composite iterative algorithm for the computation of the powering function (X^Y), for a floating point operand X and an integer exponent Y, describing the algorithm and its error analysis.

2.1 Overview

The algorithm is based on the well-known identity

$$X^Y = e^{Y \ln(X)}$$

Considering a floating-point[2] input operand $X = M_x 2^{E_x}$, with M_x the n-bit significand and E_x the exponent, and an integer b-bit input operand Y:

$$\begin{aligned} X^Y &= e^{Y \ln(M_x 2^{E_x})} = e^{Y(\ln(M_x) + \ln(2^{E_x}))} \\ &= e^{Y \ln(M_x)} e^{Y \ln(2^{E_x})} = e^{Y \ln(M_x)} e^{\ln(2^{Y E_x})} \\ &= e^{Y \ln(M_x)} 2^{Y E_x} \end{aligned} \quad (1)$$

According to (1), the powering function can be calculated by a sequence of operations consisting of the logarithm of the significand M_x, a multiplication by Y and the exponential of the resulting product. The form of the result is a significand $e^{Y \ln(M_x)}$ and an exponent $Y E_x$, typical of a floating-point operand[3].

For an efficient implementation of the powering function, the computation of the operations involved must be overlapped, which requires a left-to-right most-significant digit first (MSDF) mode of operation and the use of a redundant number system.

A problem of the proposed algorithm is the range of the digit-recurrence exponential, which is $(-\ln 2, \ln 2)$, while

---

[2]The algorithm can be adapted to a normalized fixed-point input operand $X = M_x$, $1 \leq M_x < 2$, resulting in $X^Y = e^{Y \ln(M_x)}$.

[3]For fixed-point computations, the value $e^{Y \ln(M_x)}$ can be shifted $Y E_x$ positions.

the argument of the exponential here is $Y \ln(M_x)$, being Y an integer. To extend the range of convergence of this algorithm, and thus guarantee the convergence of the overall proposed method, we use the following mathematical identity [8]:

$$e^\beta = e^{\beta \log_2(e) \ln 2}$$

If $\beta \log_2(e)$ is computed, and its integer I and fractional part F are extracted, e^β becomes:

$$\begin{aligned} e^\beta &= e^{(I+F) \ln 2} = e^{I \ln 2 + F \ln 2} \\ &= e^{\ln(2^I)} e^{F \ln 2} = 2^I e^{F \ln 2} \end{aligned}$$

Since in this case $\beta = Y \ln(M_x)$:

$$e^{Y \ln(M_x)} = 2^I e^{F \ln 2},$$

and therefore, according to (1),

$$X^Y = e^{F \ln 2} 2^{(I + Y E_x)}, \quad (2)$$

with I and F the integer and the fractional part of $Y \ln(M_x) \log_2(e)$, resulting in a bounded argument for the exponential. The product $Y_L = Y \log_2(e)$ can be computed by using a $(b \times n)$-bit multiplier in parallel with the computation of the first iteration of the logarithm.

In summary, as illustrated in Figure 1 for single-precision computations with $r = 128$, our algorithm for the computation of the powering function consists of the following steps:

1. Computing the logarithm of the input significand, $\ln(M_x)$, by employing a high-radix digit-recurrence algorithm [13].

2. Computing the product $Y_L \ln(M_x)$ by using a a serial left-to-right carry-free (LRCF) multiplication scheme [3], with $Y_L = Y \log_2(e)$ (obtained in parallel with the first iteration of the logarithm). The integer I and the fractional F parts of $Y_L \ln(M_x)$ are extracted serially.

Precision	N_e	N_l	n_e	n_l	n_{m0}	n_{m1}	n_{m2}
SP ($n = 24$)	4	5	27	35	27	33	34
DP ($n = 53$)	9	10	57	65	57	64	64

Table 1. Parameters for powering computation ($r = 128$)

3. Computing the product $H = F \ln 2$ by using a LRCF multiplication scheme [3].

4. Computing the exponential $M_z = e^{F \ln 2}$, by employing an on-line high-radix algorithm [14].

The exponent E_z can be computed in parallel by an integer multiply/add unit. The output of the algorithm is the floating-point normalized result:

$$X^Y = M_z 2^{E_z} = e^{F \ln 2} 2^{(I+YE_x)} \quad (3)$$

The use of redundancy results in $e^{F \ln 2} \in (0.5, 2)$, and therefore a normalization of the final result may be necessary. However, the condition $F \ln 2 < 0$ can be determined in advance to the last iterations of the exponential, and the final normalization can be performed with no extra delay.

The overall latency of the algorithm, as shown in Figure 1, can be estimated as

$$latency = N_e + (\delta + 1) + 3, \quad (4)$$

with N_e the number of iterations of the on-line exponential, $(\delta + 1)$ cycles to accommodate the on-line delay, 2 cycles due to the two LRCF multiplications, and an extra cycle due to the computation of the integer part I of $Y_L \ln(M_x)$. I can be obtained in a single cycle because it is guaranteed to have b bits, since Y is a b-bit integer and $r = 2^b$.

2.2 Error analysis

The final error in the algorithm consists of the accumulation of the errors due to the cascaded implementation of a set of operations, and must be bounded by 2^{-n}.

Let ϵ_l be the error in the computation of the logarithm of the input significand ($\ln(M_x) + \epsilon_l$), and ϵ_{m0} the error due to the finite arithmetic in the computation of the product Y_L ($Y \log_2(e) + \epsilon_{m0}$). When the first LRCF multiplication is performed, the associated error is:

$$\epsilon_{LRCF1} = \epsilon_l Y \log_2(e) + \epsilon_{m0} \ln(M_x) + \epsilon_l \epsilon_{m0} + \epsilon_{m1},$$

with ϵ_{m1} the error due to the LRCF multiplication scheme in the computation of the product.

The integer I and fractional F parts of the product $Y \log_2(e) \ln(M_x)$ are extracted serially, with no error affecting the integer part. The next operation to be performed

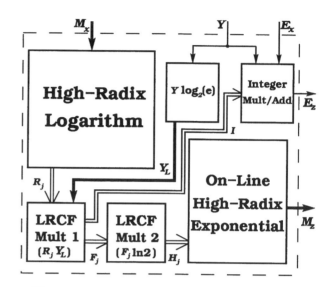

Figure 2. Block diagram of the proposed architecture

is the computation of the product $H = F \ln 2$. The error associated with this computation is:

$$\epsilon_H = \epsilon_{LRCF1} \ln 2 + \epsilon_{m2},$$

with ϵ_{m2} the error due to the LRCF multiplication scheme in the computation of the product.

Finally, the exponential e^H is computed, and therefore the obtained result is:

$$e^H e^{\epsilon_{LRCF1} \ln 2 + \epsilon_{m2}} + \epsilon_e,$$

with ϵ_e the error in the computation of the exponential.

The difference between the exact result and the obtained result must be bounded:

$$|e^H - e^H e^{\epsilon_{LRCF1} \ln 2 + \epsilon_{m2}} - \epsilon_e| < 2^{-n}$$

Taking into account that $e^H \in (0.5, 2)$:

$$|1 - e^{(\epsilon_l Y \log_2(e) + \epsilon_{m0} \ln(M_x) + \epsilon_l \epsilon_{m0} + \epsilon_{m1}) \ln 2 + \epsilon_{m2}} - \epsilon_e| < 2^{-n-1}$$

For a precision of n bits and a radix $r = 2^b$, a set of minimum values for ϵ_l, ϵ_{m0}, ϵ_{m1}, ϵ_{m2} and ϵ_e must be determined to guarantee a final result accurate to n bits. The values of the error parameters set the precision to be reached in each stage, n_l, n_e, n_{m1}, n_{m2}, and the accuracy n_{m0} of the product $Y \log_2(e)$. These parameters determine a minimum number of iterations of the logarithm (N_l) and the exponential (N_e) to be performed. As shown in Figure 1, the number of iterations of the LRCF multiplications to be performed is the same as those of the logarithm, N_l, because all the information must reach the exponential stage. The parameters N_l, N_e, n_l, n_e, n_{m1}, n_{m2} and n_{m0} set the size of the look-up tables, adders and multipliers to be used.

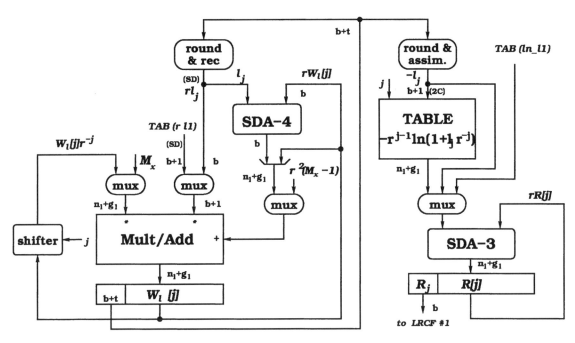

Figure 3. Block diagram of the high-radix logarithm stage

Moreover, g_l, g_e, g_{m1} and g_{m2} guard bits must be employed to guarantee that in each stage of the powering algorithm the iteration errors do not affect the achievement of the required precisions n_l, n_e, n_{m1} and n_{m2}.

The critical parameter to be first minimized is ϵ_e, since it is directly related to the required precision n_e to be reached in the exponential stage, and therefore to N_e, the number of iterations of the on-line exponential to be performed[4].

For radix $r = 128$, the set of minimum values for the considered parameters is shown in Table 1 for single-precision ($n = 24$) and double-precision ($n = 53$) computations. N_e and N_l are given in cycles, and n_e, n_l, n_{m0}, n_{m1} and n_{m2} are given in number of bits.

3 Implementation

In this section a sequential architecture is proposed for the implementation of our high-radix powering algorithm. We give a detailed explanation of the main computations involved: (i) high-radix logarithm, (ii) high-radix LRCF multiplication, and (iii) on-line high-radix exponential, and then outline the main features of the logic blocks employed. Figure 2 shows the block diagram of the proposed architecture. Single thick lines denote long-word (around n bits) operands/variables in parallel form, single thin lines denote short-word (up to 11 bits) operands/variables in parallel form, and double lines denote single-digit (b bits) variables (R_j, I, Fj and H_j).

---

[4]Note that N_e is the only parameter that affects the latency of the algorithm, since $\delta = 2$ for any n and any $r \geq 8$, as shown in [14].

3.1 High-Radix Logarithm

A high-radix digit-recurrence logarithm is described in detail in [13], although in the algorithm used here some modifications and optimizations have been made according to the operation flow in the powering computation. A slightly different notation from the used in [13] is employed here. The block diagram of the high-radix logarithm stage is shown in Figure 3.

The recurrences for performing the multiplicative normalization of the input operand M_x and computing the logarithm digits are

$$\begin{aligned} W_l[j+1] &= r(W_l[j] + l_j + l_j W_l[j] r^{-j}) \\ R[j+1] &= rR[j] - r^{j-1}\ln(1 + l_j r^{-j}) - rR_j \end{aligned} \quad (5)$$

with $j \geq 1$, $W_l[1] = r(M_x - 1)$, $R[1] = 0$ and $R_1 = 0$. For a result precision of n_l bits, a total number of $N_l = \lceil n_l/b \rceil$ iterations are necessary. The scaled recurrence $R[j]$ has been defined as

$$R[j] = r^{j-2} L[j] \quad (6)$$

in order to extract a radix-r digit R_j per iteration from the same bit-positions in all iterations.

The selection of the digits l_j in iterations $j \geq 2$ is done by rounding an estimate of the residual obtained by truncating $W_l[j]$ to t fractional bits. The selection function is

$$l_j = -round(\hat{W}_l[j]) \quad (7)$$

The sign of the digit l_j is defined as opposite of the sign of $W_l[j]$ in order to satisfy a bound on the residual, and thus assuring the convergence. The digit set for the coefficients l_j is $\{-(r-1), \ldots, -1, 0, 1, \ldots, (r-1)\}$.

Iteration $j = 1$ does not converge with selection by rounding, and therefore the selection of l_1 is performed by table look-up. This table is addressed by the $b + 1$ most significant bits of the input operand M_x, and the selection is done in such a way that the value of $|l_2|$ is bounded according to the convergence conditions [13]. However, this results in an over-redundant digit l_1 ($b + 1$ bits), increasing by one bit the size of the multiplier operand. The convergence conditions also determine a minimum value of $t = 2$ and the radix $r \geq 8$.

A multiply/add unit is used for the computation of the residual recurrence, unlike in the algorithm proposed in [13], where a separated SD multiplier and SDA4 were used. The digit l_1 is stored in the look-up table $TAB(rl_1)$ already in SD-4 recoded form, to reduce the delay of the path containing the table and the multiply/add unit.

The logarithm constants are stored in a look-up table whose size grows exponentially with the radix. However, an approximation $-l_j r^{-1}$ can be used in iterations $j \geq \lceil N_l/2 \rceil + 1$ in order to reduce the overall hardware requirements of the algorithm.

The use of redundant representation is mandatory in our algorithm for powering computation, due to the left-to-right operation flow, and results in faster execution times by making the additions independent of the precision.

3.2 LRCF Multiplication

The left-to-right carry-free (LRCF) multiplication, introduced in [3], produces the product digits from a redundant set in a most-significant-digit-first (MSDF) manner and performs the conversion on-the-fly of the product generated in a redundant form to the conventional form without using a carry-propagate adder and without additional delay. The resulting implementation is fast and regular and is very well suited for VLSI implementations.

We adapt the LRCF multiplication to carry out the intermediate multiplications in the high-radix powering, utilizing its left-to-right operation flow, with redundant representation of the operands and recoding to a high-radix of the multiplier operand. However, since the high-radix exponential is of the on-line type, the digits produced by LRCF multiplier are used without conversion. The block diagram of the two LRCF multiplication stages is shown in Figure 4.

The adapted LRCF algorithm has two operands A and D. A is the radix-2 representation of the multiplicand a, such that $a = \sum_{i=1}^{n_m} A_i 2^{-i}$, with $A_i \in \{0, 1\}$. D the recoded radix-r representation of the multiplier d, such that $d = \sum_{i=1}^{n_m/b} D_i r^{-i}$, with $D_i \in \{-r/2, \ldots, r/2\}$ and $r = 2^b$.

The recurrence produces a sequence of two accumulated

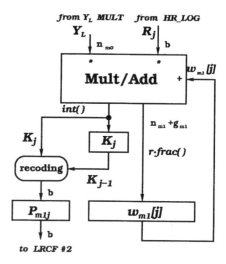

(a) LRCF multiplication 1

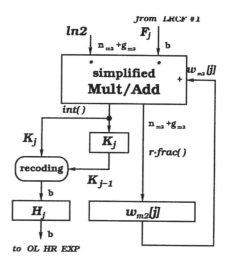

(b) LRCF multiplication 2

Figure 4. Block diagram of the LRCF multiplication stages

products w and p as follows:

$$\begin{aligned} w[j+1] &= r(fraction(w[j] + aD_{j+1})) \\ K_{j+1} &= integer(w[j] + aD_{j+1}) \\ p[j+1] &= p[j] + K_{j+1}r^{-j-1} \end{aligned} \quad (8)$$

with $j = 1, \ldots, n_m/b$ and initial values $w[0] = p[0] = 0$. The digits of the multiplier are used from most to least significant, unlike conventional multiplication schemes which use a right-to-left mode. After n_m/b steps, $p[n_m/b]$ is the most significant part of the product while $w[n_m/b]$ is the least significant part.

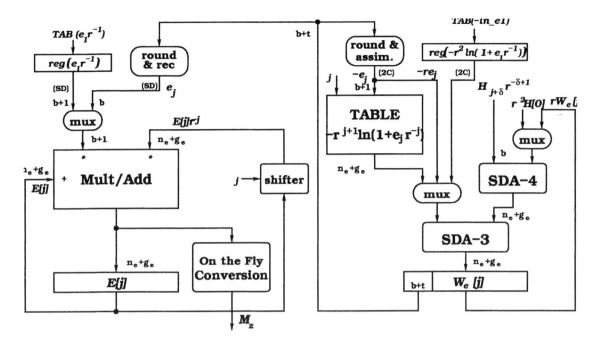

Figure 5. Block diagram of the on-line high-radix exponential stage.

A fast implementation of the LRCF multiplication scheme requires: (i) use of a redundant adder (either CS or SD) to compute the residual $w[j]$, and (ii) since the maximum value of K_j in (8) is in general larger[5] than $(r-1)$, a recoding of K_j and K_{j+1} into P_j in the range $[-(r-1),(r-1)]$ is necessary. That is, the resulting digit will be $P_j = f(K_j, K_{j+1})$.

Regarding the implementation proposed in [3], the main modification in the LRCF multiplication stages in the architecture for powering computation is the use of a multiply/add unit for computing the recurrences, instead of separated redundant multiplier and adder.

3.3 On-Line High-Radix Exponential

The on-line high-radix exponential is described in detail in [14], although a slightly different notation is used here. The block diagram of the on-line high-radix exponential stage is shown in Figure 5.

The exponential is computed on the basis of partial information:
$$H[0] = \sum_{j=1}^{\delta} H_j r^{-j}, \quad (9)$$

with δ the *on-line delay* and H_j the radix-r digits of the input operand H.

The recurrences for performing the additive normalization of H and computing the exponential are

$$\begin{aligned} W_e[j+1] &= r(W_e[j] - r^j \ln(1 + e_j r^{-j}) + H_{j+\delta} r^{-\delta}) \\ E[j+1] &= E[j](1 + e_j r^{-j}) \end{aligned} \quad (10)$$

with $j \geq 1$, $W_e[1] = rH[0]$ and $E[1] = 1$. For a result precision of n_e bits, a total number of $N_e = \lceil n_e/b \rceil$ iterations are necessary. An approximation $-e_j r$ to the logarithm constants can be used in iterations $j \geq \lceil N_e/2 \rceil + 1$ reducing the overall hardware requirements of the algorithm.

The selection of digits e_j is done in iterations $j \geq 2$ by rounding an estimate of the residual:
$$e_j = round(\hat{W}_e[j]), \quad (11)$$

with $e_j \in \{-(r-1), \ldots, -1, 0, 1, \ldots, (r-1)\}$ and $\hat{W}_e[j]$ obtained by truncating $W_e[j]$ to t fractional bits.

The use of selection by table look-up is required in the first iteration. The table is addressed by the b most significant bits of the input operand H_1, and the selection is performed so that $-(r-3) \leq e_2 \leq (r-2)$, which results in an overredundant first digit e_1. The convergence conditions also determine a minimum value of $t = 2$, an on-line delay $\delta = 2$, and a bound on the radix $r \geq 8$.

The look-up tables used in the first iteration are addressed one cycle in advance, since $H[0]$ is already known while $H_{\delta+1}$ is being computed, and the obtained values can be stored in the registers reg_e_1 and $reg\_\ln e_1$.

The conversion from redundant to conventional representation of the final result is performed using an on-the-fly method [4].

---
[5]The range of K_j is $|K_j| < 3r/2$.

Precision	latency	cycle time (τ)	exec. time (τ)	LOG area (fa)	EXP area (fa)	total area (fa)
SP ($n = 24$)	4+6=10	10.0	100.0	1886	1202	4138
DP ($n = 53$)	9+6=15	10.0	150.0	4496	3990	10131

Table 2. Execution time and total area for the proposed architecture ($r = 128$)

3.4 Implementation details

The main features of the proposed architecture are:

- All variables are in redundant representation (signed-digit) to allow faster execution of iterations, since the additions become independent of the precision.

- All products of the type $Q[j]r^{-j}$ and $q_j r^{-j}$ (with $q_j = l_j$ or e_j) are performed as shifts, since $r = 2^b$.

- SDAα is a *signed-digit binary adder* with α input bit-vectors. The addition of two SD operands requires a SDA4 adder, while SDA3 adders can be used for accumulating a SD operand and an operand in two's complement (2C) representation.

- An internal recoding from SD radix-2 to SD radix-4 representation of the multiplier operand is performed in the multiply/add units to reduce by half the number of partial products to be accumulated.

- The *round&rec* units take the $(b + t)$ most significant bits of the residuals $W_l[j]$ and $W_e[j]$ ($t = 2$ in both cases) and produce a SD-4 representation of the digits l_j and e_j, to be used as multipliers in the multiply/add units.

- The *round&assim* units take the same $(b + t)$ input words, but produce a two's complement representation of the coefficient with opposite sign ($-l_j$ or $-e_j$), by rounding the input word and performing its assimilation to non-redundant representation. This 2C representation is used for addressing the look-up tables storing the logarithm constants and as a input for the barrel shifters producing $-l_j r^{-1}$ or $-e_j r$.

4 Evaluation and comparison

In this section we present estimates of the execution time and the area costs of the proposed architecture, for single and double-precision floating-point computations ($n = 24$ and $n = 53$ bits) and radix $r = 128$. These estimates are based on an approximate model for the cost and delay of the main logic blocks used [5, 11, 13].

The actual delays and area cost depend on the technology used and on the actual implementation. However, this technology-independent model provides a good first-order approximation to the actual execution time and area values.

Table 2 shows the execution time and area cost estimates of the proposed architecture for single and double-precision FP computations, with $r = 128$. The units employed are the delay τ of a complex gate, such a full-adder, and the area of a 1-bit full-adder (fa).

The latency, as explained in (4), is $N_e + 6$ cycles, since $\delta = 2$ in this case, and therefore we have 10 and 15 cycles respectively for single and double-precision[6]. The cycle time of 10.0τ corresponds in both cases to the high-radix logarithm stage, where the critical path is the one composed of the initial table look-up for selecting the digit l_1, a multiplexer, the multiply/add unit and the register W_l. The total execution times can therefore be estimated as 100τ and 150τ, as shown in Table 2.

The total area of our architecture consists of the hardware requirements of the individual stages (high-radix logarithm, LRCF multiplications and on-line high-radix exponential, shown in Figures 3 to 5), plus the integer ($b \times n_{exp}$)-bit multiply/add unit to compute the exponent E_z, an extra ($b \times n_{m0}$)-bit multiplier required to compute $Y_L = Y \log_2(e)$, the control logic, and a b-bit CPA adder to assimilate H_1 into conventional representation before addressing the initial tables in the on-line high-radix exponential stage.

The total table size is $46Kbits$ ($5.75KB$) and $161Kbits$ ($20KB$), respectively, for single and double-precision computations, with a contribution to the total area, according to the model used, of $1673 fa$ and $5660 fa$. The total area estimates, as shown in Table 2, are $4138 fa$ and $10131 fa$.

Comparison with high-radix CORDIC

For the sake of reference, we give now execution time and area estimates for a high-radix vectoring CORDIC implementation [1]. This algorithm has been chosen as a reference because it allows the computation of two-variable functions (the modulus of a 2D vector). To allow for a fair comparison, the same assumptions are made for both algorithms, a radix-128 implementation is considered for the CORDIC[7], and we consider a fixed-point implementation of our powering algorithm for $n = 32$.

Table 3 shows the latency, cycle time, execution time and total area for our powering algorithm and the radix $r = 128$ CORDIC implementation. When computing the modulus, CORDIC [1] requires 5 microrotations, 2 scaling operations

---

[6]Note that $N_e = 4$ and 9 for SP and DP, as shown in Table 1.
[7]The algorithm proposed in [1] performs $n = 32$-bit computations, using radix $r = 512$, with a small radix $R = 32$ only for the first CORDIC microrotation, and two scaling operations to guarantee the convergence.

Scheme	latency	cycle time	exec. time	area(fa)
powering (r=128)	11	10.0τ	110τ	5400
CORDIC (r=128)	12	10.0τ	120τ	6000

Table 3. Comparison with high-radix CORDIC vectoring algorithm ($n = 32$ bits)

(taking 3 cycles due to the complexity of the second scaling), and 5 extra iterations for compensating the scaling factor. This results in a latency of 12 cycles ($5 + 3 + 4$, since the first extra iteration can be overlapped with the last microrotation). According to the approximate model used, the critical path has a delay of 10τ, and therefore the execution time can be estimated as 120τ. The total area of this implementation is about $6000 fa$, with a table size of $142 Kb$.

We can conclude that with our algorithm, the powering function can be computed efficiently in dedicated hardware with cost and performance comparable to those of the evaluation of other elementary functions.

5 Summary

A new composite algorithm for the computation of the powering function has been presented in this paper. The algorithm consists of computing X^Y as $e^{Y \ln(M_x)} 2^{Y E_x}$, through a sequence of overlapped operations: logarithm, multiplication and exponential.

The computation of the logarithm is carried out by a high-radix digit-recurrence unit, with selection by rounding. The intermediate multiplications are computed by high-radix left-to-right carry-free (LRCF) multipliers. Finally, the exponential is computed by an on-line high-radix unit with selection by rounding. The computation of the exponent of the result is carried out in parallel by an integer multiply/add unit.

A sequential architecture implementing our algorithm has been proposed. Estimates of the execution times and hardware requirements have been obtained, based on an approximate model for the delay and the area of the main components, for single and double-precision floating-point computations with radix $r = 128$.

A comparison with a high-radix CORDIC architecture shows that the execution times and area costs of the unit computing the powering are similar to those of other iterative algorithms for the computation of elementary functions.

Acknowledgement

This work was developed while J.-A. Piñeiro was with the University of California, Los Angeles (UCLA), USA. J.-A. Piñeiro and J. D. Bruguera were partially supported by the Spanish Ministry of Science and Technology (MCyT–FEDER) under contract TIC2001-3694-C02.

References

[1] E. Antelo, T. Lang, and J. D. Bruguera. Very-High Radix CORDIC Vectoring with Scalings and Selection by Rounding. In *Proc. 14th Symp. Computer Arithmetic*, pages 204–213, April 1999.

[2] W. Cody and W. Waite. *Software Manual for the Elementary Functions*. Prentice-Hall, 1980.

[3] M. D. Ercegovac and T. Lang. Fast Multiplication Without Carry Propagate Addition. Technical report, Computer Science Dept., UCLA, 1987.

[4] M. D. Ercegovac and T. Lang. *Division and Square Root: Digit Recurrence Algorithms and Implementations*. Kluwer Academic Publishers, 1994.

[5] M. D. Ercegovac, T. Lang, J.-M. Muller, and A. Tisserand. Reciprocation, Square Root, Inverse Square Root, and Some Elementary Functions Using Small Multipliers. *IEEE Trans. on Computers*, 49(7):628–637, 2000.

[6] J. F. Hart et al. *Computer Approximations*. New York: John Wiley and Songs, 1968.

[7] D. Harris. A Powering Unit for an OpenGL Lighting Engine. In *Proc. 35th Asilomar Conference on Signals, Systems, and Computers*, pages 1641–1645, 2001.

[8] E. G. Kogbetliantz. Computation of e^N for $-\infty < n < +\infty$ Using an Electronic Computer. *IBM Journal Res. and Develop*, pages 110–115, 1957.

[9] J.-M. Muller. *Elementary Functions. Algorithms and Implementation*. Birkhauser, 1997.

[10] S. F. Oberman. Floating Point Division and Square Root Algorithms and Implementation in the AMD-K7 Microprocessor. In *Proc. 14th Symp. Computer Arithmetic (ARITH14)*, pages 106–115, April 1999.

[11] J.-A. Piñeiro and J. D. Bruguera. High-Speed Double-Precision Computation of Reciprocal, Division, Square Root and Inverse Square Root. *IEEE Transactions on Computers*, 51(12):1377–1388, 2002.

[12] J.-A. Piñeiro, J. D. Bruguera, and J.-M. Muller. Faithful Powering Computation using Table Look-up and Fused Accumulation Tree. In *Proc. 15th Intl. Symposium on Computer Arithmetic (ARITH15)*, pages 40–47, 2001.

[13] J.-A. Piñeiro, M. D. Ercegovac, and J. D. Bruguera. High-Radix Logarithm with Selection by Rounding. In *Proc. IEEE ASAP'02 Conference*, pages 101–110, 2002.

[14] J.-A. Piñeiro, M. D. Ercegovac, and J. D. Bruguera. On-Line High-Radix Exponential with Selection by Rounding. In *Proc. IEEE ISCAS'2003 Conference*, 2003.

[15] M. J. Schulte and J. E. Stine. Approximating elementary functions with symmetric bipartite tables. *IEEE Transactions on Computers*, 48(8):842–847, 1999.

[16] N. Takagi. Powering by a Table Look–up and a Multiplication with Operand Modification. *IEEE Trans. Computers*, 47(11):1216–1222, 1998.

On-line multiplication in real and complex base

Christiane Frougny
LIAFA, CNRS UMR 7089
2 place Jussieu, 75251 Paris Cedex 05, France
and Université Paris 8
Christiane.Frougny@liafa.jussieu.fr

Athasit Surarerks
Computer Engineering Department
Faculty of Engineering
Chulalongkorn University
Phayathai road, Bangkok 10330, Thailand
athasit@cp.eng.chula.ac.th

Abstract

Multiplication of two numbers represented in base β is shown to be computable by an on-line algorithm when β is a negative integer, a positive non-integer real number, or a complex number of the form $i\sqrt{r}$, where r is a positive integer.

1 Introduction

On-line arithmetic, introduced in [24], is a mode of computation where operands and results flow through arithmetic units in a digit serial manner, starting with the most significant digit. To generate the first digit of the result, the δ first digits of the operands are required. The integer δ is called the *delay* of the algorithm. This technique allows the pipelining of different operations such as addition, multiplication and division. It is also appropriate for the processing of real numbers having infinite expansions: it is well known that when multiplying two real numbers, only the left part of the result is significant. On-line arithmetic is used for special circuits such as in signal processing, and for very long precision arithmetic. One of the interests of on-line computable functions is that they are continuous for the usual topology on the set of infinite sequences on a finite digit set.

To be able to perform on-line computation, it is necessary to use a redundant number system, where a number may have more than one representation (see [24, 5]). An example of such a system is the so-called signed-digit number system. It is composed of an integer base $\beta \geq 2$ and a signed-digit set of the form $\{-a, \ldots, a\}$, with $\beta/2 \leq a \leq \beta - 1$. In this sytem addition can be performed in constant time in parallel [1, 4]. On-line multiplication is also feasible [24, 5]. Parallel addition is used internally in the multiplication algorithm.

On-line algorithms for addition, subtraction, multiplication, division and square-root in integer base are well studied, see [2, 6, 19].

In this paper we study the multiplication in base β when β is a negative integer, or a non-integer real number, or a complex number of the form $i\sqrt{r}$.

It is known that any real number can be represented in negative integer base without a sign [12, 13, 15], and that, with a signed-digit set, addition is computable in constant time in parallel, and is computable by an on-line finite state automaton [7]. We show that the on-line multiplication used in the signed-digit number system can be applied in negative base. Negative base is related to some complex number systems, see below.

When the base β is a real number > 1, by the greedy algorithm of Rényi [23], one can compute a representation in base β of any real number belonging to the interval $[0, 1]$, called its β-*expansion*, and where the digits are elements of the canonical digit set $A_\beta = \{0, \ldots, \lfloor\beta\rfloor\}$ if β is not an integer, or of $A_\beta = \{0, \ldots, \beta - 1\}$ if β is an integer. In such a representation, when β is not an integer, not all the patterns of digits are allowed (see [21] for instance). For instance in base $\beta = (1 + \sqrt{5})/2$ the golden ratio, the β-expansion of the number $x = 3 - \sqrt{5}$ is $1001000\ldots$. The pattern 11 is forbidden. Different β-representations of x are $0111000\ldots$, or $100(01)(01)(01)\ldots$ for instance. This system is thus naturally redundant.

Representation of numbers in non-integer base is encountered for instance in coding theory, see [10], and in the modelization of quasicrystals, see [3].

In [8] we have studied the problem of the conversion in a real base β from a digit set $D = \{0, \ldots, d\}$ with $d \geq \lfloor\beta\rfloor$ to the canonical digit set A_β, without changing the numerical value. Addition and multiplication by a fixed positive integer are particular cases of digit set conversion. We have proved that the digit set conversion is on-line computable.

Moreover, if the base β is a Pisot number[1] this conversion is realizable by an on-line finite automaton. This means that in the Pisot case, digit set conversion needs only a finite storage memory, independent of the size of the data. However, addition is not computable in constant time in parallel.

In this work, we prove that multiplication of real numbers represented in real base β is on-line computable with a certain delay explicitly computed.

Our algorithm can be applied to the particular case that β is an integer. With digit set $\{0, \ldots, \beta\}$, this is the Carry-Save representation. The delay of our on-line multiplication algorithm in the Carry-Save representation is greater than the delay of on-line multiplication in the signed-digit representation, but the Carry-Save representation takes less memory.

We then consider the Knuth complex number system, which is composed of the base $\beta = i\sqrt{r}$, with $r \geq 2$ an integer, and canonical digit set $\{0, \ldots, r-1\}$. Every complex number has a representation in this system [12]. This allows a unified treatment of the real and imaginary parts of a complex number. We show that in the Knuth number system with a signed-digit set $\{-a, \ldots, a\}$, with $r/2 \leq a \leq r-1$, multiplication is on-line computable. It is known that addition is computable in constant time in parallel, and is computable by an on-line finite state automaton [7]. The case $\beta = 2i$ together with the signed-digit set $\{-2, \ldots, 2\}$, has been considered in [17, 16] with practical applications.

Other complex numeration systems have been considered. For instance the Penney complex number system consists of the complex base $\beta = -1 + i$ and the canonical digit set $\{0, 1\}$. Every complex number is representable in this system, and the representation is unique and finite for the Gaussian integers [22]. It is shown in [25] that in base $-1+i$ with signed-digit set $\{-1, 0, 1\}$ multiplication is on-line computable. The techniques are different from the ones used in the Knuth number system.

2 Preliminaries

2.1 On-line computability

Let A and B be two finite digit sets, and denote by $A^\mathbb{N}$ the set of infinite sequences of elements of A. Let

$$\begin{aligned} \varphi : A^\mathbb{N} &\to B^\mathbb{N} \\ (a_j)_{j\geq 1} &\mapsto (b_j)_{j\geq 1} \end{aligned}$$

The function φ is said to be *on-line computable with delay* δ if there exists a natural number δ such that, for each $j \geq 1$ there exists a function $\Phi_j : A^{j+\delta} \to B$ such that $b_j =$ $\Phi_j(a_1 \cdots a_{j+\delta})$, where $A^{j+\delta}$ denotes the set of sequences of length $j + \delta$ of elements of A. This definition extends readily to functions of several variables.

It is well known that some functions are not on-line computable, like addition in the binary system with canonical digit set $\{0, 1\}$. Addition is considered as a conversion χ from $\{0, 1, 2\}$ to $\{0, 1\}$. Denote by v^ω the infinite concatenation $vvv\ldots$, and by v^n the word v concatenated n times. Since $\chi(01^n 20^\omega) = 10^\omega$ and $\chi(01^n 0^\omega) = 01^n 0^\omega$ for any $n \geq 1$, one sees that the most significant digit of the result depends on the least significant digits of the input.

Recall that a distance ρ can be defined on $A^\mathbb{N}$ as follows: let $v = (v_j)_{j\geq 1}$ and $w = (w_j)_{j\geq 1}$ be in $A^\mathbb{N}$, $\rho(v, w) = 2^{-r}$ where $r = \min\{j \mid v_j \neq w_j\}$ if $v \neq w$, $\rho(v, w) = 0$ otherwise. The set $A^\mathbb{N}$ is then a compact metric space. This topology is equivalent to the product topology. Then any function from $A^\mathbb{N}$ to $B^\mathbb{N}$ which is on-line computable with delay δ is 2^δ-Lipschitz, and is thus uniformly continuous [8].

Let D be a digit set. We say that *multiplication is on-line computable with delay δ in base β on the digit set D* if there exists a function

$$\begin{aligned} \mu : D^\mathbb{N} \times D^\mathbb{N} &\to D^\mathbb{N} \\ ((x_j)_{j\geq 1}, (y_j)_{j\geq 1}) &\mapsto (p_j)_{j\geq 1} \end{aligned}$$

such that

$$\sum_{j\geq 1} p_j \beta^{-j} = \sum_{j\geq 1} x_j \beta^{-j} \times \sum_{j\geq 1} y_j \beta^{-j}$$

which is on-line computable with delay δ. Note that *apriori*, because of redundancy, the result of such a process is not unique, but the algorithms we shall consider later on are deterministic, and thus compute a function.

In the following, we will make the assumption that the operands begin with a run of δ zeroes. This allows to ignore the delay inside the computation.

2.2 Number representation

A survey on numeration systems can be found in [14, Chapter 7]. Let D be a finite digit set of real or complex digits and let β be a real or complex number such that $|\beta| > 1$. A *β-representation* of a real or complex number x with digits in D is a finite or a right infinite sequence $(x_j)_{j\leq n}$ with $x_j \in D$ such that $x = \sum_{j=n}^{-\infty} x_j \beta^j$. It is denoted by

$$(x_n \cdots x_0 \cdot x_{-1} x_{-2} \cdots)_\beta.$$

In this paper we consider only representations of the form $(x_j)_{j\geq 1} \in D^\mathbb{N}$ representing a number x equal to $\sum_{j\geq 1} x_j \beta^{-j}$. When a representation ends with infinitely many zeroes, it is said to be *finite*, and the zeroes are usually omitted.

---

[1] A Pisot number is an algebraic integer such that its algebraic conjugates are strictly less than 1 in modulus. The golden ratio and the natural integers are Pisot numbers.

2.3 Signed-digit number system

The base is a positive integer $\beta > 1$. With a signed-digit set of the form $S = \{-a, \ldots, a\}$, $\beta/2 \leq a \leq \beta - 1$ the representation is redundant and addition can be performed in constant time in parallel, and is computable by an on-line finite automaton, [1, 4, 19].

2.4 Negative base numeration systems

Let the base be a negative integer $\beta < -1$. It is well known (see [12, 13, 15]) that any real number can be represented without a sign in base β with digits from the canonical digit set $A = \{0, \ldots, |\beta| - 1\}$. With a signed-digit set of the form $T = \{-a, \ldots, a\}$, with $|\beta|/2 \leq a \leq |\beta| - 1$ the representation is redundant and addition can be performed in constant time in parallel, and is computable by an on-line finite automaton [7].

2.5 Representations in real base

Let β be a real number > 1, generally not an integer. Any real number $x \in [0, 1]$ can be represented in base β by the following greedy algorithm, [23] : denote by $\lfloor . \rfloor$ and by $\{.\}$ the integral part and the fractional part of a number. Let $r_0 = x$ and for $j \geq 1$ let $x_j = \lfloor \beta r_{j-1} \rfloor$ and $r_j = \{\beta r_{j-1}\}$. Thus $x = \sum_{j \geq 1} x_j \beta^{-j}$, where the digits x_j are elements of the *canonical* digit set $A_\beta = \{0, \ldots, \lfloor \beta \rfloor\}$ if $\beta \notin \mathbb{N}$, $A_\beta = \{0, \ldots, \beta - 1\}$ otherwise. The sequence $(x_j)_{j \geq 1}$ of $A_\beta^{\mathbb{N}}$ is called the β-*expansion* of x. When β is not an integer, a number x may have several different β-representations on A_β: this system is naturally redundant. The β-expansion obtained by the greedy algorithm is the greatest one in the lexicographic order.

It is shown in [8] that addition in real base is on-line computable. When the base is a Pisot number, addition is computable by an on-line finite automaton.

2.6 Knuth number system

Here the base is a complex number of the form $\beta = i\sqrt{r}$, where r is an integer ≥ 2. Any complex number is representable in base β with digits in the canonical digit set $A = \{0, \ldots, r - 1\}$ (see [12, 11, 9]). If r is a square then every Gaussian integer has a unique finite representation of the form $a_k \cdots a_0 \cdot a_{-1}, a_i \in A$.

Since $\beta^2 = -r$, we have

$$z = \sum_{j \geq 1} a_j \beta^{-j} = \sum_{k \geq 1} a_{2k}(-r)^{-k} + i\sqrt{r} \sum_{k \geq 0} a_{2k+1}(-r)^{-k-1}.$$

Thus,

$$\Re(z) = x = \sum_{k \geq 1} a_{2k}(-r)^{-k}$$

and

$$\Im(z) = y = \sqrt{r} \sum_{k \geq 0} a_{2k+1}(-r)^{-k-1}.$$

So the β-representation of z can be obtained by intertwinning the $(-r)$-representation of x and the $(-r)$-representation of $y/\sqrt{r}$.

Most studied cases are $\beta = 2i$ and $A = \{0, \ldots, 3\}$, strongly related to base -4, and $\beta = i\sqrt{2}$ and $A = \{0, 1\}$ ([12, 13, 20, 17, 16]).

With a signed-digit set of the form $R = \{-a, \ldots, a\}$, $r/2 \leq a \leq r - 1$, addition is computable in constant time in parallel, and is computable by an on-line finite state automaton [7].

3 On-line multiplication algorithm in integer base

3.1 Classical on-line multiplication algorithm

First we recall the classical algorithm for on-line multiplication in the signed-digit number system, see [24, 5]. We give our own presentation.

THEOREM 1 *Multiplication of two numbers represented in integer base $\beta > 1$ with digits in $S = \{-a, \ldots, a\}$, $\beta/2 \leq a \leq \beta - 1$, is computable by an on-line algorithm with delay δ, where δ is the smallest positive integer such that*

$$\frac{\beta}{2} + \frac{2a^2}{\beta^\delta(\beta - 1)} \leq a + \frac{1}{2}. \quad (1)$$

Proof. Denote by X_j the partial sum $\sum_{1 \leq i < j} x_i \beta^{-i}$ (and respectively Y_j and P_j), and denote by round(z) the closest integer to z

Classical on-line multiplication algorithm M_{SD}.
Input: two sequences $x = (x_j)_{j \geq 1}$ and $y = (y_j)_{j \geq 1}$ of $S^{\mathbb{N}}$ such that $x_1 = \cdots = x_\delta = 0$ and $y_1 = \cdots = y_\delta = 0$.
Output: a sequence $p = (p_j)_{j \geq 1}$ of $S^{\mathbb{N}}$ such that $\sum_{j \geq 1} p_j \beta^{-j} = \sum_{j \geq 1} x_j \beta^{-j} \times \sum_{j \geq 1} y_j \beta^{-j}$.
begin
1. $p_1 \leftarrow 0, \ldots, p_\delta \leftarrow 0$
2. $W_\delta \leftarrow 0$
3. $j \leftarrow \delta + 1$
4.   **while** $j \geq \delta + 1$ **do**
5. $W_j \leftarrow \beta(W_{j-1} - p_{j-1}) + y_j X_j + x_j Y_{j-1}$
6. $p_j \leftarrow$ round(W_j)
7. $j \leftarrow j + 1$
end

First let us prove by induction that for any $j \geq \delta$

$$W_j \beta^{-j} = X_j Y_j - P_{j-1}.$$

By Line 5 of the algorithm,

$$W_j\beta^{-j} = \beta^{-j+1}(W_{j-1} - p_{j-1}) + \beta^{-j}(y_j X_j + x_j Y_{j-1}).$$

By induction hypothesis, $W_j\beta^{-j} = X_{j-1}Y_{j-1} - P_{j-2} - \beta^{-j+1}p_{j-1} + \beta^{-j}(y_j X_j + x_j Y_{j-1})$, and the result follows from the fact that $X_j = X_{j-1} + x_j\beta^{-j}$, and the similar relations for Y_j and P_j.

Thus at step $n \geq \delta$, $X_n Y_n = \beta^{-n}W_n + P_{n-1} = \beta^{-n}(W_n - p_n) + P_n$. Since $|W_n - p_n| \leq \frac{1}{2}$,

$$|X_n Y_n - P_n| \leq \frac{\beta^{-n}}{2}$$

and the algorithm is convergent. The sequence $p_1 \cdots p_n$ is a β-representation of the most significant half of the product $X_n Y_n$.

Now it remains to prove that the digits p_j's computed in Line 6 of the algorithm are in the digit set S. It is enough to show that $|W_j| \leq a + \frac{1}{2}$.
From Line 5 and the fact that $|X_j|$ and $|Y_{j-1}|$ are less than $\frac{a}{\beta^\delta(\beta-1)}$ follows that

$$|W_j| < \frac{\beta}{2} + \frac{2a^2}{\beta^\delta(\beta-1)} \leq a + \frac{1}{2}$$

by (1). ∎

Note that additions and multiplications by a digit in Line 5 are performed in constant time in parallel.

COROLLARY 1 *The delay δ for Algorithm M_{SD} takes the following values. If $\beta = 2$ and $a = 1$, $\delta = 2$. If $\beta = 3$ and $a = 2$, $\delta = 2$. If $\beta = 2a \geq 4$ then $\delta = 2$. If $\beta \geq 4$ and if $a \geq \lfloor \beta/2 \rfloor + 1$, $\delta = 1$.*

3.2 On-line multiplication in negative base

Now we consider the case where the base is a negative integer $\beta < -1$ and the digit set is $T = \{-a, \ldots, a\}$, $|\beta|/2 \leq a \leq |\beta| - 1$.

PROPOSITION 1 *Multiplication of two numbers represented in negative base $\beta < -1$ and digit set $T = \{-a, \ldots, a\}$, $|\beta|/2 \leq a \leq |\beta| - 1$, is computable by the classical on-line algorithm M_{SD} with delay δ, where δ is the smallest positive integer such that*

$$\frac{|\beta|}{2} + \frac{2a^2}{|\beta|^\delta(|\beta|-1)} \leq a + \frac{1}{2}. \quad (2)$$

Proof. At step $n \geq \delta$ of the algorithm we get that $X_n Y_n - P_n = \beta^{-n}(W_n - p_n)$. Since $|W_n - p_n| \leq \frac{1}{2}$,

$$|X_n Y_n - P_n| \leq \frac{|\beta|^{-n}}{2}.$$

To show that p_j is in T, we have to show that $|W_j| \leq a + \frac{1}{2}$. From Line 5 and the fact that $|X_j|$ and $|Y_{j-1}|$ are less than $\frac{a}{|\beta|^\delta(|\beta|-1)}$ follows that

$$|W_j| < \frac{|\beta|}{2} + \frac{2a^2}{|\beta|^\delta(|\beta|-1)} \leq a + \frac{1}{2}$$

by (2). ∎

4 On-line multiplication in real base

Let $D = \{0, \ldots, d\}$ be a digit set containing A_β, that is, $d \geq \lfloor \beta \rfloor$.

THEOREM 2 *Multiplication of two numbers represented in base β with digits in D is computable by an on-line algorithm with delay δ, where δ is the smallest positive integer such that*

$$\beta + \frac{2d^2}{\beta^\delta(\beta-1)} \leq d + 1. \quad (3)$$

Proof. Clearly a number δ satisfying (3) exists, because $d \geq \lfloor \beta \rfloor$.

Real base on-line multiplication algorithm $M_\mathbb{R}$.
Input: two sequences $x = (x_j)_{j\geq 1}$ and $y = (y_j)_{j\geq 1}$ of $D^\mathbb{N}$ such that $x_1 = \cdots = x_\delta = 0$ and $y_1 = \cdots = y_\delta = 0$ [2].
Output: a sequence $p = (p_j)_{j\geq 1}$ of $D^\mathbb{N}$ such that $\sum_{j\geq 1} p_j \beta^{-j} = \sum_{j\geq 1} x_j \beta^{-j} \times \sum_{j\geq 1} y_j \beta^{-j}$.
begin
1. $p_1 \leftarrow 0, \ldots, p_\delta \leftarrow 0$
2. $W_\delta \leftarrow 0$
3. $j \leftarrow \delta + 1$
4.   while $j \geq \delta + 1$ do
5. $W_j \leftarrow \beta(W_{j-1} - p_{j-1}) + y_j X_j + x_j Y_{j-1}$
6. $p_j \leftarrow \lfloor W_j \rfloor$
7. $j \leftarrow j + 1$
end

As above, at step $n \geq \delta$, $X_n Y_n - P_n = \beta^{-n}(W_n - p_n)$. Since $0 \leq W_n - p_n < 1$ we get

$$0 \leq X_n Y_n - P_n < \beta^{-n}$$

and the algorithm is convergent.

It remains to prove that the p_j's are in D, i.e. $0 \leq p_j \leq d$. It is enough to show that $W_j < d + 1$. From Line 5 and the fact that X_j and Y_{j-1} are less than $\frac{d}{\beta^\delta(\beta-1)}$ follows that

$$W_j < \beta + \frac{2d^2}{\beta^\delta(\beta-1)} \leq d + 1$$

by (3). ∎

---
[2]This implies that $\sum_{j\geq 1} x_j \beta^{-j}$ and $\sum_{j\geq 1} y_j \beta^{-j}$ are in $[0, 1]$.

EXAMPLE 1 *Let $\beta = \varphi = (1 + \sqrt{5})/2$ be the golden ratio. Then the canonical digit set is $A_\varphi = \{0, 1\}$. Multiplication on A_φ is on-line computable with delay $\delta = 5$ by (3). This delay does not seem to be optimal, we conjecture that 4 is optimal. We give in Table 1 below the detail of a computation with $x = y = .0^5 10101$. The numerical value of x is equal to $\varphi^{-5}(\varphi^{-1} + \varphi^{-3} + \varphi^{-5})$. The result is $p = .0^{10}101000100001$. Computations are represented in base φ, in a symbolic way.*

j	$(W_j)_\varphi$	p_j
6	.000001	0
7	.00001	0
8	.0010001001	0
9	.010001001	0
10	.101000100001	0
11	1.01000100001	1
12	.1000100001	0
13	1.000100001	1
14	.00100001	0
15	.0100001	0
16	.100001	0
17	1.00001	1
18	.0001	0
19	.001	0
20	.01	0
21	.1	0
22	1.0	1
23	.0	0

Table 1. On-line multiplication in base φ with delay 5

Note that the result is not the greedy β-expansion in general. For instance with $x = .0^5 10100101010101$ and $y = .0^5 010100101010101$, Algorithm M_R gives the result $p = .0^{11}1010000110000101010001001$, which contains the forbidden pattern 11.

EXAMPLE 2 *Let $\beta = (3 + \sqrt{5})/2$ be the square of the golden ratio. Then the canonical digit set is $A_\beta = \{0, 1, 2\}$. Multiplication on A_β is on-line computable with delay $\delta = 3$. This delay is optimal for our algorithm : suppose that the delay 2 is achievable, and take $x = y = .002222$. The result given by Algorithm M_R would then be $p = .00010301011011$ which is not on the alphabet A_β.*

5 Carry-Save *versus* Signed-Digit

In this section β is an integer > 1. Take $D = \{0, \ldots, \beta\}$, then the representation on D is redundant. We call it the Carry-Save representation, because it is used in computer arithmetic under that name in the case that $\beta = 2$ for internal additions in multipliers, see [18].

By the real base algorithm M_R, multiplication in base 2 on $\{0, 1, 2\}$ is on-line computable with delay $\delta = 3$. This delay is optimal, as shown by the following example. Suppose that the delay is 2, and take $x = .00222$ and $y = .00212$. The result computed by the algorithm would be equal to $p = .0001301$.

If $\beta \geq 3$, multiplication on $D = \{0, \ldots, \beta\}$ is on-line computable with the optimal delay $\delta = 2$.

Relation (3) is never satisfied for β integer and $d = \beta - 1$, which is not surprising because it is known that multiplication in integer base on the canonical digit set is not on-line computable.

Internal additions and multiplications by a digit in Algorithm M_R can be performed in parallel when β is an integer. This is well known when $\beta = 2$. We give below the algorithm for addition in the general case.

PROPOSITION 2 *Addition in the Carry-Save representation can be performed in constant time in parallel.*

Proof. Input: $x_{n-1} \cdots x_0$ and $y_{n-1} \cdots y_0$ with x_i and y_i in $D = \{0, \ldots, \beta\}$ for $0 \leq i \leq n - 1$.
Output: $s_n \cdots s_0$ with s_i in D such that
$$\sum_{0 \leq i \leq n} s_i \beta^i = \sum_{0 \leq i \leq n-1} x_i \beta^i + \sum_{0 \leq i \leq n-1} y_i \beta^i.$$

begin
1.   In parallel for $0 \leq i \leq n - 1$ **do**
2. $z_i \leftarrow x_i + y_i$
3.   **if** $z_i = 2\beta$ **then** $\{ c_{i+1} \leftarrow 2; r_i \leftarrow 0\}$
4.   **if** $z_i = 2\beta - 1$ **then**
        if $z_{i-1} \geq \beta$ **then**
            $\{ c_{i+1} \leftarrow 2; r_i \leftarrow -1\}$
        else $\{ c_{i+1} \leftarrow 1; r_i \leftarrow \beta - 1\}$
5.   **if** $z_i = 2\beta - k$ ($2 \leq k \leq \beta - 1$) **then**
        $\{ c_{i+1} \leftarrow 1; r_i \leftarrow \beta - k\}$
6.   **if** $z_i = \beta$ **then** $\{ c_{i+1} \leftarrow 1; r_i \leftarrow 0\}$
7.   **if** $z_i = \beta - 1$ **then**
        if $z_{i-1} \geq \beta$ **then**
            $\{ c_{i+1} \leftarrow 1; r_i \leftarrow -1\}$
        else $\{ c_{i+1} \leftarrow 0; r_i \leftarrow \beta - 1\}$
8.   **if** $0 \leq z_i \leq \beta - 2$ **then**
        $\{ c_{i+1} \leftarrow 0; r_i \leftarrow z_i\}$
9. $s_i \leftarrow c_i + r_i$
10. $s_n \leftarrow c_n$
end
Clearly,
$$\sum_{0 \leq i \leq n} s_i \beta^i = \sum_{0 \leq i \leq n-1} x_i \beta^i + \sum_{0 \leq i \leq n-1} y_i \beta^i.$$

One has to prove that the digits s_i's are elements of D. Since $-1 \leq r_i \leq \beta - 1$ and $0 \leq c_i \leq 2$ then $-1 \leq s_i \leq \beta + 1$. The case $s_i = -1$ can happen only if $r_i = -1$ and $c_i = 0$, which is impossible.
The case $s_i = \beta + 1$ can happen only if $r_i = \beta - 1$ and $c_i = 2$, which is impossible as well. ∎

The delay for multiplication in the signed-digit representation is better than in the Carry-Save representation, but note that the digit-set in the signed-digit representation has cardinality $2a + 1$, to be compared to $\text{card}(D) = \beta + 1$, and the digits in D being nonnegative take less memory to be stored.

6 On-line multiplication in the Knuth complex number system

THEOREM 3 *Multiplication of two complex numbers represented in base $\beta = i\sqrt{r}$, with r an integer ≥ 2, and digit set $R = \{-a, \ldots, a\}$, $r/2 \leq a \leq r - 1$, is computable by an on-line algorithm with delay δ, where δ is the smallest odd integer such that*

$$\frac{r}{2} + \frac{4a^2}{r^{\frac{\delta-1}{2}}(r-1)} \leq a + \frac{1}{2}. \quad (4)$$

Proof. On-line multiplication algorithm M_C.
Input: two sequences $x = (x_j)_{j \geq 1}$ and $y = (y_j)_{j \geq 1}$ of $R^{\mathbb{N}}$ such that $x_1 = \cdots = x_\delta = 0$ and $y_1 = \cdots = y_\delta = 0$.
Output: a sequence $p = (p_j)_{j \geq 1}$ of $R^{\mathbb{N}}$ such that $\sum_{j \geq 1} p_j \beta^{-j} = \sum_{j \geq 1} x_j \beta^{-j} \times \sum_{j \geq 1} y_j \beta^{-j}$.
begin
1. $p_1 \leftarrow 0, \ldots, p_\delta \leftarrow 0$
2. $W_\delta \leftarrow 0$
3. $j \leftarrow \delta + 1$
4.   **while** $j \geq \delta + 1$ **do**
5. $W_j \leftarrow \beta(W_{j-1} - p_{j-1}) + y_j X_j + x_j Y_{j-1}$
6. $p_j \leftarrow \text{sign}(\Re(W_j)) \lfloor |\Re(W_j)| + \frac{1}{2} \rfloor$
7. $j \leftarrow j + 1$
end

The digit p_j will belong to R if $|\Re(W_j)| < a + \frac{1}{2}$. By Line 6, for all j, $\Re(|W_j - p_j|) \leq \frac{1}{2}$ and $\Im(W_j - p_j) = \Im(W_j)$. Thus, by Line 5,

$$|\Re(W_j)| \leq \sqrt{r}|\Im(W_{j-1})| + a(|\Re(X_j) + \Re(Y_{j-1})|)$$

and

$$|\Im(W_j)| \leq \frac{\sqrt{r}}{2} + a(|\Im(X_j) + \Im(Y_{j-1})|).$$

First suppose that δ is odd. Then

$$|\Re(X_j)| < \frac{a}{r^{\frac{\delta-1}{2}}(r-1)}$$

and

$$|\Im(X_j)| < \sqrt{r}\frac{a}{r^{\frac{\delta+1}{2}}(r-1)}$$

and the same holds true for Y_{j-1}. Thus

$$|\Re(W_j)| \leq \frac{r}{2} + \frac{4a^2}{r^{\frac{\delta-1}{2}}(r-1)} < a + \frac{1}{2}$$

by (4).

Suppose now that a better even delay δ' could be achieved. Then

$$|\Re(X_j)| < \frac{a}{r^{\frac{\delta'}{2}}(r-1)}$$

and

$$|\Im(X_j)| < \sqrt{r}\frac{a}{r^{\frac{\delta'}{2}}(r-1)}$$

thus

$$|\Re(W_j)| < \frac{r}{2} + \frac{2a^2(r+1)}{r^{\frac{\delta'}{2}}(r-1)}.$$

This delay will work if

$$\frac{r}{2} + \frac{2a^2(r+1)}{r^{\frac{\delta'}{2}}(r-1)} \leq a + \frac{1}{2}. \quad (5)$$

Suppose that the delay in (4) is of the form $\delta = 2k + 1$ and the delay in (5) is of the form $\delta' = 2k'$, and set

$$C = \frac{(r-1)(2a+1-r)}{4a^2}.$$

Then k is the smallest positive integer such that

$$k > \frac{\log(2/C)}{\log(r)}$$

and k' is the smallest positive integer such that

$$k' > \frac{\log((r+1)/C)}{\log(r)}$$

and obviously $k < k'$.
Since for $n \geq \delta$

$$X_n Y_n - P_n = \beta^{-n}(W_n - p_n),$$

$$|\Re(W_n - p_n)| \leq 1/2$$

and

$$|\Im(W_n - p_n)| = |\Im(W_n)| \leq \frac{\sqrt{r}}{2} + \sqrt{r}\frac{2a^2}{r^{\frac{\delta+1}{2}}(r-1)}$$

the algorithm is convergent, and $p_1 \cdots p_n$ is a β-representation of the most significant half of $X_n Y_n$. ∎

COROLLARY 2 *The delay δ for Algorithm M_C takes the following values. If $r = 2$ and $a = 1$, $\delta = 7$. If $r = 8$ or $r = 9$ and $a = r - 1$ then $\delta = 3$. If $r = 10$ and $a \geq 7$ then $\delta = 3$. In the other cases, for $r \leq 10$ the delay is $\delta = 5$.*

EXAMPLE 3 *Let $\beta = 2i$ and $R = \{-2, -1, 0, 1, 2\}$. By Corollary 2 the delay δ is equal to 5. Let $x = .0^5 1\bar{2}0\bar{1}201$ and $y = .0^5 1\bar{1}00121$. The result is $p = .0^{10}1111\bar{1}\bar{1}2\bar{1}\bar{1}\ldots$*

j	$(W_j)_{2i}$	p_j
6	.000001	0
7	.0001112	0
8	.001112	0
9	.01112$\bar{1}$1	0
10	.11110000$\bar{1}$2	0
11	1.1110120$\bar{2}$	1
12	1.11$\bar{1}$1$\bar{1}$2$\bar{1}$$\bar{1}$$\bar{1}$$\bar{1}$21	1
13	1.1$\bar{1}$1$\bar{1}$2$\bar{1}$$\bar{1}$$\bar{1}$$\bar{1}$21	1
14	1.$\bar{1}$1$\bar{1}$2$\bar{1}$$\bar{1}$$\bar{1}$$\bar{1}$21	1
15	$\bar{1}$.1$\bar{1}$2$\bar{1}$$\bar{1}$$\bar{1}$$\bar{1}$21	$\bar{1}$
16	1.$\bar{1}$2$\bar{1}$$\bar{1}$$\bar{1}$$\bar{1}$21	1
17	$\bar{1}$.2$\bar{1}$$\bar{1}$$\bar{1}$$\bar{1}$21	$\bar{1}$
18	2.$\bar{1}$$\bar{1}$$\bar{1}$$\bar{1}$21	2
19	$\bar{1}$.$\bar{1}$$\bar{1}$$\bar{1}$21	$\bar{1}$
20	$\bar{1}$.$\bar{1}$$\bar{1}$21	$\bar{1}$

Table 2. On-line multiplication in base $2i$ with delay 5

Acknowledgements

We are pleased to thank the anonymous referee who indicated us references [17] and [16].

References

[1] A. Avizienis, Signed-digit number representations for fast parallel arithmetic. *IRE Transactions on electronic computers* **10** (1961), 389–400.

[2] J.-C. Bajard, J. Duprat, S. Kla and J.-M. Muller, Some operators for on-line radix 2 computations. *J. of Parallel and Distributed Computing* **22** (1994), 336–345.

[3] Č. Burdík, Ch. Frougny, J.-P. Gazeau and R. Krejcar, Beta-integers as natural counting systems for quasicrystals. *J. of Physics A: Math. Gen.* **31** (1998), 6449–6472.

[4] C.Y. Chow and J.E. Robertson, Logical design of a redundant binary adder. *Proc. 4th Symposium on Computer Arithmetic*, I.E.E.E. Computer Society Press (1978), 109–115.

[5] M.D. Ercegovac, On-line arithmetic: An overview. *Real time Signal Processing VII* **SPIE 495** (1984), 86–93.

[6] M.D. Ercegovac, An on-line square-rooting algorithm. *Proc. 4th Symposium on Computer Arithmetic*, I.E.E.E. Computer Society Press (1978).

[7] Ch. Frougny, On-line finite automata for addition in some numeration systems. *Theoretical Informatics and Applications* **33** (1999), 79–101.

[8] Ch. Frougny, On-line digit set conversion in real base. *Theoret. Comp. Sci.* **292** (2003), 221–235.

[9] W. Gilbert, Radix representations of quadratic fields. *J. Math. Anal. Appl.* **83** (1981), 264–274.

[10] W.H. Kautz, Fibonacci codes for synchronization control. *I.E.E.E. Trans. Infor. Th.* **11** (1965), 284–292.

[11] I. Kátai and J. Szabó, Canonical number systems. *Acta Sci. Math.* **37** (1975), 255–280.

[12] D.E. Knuth, An imaginary number system. *C.A.C.M.* **3** (1960), 245–247.

[13] D.E. Knuth, *The Art of Computer Programming, vol. 2: Seminumerical Algorithms*, 2nd ed., Addison-Wesley, 1988.

[14] M. Lothaire, *Algebraic Combinatorics on Words*, Cambridge University Press, 2002.

[15] D.W. Matula, Basic Digit Sets for Radix Representation. *J.A.C.M.* **29** (1982), 1131–1143.

[16] R. McIlhenny, Complex number on-line arithmetic for reconfigurable hardware : algorithms, implementations, and applications. PhD Dissertation, Computer Science Department, UCLA, 2002.

[17] R. McIlhenny and M.D. Ercegovac, On-line algorithms for complex number arithmetic. *Proc. 32nd Asilomar Conference on Signals, Systems, and Computers* (1998).

[18] J.-M. Muller, *Arithmétique des ordinateurs*, Masson, 1989.

[19] J.-M. Muller, Some characterizations of functions computable in on-line arithmetic. *I.E.E.E. Trans. on Computers* **43** (1994), 752–755.

[20] A. M. Nielsen and J.-M. Muller, Borrow-Save Adders for Real and Complex Number Systems. In Proceedings of the Conference *Real Numbers and Computers*, Marseilles, 1996, 121–137.

[21] W. Parry, On the β-expansions of real numbers. *Acta Math. Acad. Sci. Hungar.* **11** (1960), 401–416.

[22] W. Penney, A "binary" system for complex numbers. *J.A.C.M.* **12** (1965), 247–248.

[23] A. Rényi, Representations for real numbers and their ergodic properties. *Acta Math. Acad. Sci. Hungar.* **8** (1957), 477–493.

[24] K. S. Trivedi and M. D. Ercegovac, On-line algorithms for division and multiplication. *I.E.E.E. Trans. on Computers* **C 26** (1977), 681–687.

[25] A. Surarerks, On-line arithmetics in real and complex base. Ph.D. Dissertation, University Paris 6, LIAFA report 2001/06, june 2001.

A VLSI Algorithm for Modular Multiplication/Division

Marcelo E. Kaihara and Naofumi Takagi
Department of Information Engineering
Nagoya University
Nagoya, 464-8603, Japan
mkaihara@takagi.nuie.nagoya-u.ac.jp

Abstract

We propose an algorithm for modular multiplication/division suitable for VLSI implementation. The algorithm is based on Montgomery's method for modular multiplication and on the extended Binary GCD algorithm for modular division. It can perform either of these operations with a reduced amount of hardware. Both calculations are carried out through iterations of simple operations such as shifts and additions/subtractions. The radix-2 signed-digit representation is employed so that all additions and subtractions are performed without carry propagation. A modular multiplier/divider based on this algorithm has a linear array structure with a bit-slice feature and carries out an n-bit modular multiplication in at most $\lfloor \frac{2(n+2)}{3} \rfloor + 3$ clock cycles and an n-bit modular division in at most $2n+5$ clock cycles, where the length of the clock cycle is constant and independent of n.

1 Introduction

With the proliferation of Internet usage, there is an increasing necessity for PCs and mobile devices, such as PDAs, of having ability to manage several security protocols. Since processing of public-key cryptosystems requires huge amount of computation, there is a growing demand for developing dedicated hardware to accelerate this.

In this paper, we propose a VLSI algorithm for modular multiplication/division with a large modulus. Modular multiplication with a large modulus is the basic operation in calculating modular exponentiation which is used to process public-key cryptosystems such as RSA [4]. One of the efficient methods for calculating the modular multiplication is by using Montgomery's multiplication algorithm [3]. Several implementations of the algorithm have been proposed [1]. On the other hand, modular division with a large modulus is used in decryption of public-key cryptosystems such as ElGamal [2]. It can be calculated by using the extended Binary GCD algorithm which is suited for binary arithmetic [5].

Since PCs and mobile devices do not seem to process more than one cryptosystem simultaneously, we combine multiplier and divider so that the hardware requirement is reduced by making large part of the circuit be shared by the two operations.

In the VLSI algorithm to be proposed, multiplication is based on Montgomery's algorithm and division is based on the extended Binary GCD algorithm. The algorithm is accelerated by introducing redundant representation in all additions/subtractions so that they are carried out in constant time independent of the length of the operands. Almost all the components in the VLSI algorithm are shared reducing considerably hardware requirements.

A modular multiplier/divider based on the algorithm has a linear array structure with a bit-slice feature and is suitable for VLSI implementation. The amount of hardware of an n-bit modular multiplier/divider is proportional to n. It performs an n-bit modular multiplication in at most $\lfloor \frac{2(n+2)}{3} \rfloor + 3$ clock cycles and an n-bit modular division in at most $2n+5$ clock cycles where the length of clock cycle is constant independent of n.

In the next section, we explain the extended Binary GCD algorithm and Montgomery's multiplication algorithm. In Section 3, we propose a VLSI algorithm for modular multiplication/division. In Section 4, we discuss several aspects about implementation. In Section 5, we present the concluding remarks.

2 Preliminaries

2.1 Extended Binary GCD Algorithm for Modular Division

Extended Binary GCD Algorithm is an efficient way of calculating modular division [5]. Consider the residue class field of integers with an odd prime modulus M. Let X and

$Y(\neq 0)$ be elements of the field. The algorithm calculates $Z(<M)$ such that $Z \equiv X/Y \pmod{M}$. It performs modular division by intertwining a procedure for finding the modular quotient with that for calculating $\gcd(Y, M)$. The algorithm is based on the following facts: if A is even and B is odd, then $\gcd(A, B) = \gcd(A/2, B)$; if A and B are both odd, then either $A + B$ or $A - B$ is divisible by 4; in this case, if $A + B$ is divisible by 4, then $\gcd(A, B) = \gcd((A+B)/2, B)$, $(A+B)/2$ is even and $|(A+B)/2| \leq \max(|A|, |B|)$; otherwise $A - B$ is divisible by 4, $\gcd(A, B) = \gcd((A-B)/2, B)$, $(A-B)/2$ is even and $|(A-B)/2| \leq \max(|A|, |B|)$.

We show the algorithm below. Note that A and B are integers and are allowed to be negative. δ represents $\alpha - \beta$, where α and β are values such that 2^α and 2^β indicates the minimums of the upper bounds of $|A|$ and $|B|$ respectively.

[Algorithm 1]
(Extended Binary GCD Algorithm)
Function: Modular Division
Inputs: $M: 2^{n-1} < M < 2^n$
$\quad\quad X, Y: 0 \leq X < M, 0 < Y < M$
Output: $Z \equiv X/Y \bmod M$
Algorithm:
$\quad A := Y; B := M; U := X; V := 0; \delta := 0;$
$\quad$**while** $A > 0$ **do**
$\quad\quad$**while** $A \bmod 2 = 0$ **do**
$\quad\quad\quad A := A/2; U := U/2 \bmod M; \delta := \delta - 1;$
$\quad\quad$**end while**
$\quad\quad$**if** $\delta < 0$ **then**
$\quad\quad\quad T := A; A := B; B := T;$
$\quad\quad\quad T := U; U := V; V := T;$
$\quad\quad\quad \delta := -\delta;$
$\quad\quad$**end if**
$\quad\quad$**if** $(A + B) \bmod 4 = 0$ **then** $q := 1;$
$\quad\quad$**else** $q := -1;$ **end if**
$\quad\quad A := (A + qB)/2; U := (U + qV)/2 \bmod M;$
$\quad$**end while**
$\quad$**if** $B = 1$ **then** $Z := V;$
$\quad$**else** /* $B = -1$ */ $Z := M - V;$ **end if**
$\quad$output Z as the result;

To calculate $U/2 \bmod M$, the algorithm examines the least significant bit of U to determine whether it is even or odd. If it is even, the algorithm performs $U/2$, otherwise it performs $(U+M)/2$. In this way, modular reduction is accomplished by a simple shift operation.

It can easily be shown that the equivalences $V \times Y \equiv B \times X \pmod{M}$ and $U \times Y \equiv A \times X \pmod{M}$ always hold. Since $\gcd(Y, M) = 1$, when $A = 0$, B is 1 or -1. Hence, in the final step $Z \times Y \equiv X \pmod{M}$ holds, and Z is the quotient of X/Y modulo M.

2.2 Montgomery's Modular Multiplication Algorithm

Montgomery introduced an efficient algorithm for calculating modular multiplication [3]. Consider the residue class ring of integers with an odd modulus M. Let X and Y be elements of the ring. Montgomery's modular multiplication algorithm calculates $Z(<M)$ such that $Z \equiv XYr^{-1} \pmod{M}$ where r is an arbitrary constant relatively prime to M. The value of r is usually set to 2^n when the calculations are performed in radix-2 with an n-bit modulus M.

The radix-2 Montgomery's multiplication algorithm is described below. We use the same notation as in the extended Binary GCD algorithm to emphasize the similitude of these algorithms.

[Algorithm 2]
(Montgomery's Multiplication Algorithm)
Function: Montgomery's Modular Multiplication
Inputs: $M: 2^{n-1} < M < 2^n$
$\quad\quad X, Y: 0 \leq X, Y < M$
Output: $Z \equiv XY2^{-n} \bmod M$
Algorithm:
$\quad A := Y; U := 0; V := X;$
$\quad$**for** $i = 1$ **to** n
$\quad\quad$**if** $A \bmod 2 = 0$ **then** $q = 0;$
$\quad\quad$**else** $q = 1;$ **end if**
$\quad\quad A := (A - q)/2; U := (U + qV)/2 \bmod M;$
$\quad$**end for**
$\quad$**if** $U \geq M$ **then** $Z := U - M;$
$\quad$**else** $Z := U;$ **end if**
$\quad$output Z as the result;

Note that U is always bounded by $2M$ throughout all iterations. Therefore, the last correction step assures that the output is correctly expressed in modulo M.

3 A VLSI Algorithm for Montgomery's Modular Multiplication and Modular Division

We propose a VLSI algorithm that performs Montgomery's modular multiplication and modular division, which is efficient in execution time and hardware requirements.

3.1 Use of a Redundant Representation

We assume that the input modulus M is an n-bit binary odd number that satisfies the condition $2^{n-1} < M < 2^n$. We also assume that the input operands X and Y and the output result Z are n-digit radix-2 signed-digit (SD2) integers in the range $(-M, M)$.

The SD2 representation uses the digit set $\{\bar{1}, 0, 1\}$ where $\bar{1}$ denotes -1. An n-digit SD2 integer $A = [a_{n-1}, a_{n-2}, \cdots a_0]$ ($a_i \in \{\bar{1}, 0, 1\}$) has the value $\sum_{i=0}^{n-1} a_i \cdot 2^i$. Addition of two SD2 numbers can be performed without carry propagation. We use the addition rules for SD2 numbers shown in table 1 [6]. The addition is accomplished by first calculating the interim sum u_i and the carry digit c_i and then performing the final sum $s_i = u_i + c_{i-1}$ for each i. To calculate s_i, we just have to check the digits a_i, b_i and their preceding ones. All the digits of the result can be computed in parallel. The negation of an SD2 number can be done simply by changing the signs of all nonzero digits in it. Subtraction can be performed through negation and addition in one step. We require a carry-propagate addition to convert an SD2 number to the binary representation.

Table 1. The rules for adding binary SD2 numbers

$a_i b_i$	$a_{i-1} b_{i-1}$	c_i	u_i
00	–	0	0
01/10	neither is $\bar{1}$	1	$\bar{1}$
01/10	at least one is $\bar{1}$	0	1
0$\bar{1}$/$\bar{1}$0	neither is $\bar{1}$	0	$\bar{1}$
0$\bar{1}$/$\bar{1}$0	at least one is $\bar{1}$	$\bar{1}$	1
11	–	1	0
$\bar{1}\bar{1}$	–	$\bar{1}$	0
1$\bar{1}$/$\bar{1}$1	–	0	0

We represent the internal variables A, B, U and V in n-digit SD2 representation so that all basic operations are carried out in constant time independent of the lengths of the operands by a combinational circuit.

In applications such as exponentiation, chained multiplications are required. To remove time-consuming SD2 to binary conversion in each multiplication, we allow the input operands X and Y as well as the output result Z be expressed in the same redundant representation so that the output can be directly fed into the inputs. Note that the operands X, Y can still be given in ordinary binary representation.

3.2 Division Mode

We follow the structure of the VLSI algorithm for modular division based on the Binary GCD algorithm [5] and further accelerate it.

This algorithm [5] performs all basic operations in constant time independent of n by a combinational circuit. This algorithm implements the 'while' loop introducing P which represents a binary number of $n + 2$ bits and indicates the minimum of the upper bounds of $|A|$ and $|B|$, i.e., $\min(2^\alpha, 2^\beta)$. Note that P has only one bit in 1 and the rest in 0. In this way, the termination condition check, $A = 0$, that may require an investigation of the whole bits of A is replaced by a check of $P = 1$ which can be carried out by just looking at the least significant bit of P, i.e. p_0. A binary number D and a flag s ($\in \{0, 1\}$) are introduced to implement δ. D has n bits of length and has the value $D = 2^{(-1)^s \cdot \delta}$. Note that this variable also has only one bit in 1 and the rest in 0. In this way, the decrement of δ, $\delta := \delta - 1$, which may require a long borrow propagation is replaced by a one-bit shift of D.

The calculation of $T/2$ modulo M is implemented by the operation $MHLV(T, M)$. It is carried out by performing $T/2$ or $(T + M)/2$ accordingly as T is even or odd. Note that only the least digit of T has to be checked to determine whether it is even or odd. The calculation of $T/4$ modulo M is implemented by the operation $MQRTR(T, M)$. It is carried out by performing the following calculations: If M (mod 4) is 1, it performs $T/4$ or $(T - M)/4$ or $(T + 2M)/4$ or $(T + M)/4$, accordingly as T (mod 4) is 0, 1, 2 or 3. If M (mod 4) is 3, it performs $T/4$ or $(T + M)/4$ or $(T + 2M)/4$ or $(T - M)/4$, accordingly as T (mod 4) is 0, 1, 2 or 3. Since M is an ordinary binary number, addition of M or $-M$ or $2M$ in $MHLV$ and $MQRTR$ is simpler than the ordinary SD2 addition. For the details of the simpler SD2 addition, see, e.g., [7].

The operation $U/2$ modulo M that is performed with the operation $A := A/2$ in Algorithm 1 when A is divisible by 2, is implemented with the operation $MHLV(U, M)$.

Since, $A := (A + B)/2$ (or $A := (A - B)/2$) is always divisible by 2, the algorithm combines this calculation with its succeeding one $A := A/2$ obtaining $A := (A + B)/4$ (or $A := (A - B)/4$) and its corresponding operation $U := (U + V)/4 (\bmod M)$ (or $U := (U - V)/4 (\bmod M)$). The latter operation is implemented by using $MQRTR(U+V, M)$ (or $MQRTR(U-V, M)$). The calculations of $A := (A + B)/4$ and $MQRTR(U+V, M)$ (or $A := (A - B)/4$ and $MQRTR(U - V, M)$) are also combined with their preceding swap of A and B and that of U and V, respectively. All the results of these basic operations are always in the range from $-M$ to M and no over-flow occurs.

In order to accelerate the calculation, for the case that A is divisible by 4, instead of performing $A/2$ and $U/2$ modulo M in two different steps, we modify the algorithm by grouping two of each operation into the calculations of $A/4$ and $U/4$ modulo M. We perform the latter calculation by using $MQRTR(U, M)$.

3.3 Multiplication Mode

We implement the while loop by using the same P as in the division case.

In Algorithm 2, A and V are initialized with the values of Y and X. U is used to store the partial products and it is initialized with the value 0. The algorithm examines the least significant bit of A to determine whether V has to be added. Then it performs a division of U by 2 modulo M and A is shifted down one position.

To accelerate the calculation, we modify this algorithm so that it processes two digits at a time. We examine the least two significant digits of A, i.e. $[a_1 a_0]$. If $[a_1 a_0] = [00]$, we perform $U/4$ modulo M and shift down A two digit positions. If $[a_1 a_0] = [10]$ or $[\bar{1}0]$, we perform $U/2$ modulo M and shift down A only one position. The 1 or $\bar{1}$ digit that is shifted into the least significant digit position is processed in the next iteration. The operations $U/4$ modulo M and $U/2$ modulo M can be accomplished by performing $MHLV(U, M)$ and $MQRTR(U, M)$ respectively.

If $[a_1 a_0] = [01]$ or $[0\bar{1}]$, we perform $MQRTR(U + V, M)$ or $MQRTR(U - V, M)$ accordingly, and we shift down A two digit positions. If $[a_1 a_0] = [1\bar{1}]$ or $[\bar{1}1]$, we convert it into $[01]$ or $[0\bar{1}]$ so that we can perform the same operations as the previous case. If $[a_1 a_0] = [\bar{1}\bar{1}]$ or $[11]$, we convert it into $[01]$ or $[0\bar{1}]$ and add -4 or 4 to A so that this case is also reduced to the previous ones.

In this way, all the operations can be accomplished with shifts, $MHLV$ and $MQRTR$, and all the results are always bounded in magnitude by M. The Montgomery's constant r is now 2^{n+2}.

To make use of the same decision rule as in the division, we initialize B with its least significant digit in $\bar{1}$. In this way, when the least significant digit of A has value 1, $A + B = 0 \mod 4$. The correction of adding -4 or 4 can be done introducing the digit $\bar{1}$ in the third least significant bit of B, i.e. b_2. The conversions and corrections are performed in the algorithm by rewrite(a_2, a_1, b_2) and the rules are summarized in table 2. Note that when $[a_1 a_0] = [11]$, these digits are replaced by $[0\bar{1}]$ and subtraction is performed. Therefore, the correction of adding 4 is performed by introducing the value $\bar{1}$ in b_2.

3.4 The VLSI Algorithm

The VLSI algorithm is presented here. In the following, $\{C1, C2\}$ means that two calculations, $C1$ and $C2$, are performed in parallel.

[Algorithm 3]
(A VLSI Algorithm for Montgomery's modular multiplication and modular division)
Function: Montgomery's Modular Multiplication and

Table 2. Conversion rule for rewrite (a_1, a_0, b_2)

a_1	a_0	b_2	a_1	a_0	meaning
$\bar{1}$	$\bar{1}$	$\bar{1}$	0	1	$-4+1$
$\bar{1}$	0	0	$\bar{1}$	0	$0-2$
$\bar{1}$	1	0	0	$\bar{1}$	$0-1$
0	$\bar{1}$	0	0	$\bar{1}$	$0-1$
0	0	0	0	0	0
0	1	0	0	1	$0+1$
1	$\bar{1}$	0	0	1	$0+1$
1	0	0	1	0	$0+2$
1	1	$\bar{1}$	0	$\bar{1}$	$4-1$

Modular Division
Inputs: $M : 2^{n-1} < M < 2^n$
$X, Y : -M < X, Y < M$
Output: mode = 0 : $Z \equiv XY2^{-(n+2)} \mod M$
mode = 1 : $Z \equiv X/Y \mod M$
Algorithm:
Step 1:
$A := Y; P := 2^{n+1}; s := 1; D := 1; M := M;$
if $mode = 0$ **then**
$B := \bar{1}; U := 0; V := X;$
else
$B := M; U := X; V := 0;$
end if
Step 2:
while $p_0 \neq 1$ **do**
  if $mode = 0$ **then** rewrite (a_1, a_0, b_2); **end if**
  if $[a_1 a_0] = 0$ **then** /* A mod 4=0 */
    $A := A >> 2; U := MQRTR(U, M);$
    if $s = 0$ **then**
      if $d_1 = 0$ **then** $D := D >> 2;$
        if $d_0 = 1$ **then** $s := 1;$ **end if**
      else $P := P >> 1; s := 1;$ **end if**
    else /* $s = 1$ */
      $D := D << 2;$
      if $p_1 = 0$ **then** $P := P >> 2;$
      else $P := P >> 1; s := 0;$ **end if**
    end if
  elseif $a_0 = 0$ **then** /* A mod 4=2 */
    $A := A >> 1; U := MHLV(U, M);$
    if $s = 0$ **then** $D := D >> 1;$
      if $d_0 = 1$ **then** $s := 1;$ **end if**
    else /* $s = 1$ */
      $D := D << 1; P := P >> 1;$
  else /* A mod 4=1 or A mod 4=3 */
    if $([a_1 a_0] + [b_1 b_0]) \mod 4 = 0$ **then** $q = 1$
    else $q = -1$ **end if**
    if $mode = 0$ **or** $s = 0$ **or** $d_0 = 1$ **then**

$A := (A + qB) >> 2;$
$U := MQRTR(U + qV, M);$
if $s = 1$ **then**
   if $mode = 0$ **and** $p_1 = 0$ **then**
      $P := P >> 2;$
   else $P := P >> 1;$
      if $p_0 = 1$ **then** $s = 0$ **end if**
   end if
   $D := D << 1;$
else /* $s = 0$ */
   $D := D >> 1;$
   if $d_0 = 1$ **then** $s := 1;$ **end if**
end if
else /* $mode = 1$ and $s = 1$ and $D > 1$ */
   $\{A := (A + qB) >> 2, B := A\};$
   $\{U := MQRTR(U + qV, M), V := U\};$
   $s := 0; D := D >> 1;$
   if $d_0 = 1$ **then** $s := 1;$ **end if**
end if
end if
end while
Step 3:
   if $mode = 0$ **and** $s = 1$ **then**
      $U := MHLV(U, M);$
   else if $mode = 1$ **and** $[b_1 b_0] \mod 4 = 3$ **then**
      $V := -V;$ **end if**
end if
Step 4:
   if $mode = 0$ **then** $Z := U;$
   else $Z := V;$ **end if**
   output Z as the result;

In division mode, i.e. mode= 1, when $A \mod 4 = 0$, A is shifted down two digits and $MQRTR(U, M)$ is performed. Note that when $P = 2$ and $a_0 = 0$, an extra 0 digit is processed together. However, since these operations only updates the values of A and U, this calculation does not affect the final result nor does increase the number of iterations needed. No special consideration has to be taken for the termination condition.

Note also that in the algorithm, δ is represented with the values of D and s. We take as convention to represent $\delta = 0$ with $D = 1$ and $s = 1$.

In Step 3, B is 1 when $B \mod 4 = 1$ and it is -1 otherwise, i.e., when $B \mod 4 = 3$. When $B = -1$, V is negated in the SD2 system.

Fig. 1 shows an example of a modular division, $-115/249 \mod 251 = -68 \mod 251 = 183$ where $n = 8$ by [Algorithm 3]. The leftmost column shows which calculations have been carried out. For example, '$(A - B)/4, A$' means that $\{ A := (A - B)/4, B := A \}$ and $\{ U := MQRTR(U - V, M), V := U \}$ have been carried out.

In multiplication mode, i.e mode=0 the flag s is set to 1 and it remains in this value until the end of Step 2.

In the case that $P = 2$, and the corresponding operation to be performed involves two digits shift, we shift P only one position to mark the end of the loop and reset the flag s to 0. At this point, $n + 2$ digits of A are processed so no extra calculations are needed. In the case that $P = 2$, and the corresponding operation to be performed involves only one digit shift, P is shifted one position and the loop finishes leaving one digit of A unprocessed. This is the same case as having $P = 4$ with operations involving two digits shifts. The flag s is left in the value 1 indicating that an extra operation is needed in Step 3. It can be shown that this unprocessed digit is always 0, so we only need to perform $MHLV(U, M)$ at the end. In this way, all the $n + 2$ digits of A are always processed and the Montgomery's constant has the value $r = 2^{n+2}$

Proposition 1: Let Y be expressed in SD2 representation with n bits of length such that $-M < Y < M$, and M be an n-bit binary number that satisfies the condition $2^{n-1} < M < 2^n$. If Algorithm 3 is used with this input and Step 2 finishes leaving the topmost significant bit of A unprocessed, this digit is always 0.

Proof: At initialization time, the value of Y is copied into A. Suppose the case that A is positive and $a_{n-1} = 1$, this digit can be transformed into $[10]$ or into $[1\bar{1}]$ when $A + B$ or $A - B$ is performed following the addition rules of SD2 numbers described in table 1. For the former case, the digits $[10]$ can in turn be transformed into $[1\bar{1}0]$. Further expansion does not occur when the most significant digit is followed by $\bar{1}$. Now, consider the case that $n - 1$ bits of A have been processed and we are about to process the next two of the remaining three bits. A can have its bits $[a_2, a_1, a_0] = 1\bar{1}0$ or $1\bar{1}\bar{1}$. No other possibilities are left because of the restriction of $|A| < M$. In the former case, A is shifted by only one position leaving the other two bits to be processed in the next iteration. In fact, these bits $1\bar{1}$ are recoded into 01 and they are processed together in the next iteration. No extra calculation is needed. In the latest case, the least significant two digits $\bar{1}\bar{1}$ of A are recoded into $0\bar{1}$ and processed together. The generated carry digit $\bar{1}$ is subtracted from A so that the most significant bit of A that has been left is cancelled and reset to 0. Similarly, when A is negative and $a_{n-1} = \bar{1}$, this digit can be transformed into $[\bar{1}1]$ and no further expansion occurs. ∎

Fig. 2 shows an example of a Montgomery's multiplication, $-115 \times 249 \times 2^{-10} \mod 251 = 137$ where $n = 8$ by [Algorithm 3]. The leftmost column shows which calculations have been carried out. For example, '$A >> 1$' means that $A >> 1$ and $U := MHLV(U, M)$ have been carried out and '$(A + B) >> 2$' means that $(A + B) >> 2$ and $U := MQRTR(U, M)$ have been carried out. In this

$mode = 1, M = [1111011]_2\ (251), X = [\bar{1}0010\bar{1}01]_{SD}\ (-115), Y = [111111\bar{1}\bar{1}]_{SD}\ (249)$

	A		B		P	D	s	U		V	
	$111111\bar{1}\bar{1}$	(249)	11111011	(251)	1000000000	0000000001	1	$\bar{1}0010\bar{1}01$	(-115)	00000000	(0)
$(A+B)/4,B$	01111101	(125)	11111011	(251)	0100000000	0000000010	1	00100010	(34)	00000000	(0)
$(A+B)/4,A$	01011110	(94)	01111101	(125)	0100000000	0000000001	1	$1000\bar{1}\bar{1}10$	(134)	00100010	(34)
$A/2,B$	00101111	(47)	01111101	(125)	0010000000	0000000001	1	$0100\bar{1}11$	(67)	00100010	(34)
$(A+B)/4,B$	00101011	(43)	00101111	(47)	0010000000	0000000001	1	01011000	(88)	$0100\bar{1}11$	(67)
$(A-B)/4,B$	$0000001\bar{1}$	(-1)	00101111	(47)	0001000000	0000000010	1	01000100	(68)	$0100\bar{1}11$	(67)
$(A-B)/4,A$	$0000\bar{1}\bar{1}00$	(-12)	$0000001\bar{1}$	(-1)	0001000000	0000000001	1	$010000\bar{1}1$	(63)	01000100	(68)
$A/4,B$	$000000\bar{1}\bar{1}$	(-3)	$0000001\bar{1}$	(-1)	0000010000	0000000100	1	$0\bar{1}\bar{1}0\bar{1}\bar{1}\bar{1}$	(-47)	01000100	(68)
$(A+B)/4,A$	$0000000\bar{1}$	(-1)	$000000\bar{1}\bar{1}$	(-3)	0000010000	0000000010	0	01000100	(68)	$0\bar{1}\bar{1}0\bar{1}\bar{1}\bar{1}$	(-47)
$(A+B)/4,B$	$0000000\bar{1}$	(-1)	$000000\bar{1}\bar{1}$	(-3)	0000010000	0000000001	1	01000100	(68)	$0\bar{1}\bar{1}0\bar{1}\bar{1}\bar{1}$	(-47)
$(A+B)/4,B$	$0000000\bar{1}$	(-1)	$000000\bar{1}\bar{1}$	(-3)	0000001000	0000000010	1	01000100	(68)	$0\bar{1}\bar{1}0\bar{1}\bar{1}\bar{1}$	(-47)
$(A+B)/4,A$	$0000000\bar{1}$	(-1)	$0000000\bar{1}$	(-1)	0000001000	0000000001	1	01000100	(68)	01000100	(68)
$(A-B)/4,B$	00000000	(0)	$0000000\bar{1}$	(-1)	0000000100	0000000010	1	00000000	(0)	01000100	(68)
$A/4,B$	00000000	(0)	$0000000\bar{1}$	(-1)	0000000001	0000001000	1	00000000	(0)	01000100	(68)
$-V$										$0\bar{1}000\bar{1}00$	(-68)

$Z = [0\bar{1}000\bar{1}00]_{SD}\ (-68)$

Figure 1. A modular division by [Algorithm 3]

$mode = 0, M = [1111011]_2\ (251), X = [\bar{1}0010\bar{1}01]_{SD}\ (-115), Y = [111111\bar{1}\bar{1}]_{SD}\ (249)$

	A		B		P	s	U		V	
	$111111\bar{1}\bar{1}$	(249)	$0000000\bar{1}$	(-1)	1000000000	1	00000000	(0)	$\bar{1}0010\bar{1}01$	(-115)
$(A+B) \gg 2$	01000110	(62)	00000101	(-5)	0010000000	1	00100010	(34)	$\bar{1}0010\bar{1}01$	(-115)
$A \gg 1$	$0010001\bar{1}$	(31)	$0000000\bar{1}$	(-1)	0001000000	1	00010001	(17)	$\bar{1}0010\bar{1}01$	(-115)
$(A-B) \gg 2$	$000\bar{1}\bar{1}000$	(8)	$0000000\bar{1}$	(-1)	0000010000	1	$0\bar{1}\bar{1}0001\bar{1}$	(33)	$\bar{1}0010\bar{1}01$	(-115)
$A \gg 2$	$000000\bar{1}10$	(2)	$0000000\bar{1}$	(-1)	0000000100	1	$010001\bar{1}1$	(71)	$\bar{1}0010\bar{1}01$	(-115)
$A \gg 1$	$0000001\bar{1}$	(1)	$0000000\bar{1}$	(-1)	0000000010	1	10100001	(161)	$\bar{1}0010\bar{1}01$	(-115)
$(A+B) \gg 2$	00000000	(0)	$0000000\bar{1}$	(-1)	0000000001	0	10001001	(137)	$\bar{1}0010\bar{1}01$	(-115)
U							10001001	(137)		

$Z = [10001001]_{SD}\ (137)$

Figure 2. A Montgomery's modular multiplication by [Algorithm 3]

example, Step 2 terminates with $s = 0$, so no extra calculations are needed.

4 Discussions

4.1 Chained Multiplications and Exponentiation

In applications such as exponentiation, chained multiplications are performed in Montgomery's representation. Observing that the result Z of the modular multiplication satisfies $|Z| < M$, it is possible to reuse the result as input operands of another modular multiplication. Note that r is an arbitrary constant relatively prime to M. In our proposed algorithm r has the value 2^{n+2}. Only one carry propagation addition is needed at the end of the whole calculation to convert the result from SD2 representation into binary number. In the case that $Z < 0$, we need to add M as a final correction step. The same correction step is applied in division mode.

Furthermore, modular multiplication/division can also be used to accelerate the calculation of modular exponentiations. That is, consider the operation $x^b (\bmod\ M)$. Let b be expressed in SD2 representation. The modular exponentiation can be calculated by examining each digit of the exponent from the topmost significant position and performing a modular squaring for each digit in 0, a modular squaring and a modular multiplication for each digit in 1 and a modular squaring and a modular division for each digit in $\bar{1}$. Since b can be recoded to reduce the number of 1s, the number of the overall operations can be considerably reduced.

4.2 Hardware Implementation

We assume to perform one pass of the computations in the 'while' loop of Step 2, i.e., one row in Fig. 1/Fig. 2, in one clock cycle.

A modular multiplier/divider based on Algorithm 3 mainly consists of 7 registers for storing A, B, P, D, U, V and M, three SD2 adders one of which is simpler, selectors, and a small control circuit. Fig. 3 shows a block diagram of the multiplier/divider.

In multiplication mode D is not used. Therefore, D can be disconnected during this mode to reduce power consumption. The circuit has a linear array structure with a bit-

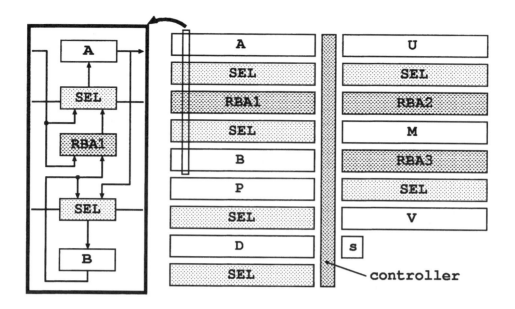

Figure 3. Block diagram of the multiplier/divider

slice feature. The amount of hardware of the modular multiplier/divider is proportional to n. Since the depth of the combinational circuit part is constant, the length of clock cycle is a constant independent of n.

4.3 Use of Two Level 1-hot Counters or Binary Counters

We can reduce the amount of hardware for keeping P and D by replacing the 1-hot counters with two-level 1-hot counters. Let n_h and n_l be integers such that $n+2 \leq n_h \cdot n_l$ is satisfied and $n_h + n_l$ is minimized, namely, $n_h \approx n_l \approx \sqrt{n}$. We replace P with n_h-bit and n_l-bit 1-hot counters P_h and P_l which keep p_h and p_l, respectively, such that $p_h \cdot n_l + p_l = P$. We replace D with D_h and D_l in the same way.

When we use P_h and P_l instead of P and use D_h and D_l instead of D, we modify the algorithm as follows. P_h and P_l are initialized so that $p_h = \lfloor (n+1)/n_l \rfloor$ and $p_l = n + 1 \mod n_l$ are satisfied. D_h and D_l are initialized so that $d_h = 0$ and $d_l = 1$. The operation $P := P >> 1$ is realized as:

1. If the rightmost bit of P_l is 1, then perform 1-bit right shift of P_h;

2. Perform 1-bit cyclic right shift of P_l.

Similarly, the operation of $P >> 2$ can be accomplished by looking at the rightmost two bits of P_l. Shift operations of D can be realized in similar ways.

The check of $p_0 = 1$ can be replaced by the check of the rightmost bits of both P_h and P_l being 1.

When we use a 1-hot counter for each counter, it requires $n + 2$ flip-flops. When we use a two-level 1-hot counter, it requires about $2\sqrt{n}$ flip-flops. We can further reduce the amount of hardware for counters by using binary counters, each of which requires about $\log_2 n$ flip-flops. Although the depth of the binary counter is not a constant, it is proportional to $\log \log n$ and is very small even when n is several hundreds. Therefore, in practice, it may be efficient to use binary counters.

When we employ binary counters, we should introduce a zero flag and perform zero detection of the counter in the previous step, i.e., in one step earlier than in [Algorithm 3] in order to avoid the increase of the clock period.

5 Concluding Remarks

We have proposed a VLSI algorithm for modular multiplication/division. We have modified the extended Binary GCD algorithm and Montgomery's modular multiplication and have accelerated them by the use of a redundant representation for internal computation.

A modular multiplier/divider based on the algorithm has a linear array structure with a bit-slice feature, and is suitable for VLSI implementation. The amount of hardware of an n-bit modular multiplier/divider is proportional to n. It performs an n-bit modular multiplication in at most $\lfloor \frac{2(n+2)}{3} \rfloor + 3$ clock cycles and an n-bit modular division in at most $2n + 5$ clock cycles, where the length of the clock cycle is constant and independent of n.

References

[1] S. E. Eldridge and C. D. Walter. 'Hardware implementation of Montgomery's modular multiplication algorithm,' *IEEE Trans. Computers*, vol. 42, no. 6, pp. 693-699, June 1993.

[2] T. ElGamal, 'A public key cryptosystem and a signature scheme based on discrete logarithms,' *IEEE Trans. Information Theory*, vol. IT-31, no. 4, pp. 469–472, July 1985.

[3] P. L. Montgomery, 'Modular Multiplication without Trial Division' *Mathematics of Computation*, vol. 44, no. 170, pp. 519-521, Apr. 1985.

[4] R. L. Rivest, A. Shamir, and L. Adleman, 'A method for obtaining digital signatures and public-key cryptosystems,' *Commun. ACM*, vol. 21, no. 2, pp. 120-126, Feb. 1978.

[5] N. Takagi, 'A VLSI Algorithm for Modular Division Based on the Binary GCD Algorithm,' *IEICE Trans. Fundamentals*, vol. E81-A, no. 5, pp. 724–728, May 1998.

[6] N. Takagi, H. Yasuura and S. Yajima, 'High-speed VLSI multiplication algorithm with a redundant binary addition tree,' *IEEE Trans. Computers*, vol. C-34, no. 9, pp. 789–796, Sep. 1985.

[7] N. Takagi and S. Yajima, 'Modular multiplication hardware algorithms with a redundant representation and their application to RSA cryptosystem,' *IEEE Trans. Computers*, vol. 41, no. 7, pp. 887–891, July 1992.

Saturating Counters: Application and Design Alternatives

Israel Koren
Department of Electrical and Computer Engineering
University of Massachusetts, Amherst, MA 01003
Email: koren@euler.ecs.umass.edu

Yaron Koren
Bear Stearns, New York, NY 10179

Bejoy G. Oomman
Genesys Testware, Fremont, CA 94539

Abstract

We define a new class of parallel counters, Saturating Counters, which provide the exact count of the inputs that are 1 only if this count is below a given threshold. Such counters are useful in, for example, a self-test and repair unit for embedded memories in a system-on-a-chip. We describe this application and present several alternatives for the design of the saturating counter. We then compare the delay and area of the proposed design alternatives.

1. Introduction

Various designs of parallel counters to be used in multiplier units and other applications have been proposed and implemented (e.g., [1, 2]). Such designs use different basic building blocks like (3,2) counters, (7,3) counters and the like [3]. An (n,k) parallel counter has n input bits and produces a k-bit binary count of its inputs that are 1. Clearly, k must satisfy $2^k - 1 \geq n$ or $k \geq \lceil \log_2(n+1) \rceil$. We define here a new type of parallel counters which we call *saturating counters*. A saturating counter needs to provide the exact count of its inputs that are 1 only if this count is below a certain threshold, denoted by T. The exact output is less important when the number of inputs that are 1 exceeds the threshold T, as long as the output indicates that the threshold has been exceeded. Such a saturating counter is needed in the design of a self-test and repair circuit for large memories embedded in a system-on-a-chip. Note that the saturating counters considered here are different from those used in certain image processing applications and in microprocessors' branch prediction units. The latter normally saturate at their maximum count of $2^k - 1$ (and, sometimes also at their minimum count of 0) and all other results must be exact.

The necessary number of output bits of a saturating counter, denoted by k, does not have to satisfy the condition $k \geq \lceil \log_2(n+1) \rceil$. Instead, the inequality which must be satisfied is $k \geq \lceil \log_2(T+1) \rceil$. In principle, T can be any number smaller than n; however, a simpler and faster implementation can be achieved when T is a power of 2. Moreover, for the application considered in this paper, if the threshold is not a power of 2 we can still employ a saturating counter with $k = i+1$ output bits, where $i = \lceil \log_2 T \rceil$. We will therefore focus in this paper on the special case of $[n, k]$ saturating counters with n inputs, a threshold of $T = 2^{k-1}$, and k output bits.

The paper is organized as follows. In Section 2 we describe the application that requires the design of a fast saturating counter. In Section 3 we present some design alternatives for saturating counters and in Section 4 we compare the delay and area of the various alternatives. Section 5 concludes the paper.

2. Self-Test and Repair for Embedded Memories in a System on a Chip

The high density and size of memory units, implemented either as separate ICs or as embedded memories, have resulted in an increasing number of manufacturing defects leading to low yields of high volume ICs. System-On-a-Chip (SOC) designs that contain megabits of embedded memory are now available from several companies. The manufacturing yield of these SOC products is strongly dependent on the yield of their embedded memory.

Spare memory rows and columns have traditionally been added to memory designs to replace defective rows, columns or individual cells. To perform such replacements, the defective rows, columns or cells must be identified first. In the past, dedicated external memory testers with fault diagnosis capabilities have been used. Following the identification of the defective cells, the chip is taken to a laser repair station and fuses are blown to replace faulty memory cells with spare memory cells [4].

To eliminate the costly memory tester from the chip manufacturing process, designers have started to incorporate Built-In Self-Test (BIST) circuitry into large memory units. Such circuitry is capable of executing memory tests to diagnose any error, which may be the result of either a manufacturing fault or a fault (intermittent or permanent) that occurs during the normal operation of the IC.

Designers of systems-on-a-chip have gone one step further, and several current designs include a Built-In Self-Test Diagnosis and Repair (BISTDR) circuit for the embedded memories in the SOC. The use of BISTDR not only enables permanent memory repair following manufacturing (hard repair), but also every time the system is powered up (soft repair). Hard repair can be done by laser blown fuses or by writing non-volatile re-configuration flip-flops, while soft repair uses only the latter [5, 6, 7].

The process starts with a self-test operation performed internally in the memory unit. Once the faulty data bits and faulty addresses have been identified, the faulty data bits are replaced with spare data bits, and faulty words are replaced with spare words. Figure 1 shows a block diagram

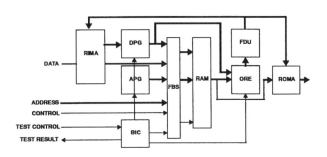

Figure 1. A block diagram of the BISTDR unit.

of the BISTDR unit, which consists of the following functional blocks: APG (Address Pattern Generator), DPG (Data Pattern Generator), ORE (Output Response Evaluator), BIC (BIST Interface Controller), FDU (Fault Diagnosis Unit), FBS (Function to BIST selector), RIMA (Repair Input Multiplexor Array) and ROMA (Repair Output Multiplexor Array).

The built-in self-repair is usually executed automatically during the power-on reset sequence of the SOC and must, therefore, be performed at system speed using the system clock. The test and repair process is done on the fly in a single cycle to avoid the need to store fault information. The Fault Diagnosis Unit (FDU) is therefore on the critical path in the BISTDR circuit since it has to identify the faulty bit(s), and make a repair decision (address or data repair) within one memory read access cycle. This repair decision is based on the number of failing data bits at the current address, and the currently available repair resources (unused spare data bits and address locations). The failing bits are determined using an array of XOR gates which compares the memory output with the expected output. This produces a bit vector whose width equals the width of the memory. A bit in this vector will have a 1 if there is a mismatch at the corresponding bit position, and a 0 otherwise.

The critical path within the FDU includes a circuit that counts the number of failures, or 1's, in this bit vector. If the number of bit failures exceeds the number of spare bits (typically no larger than 8), the memory is not repairable. Therefore, it is sufficient to know the failing bit count accurately only when it is less than the number of spares available. The fast saturating counter we design is given a threshold T, where T is a power of 2 equal to or slightly larger than the number of available spares. The number of inputs of the required saturating counter, n, is the width of the memory. Unlike stand-alone memory chips, embedded memories in SOC designs have no restriction on the data width due to pad limitations. Thus, embedded memories of width of up to 1024 are commonly used in SOCs. There is therfore a need for saturating counters for as many as 1024 input bits.

3. Saturating Counters - Design Alternatives

An $[n, k]$ saturating counter has n input bits denoted by $a_1, a_2, \cdots, a_n$, k output bits denoted by $s_{k-1}, s_{k-2}, \cdots, s_0$ and a corresponding threshold of $T = 2^{k-1}$. The output satisfies

$$(s_{k-1}, s_{k-2}, \cdots, s_0)_2 = \sum_{i=1}^{n} a_i$$

$$\text{if } \sum_{i=1}^{n} a_i \leq 2^{k-1}$$

while

$$(s_{k-1}, s_{k-2}, \cdots s_0)_2 \in [2^{k-1}+1, 2^k - 1]$$

$$\text{if } \sum_{i=1}^{n} a_i > 2^{k-1}$$

For example, a $[1024, 4]$ saturating counter has 1024 inputs, a threshold of 8, and produces four output bits satisfying

$$(s_3, s_2, s_1, s_0)_2 = \sum_{i=1}^{1024} a_i$$

if there are at most eight input bits which equal 1, and $(s_3, s_2, s_1, s_0)_2 \in [9, 15]$ if there are nine or more input bits which equal 1.

A complete Wallace tree for 1024 inputs produces nine output bits and requires 16 levels of (3,2) counters. A straightforward way to implement a $[1024, 4]$ saturating counter is to use (3,2) counters in the columns with weights $2^2, 2^1$ and 2^0 but use only OR gates in the column with weight 2^3. This implementation, shown in Table 1, requires 11 levels of (3,2) counters plus one level of an OR gate, assuming that two levels of OR operations in column 2^3 can be completed in parallel to the operation of a single level of (3,2) counters in the $2^2, 2^1$ and 2^0 columns. Table 1 shows, for each level of the tree, the number of (3,2) or (2,2) counters required in every column, and the resulting number of intermediate results in every column. For example, in the second level of the tree, 114 (3,2) counters are used in the 2^0 column, producing 114 intermediate bits of weight 2^0 and 114 bits of weight 2^1, which are added to the 115 bits generated directly in the 2^1 column. The notation 9+2(OR$_4$) in the 2^3 column means that two levels of OR gates are used, 9 in the first level and 2 in the second.

Note that the implementation depicted in Table 1 will produce a result of 8 if the number X of input bits which equal 1 satisfies $X \bmod 8 = 0$, e.g., 16, 32 and so on. If such a situation is not allowed, a threshold of $T = 16$ can be selected. However, for the application at hand the probability of such an event occurring was deemed to be negligible. The average expected number of defective memory cells in a single row is less than 4, with a standard deviation of less than 2, making the probability of 16 defective cells in one row practically zero.

In [1] Jones and Swartzlander have compared the design of parallel counters using only (3,2) or (2,2) counters to designs using more complex counters like (7,3), (15,4) and (31,5). They have analyzed the delay and area of different implementations and concluded that designs based on (3,2) and (2,2) counters only are generally superior. We therefore decided not to experiment with counters like (7,3), (15,4) and (31,5). However, in recent years (4;2) compressors [3] have become common in parallel multiplier designs, and very efficient implementations for them have been proposed (e.g., [8]). Consequently, we studied the possibility of using (4;2) compressors instead of (3,2) counters in one or more levels of the saturating counter. Table 2 shows that if (4;2) compressors are used in levels 1 through 5, the total number of levels is reduced from 12 to 9. (4;2) compressors, though, have a higher delay than (3,2) counters. However, if the delay of a (4;2) compressor is only about 50% larger than the delay of a (3,2) counter, the overall delay of the [1024,4] saturating counter still decreases when (4;2) compressors are used. Detailed delay comparisons are reported in the next section.

Tables 1 and 2 were generated using an online saturating counter simulator which is available at [9].

3.1. $(\{m, i, j\}, 3)$ units

Re-examining Tables 1 and 2, one can notice that the last few stages achieve only a small reduction in the number of bits but incur a high delay. One could replace the last four stages in Table 1, which reduce the number of bits from (5,5,2,1) to (1,1,1,1), by a look-up table with 2^{5+5+2} inputs and 4 outputs. However, a simpler and probably faster (for most technologies) solution exists which takes advantage of the saturating nature of the counter. This solution uses a special 3-column $(\{5,5,2\},3)$ unit, as shown in Table 3. If we wish to apply the same approach to the [1024,4] saturating counter which uses (4;2) compressors (see Table 2), a $(\{13,8,3\},3)$ unit could be used.

2^3	2^2	2^1	2^0	Level
			341(3,2)	
		341	342	1
		113(3,2)	114(3,2)	
	113	114+115	114	2
	113	229	114	
	37(3,2)	76(3,2)	38(3,2)	
37	76+39	38+76	38	3
37	115	115	38	
9+2(OR$_4$)	38(3,2)	38(3,2)	12(3,2)	
4+38	38+39	12+39	14	4
42	77	51	14	
10+3(OR$_4$)	25(3,2)	17(3,2)	4(3,2)	
3+25	17+27	4+17	6	5
28	44	21	6	
7+1(OR$_4$)	14(3,2)	7(3,2)	2(3,2)	
4+14	7+16	2+7	2	6
18	23	9	2	
4+1(OR$_4$)	7(3,2)	3(3,2)	1(2,2)	
3+7	3+9	1+3	1	7
10	12	4	1	
2+1(OR$_4$)	4(3,2)	1(3,2)		
1+4	1+4	2	1	8
5	5	2	1	
1+0(OR$_4$)	1(3,2)	1(2,2)		
2+1	3+1	1	1	9
3	4	1	1	
0(OR$_4$)	2(2,2)			
3+2	2	1	1	10
5	2	1	1	
1(OR$_4$)	1(2,2)			
2+1	1	1	1	11
3	1	1	1	
1(OR$_4$)				12
1	1	1	1	

Table 1. A [1024, 4] saturating counter using (3,2) and (2,2) counters. It uses 892 (3,2) counters, 5 (2,2) counters and 49 4-input OR gates.

An $(\{m, i, j\}, 3)$ unit, shown in Figure 2, is a saturating parallel counter which receives m inputs of weight 2^{k-1}, i inputs of weight 2^{k-2} and j inputs of weight 2^{k-3}. It produces three outputs of weights 2^{k-1}, 2^{k-2} and 2^{k-3} where $2^{k-1} = T$ is the threshold of the saturating counter.

We restrict our discussion to the case where $2 \leq j \leq 3$, for which the maximum carry from the position of weight 2^{k-3} to the position of weight 2^{k-2} is 1. Thus, we have $i + 1$ bits of weight 2^{k-2} to be added. $s_{k-3} = y_1 \oplus \cdots \oplus y_j$ can be replaced, if

2^3	2^2	2^1	2^0	Level
			205(4;2)	
		410	205	1
		90(4;2)	41(4;2)	
	90+90	91+41	41	2
	180	132	41	
	41(4;2)	28(4;2)	8(4;2)	
	1(3,2)			
82+1	42+28	28+8	9	3
83	70	36	9	
20+5(OR$_4$)	15(4;2)	7(4;2)	2(4;2)	
	1(2,2)	1(3,2)		
30+1+8	16+7+1	8+2	2	4
39	24	10	2	
9+3(OR$_4$)	5(4;2)	2(4;2)	1(2,2)	
3+10	6+2	2+1	1	5
13	8	3	1	
3+1(OR$_4$)	2(3,2)	1(3,2)		
	1(2,2)			
1+3	3+1	1	1	6
4	4	1	1	
1(OR$_4$)	2(2,2)			
1+2	2	1	1	7
3	2	1	1	
	1(2,2)			
3+1	1	1	1	8
4	1	1	1	
1(OR$_4$)				9
1	1	1	1	

Table 2. A $[1024, 4]$ saturating counter using (4;2) compressors, (3,2) and (2,2) counters. It uses 444 (4;2) compressors, 5 (3,2) counters, 6 (2,2) counters and 44 4-input OR gates.

2^3	2^2	2^1	2^0	Level
			341(3,2)	
		341	342	1
		113(3,2)	114(3,2)	
	113	114+115	114	2
	113	229	114	
	37(3,2)	76(3,2)	38(3,2)	
37	76+39	38+76	38	3
37	115	115	38	
9+2(OR$_4$)	38(3,2)	38(3,2)	12(3,2)	
4+38	38+39	12+39	14	4
42	77	51	14	
10+3(OR$_4$)	25(3,2)	17(3,2)	4(3,2)	
3+25	17+27	4+17	6	5
28	44	21	6	
7+1(OR$_4$)	14(3,2)	7(3,2)	2(3,2)	
4+14	7+16	2+7	2	6
18	23	9	2	
4+1(OR$_4$)	7(3,2)	3(3,2)	1(2,2)	
3+7	3+9	1+3	1	7
10	12	4	1	
2+1(OR$_4$)	4(3,2)	1(3,2)		
1+4	1+4	2	1	8
5	5	2	1	
1(OR$_4$)	({5,5,2},3)			9
1	1	1	1	

Table 3. A $[1024, 4]$ saturating counter using a $(\{5, 5, 2\}, 3)$ unit. It uses 891 (3,2) counters, one (2,2) counter, 43 4-input OR gates and a $(\{5, 5, 2\}, 3)$ unit.

$z_1 + \cdots + z_m$	$x_1 \cdots x_i\, x_{i+1}$	s_{k-1}	s_{k-2}
1	ϕ	1	ϕ
0	All 0's	0	0
0	A single 1	0	1
0	Two or more 1's	1	ϕ

The resulting Boolean expressions are:

$$s_{k-2} = x_1 + \cdots + x_i + x_{i+1} \quad \text{and}$$

$$s_{k-1} = z_1 + \cdots + z_m + (x_1 x_2 + x_1 x_3 + \cdots + x_i x_{i+1})$$

We can substitute x_{i+1} into the expressions for s_{k-2} and s_{k-1}. This would result in product terms with up to three literals in s_{k-1}, i.e., fan-in $\leq$

needed, by $s_{k-3} = y_1 + \cdots + y_j$. For $j \leq 3$ this simplification is not needed since the delay of the two (or less) XOR gates will be smaller than the delay of the gates required to generate s_{k-1}. The contribution of $y_1, \cdots, y_j$ to s_{k-2} can be represented by an additional input of weight 2^{k-2}, i.e.,

$$x_{i+1} = \begin{cases} y_1 y_2 & \text{for } j = 2 \\ y_1 y_2 + y_1 y_3 + y_2 y_3 & \text{for } j = 3 \end{cases}$$

The truth table for the $(\{m, i, j\}, 3)$ unit is:

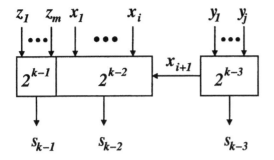

Figure 2. An ($\{m, i, j\}, 3$) unit.

3. Notice that the simplified Boolean equation for s_{k-2} may in fact produce $s_{k-2}=1$ even if the correct value is 0, but only if $s_{k-1}=1$. Therefore, the probability of producing an output of 8, when the number X of input bits which equal 1 satisfies $X \bmod 8 = 0$ and $X > 8$ (e.g., $X=16$), is lower than the corresponding probability for the saturating counters of the types depicted in Tables 1 and 2.

To calculate the delay and area of the proposed ($\{m, i, j\}, 3$) unit, some further analysis is required. The total number of signals in an implementation of s_{k-1} after the first level of gates (OR gates for the z inputs and AND gates for the remaining terms) is

$$N_2 = \binom{i+1}{2} + m \, div \, fi + m \, mod \, fi$$

if $j = 2$, and

$$N_2 = \binom{i+1}{2} + 2i + m \, div \, fi + m \, mod \, fi$$

if $j = 3$, where fi is the maximum fan-in allowed.

The table below shows the number of signals (after the first level of gates, i.e., OR gates for the z inputs and AND gates for the $x_p x_q$ and $x_p y_s y_t$ terms) and the number of logic levels for the special case of fan-in $fi = 4$, $m \leq 4$ and $j = 2$.

i	2	3	4	5	6	7	8
N_2	4	7	11	16	22	29	37
# of levels	2	3	3	3	4	4	4

For fan-in $fi \geq 4$ the total number of gate levels is therefore $1 + \lceil \log_4 N_2 \rceil$.

The exact benefit of using an ($\{m, i, j\}, 3$) unit instead of several levels of (3,2) and (2,2) counters is highly dependent on its circuit implementation. For simplicity, we will assume for the numerical results summarized in the next section that the ($\{m, i, j\}, 3$) unit is implemented using basic logic gates with a delay of Δ_G for an OR or AND gate with fan-in=fi or less. The (3,2) and (2,2) counters are implemented using 2-input XOR gates whose delay is denoted by Δ_{XOR}.

4. Numerical Results

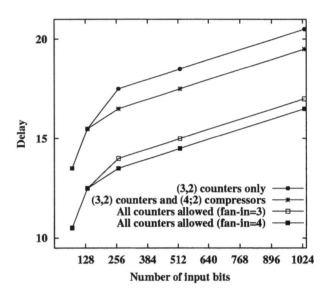

Figure 3. Delay comparison (the delay unit is Δ_{XOR}).

Figure 3 compares the delay of an $[n,4]$ saturating counter (for n=72, 136, 264, 520 and 1032) implemented in four different ways: using (3,2) and (2,2) counters only, allowing the use of (4;2) compressors as well, and allowing all types of counters including the special ($\{m, i, j\}, 3$) unit. The latter has two implementations, one with fan-in $fi = 4$ and another with $fi = 3$. Only the basic design, which is restricted to the use of (3,2) and (2,2) counters, is unique. The remaining three designs have multiple possible implementations and the delay of the fastest implementation for that type is shown. The delays in Figure 3 are measured in terms of Δ_{XOR} under the assumptions that $\Delta_G = 0.5\Delta_{XOR}$, $\Delta_{(2,2)} = \Delta_{XOR}$, $\Delta_{(3,2)} = 2\Delta_{XOR}$ and $\Delta_{(4:2)} = 3\Delta_{XOR}$. The results indicate that the use of (4;2) compressors reduces the delay by no more than one Δ_{XOR} in the given range of inputs, but using an $(\{m,i,j\},3)$ unit in addition further reduces the delay by 3 Δ_{XOR} or 2.5 Δ_{XOR} for fan-in $fi = 4$ or 3, respectively.

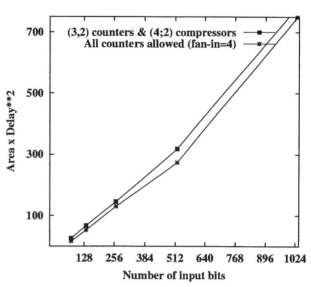

Figure 5. Area $\times$ Delay^2 comparison.

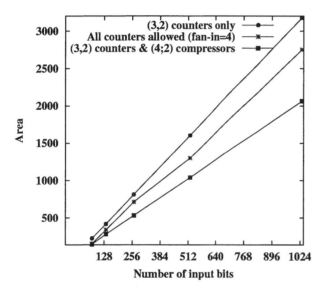

Figure 4. Area comparison (the area unit is A_{XOR}).

Figure 4 shows the estimated area for three of the design alternatives. The area is measured in terms of A_{XOR} under the assumptions that $A_G = 0.5A_{XOR}$, $A_{(2,2)} = 1.5A_{XOR}$, $A_{(3,2)} = 3.5A_{XOR}$ and $A_{(4:2)} = 4.5A_{XOR}$. The area of an $(\{m,i,j\},3)$ unit is determined by the number of gates required. Figure 4 shows that the reduction in total area due to the use of (4;2) compressors increases with the number of inputs (under the above-mentioned area ratios assumption). The use of an $(\{m,i,j\},3)$ unit may increase the total area but the area will still be lower than that of the basic design using only (3,2) and (2,2) counters.

To make a decision regarding the use of an $(\{m,i,j\},3)$ unit, the designer should consider the delay as well as the area. A measure like *Area $\times$ Delay^2* can help, and Figure 5 shows that the use of an $(\{m,i,j\},3)$ unit is beneficial.

As mentioned above, a design using (4;2) compressors and an $(\{m,i,j\},3)$ unit is not unique and therefore one can trade off area and delay. Figure 6 illustrates such a tradeoff for n=520 and 1032; where the basic designs (using only (3,2) and (2,2) counters) are also shown for reference. Note that a reasonable reduction in the area can be achieved with some increase in delay. Further increases in the delay will only marginally reduce the area, and thus are not advisable.

5. Conclusion

Saturating counters have been defined and several design alternatives have been presented and evaluated. The motivation for this study was the need to design such a counter as part of a self test and repair unit for an embedded memory in a system on a chip. The saturating counter that has been implemented uses (3,2) counters and an

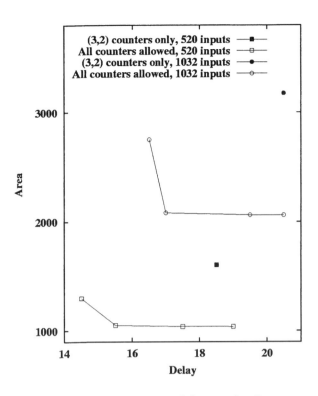

Figure 6. Area vs. delay tradeoff.

($\{m, i, j\}, 3$) unit. It has been implemented using the Perfect SAGE standard cell library for 0.15micron TSMC CMOS from Artisan. A preliminary design which did not use an ($\{m, i, j\}, 3$) unit did not satisfy the timing requirements.

References

[1] R. F. Jones and E. E. Swartzlander, "Parallel Counter Implementation," *Journal of VLSI Signal Processing,* pp. 223-232, 1994.

[2] E. E. Swartzlander, "Parallel Counters," *IEEE Trans. on Computers,* Vol. C-22, pp. 1021-1024, 1973.

[3] I. Koren, *Computer Arithmetic Algorithms,* 2nd edition, A K Peters, Natick, MA, 2002.

[4] I. Koren and Z. Koren, "Defect Tolerant VLSI Circuits: Techniques and Yield Analysis," *Proceedings of the IEEE,* Vol. 86, pp. 1817-1836, Sept. 1998.

[5] H. C. Ritter and B. Muller, "Built-In Test Processor for Self-Testing Repairable Random Access Memories," *Proceedings of the International Test Conference,* 1987, pp. 1078-1084.

[6] R. Treur and V. K. Agarwal, "Built-In Self-Diagnosis for Repairable Embedded RAMs," *IEEE Design & Test of Computers,* June 1993, pp. 24-33.

[7] Y. Nagura *et. al,* "Test cost reduction by At-speed BISR for embedded DRAMS," *Proceedings of the International Test Conference,* 2001, pp. 182-186.

[8] N. Ohkubo, M. Suzuki, *et. al.,* "A 4.4-ns CMOS 54 × 54-b Multiplier Using Pass-Transistor Multiplexor," *IEEE Journal of Solid-State Circuits,* vol.30, pp. 251–256, Mar. 1995.

[9] http://www.ecs.umass.edu/ece/koren/arith/simulator/SatCount/

Session 10:
Number Systems

Chair: Colin Walter

Error-Free Arithmetic for Discrete Wavelet Transforms using Algebraic Integers

K. A. Wahid, V. S. Dimitrov and G. A. Jullien

ATIPS Laboratory, Dept. of Electrical and Computer Engineering
University of Calgary, Calgary, Alberta, Canada T2N 1N4
wahid@atips.ca, dimitrov@atips.ca, jullien@atips.ca

Abstract

In this paper a novel encoding scheme is introduced with applications to error-free computation of Discrete Wavelet Transforms (DWT) based on Daubechies wavelets. The encoding scheme is based on an algebraic integer decomposition of the wavelet coefficients. This work is a continuation of our research into error-free computation of DCTs and IDCTs, and this extension is timely since the DWT is part of the new standard for JPEG2000. This encoding technique eliminates the requirements to approximate the transformation matrix elements by obtaining their exact representations. As a result, we achieve error-free calculations up to the final reconstruction step where we are free to choose an approximate substitution precision based on a hardware/accuracy trade-off.

1. Introduction

Low bit-rate image compression is essential for the transmission and storage of digital images. A number of techniques for image coding have been proposed, and, due to some very attractive characteristics, the wavelet transform has proven to be very useful in this regard. Recently there has been a great deal of interest shown in wavelet transforms, especially in the field of image and video processing; this is due, in part, to their flexibility in representing image signals and the ability to adapt to human visual characteristics [1]. The Discrete Wavelet Transform has also been used in the most recent image compression standard, JPEG2000. The wavelet representation provides a multiresolution expression of a signal with localization in both time and frequency and, because this multiresolution analysis is very similar to the way in which the human visual system interprets images, it proves to be an efficient way to code image and video signals [2].

There have been several wavelet filters proposed for applications in image processing. In this paper we will concentrate on the use of Daubechies Discrete Wavelets, which use irrational numbers in the formulation of the coefficients. Thus the use of any conventional number representation introduces approximation errors at the very beginning of the process. These errors propagate through the wavelet transform computation and degrade the quality of image reconstruction. In this paper, our aim is to eliminate these errors from the main computational part of the transform by the use of algebraic integer representation of the filter coefficients.

2. Algebraic integers

Algebraic integers are defined by real numbers that are roots of monic polynomials with integer coefficients [3]. As an example, let $\omega = e^{\frac{2\pi j}{16}}$ denote a primitive 16th root of unity over the ring of complex numbers. Then ω satisfies the equation $x^8 + 1 = 0$. If ω is adjoined to the rational numbers, then the associated ring of algebraic integers is denoted by $Z[\omega]$. The ring $Z[\omega]$ can be regarded as consisting of polynomials in ω of degree 7 with integer coefficients. The elements of $Z[\omega]$ are added and multiplied as polynomials, except that the rule $\omega^8 = -1$ is used in the product to reduce the degree of powers of ω to below 8. For an integer M, $Z[\omega]_M$ is used to denote the elements of $Z[\omega]$ with coefficients between $-\frac{M}{2}$ and $\frac{M}{2}$.

A real number x in $Z[\omega]$ can be written in the form

$$x = a_0 + \sqrt{2 + \sqrt{2}}\, a_1 + \sqrt{2}\, a_2 + \sqrt{2 - \sqrt{2}}\, a_3 \quad (1)$$

The ring of all such elements is denoted by $Z[\sqrt{2 + \sqrt{2}}]$. If $\theta = \sqrt{2 + \sqrt{2}}$, then θ is a root of the polynomial $x^4 - 4x^2 + 2$ and the elements of $Z[\theta]$ have a polynomial form, where the relation $\theta^4 = 4\theta^2 - 2$ is used to reduce powers of θ above three. The elements of $Z[\theta]$ are used to process separately the real and imaginary part of $Z[\omega]$.

In summary, algebraic integers of an extension of degree n can be assumed to be of the form

$$a_0\omega_0 + a_1\omega_1 + \ldots + a_{n-1}\omega_{n-1} \quad (2)$$

where $\{\omega_0, \omega_1, \ldots, \omega_{n-1}\}$ is called the algebraic-integer *basis* and the coefficients a_i are integers.

The idea of using algebraic integers in DSP applications was first explored by Cozzens and Finkelstein [4]. In their work, the algebraic integer number representation, in which the signal sample is represented by a set of (typically four to eight) small integers, and combined with the Residue Number System (RNS) to produce processors composed of simple parallel channels [5]. In this procedure, the algebraic integer representation was used to approximate complex input signals.

Our group initially introduced algebraic integer coding to provide low complexity error-free computation of the DCT and IDCT [6]. More recently, the 'cas' function of the Discrete Hartley Transform [7], was shown to be amenable to an algebraic integer encoding scheme. The application of the DCT transform is in the field of video compression for low bandwidth transmission, and so the extension of algebraic integer encoding to wavelet transforms is quite timely in this regard.

In order to better introduce the concept for error-free implementation of Daubechies wavelets, we will first provide a quick review of the use of Algebraic Integers for the Discrete Cosine Transform.

3. Discrete Cosine Transform (DCT)

We adopt the following definitions for the 1-D DCT and IDCT as shown in eqn. (3):

$$F(k) = 2\sum_{n=0}^{N-1} x(n)\cos\frac{(2n+1)k}{2N}\pi; \quad (0 \le k \le N-1)$$

$$x(n) = \frac{1}{N}\sum_{k=0}^{N-1} \overline{F(k)}\cos\frac{(2n+1)k}{2N}\pi; \quad (0 \le n \le N-1) \quad (3)$$

where $x(n)$ is a real sequence of length N, and

$$\overline{F(k)} = \begin{cases} \frac{F(0)}{2} & k=0 \\ F(k) & \text{otherwise} \end{cases}$$

The two dimensional DCT, used extensively in image processing, can be viewed as a simple extension of the one dimensional case.

The elements of the transform matrix for the DCT are real numbers of the form $\cos n\pi/2N$, where n is an integer. Rather than applying the classical procedure of using approximations to these elements, the algebraic integer encoding scheme processes numbers of this form using exact representations. The first non-zero angle for the 8 point DCT is $\pi/16$. Denote $z = 2\cos\pi/16$ and consider the polynomial expansion:

$$f(z) = \sum_{i=0}^{7} a_i \cdot z^i \quad (4)$$

where a_i are integers. This error-free representation is provided in Table 1 for $2\cos n\pi/16$, $0 \le n < 8$.

Table 1: Algebraic integer representations for the 8-point DCT.

	a_0	a_1	a_2	a_3	a_4	a_5	a_6	a_7
$2\cos 0\pi/16$	2	0	0	0	0	0	0	0
$2\cos \pi/16$	0	1	0	0	0	0	0	0
$2\cos 2\pi/16$	-2	0	1	0	0	0	0	0
$2\cos 3\pi/16$	0	-3	0	1	0	0	0	0
$2\cos 4\pi/16$	2	0	-4	0	1	0	0	0
$2\cos 5\pi/16$	0	5	0	-5	0	1	0	0
$2\cos 6\pi/16$	-2	0	9	0	-6	0	1	0
$2\cos 7\pi/16$	0	-7	0	14	0	-7	0	1

The multiplication between a real number and any coefficient above can thus be implemented with at most 2 shifts and 1 addition. An important feature for any given cosine element, is that at least every other coefficient of the algebraic integer representation is zero. In order to simplify the final reconstruction, one has to choose the parameter with the smallest number of non-zero digits in its canonic-signed digit representation within the dynamic range given. For 8-bit and 16-bit dynamic range, $z = \cos\frac{\pi}{16}$ turns out to be the best choice.

4. Discrete Wavelet Transform (DWT)

Wavelets are functions defined over a finite interval and having an average value of zero [8]. The basic idea of the wavelet transform is to represent any arbitrary function $x(t)$ as a superposition of a set of such wavelets or basis functions. These basis functions or *baby wavelets* are obtained from a single prototype wavelet called the *mother wavelet*, by dilations or contractions (scaling) and translations (shifts). The Wavelet Transform of a finite length signal $x(t)$ is given by the following eqn. (5).

$$\psi(\tau, s) = \frac{1}{\sqrt{s}} \int x(t) \Theta \psi^\circ \left(\frac{t-\tau}{s}\right) dt \qquad (5)$$

Here, τ = transition and s = scaling. The basic idea of the multiresolution analysis of the DWT is to decompose the fundamental signal space V_O into orthogonal subspaces. That is:

$$V_O = V_J \oplus W_J \oplus W_{J-1} \oplus \ldots \oplus W_1 \qquad (6)$$

where, $V_J \supset V_{J+1}$ and $V_J = V_{J+1} \oplus W_{J+1}$.

So, the DWT splits the original signal V_O into a high-pass and a low-pass signal, which corresponds to the subspaces W_1 and V_1 respectively. Then the low-pass signal is further split into a high-pass and a low-pass signal, which corresponds to the subspaces W_2 and V_2 respectively. This splitting process iterates until the original signal space is decomposed into V_J and $W_J, W_{J-1}, \ldots, W_1$. Figure 1 shows the 2-level decomposition of the Tree image.

a)

b)

Figure 1. a) Original "Tree" image; b) 2-level decomposition of DWT

5. Algebraic integer (AI) encoding of Daubechies wavelets

There have been several wavelet filters, such as Haar, Symlets, Gaussian, Mexican hat etc., defined over the past few years for compression algorithms. Wavelets have also been proposed for lossless compression [9]. In this paper we will restrict ourselves to Daubechies wavelets. This class of wavelets includes members ranging from highly localized to highly smooth and also provides excellent performance in image compression applications [10]. Daubechies wavelet coefficients are based on computing wavelet coefficients C_n (where, $n = 0, 1, 2, \ldots, N-1$ and N is the number of coefficients) to satisfy the following conditions [11]:

1. The conservation of area under $x(t)$: $\sum_n C_n = 2$

2. The accuracy conditions: $\sum_n (-1)^n n^m C_n = 0$ (where $m = 0, 1, 2, \ldots, p-1$ and $p = \frac{N}{2}$)

3. The perfect reconstruction conditions: $\sum_n C_n^2 = 2$ and $\sum_n C_n \cdot C_{n+2m} = 0$.

Then the low-pass filter is $h(n) = \frac{C_n}{2}$ and the high-pass filter is $g(n) = (-1)^{n+1} h(n-N-1)$.

Daubechies coefficients range from Daubechies-2 (in short, DAUB2 which has 2 coefficients) to Daubechies-20 (DAUB20, 20 coefficients). Among them DAUB4 has been mostly used in image processing algorithms. The DAUB4 coefficients in closed form are as follows [8]:

$$C_0 = \frac{(1+\sqrt{3})}{4\sqrt{2}} \quad C_1 = \frac{(3+\sqrt{3})}{4\sqrt{2}}$$
$$C_2 = \frac{(3-\sqrt{3})}{4\sqrt{2}} \quad C_3 = \frac{(1-\sqrt{3})}{4\sqrt{2}} \qquad (7)$$

5.1. AI encoding of DAUB4

Taking a variable $z = \sqrt{3}$ we can clearly express all the coefficients (scaled by $4\sqrt{2}$) of eqn. (7) as a first degree polynomial in z with integer coefficients, as follows:

$$C''_0 = 1+z \quad C''_1 = 3+z$$
$$C''_2 = 3-z \quad C''_3 = 1-z \quad (8)$$

Now, consider the polynomial: $f(z) = a_0 + a_1 z$ where a_i are integers. Then z corresponds to the following particular choice of a_1 : (0, 1). So for all the DAUB4 coefficients we have the exact codes given in Table 2. By manipulating these polynomial representations of the coefficients, instead of approximate representations of the coefficients themselves, we have eliminated any errors in the calculations until the final reconstruction step. The input data of the corresponding pixels are coded with integers and the multiplications we require are very simple and, most importantly, parallel. In the worst case, we need 1 addition/subtraction operation and no multiplications. Both forward and inverse mappings can be performed using the same polynomial expansion.

Table 2: Exact representation of DAUB4 coefficients (using 1st Degree)

	a_0	a_1
C'_0	1	1
C'_1	3	1
C'_2	3	-1
C'_3	1	-1

Another representation of DAUB4 can be achieved using a 2nd degree polynomial, as given in eqn. (9).

$$C'_0 = 1+z \quad C'_1 = z+z^2$$
$$C'_2 = -z+z^2 \quad C'_3 = 1-z \quad (9)$$

In this case the polynomial will be:

$$f(z) = a_0 + a_1 z + a_2 z^2 \quad (10)$$

The exact codes of the coefficients are given in Table 3.

Table 3: Exact representation of DAUB 4 coefficients (using 2nd Degree)

	a_0	a_1	a_2
C'_0	1	1	0
C'_1	0	1	1
C'_2	0	-1	1
C'_3	1	-1	0

5.2. AI encoding of DAUB6

The same technique can also be applied for Daubechies-6 coefficients. DAUB6 coefficients in closed form are mapped as follows [9].

$$C_0 = \frac{(1+\sqrt{10}+\sqrt{5+2\sqrt{10}})}{16\sqrt{2}}$$
$$C_1 = \frac{(5+\sqrt{10}+3\sqrt{5+2\sqrt{10}})}{16\sqrt{2}}$$
$$C_2 = \frac{(10-2\sqrt{10}+2\sqrt{5+2\sqrt{10}})}{16\sqrt{2}}$$
$$C_3 = \frac{(10-2\sqrt{10}-2\sqrt{5+2\sqrt{10}})}{16\sqrt{2}} \quad (11)$$
$$C_4 = \frac{(5+\sqrt{10}-3\sqrt{5+2\sqrt{10}})}{16\sqrt{2}}$$
$$C_5 = \frac{(1+\sqrt{10}-\sqrt{5+2\sqrt{10}})}{16\sqrt{2}}$$

Now considering $z = \sqrt{5+2\sqrt{10}}$, we will obtain the exact codes for DAUB6 coefficients (scaled by $32\sqrt{2}$) which are given in eqn. (12) and summarized in Table 4. The addition in this ring, $Z[\sqrt{5+2\sqrt{10}}]$, is componentwise and multiplication is equivalent to a polynomial multiplication modulo $z^4 - 10z^2 - 15 = 0$. For the final reconstruction we can use Horner's rule [12]. Then eqn. (10) can be rewritten as $f(z) = (a_2 z + a_1)z + a_0$.

$$C'_0 = -3+2z+z^2 \quad C'_1 = 5+6z+z^2$$
$$C'_2 = 30+4z-2z^2 \quad C'_3 = 30-4z-2z^2 \quad (12)$$
$$C'_4 = 5-6z+z^2 \quad C'_5 = -3-2z+z^2$$

Here we see that, unlike the DCT, where we required a polynomial of degree 7 [6], for the DWT we have a polynomial of only 2nd degree. In addition, the value of the coefficients are small; the largest coefficient for DAUB 4 is 1, and for DAUB 6, it is 30. So, in the worst case, we will need 2 additions / subtractions and 1 multiplication. Using a signed digit representation for the DAUB6 coefficients, no coefficient will require more than 2 non-zero digits.

Table 4: Exact representation for DAUB6 coefficients

	a_0	a_1	a_2
C'_0	-3	2	1
C'_1	5	6	1
C'_2	30	4	-2
C'_3	30	-4	-2
C'_4	5	-6	1
C'_5	-3	-2	1

6. Final reconstruction step

For the computation of a two-dimensional DWT or IDWT, we need to recover the integer part of the result and the most significant bit of the fractional part, in order to allow correct rounding. Since the final result is in an error free format, we can easily estimate the precision we need to guarantee sufficient accuracy. As an example, if the input and output data are to be represented within 16-bits, then the representation of z, as provided in eqn. (13) for DAUB4 and in eqn. (14) for DAUB6),

$$z \approx 10.0\bar{1}000\bar{1}00\bar{1}0 = 2 - 2^{-2} - 2^{-6} - 2^{-9} \quad (13)$$

$$z \approx 100.\bar{1}0\bar{1}000\bar{1}0\bar{1}0000 = 4 - 2^{-1} - 2^{-3} - 2^{-7} - 2^{-9} \quad (14)$$

is sufficient. Now, taking different bit-lengths, one can use Booth encoding and easily find the errors for different precisions. These signed-digit encoding errors for different word lengths are provided in Table 5.

Table 5: Bit encoding errors for DAUB4 and DAUB6

No of Bits	Error (%)	
	DAUB4	DAUB6
8	9.13×10^{-2}	5.69×10^{-2}
12	6.67×10^{-3}	6.16×10^{-3}
16	6.00×10^{-4}	2.69×10^{-4}
24	1.79×10^{-6}	1.73×10^{-7}
32	6.97×10^{-9}	2.02×10^{-8}

7. Comparison study

To demonstrate the improvement in image reconstruction quality in terms of comparable hardware cost we will take a sample 8×8 data input and use a DAUB4 wavelet transform. The processing steps consist of multiplying this data matrix with the DAUB4 forward transform matrix of (15) (using an assumption of periodicity [11]). The structure of the matrix uses coefficients C_0, C_1, C_2 and C_3 as a smoothing filter and the coefficients C_3, $-C_2$, C_1 and $-C_0$ as a non-smoothing filter.

$$\begin{bmatrix} C_0 & C_1 & C_2 & C_3 & 0 & 0 & 0 & 0 \\ C_3 & -C_2 & C_1 & -C_0 & 0 & 0 & 0 & 0 \\ 0 & 0 & C_0 & C_1 & C_2 & C_3 & 0 & 0 \\ 0 & 0 & C_3 & -C_2 & C_1 & -C_0 & 0 & 0 \\ 0 & 0 & 0 & 0 & C_0 & C_1 & C_2 & C_3 \\ 0 & 0 & 0 & 0 & C_3 & -C_2 & C_1 & -C_0 \\ C_2 & C_3 & 0 & 0 & 0 & 0 & C_0 & C_1 \\ C_1 & -C_0 & 0 & 0 & 0 & 0 & C_3 & -C_2 \end{bmatrix} \quad (15)$$

The DWT is invertible and orthogonal - the inverse transform, when viewed as a matrix, is simply the transpose of the forward transform matrix.

Using a fixed-point (FP) binary implementation, we will need 4 multiplications and 3 additions to compute each output data point (since, in (15), there are only 4 coefficients in each row and everything else is zero); but, using the algebraic integer (AI) representation (Table 3), we will need only 1 multiplication and 6 additions. Using a full adder as the atom for the hardware cost, and making a simple assumption of a hardware cost of n for an n-bit word (this will be a best case comparison for the fixed-point binary implementation) the results are summarized in Table 6 for the standard "Lena" image, along with a measure of image reconstruction quality, Peak Signal-Noise Ratio (PSNR).

Figure 2 plots the results from Table 6 to show the improvement in PSNR for both the FP and AI implementations as a function of hardware cost. The advantage of using the AI approach is clear.

An interesting comparison is to select similar hardware costs (here, hardware cost is measured in terms of adders) and then compare the reconstruction performance. An example of this is to select the 10-bit fixed point implementation and compare the results to that of the 18-bit DAUB4 algebraic integer implementation which have similar hardware costs. In this case the difference in the PSNR is almost 53 dB.

Table 7 provides a hardware vs. performance cost comparison between a fixed-point and an algebraic integer DAUB6 implementation for the "Lena" image. In this case we note that the performance difference is not as dramatic as for the DAUB4 case, but we still see an almost 20dB improvement for the 10-bit fixed point implementation. Overall, the DAUB6 will provide superior image reconstruction to DAUB4, and so this performance improvement is important.

Table 6: Comparison of hardware costs between fixed-point and AI for DAUB4

No of Bits	Hardware Complexity		PSNR (dB)	
	FP	AI	FP	AI
8	152	80	42.89	50.49
10	230	104	54.53	62.53
12	324	132	65.67	74.57
14	434	164	81.78	86.37
16	560	200	88.79	92.68
18	702	240	100.40	106.92
20	860	284	115.39	118.43
22	1034	332	129.21	133.52
24	1224	384	137.00	144.83

Figure 3 shows the image construction of the "Lena" image for a 8-bit fixed point vs. a 14-bit AI for DAUB4. In this experiment the original image was encoded using the DAUB4 wavelet transform and the decoded using the inverse transform. No compression was performed, so the image quality degradation is purely due to arithmetic quantization effects. The difference in PSNR is about 44dB (from Table 6) and the level of improvement is quite noticeable from the image reconstruction results.

Table 7: Comparison of Hardware Costs between Fixed-Point and AI for DAUB6

No of Bits	Hardware Complexity		PSNR (dB)	
	FP	AI	FP	AI
8	232	200	40.11	49.87
10	350	240	55.97	62.39
12	492	288	64.57	74.57
14	658	344	76.42	86.61
16	848	408	90.74	98.56
18	1062	480	103.68	108.60
20	1300	560	116.90	122.74
22	1562	648	132.85	134.51
24	1848	744	137.65	146.82

8. Conclusions

In this paper, we have proposed a new approach for the efficient and error-free computation of Daubechies wavelet transforms used in image processing applications. The approach is based on encoding the basis set using algebraic integers. The algebraic integer quantization not only reduces the number of arithmetic operations, but also reduces the dynamic range of the computations. The Final Reconstruction Step generates some rounding errors but these errors are only introduced at the end of the calculation, not distributed through the calculation, as is the case for a fixed-point binary implementation. We have shown that significant improvements are possible, for similar hardware costs, between a fixed-point binary and algebraic integer implementation. These improvements are demonstrated in an image reconstruction example of a 8-bit fixed-point binary vs. 14-bit algebraic integer implementation.

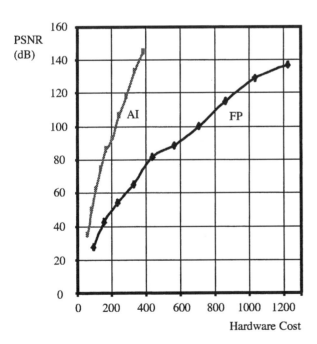

Figure 2. Comparison of AI and Fixed-Point implementations of DAUB4

 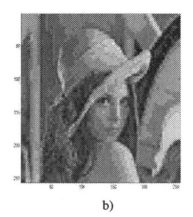 

a) b) c)

Figure 3. DAUB4 encoding and decoding results for the same hardware complexity: a) original "Lena" image; b) binary implementation; c) algebraic integer implementation

9. References

[1] H. Meng, Z. Wang and G. Lui, "Performance of the Daubechies Wavelet Filters compared with other Orthogonal Transforms in Random Signal Processing", 5th International Conference on Signal Processing, WCCC-ICSP, vol. 1, pp. 333-336, 2000.

[2] J. H. Park, K. O. Kim and Y. K. Yang, "Image Fusion using Multiresolution Analysis", IGARSS, vol. 2, pp. 864-866, 2001.

[3] Richard Dedekind, "Theory of Algebraic Integers", Translated and introduced by John Stillwell, September, 1996.

[4] J. H. Cozzens and L. A. Finkelstein, "Computing the Discrete Fourier Transform using Residue Number Systems in a Ring of Algebraic Integers", IEEE Transactions on Information Theory, vol. 31, pp. 580-588, 1985.

[5] R. A. Games, D. Moulin, S. D. O'Neil and J. Rushanan, "Algebraic Integer Quantization and Residue Number System Processing", IEEE International Conference on Acoustics, Speech and Signal processing, pp. 948-951, May 1989.

[6] V. S. Dimitrov, G. A. Jullien and W. C. Miller, "A New DCT Algorithm Based on Encoding Algebraic Integers", ICASSP, pp. 1377-1380, 1998.

[7] R. Baghaie and V. Dimitrov, "Systolic Implementation of Real-valued Discrete transforms via Algebraic Integer Quantization", An International Journal on Computers and Mathematics with Applications, vol. 41, pp. 1403-1416, 2001.

[8] A. Primer, "Introduction to Wavelets and Wavelet Transforms", Prentice Hall, 1998, ISBN: 0-13-489600-9.

[9] K. Komatsu and K. Sezaki, "Optimum Quantization Step Size for Integer Lossless Wavelet Coefficients", The 6th World Multiconference on Systemics, Cybernetics and Informatics, 2002.

[10] J. P. Andrew, P. O. Ogunbona and F. J. Paoloni, "Comparison of Wavelet Filters and Subband Analysis Structure for Still Image Compression", ICASSP, vol. 5, pp. 589-592, 1994.

[11] I. Daubechies, "Communications on Pure and Applied Mathematics", vol. 41, pp. 909-996, 1988.

[12] D. Knuth, "The Art of Computer Programming", vol. 2 - Seminumerical Algorithms, 3rd edition, Addison Wesley, 1981.

10. Acknowledgements

The authors acknowledge financial support from the Alberta Informatics Circle of Research Excellence (iCORE), the Natural Sciences and Engineering Council (NSERC) of Canada, the Micronet Network of Centres of Excellence and Gennum Corporation. The authors are indebted to the Canadian Microelectronics Corporation (CMC) for their equipment and software loan program and fabrication services.

On Computing Addition Related Arithmetic Operations via Controlled Transport of Charge

Sorin Cotofana, Casper Lageweg and Stamatis Vassiliadis
Computer Engineering Laboratory,
Delft University of Technology,
Delft, The Netherlands
{Sorin,Casper,Stamatis}@CE.ET.TUDelft.NL

Abstract

In this paper we investigate the implementation of basic arithmetic functions, such as addition and multiplication, in Single Electron Tunneling (SET) technology. First, we describe the SET equivalents of Boolean CMOS gates and Threshold logic gates. Second, we propose a set of building blocks, which can be utilized for a novel design style, namely arithmetic operations performed by direct manipulation of the location of individual electrons within the system. Using this new set of building blocks, we propose several novel approaches for computing addition related arithmetic operations via the controlled transport of charge (individual electrons). In particular, we prove the following: n-bit addition can be implemented with a depth-2 network built with $O(n)$ circuit elements; n-input parity can be computed with a depth-2 network constructed with $O(n)$ circuit elements and the same applies for $n\lfloor \log n \rfloor$ counters; multiple operand addition of m n-bit operands can be implemented with a depth-2 network using $O(mn)$ circuit elements; and finally n-bit multiplication can be implemented with a depth-3 network built with $O(n)$ circuit elements.

1 Introduction

Feature size reduction in microelectronic circuits has been an important contributing factor to the dramatic increase in the processing power of computer arithmetic circuits. However, it is generally accepted that sooner or later MOS based circuits cannot be reduced further in (feature) size due to fundamental physical restrictions [15]. Therefore, several emerging technologies are currently being investigated [13]. Single Electron Tunneling (SET) [8] is one such technology candidate and offers greater scaling potential than MOS as well as ultra-low power consumption. Additionally, recent advances in silicon based fabrication technology (see for example [14]) show potential for room temperature operation. However SET devices display a switching behavior that differs from traditional MOS devices. This provides new possibilities and challenges for implementing digital circuits.

SET technology introduces the quantum tunnel junction as a new circuit element for (logic) circuits. The tunnel junction can be thought of as a "leaky" capacitor, such that the "leaking" can be controlled by the voltage across the tunnel junction. Although this behavior at first glance appears similar to that of a diode, the difference stands in the scale at which switching occurs. Charge transport though a tunnel junction can only occur in quantities of a single electron at a time. Additionally, given the feature sizes anticipated for such circuits, the transport of a single electron can have a significant effect on the voltage across a tunnel junction. This implies that transporting a few electrons through a tunnel junction will inhibit further charge transport, making it possible to control the transport of charge in discrete and accurate quantities.

The ability to control the transport of individual electrons in SET technology introduces a broad range of new possibilities and challenges for implementing computer arithmetic circuits. In this paper we investigate the computation of addition related arithmetic operations in SET technology by controlling the transport of individual electrons. First, we briefly present the SET equivalent of two conventional design styles, namely the equivalents of CMOS and threshold logic gates. Second, we propose a set of building blocks, which can be utilized for charge controlled computations. Third, using the new set of building blocks, we propose several novel approaches for computing addition related arithmetic functions, e.g., addition, parity, counting, multiplication, via the controlled transport of charge. Related to these new schemes we prove that the following holds true:

- The addition/subtraction of two n-bit operands can be computed with a depth-2 network composed out of

$3n + 1$ circuit elements[1].

- The n-parity function can be computed with a depth-2 network constructed with $n + 1$ circuit elements.

- The $n|\log n$ counter can be implemented with a depth-2 network constructed with $n + \log n$ circuit elements.

- The m n-bit multiple operand addition can be implemented using at most $n(m + 1) + \lceil \log m \rceil + 1$ circuit elements.

- The multiplication of two n-bit operands can be computed with a depth-3 network with $4n - 1$ circuit elements.

The remainder of this paper is organized as follows: Section 2 briefly presents some SET background theory, explaining the basic switching behavior appearing in SET circuits. Section 3 presents the SET equivalent of the CMOS design style. Section 4 presents the implementation of threshold gates in SET technology. In Section 5 we propose a set of new building blocks for controlled charge transport. Section 6 proposes new schemes for the calculation of addition and multiplication via the controlled transport of single electrons. Finally, Section 7 concludes the paper with some final remarks.

2 Background

Single Electron Tunneling technology introduces the quantum tunnel junction as a new circuit element. A tunnel junction consist of two conductors separated by an extremely thin insulating layer. The insulating layer acts as an energy barrier which inhibits charge transport under normal (classical) physics laws. However, according to quantum physics theory, charge transport of individual electrons through this insulating layer can occur if this results in a reduction of the total energy present in the circuit. The transport of charge through a tunnel junction is referred to as *tunneling*, while the transport of a single electron is referred to as a *tunnel event*. Electrons are considered to tunnel through a tunnel junction strictly one after another.

Rather then calculating for each tunnel junction if a hypothetical tunnel event results in a reduction of the circuit's energy, we can calculate the critical voltage V_c, which is the voltage threshold needed across the tunnel junction to make a tunnel event through this tunnel junction possible. For calculating the critical voltage of a junction, we assume a tunnel junction with a capacitance of C_j. The remainder of the circuit, as viewed from the tunnel junction's perspective,

[1] By circuit element we mean in this context any of the building block presented in Section 5.

has an equivalent capacitance of C_e. Given the approach presented in [5], we calculate V_c for the junction as

$$V_c = \frac{e}{2(C_e + C_j)}. \quad (1)$$

In the equation above, as well as in the remainder of this discussion, we refer to the charge of the electron as $e = 1.602 \cdot 10^{-19}$ C. Strictly speaking this is incorrect, as the charge of the electron is of course negative. However, it is more intuitive to consider the electron as a positive constant for the formulas which determine if a tunnel event will take place or not. We will of course correct for this when we discuss the direction in which the tunnel event takes place. Generally speaking, if we define the voltage across a junction as V_j, a tunnel event will occur through this tunnel junction if and only if $|V_j| \geq V_c$. If tunnel events cannot occur in any of the circuit's tunnel junctions, i.e., $|V_j| < V_c$ for all junctions in the circuit, the circuit is in a *stable state*. For our research we focus on circuits where a limited number of tunnel events may occur, resulting in a stable state. Each stable state determines a new output value resulting from the distribution of charge throughout the circuit.

Assuming that a tunnel event is possible, the orthodox theory for single electron tunneling (see for example [5] for a more extensive introduction) states that tunneling is a stochastic process, in which the rate at which tunnel events occur at $0K$ temperature is

$$\Gamma = \frac{|V_j - V_c|}{eR_t} \quad (2)$$

where R_t is the tunnel resistance (usually $\approx 10^5 \Omega$). Note that a non-$0K$ temperature implies a lower event rate. Assuming that an individual tunnel event can be described as a Poisson process, we can calculate the required delay t for a single tunnel event to occur for a given error chance P_{err} as:

$$t = \frac{ln(P_{err})eR_t}{|V_j - V_c|}. \quad (3)$$

Given that the minimum amount of transportable charge consists of a single electron, there exists a minimum energy threshold, called the Coulomb energy, which must be present in the circuit so that the transport of a single electron reduces the total amount of energy in the system. Resulting, in order to utilize the electron tunneling phenomenon, all other types of energy must be much smaller then the Coulomb energy. For thermal energy, this implies that, if we intend to add or remove charge to a circuit node by means of tunnel events, the total capacitance attached to such circuit nodes must be less then $900aF$ for $1K$ temperature operation, or less then $3aF$ for $300K$ (room temperature) operation [10]. This represents a major SET fabrication technology hurdle as even for cryostat temperature operation very small circuit features are required to implement such small capacitors. Another major technology challenge comes from the fact that thus far all experimental circuits have displayed a random offset charge (random charge present on circuit nodes), which is assumed to be the result of trapped charge particles in the tunnel junctions themselves or in the substrate. This random charge results in a random additional voltage across tunnel junctions, which can cause errors in their switching behavior. At the same time there are indications [8] that the offset charge problem may reduce or even disappear entire for the nanometer-scale feature size circuits required for room temperature operations. Given this and the fact that in our investigation we focus on the efficient utilization of the SET behavioral properties we ignore the aspects related to offset charge and its potential influence on SET based computational structures.

3 CMOS Like SET Gates

One of the first SET circuits examined in literature is the SET transistor (see [7] for an early review paper). The SET transistor consists of two tunnel junctions in series, with a capacitor attached to the interlaying circuit node, as is displayed in Figure 1. The resulting 3-terminal structure can be seen as being similar to a MOS transistor, such that the gate voltage V_g can control the transport of charge through the tunnel junctions (current I_d).

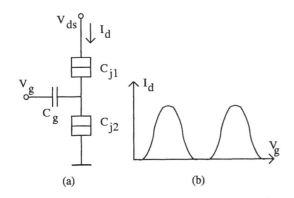

Figure 1. The SET transistor (a) circuit and (b) transfer function.

However, unlike the MOS transistor, the current I_d through the SET transistor has a periodic response to the input voltage V_g. By adding a capacitively coupled bias voltage to the interlaying node, the transfer function of the SET transistor can be translated over the V_g axis.

When two SET transistors with different biasing voltages are combined in a single circuit, we arrive at the CMOS-type inverter structure proposed in [6], as displayed in Figure 2. In the Figure, the upper and lower SET transistor behave as a p-type and an n-type MOS transistor, respectively. Additionally it was suggested in [6] that using p-type and n-type SET transistors as a basis, one can now convert existing CMOS cell libraries to SET technology.

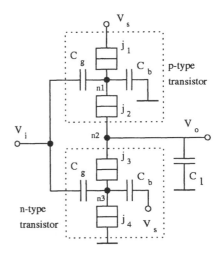

Figure 2. CMOS-type SET inverter.

The main disadvantage of the approach described above is that the current transport though an "open" transistor still consists of a large number of individual electrons "dripping" through the tunnel junctions. This is obviously a far slower process then the transport of just one single electron through the same junction, and consequently does not use the SET technology to its full potential. Therefore, the next step would be to limit the charge transport through an open transistor to just 1 electron. This results in the principle of Single Electron Encoded Logic (SEEL), in which the Boolean logic values 0 and 1 are encoded as a net charge of 0 and 1e on the circuit's output node.

However, when the SEEL approach is applied to converted CMOS cells with multiple p-type or n-type transistors in series, the circuits will no longer operate correctly, as clarified by the following example. Assume a series of 2 p-type transistors, of which the one bordering the load capacitor is open while the other one is closed. This situation will result in the removal of 1 electron from the load capacitor, resulting in an incorrect "high" output. Only the inverter circuit itself will operate correctly under a single electron encoded logic regime. This implies that CMOS type SET logic must encode the Boolean logic values 0 and 1 as "few" and "many" electron charges. We can therefore conclude that CMOS-type SET logic cannot efficiently utilize the SET features. To circumvent this problem we introduce in the next section SET based threshold logic gates, which can operate according to the SEEL paradigm.

4 Threshold Logic Gates

Threshold logic gates are devices able to compute any linearly separable Boolean function given by:

$$F(X) = sgn\{\mathcal{F}(X)\} = \begin{cases} 0 & \text{if } \mathcal{F}(X) < 0 \\ 1 & \text{if } \mathcal{F}(X) \geq 0 \end{cases} \quad (4)$$

where $\mathcal{F}(X) = \sum_{i=1}^{n} \omega_i x_i - \psi$, x_i are the n Boolean inputs, and w_i are the corresponding n integer weights. The linear threshold gate performs a comparison between the weighted sum of the inputs $\Sigma_{i=1}^{n} \omega_i x_i$ and the threshold value ψ. If the weighted sum of inputs is *greater than or equal to* the threshold, the gate produces a logic 1. Otherwise the output is a logic 0. Threshold logic gates are inherently more powerful then standard Boolean gates [12].

As stated in Section 2, a SET tunnel junction requires a minimum voltage $|V_j| \geq V_c$ in order for a tunnel event to occur. This critical voltage V_c acts as a naturally occurring threshold ψ with which the junction voltage V_j is compared. If we add capacitively coupled inputs to the circuit nodes on either side of the tunnel junction, the inputs will make a positively or negatively weighted contribution to the voltage across this junction (depending on the sign definition of V_j). Similarly, we can add a capacitively coupled biasing voltage in order to adjust the threshold to the desired value. This approach resulted in a generic threshold gate implementation [2] as displayed in Figure 3.

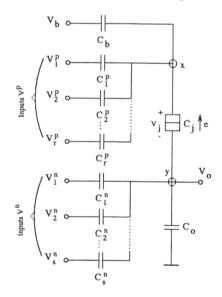

Figure 3. The n-input linear threshold gate.

In this figure, the input signals $V^p = \{V_1^p, V_2^p, \ldots, V_r^p\}$ are weighted by their corresponding capacitors $C^p = \{C_1^p, C_2^p, \ldots, C_r^p\}$ and added to the voltage across the tunnel junction. The input signals $V^n = \{V_1^n, V_2^n, \ldots, V_r^n\}$ are weighted by their corresponding capacitors $C^n = \{C_1^n, C_2^n, \ldots, C_r^n\}$ and subtracted from the voltage across the tunnel junction. The biasing voltage V_b, weighted by the capacitor C_b, is used to adjust the gate threshold to the desired value ψ. If $sgn\{V_j - V_c\}$, a single electron is transported from node y to node x, which results in a high output.

The generic threshold gate described above can be used to implement any threshold function (including the standard Boolean logic gates). However, due to the passive nature of this circuit we must apply sufficient buffering between different threshold gates in order to alleviate feedback effects as well as to maintain correct logic levels [3]. The CMOS style inverter described in Section 3 can act as a SEEL buffer [1]. A similar non-inverting buffer can also be derived from the inverter by removing both bias capacitors C_b and choosing a different set of capacitor values, resulting in an even smaller buffer.

Thus by utilizing the SET based threshold gate approach all the Boolean and/or Threshold logic schemes for the computation of arithmetic functions can be potentially implemented with no major change in the paradigm. Moreover by encoding Boolean values in a net charge of 0 or $1e$, the SET threshold logic makes efficient use of the SET technology and potentially provides the premises for ultra-low power consumption computations.

However, given that in SET technology it is possible to control the transport of individual electrons, we can further improve efficiency if we can encode n-bit operands as a number of electrons stored at a specific circuit location and perform arithmetic operations via the controlled transport of single electrons. Before exploring this novel concept further we propose in the next section a set of new building blocks to constitute the fundament for computing arithmetic operations via controlled transport of charge.

5 Building Blocks for Electron Counting

In this section we propose two basic blocks: one can be utilized to move electrons within a SET circuit and the other one can be utilized to implement periodic symmetric functions. The novel circuit blocks introduced in the sequel are then utilized in Section 6 as a basis for constructing larger structures for addition related arithmetic operations, e.g., addition, parity, counting, and multiplication.

5.1 The MVke Block

The $MVke$ block displayed in Figure 4 is a basic block with which a variable number of electrons can be added to or removed from a charge reservoir. Typically, a charge reservoir is a circuit node that is capacitively coupled to ground. A charge reservoir with a capacitance C_r containing a charge of $V \times e$ is therefore equivalent to a voltage source $U = \frac{V \times e}{C_r}$. The $MVke$ block operates as follows: if the Boolean control signal $E = 1$, a charge of $V \times k \times e$ is moved to the electron reservoir when the block is triggered by a clock pulse (CLK), where k is a positive integer constant and V is an integer (variable) value. Note that V could either be another charge reservoir containing $V \times e$ electrons or an equivalent voltage source. For positive V values the $MVke$ block is in "add" mode (adding charge to the reservoir) while for negative V values the $MVke$ block is in "remove" mode (removing charge from the reservoir).

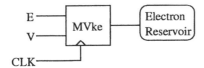

Figure 4. The MVke block.

Earlier experiments have demonstrated [4, 11] that the SET turnstile circuit, originally proposed by [9], can be modified such that it can control the transport of charge to and from a charge reservoir. However, the (modified) turnstile can only move one electron per clock pulse fact that precludes its direct utilization as an $MVke$ block. Given that the SET transistor operates as a controlled switch it can be used to extend the (modified) turnstile circuit capabilities such as it can let tunnel a larger amount of electrons per clock signal. A possible implementation of the $MVke$ block, based on the SET transistor and the operating principle of the turnstile circuit, is displayed in Figure 5.

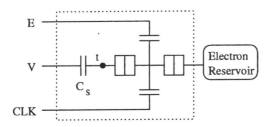

Figure 5. Possible $MVke$ block implementation.

The circuit operates as follows. If a clock pulse arrives, the SET transistor is opened if and only if $E = 1$. When the transistor opens, $V \times k \times e$ charge will be added to or removed from the electron reservoir due to charge pulling or pushing effect of V. As a result of this charge transport, an opposite charge $-V \times k \times e$ will be stored on node 't'. The voltage resulting from this charge will cancel the effect of voltage source V, inhibiting further charge transport. Given that the capacitor C_s acts as a weight factor for V, the desired constant value k can be adjusted by changing the value of C_s.

5.2 The PSF Block

A Boolean symmetric function $F_s(x_0, x_1, \ldots, x_{n-1})$ is a Boolean function for which the output depends on the sum of the inputs $X = \sum_{i=0}^{n-1} x_i$. A Periodic Symmetric Function (PSF) $F_p(X)$ is a symmetric function for which $F_p(X) = F_p(X + T)$, where T is the period. Any PSF can be completely characterized by T, the value of its period, and a,b, the values of X corresponding with the first positive transition and the first negative transition, as displayed in Figure 6. Efficient implementation of periodic symmetric functions is quite important as many functions involved in computer arithmetic computations, e.g., parity, belong to this class of functions.

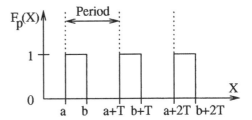

Figure 6. Period symmetric function $F_p(X)$.

Given the periodic transfer function of the SET transistor, as displayed in Figure 1, we can design a *PSF* block that can compute any PSF using a single SET transistor as a basis. The period of the SET transistor's transfer function can be adjusted to T by varying the value of the gate capacitor C_g. Likewise, the drain-source voltage V_{ds} determines the part of the function period in which $I_d > 0$, i.e., the length of the $[a, b]$ interval. Finally, a capacitively coupled bias voltage similar to that used for the CMOS-type inverter can translate the transfer function over the X axis in order to place the $[a, b]$ intervals in the required positions.

6 Electron Counting Based Arithmetic

In Sections 3, 4 we have demonstrated the feasibility of implementing classical Boolean and Threshold logic gates in SET technology. Although we can achieve the encoding of Boolean variables as a net charge of 0 and 1e, this still does not use the full potential of SET. Given that we can control the transport of individual electrons, we have the possibility of encoding integer values X directly as a net charge Xe. Once integer values have been encoded as a number of electrons, we can perform arithmetic operations directly in electron charges. This reveals a broad range of novel computational schemes, which we will generally refer to as electron counting.

Within the context of this novel electron counting paradigm we investigate in the sequel SET networks for addition related arithmetic operations. We are mainly concerned in establishing the limits of such SET based circuit designs, thus we are interested in establishing theoretical bounds for delay and size, measured in terms of circuit elements[2], of the proposed implementations.

6.1 Addition & Subtraction

In this subsection we assume binary encoded n-bit operands, $A = (a_0, a_1, \ldots, a_{n-1})$ and $B = (b_0, b_1, \ldots, b_{n-1})$, and propose an electron counting scheme to compute the result of their addition/subtraction. The basic idea behind the method is first to convert the operands from digital to charge representation, add/subtract them in charge format, and convert the result back to binary digital representation.

Assuming binary operands, the first step in any electron counting process will be to convert a binary integer value X to its discrete analog equivalent Xe using a Digital to Analog Converter (DAC) which follows the general organization of the one introduced in [4]. As described in Section 5.1 the $MVke$ circuit (depicted in Figure 4) can be utilized to add/remove a number of electrons to/from a charge reservoir. When multiple such $MVke$ blocks operate in parallel on the same charge reservoir, electrons can be added to the reservoir in parallel. More specific, to convert an operand $X = (x_0, x_1, \ldots, x_{n-1})$, each bit x_i, $i = 0, 1, \ldots, n-1$ is connected to the E input of an $MVke$ block that has the V input hardwired to a bias potential that induces a $V \times k$ value equal with 2^i. Therefore, the operand X can be encoded as $\sum_{i=0}^{n-1} x_i 2^i e$ at the cost of n $MVke$ blocks in "add" mode. Thus this new DAC scheme has an $O(n)$ asymptotic complexity in terms of circuit elements.

Given the $MVke$-DAC encoding scheme described above, the addition and subtraction operations can be implemented in a straightforward manner. The addition of two n-bit operands A and B can be embedded in the conversion process if the operands are converted into charge format, via a total of $2n$ $MVke$ blocks in "add" mode that share a single charge reservoir. Similar, the subtraction operation, i.e., $A - B$, can also be embedded in the conversion process for the same cost. In this case, the $MVke$ blocks converting B operate in "remove" mode, encoding B as $-Be$, while still operating on the same electron reservoir.

Once the result corresponding to the addition/subtraction is available in the charge reservoir as a charge Ye, where

---

[2]By circuit element we mean in this context any of the building blocks presented in Section 5. We also assume that all the building blocks have the same cost, measured in terms of tunnel junctions and capacitors, and the same delay. More detailed computations can be also made in order to evaluate the proposed networks in terms of tunnel junctions and capacitors but such computations are beyond the scope of the paper. We preferred to use the generic concept of circuit element to keep the discussion implementation independent and to simplify the derivations.

$Y = A + B$ or $Y = A - B$, we need to convert this result back to a digital format in order to finalize the computation process. To achieve this an Analog to Digital Conversion (ADC) process is required. In the following we propose an ADC circuit that is taking advantage of the periodic transfer function of the SET transistor.

If N is the maximum number of extra electrons that can be present in the result electron reservoir, $m = 1 + \lceil \log N \rceil$ bits are required to represent this value in binary format. Then, following the base 2 counting rules, any ADC output bit s_i, $i = 0, 1, \ldots, \lceil \log N \rceil$ is equal to 1 inside an interval that includes 2^i consecutive integers, every 2^{i+1} integers, and 0 otherwise. Thus each bit s_i can be described by a periodic symmetric function with period 2^{i+1}. Then each output bit s_i can be computed by a PSF block that had been adjusted in order to have a transfer function that copies the periodic symmetric function required for the bit position i.

Thus we can implement an m-bit ADC using m PSF blocks (the PSF applied at bit position i is tuned to exhibits the periodic transfer function corresponding to that s_i bit) that operate in parallel on an electron reservoir. Given that we are addressing the particular case of n-bit operand addition, such that $m = n + 1$, then the cost of the required ADC circuit is in the order of $O(n)$.

Summarizing, the electron counting based addition/subtraction of two n-bit operands can be implemented with a depth-2 SET network built with $3n + 1$ circuit elements, therefor with an $O(n)$ asymptotic complexity measured in terms of circuit elements.

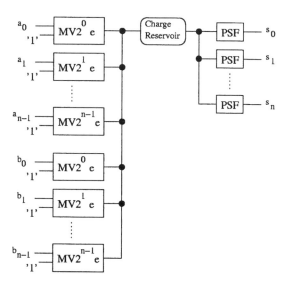

Figure 7. Organization of n-bit addition/subtraction circuit.

The overall organization of the circuit is depicted in Figure 7. We note here that in the Figure, the value k of the $MVke$ blocks has been drawn inside the block to suggest that it was implemented by properly adjusting the C_S value, while all inputs V have been fixed to the equivalent of a charge reservoir with $1e$ charge.

Even though the proposed scheme was meant for addition/subtraction it has a broader scope and some of its alternative utilizations are discussed in the following:

- **n-bit parity function:** The scheme can be applied for $O(1)$ delay computation of the n-bit parity function as follows: The n inputs are connected via $MVke$ blocks (all the V inputs are hardwired to a bias potential that induces $V \times k = 1$) to a charge reservoir that provides input information to one PSF, molded to have the transfer function equal to 1 inside an interval that includes 2^0 consecutive integers, every 2^1 integers, and 0 otherwise, i.e., corresponding to parity. Thus the n-parity function can be computed with a depth-2 network constructed with $n+1$ circuit blocks. When compared to Boolean and Threshold logic based schemes (an AND-OR implementation of the n-input parity requires 2^{n-1} n-input AND gates and one 2^{m-1}-input OR gate) this is a substantial improvement.

- **$n|\log n$ counters:** To implement an $n|\log n$ counter we just have to augment the n-bit parity network with $\log n - 1$ additional PSF blocks. Thus the $n|\log n$ counter can be implemented with a depth-2 network constructed with $n + \log n$ circuit elements.

- **Multiple operand addition:** The addition scheme can be easily extended to support multiple operand addition by connecting more than 2 DAC circuits to the same electron reservoir and adjusting the ADC converter in order to be able to convert values up to $m \times (2^n - 1)$, where m is the number of operands and n their bit length. Thus the m n-bit multiple operand addition can be implemented using $m \times n$ $MVke$ blocks and at most $n + \lceil \log m \rceil + 1$ PSF blocks.

6.2 Multiplication

In this section we propose an electron counting multiplication scheme that follows to some extent the paradigm we introduced for addition. Assume we have the input operands $A = (a_0, a_1, \ldots, a_{n-1})$ and $B = (b_0, b_1, \ldots, b_{n-1})$ and we want to compute $P = A \times B$.

One way to do the multiplication is to utilize the multiple operand addition scheme presented at the end of the previous section. To utilize that scheme we have to calculate first all the partial products $a_i b_j$, $i = 0, 1, \ldots, n-1$, $j = 0, 1, \ldots, n-1$ with n^2 2-input AND gates. Subsequently, each row of partial products is connected to a DAC structure with the $MVke$ block input V hardwired to a potential that reflects the correct weight for the partial product it processes. This implies that n DAC circuits are now connected to the charge reservoir and that the ADC converter is adjusted in order to be able to convert values up to $n \times (2^n - 1)$. Thus the multiplication can be implemented with a depth-3 network constructed with n^2 2-input AND gates, n^2 Mke blocks, and $2n - 1$ PSF blocks. This implies that the overall asymptotic complexity of the multiplication circuit is in the order of $O(n^2)$ circuit elements.

In the sequel we introduce a different technique that makes use of the ability to transport a variable number of electrons to/from a charge reservoir exhibited by the $MVke$ structure discussed in Section 5.1 and depicted in Figure 4. Such a block can transport $V \times k$ electrons when k is a built-in constant (can be changed via a circuit parameter, i.e., C_S value) and V is a variable specified by the content of a charge reservoir.

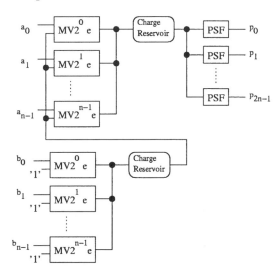

Figure 8. Organization of n-bit multiplication circuit.

The basic idea behind the scheme is again to add a charge Pe to a charge reservoir and to utilize an ADC structure to obtain the binary representation of the product P. The general organization of the proposed multiplication circuit is depicted in Figure 8. Again, the value k of the $MVke$ blocks has been drawn inside the blocks themselves to suggest that that k value was implemented inside the block by properly adjusting the C_S value. The scheme is utilizing a clock for synchronization purposes[3] and the computation process can be described as follows: First, on the positive clock value, a number of electrons corresponding to the value of the B operand, i.e., $\sum_{i=0}^{n-1} b_i 2^i$, are added to

---
[3] We assume here a level triggered behavior but the scheme can work with edge triggered policy as well.

the corresponding charge reservoir. This is achieved with n $MVke$ blocks each of them assuming as inputs the b_i bit and having the V input hardwired to the equivalent of a charge reservoir with $1e$ charge, such that $V \times k = 2^i$. Second, on the negative clock value, a charge of $A \times Be$ is added to the other charge reservoir. This is achieved with n $MVke$ blocks assuming as inputs the a_i bits and the analog value present on the charge reservoir processed in the previous computation step. As each $MVke$ block in this stage contributes $a_i \times 2^i \times B$ electrons, a final charge of $\sum_{i=0}^{n-1} a_i 2^i \times Be$, i.e, $A \times Be$ is present in the output charge reservoir when the second step is completed. Last, the value on the output charge reservoir is converted to a binary encoded value representing the product with $2n - 1$ PSF blocks.

This new scheme still implies a depth-3 network but requires $2n$ $MVke$ blocks and $2n - 1$ PSF blocks. Thus the new scheme reduces the cost in terms of $MVke$ blocks from $O(n^2)$ to $O(n)$ and therefore, the overall asymptotic complexity is also reduced to $O(n)$ as no partial products have to be explicitly generated.

7 Conclusions

Single Electron Tunneling (SET) technology offers a potential for (sub)nanometer feature size scaling, room temperature operation, as well as ultra-low power consumption. However, it displays a switching behavior that differs from traditional MOS devices. This provides new possibilities and challenges for implementing computer arithmetic circuits. In this line of reasoning we investigated the implementation of basic arithmetic functions, such as addition and multiplication, in SET technology. First, we described the SET equivalent of two conventional design styles, namely the equivalents of CMOS and threshold logic gates. Second, we proposed a set of building blocks, which can be utilized for a novel design style, namely arithmetic operations performed by direct manipulation of the location of individual electrons within the system. Third, using the new set of building blocks, we proposed several novel approaches for computing arithmetic functions, e.g., addition, parity, counting, multiplication, via the controlled transport of individual electrons. Related to these new electron counting schemes we proved that the following holds true: the addition/subtraction of two n-bit operands can be computed with a depth-2 network composed out of $3n + 1$ circuit elements; the n-parity function can be computed with a depth-2 network constructed with $n + 1$ circuit elements; $n|\log n$ counter can be implemented with a depth-2 network constructed with $n + \log n$ circuit elements; m n-bit multiple operand addition can be implemented using at most $nm + n + \lceil \log m \rceil + 1$ circuit elements; and finally the multiplication of two n-bit operands can be computed with a depth-3 network with $4n - 1$ circuit elements.

References

[1] C.Lageweg, S.Cotofana, and S.Vassiliadis. Static Buffered SET Based Logic Gates. In *2nd IEEE Conference on Nanotechnology (NANO)*, August 2002.

[2] C.Lageweg and S.Cotofana and S.Vassiliadis. A Linear Threshold Gate Implementation in Single Electron Technology. In *IEEE Computer Society Workshop on VLSI*, pages 93–98, April 2001.

[3] C.Lageweg and S.Cotofana and S.Vassiliadis. Achieving Fanout Capabilities in Single Electron Encoded Logic Networks. In *6th International Conference on Solid-State and IC Technology (ICSICT)*, October 2001.

[4] C.Lageweg and S.Cotofana and S.Vassiliadis. Digital to Analog Conversion Performed in Single Electron Technology. In *1st IEEE Conference on Nanotechnology (NANO)*, October 2001.

[5] C.Wasshuber. *About Single-Electron Devices and Circuits*. PhD thesis, TU Vienna, 1998.

[6] J.R.Tucker. Complementary Digital Logic based on the "Coulomb Blockade". *Journal of Applied Physics*, 72(9):4399–4413, November 1992.

[7] K.K.Likharev. Correlated Discrete Transfer of Single Electrons in Ultrasmall Tunnel Junctions. *IBM Journal of Research and Development*, 32(1):144–158, January 1988.

[8] K.K.Likharev. Single-Electron Devices and Their Applications. *Proceeding of the IEEE*, 87(4):606–632, April 1999.

[9] L.J.Geerligs, V.F.Anderegg, P.A.M.Holweg, J.E.Mooij, H.Poitier, D.Esteve, C.Urbina, and M.H.Devonet. Frequency-Locked Turnstile Device for Single Electrons. *Physical Review Letters*, 64(22):2691–2694, May 1990.

[10] M.Goossens. *Analog Neural Networks in Single-Electron Tunneling Technology*. PhD thesis, Delft University of Technology, 1998.

[11] K. M.Kirihara. Asymmetric Single Electron Turnstile and Its Electronic Circuit Applications. *IEICE Transactions on Electronics*, E81-C(1):57–61, January 1998.

[12] S.Muroga. *Threshold Logic and its Applications*. Wiley and Sons Inc., 1971.

[13] Technology roadmap for nanoelectronics. Downloadable from website http://www.cordis.lu/esprit/src/melna-rm.htm, 1999. Published by the Microelectronics Advanced Research Initiative (MELARI NANO), European Commission.

[14] Y.Ono, Y.Takahashi, K.Yamazaki, M.Nagase, H.Namatsu, K.Kurihara, and K.Murase. Fabrication Method for IC-Oriented Si Single-Electron Transistors. *IEEE Transactions on Electron Devices*, 49(3):193–207, March 2000.

[15] Y.Taur, D.A.Buchanan, W.Chen, D.Frank, K.Ismail, S.Lo, G.Sai-Halasz, R.Viswanathan, H.Wann, S.Wind, and H.Wong. CMOS Scaling into the Nanometer Regime. *Proceeding of the IEEE*, 85(4):486–504, 1997.

The Interval Logarithmic Number System

Mark G. Arnold and Jesus Garcia
Lehigh University
CSE Department
Bethlehem, PA 18015, USA
{marnold,jegj}@eecs.lehigh.edu

Michael J. Schulte
University of Wisconsin-Madison
Department of ECE
Madsion, WI 53706, USA
schulte@engr.wisc.edu

Abstract

This paper introduces the Interval Logarithmic Number System (ILNS), in which the Logarithmic Number System (LNS) is used as the underlying number system for interval arithmetic. The basic operations in ILNS are introduced and an efficient method for performing ILNS addition and subtraction is presented. The paper compares ILNS to Interval Floating Point (IFP) for a few sample applications. For applications like the N-body problem, which have a large percentage of multiplies, divides and square roots, ILNS provides much narrower intervals than IFP. In other applications, like the Fast Fourier Transform, where addition and subtraction dominate, ILNS and IFP produce intervals having similar widths. Based on our analysis, ILNS is an attractive alternative to IFP for application that can tolerate low to moderate precisons.

1. Introduction

Floating-point arithmetic is the dominant choice for scientific and graphics applications; however, other alternatives exist. These alternatives are desirable for niche applications that have special requirements. An alternative with very similar properties to floating point is the Logarithmic Number System (LNS), which can be more cost effective for applications that require low-to-moderate precision. Advantages of LNS include: low power consumption [23], efficient matrix operations [8], and low-cost multiplies, divides, powers, square roots, etc., all of which are useful in massively parallel scientific simulation [19].

Interval arithmetic is an alternative used instead of conventional floating point to give reliable bounds on roundoff, approximation and inexact-input errors. Interval arithmetic is often used in scientific code to explore the effect of measurement error on the final result of a computation. For example, astronomical applications of interval arithmetic have included calculating the effects of experimental error to obtain Newton's gravitational constant to a higher degree of accuracy [12], and analyzing the stability of solar systems [15].

This paper proposes combining LNS and interval arithmetic. To the authors' knowledge, this combination of interval arithmetic and LNS has not been studied previously. This may primarily be due to the perception that interval arithmetic, which is often used for high-precision applications, is incompatible with the low or moderate precision applications where LNS offers significant advantages. We have identified at least two applications (the N-body problem [19] and computer graphics [29]), where this combination of interval arithmetic and LNS arithmetic may offer significant advantages. ILNS is especially attractive in application-specific systems that allow for specialized hardware designs. The simple algorithms chosen in this paper to illustrate the benefits of ILNS frequently are implemented using application-specific systems, rather than general-purpose processors. Future research will investigate algorithms that use interval arithmetic to achieve convergence by interval methods, and the advantages that ILNS yields for them.

The remainder of this paper is organized as follows. Section 2 gives an overview of LNS. Section 3 discusses interval arithmetic and shows how LNS arithmetic can be used to implement interval LNS. Section 4 describes routines that are used to simulate algorithms with conventional and logarithmic interval arithmetic. Section 5 describes how various algorithms perform using these two arithmetics. Section 6 discusses an implementation of the proposed system. Section 7 summarizes the effects observed in the simulations and gives conclusions.

2. The Logarithmic Number System

LNS represents a real number, x, using a sign bit, $sign(x)$, and the fixed point, base-b logarithm of the absolute value of that real number, such that $\bar{x} = \text{round}(\log_b |x|, \text{mode})$, rounded using some mode (e.g., towards nearest, towards $\pm\infty$, or by special LNS modes [4]). The representation of a quotient, q, is simply the difference of the logarithms ($\bar{q} = \bar{x} - \bar{y}$) and the exclusive OR of the sign bits (sign(q) = sign(x)$\oplus$sign(y)). A similar approach holds for products by adding logarithms and for square roots by right shifting the logarithm. When the operands are exact and the results do not overflow, the LNS product or quotient is exact, because only fixed point operations are involved on the logarithms. This is an advantage compared to floating point, where only a small subset of products and quotients of exact operands produce exact results. For example, if $b = 2$ with the inputs $x = -8$ and $y = -1/\sqrt{2}$, we have $\bar{x} = 3$, sign(x) = 1, $\bar{y} = -0.5$ and sign(y) = 1, which are exact representations. The quotient, $q = 8\sqrt{2}$, is represented as $\bar{q} = 3.5$ and sign(q) = 0, which is an exact representation in this base-2 LNS.

LNS addition and subtraction [14] are more complicated and their results are seldom exact. Addition can be rewritten as:
$$x + y = y \cdot (1 + x/y). \quad (1)$$
The advantage of (1) is that a function of two variables ($x + y$) has been reduced to a function of one variable (increment of the quotient). In order to increment with LNS, the hardware needs an implementation of the *addition logarithm*,
$$s_b(z) = \log_b(1 + b^z) \quad (2)$$
and (for applications involving adding numbers of opposite signs) the *subtraction logarithm*,
$$d_b(z) = \log_b |1 - b^z|. \quad (3)$$
From (1), when the signs of the operands are the *same*,
$$\begin{aligned}\log(|x| + |y|) &= \log_b(|y| \cdot (1 + |x|/|y|)) \\ &= \bar{y} + \log_b(1 + b^{(\bar{x}-\bar{y})}) \\ &= \bar{y} + s_b(\bar{x} - \bar{y}). \end{aligned} \quad (4)$$
When the signs of the operands *differ*, a similar approach holds:
$$\log||x| - |y|| = \bar{y} + d_b(\bar{x} - \bar{y}). \quad (5)$$
The result of s_b or d_b must be rounded in some way [4] to make it representable in the finite $b = 2$ LNS.

For example, $x = -8$ and $y = -1/\sqrt{2}$, gives $\bar{x} - \bar{y} = 3.5$ and sign(x) = sign(y), Thus, $s_2(3.5) = 3.6221934162022...$, which must be rounded. The computed result of (4), $\log_2(|x| + |y|) = -0.5 + s_2(3.5) \approx 3.122... \approx \log_2(8.707...)$, is only an approximation of the exact value $\log_2(8 + 1/\sqrt{2})$.

3. Interval Arithmetic

Interval arithmetic specifies a precise method for performing arithmetic operations on intervals [21]. An interval-arithmetic operation takes intervals as input operands and produces an interval as the output result. The output interval is defined by lower and upper endpoints, such that the true result is guaranteed to lie in this interval. The width of the interval (i.e., the distance between the two endpoints) indicates the accuracy of the result or the certainty with which the result is known.

Interval arithmetic is implemented by using some underlying real number system and its associated arithmetic operations. For example, the underlying real number system could be exact rationals (as available in LISP) [27] that require an unbounded amount of memory per number. In this case, the resulting intervals are as narrow (precise) as the algorithm (involving the four basic operations on intervals) allows for the given inputs.

More commonly, the underlying real number system is chosen to be one that is easier to implement in hardware. Often, such systems have a constant, rather than unbounded, datum size that is determined when the hardware is fabricated (i.e., the word size). For example, the fixed-point number system has the advantage that overflow-free addition and subtraction are exact using simple integer ALUs, but suffers from the fact that roundoff error occurs for multiplication and division. Another disadvantage of fixed point interval arithmetic is its limited dynamic range [24].

Although specialized hardware for variable-precision floating-point interval arithmetic [26] has been suggested, the most common number system for implementing interval arithmetic has been conventional floating point, as exemplified by the IEEE-754 standard [13]. This paper refers to this implementation as Interval Floating Point (IFP). A description of IFP and its implementation using IEEE-754 arithmetic is provided in [11]. IFP offers greater dynamic range than interval fixed point for the same constant word size, but suffers from roundoff errors, in many cases, for all four basic arithmetic operations (+, -, *, /).

This paper introduces the Interval Logarithmic Number System (ILNS), which uses LNS to perform interval arithmetic. LNS has approximately the same dynamic range as floating point, but LNS has the advantage that roundoff never occurs in the operations of multiplication and division. This suggests ILNS may offer interval widths closer to the theoretical minimum for the same constant word size as IFP. ILNS is most useful for applications that need low-to-moderate precision and that have a predominance of multiplications, divisions, square roots and other powering operations. Although only a subset of applications meet these criteria, for those that do, ILNS may offer significant advantages.

Interval arithmetic was originally proposed as a tool for bounding roundoff errors in numerical computations [21]. It can also used to determine the effects of approximation errors and errors that occur due to non-exact inputs [1]. Interval arithmetic is especially useful for scientific computations, in which data is uncertain or can take a range of values. For example, Kreinovich and Bernat have used interval arithmetic to investigate the stability of the solar system [15].

In the discussion to follow, intervals are denoted by capital letters and real numbers are denoted by small letters. The lower and upper endpoints of an interval X are denoted as x_l and x_u, respectively. As defined in [21] and [1], a closed interval $X = [x_l, x_u]$ consists of the set of real numbers between and including the two endpoints x_l and x_u (i.e. $X = \{x : x_l \leq x \leq x_u\}$). An real number x is equivalent to the degenerate interval $[x, x]$.

The ILNS representation, $\bar{X}$, of an interval $X > 0$ is itself an interval involving the fixed-point logarithmic representations of the endpoints: $\bar{X} = [\bar{x}_l, \bar{x}_u]$. When an algorithm allows $x_l < 0$, the ILNS representation must also include sign(x_l) and sign(x_u).

The width and midpoint of an interval X are defined as:

$$\text{width}(X) = x_u - x_l \quad (6)$$
$$\text{midpoint}(X) = (x_l + x_u)/2 \quad (7)$$

The endpoints of a result interval may not be representable in the underlying number system. In this case, the endpoints have to be rounded outwards: the lower towards $-\infty$, which is denoted as $\triangledown()$, and the upper towards $+\infty$, which is denoted as $\triangle()$. This guarantees that the *true* result is still contained in the new interval, but introduces an undesired uncertainty, by making the interval wider, as defined by (6).

Conventional interval addition and subtraction are defined as [21]:

$$X + Y \approx R = [\triangledown(x_l + y_l), \triangle(x_u + y_u)] \quad (8)$$
$$X - Y \approx R = [\triangledown(x_l - y_u), \triangle(x_u - y_l)] \quad (9)$$

The ILNS implementation of addition is:

$$\bar{R} = [\bar{y}_l + \triangledown(f_l(\bar{x}_l - \bar{y}_l)), \bar{y}_u + \triangle(f_u(\bar{x}_u - \bar{y}_u))], \quad (10)$$

where f_l is either s_b or d_b, depending on whether sign(x_l) = sign(y_l). Likewise, f_u is either s_b or d_b, depending on whether sign(x_u) = sign(y_u). Certain algorithms, like Euclidean-distance calculations, guarantee positive signs, and therefore $f_l = f_u = s_b$.

Conventional interval multiplication [21] is defined as:

$$X \cdot Y \approx R = [\triangledown(\min(x_l y_l, x_l y_u, x_u y_l, x_u y_u)), \quad (11)$$
$$\triangle(\max(x_l y_l, x_l y_u, x_u y_l, x_u y_u))].$$

Alternatively, the interval endpoints of X and Y that give the correct interval product can be determined by examining their sign bits. Since when $x_l > 0$, x_u must also be positive, there are three possible cases for sign(x_l) and sign(x_u) This results in a total of nine cases [22]. With this technique, only two multiplications are required, unless both X and Y cross zero. Note that unlike IFP, ILNS does not require rounding since it is implemented using addition.

Interval division is defined as:

$$X/Y \approx R = [\triangledown(\min(x_l/y_l, x_l/y_u, x_u/y_l, x_u/y_u)), \quad (12)$$
$$\triangle(\max(x_l/y_l, x_l/y_u, x_u/y_l, x_u/y_u))]$$

if Y does not contain zero. Again, this can be implemented with no rounding in ILNS, since division is implemented by subtracting logarithms. If Y contains zero, the resulting interval is infinite. To allow for division by an interval that contains zero, extended interval arithmetic is required [10]. Extended interval arithmetic specifies results for division by an interval that contains zero and for operations on plus and minus infinity, as provided by IEEE-754 [13]. ILNS can be extended to deal with this in an analogous way [2]. Like the product, each endpoint of the quotient can be determined by a single division after examining the sign bits [21].

Interval arithmetic is also defined for the elementary functions. If an elementary function $f(x)$ is monotonically increasing on $X = [x_l, x_u]$, the resulting interval is $f(X) = [f(x_l), f(x_u)] \approx [\triangledown(f(x_l)), \triangle(f(x_u))]$. For example, when $x_l \geq 0$, the square-root function can be computed as

$$\sqrt{X} \approx [\triangledown(\sqrt{x_l}), \triangle(\sqrt{x_u})]. \quad (13)$$

4. Interval Arithmetic Emulation

To investigate the numerical differences between ILNS and IFP, a library of basic interval operations was written for ILNS and IFP addition, subtraction, multiplication, division, and square root. Algorithms using this library perform each arithmetic operation on interval data. The underlying number representation chosen greatly affects interval results, since every time an endpoint of a result cannot be exactly represented, it must be rounded. The frequency of this rounding and the size of the roundoff error determine the width of the interval results.

The IEEE-754 standard [13] for floating point represents values in a 32-bit single-precision format, where the MSB is the sign bit (sign(x)), the following eight bits represent the exponent (e), and the remaining 23 bits represent the mantissa (m). The mantissa is a fixed-point value, with a binary point to the left of the mantissa and an implicit 1 to the left of the binary point. The value, x, represented by the IEEE-754 single-precision format is:

$$x = (1 - 2\,\text{sign}(x))2^{e-127}(1 + m), \quad (14)$$

except in the subnormal region. The maximum roundoff error in FP is given by

$$\epsilon_{FP} = 2^{e-127} \cdot 2^{-23} = 2^{e-150}, \quad (15)$$

which is relative to the exponent, but not to the mantissa. Thus, the maximum relative error increases as the mantissa decreases.

The selected LNS implementation also uses a 32-bit format. The base is 2, the MSB is the sign bit, and a 31-bit fixed-point exponent l follows it. The first 8 bits in the exponent are the integer part, and the last 23 bits come after the binary point. This allows a similar dynamic range and also similar accuracy [4] compared to single-precision floating point. The value, u, represented by the LNS format is

$$u = (1 - 2\,\text{sign}(u))2^{l-128}, \quad (16)$$

where in the notation of the previous sections, the LNS representation is $\bar{u} = 128 - l$ and $\text{sign}(u)$. (16) provides constant relative precision. The maximum roundoff error is given by

$$\epsilon_{LNS} = 2^{l-128}(2^{2^{-23}} - 1) \approx 2^{l-151.5}, \quad (17)$$

which is proportional to the value of u. The value *zero* is considered a special case, and is represented with a reserved value. When zero is used in an operation, it is detected and the result is exact. Denormal and infinite values, although possible in LNS [2], were not implemented for this research.

Emulating IFP operations is straightforward. The IEEE-754 compliant FPU in any modern general-purpose processor allows setting the rounding mode toward $+\infty$ or $-\infty$, as required by the interval operations.

To emulate an LNS ALU, LNS values are stored using the 32-bit sign and exponent representation. Multiplication and division are exact (unless overflow occurs) and can be implemented with the simple fixed-point addition or subtraction of the logarithmic representations. Operations that are not exact require outward rounding. Assuming an exact representation of the logarithm of the absolute value of the result (which, in general, would require infinite precision), the result is truncated to 23 bits. Positive values rounded toward $+\infty$ and negative values rounded toward $-\infty$ need to have 2^{-23} added to the truncated logarithm. This is accomplished by adding 1 to the LSB, to produce the correctly rounded result. In the remaining cases, truncation is enough. The assumption of an exact result is justified for each operation in the next two paragraphs.

In LNS, the square root is equivalent to dividing the exponent by 2, and is therefore implemented using a 1-bit right shift. The exact result can be represented with just one extra precision bit. There is a 50% probability that the endpoints for (13) are exact and do not require rounding. Rounding is straight-forward, since only the the single bit that was shifted out and the rounding direction need to be examined.

Addition and subtraction in LNS are more complicated. Of the steps in (4) and (5), only the computations of s_b and d_b are subject to roundoff errors. To compute the result, (2) and (3) are computed as shown. In the ILNS library routines, the separate steps involving exponentials and logarithms in (2) and (3) are computed in double-precision floating point, and the final result is rounded as required (depending on which interval endpoint is being calculated). Double precision is sufficiently more accurate than the precision used for the simulation that such double-precision results rounded towards $+\infty$ or $-\infty$ should form the correct interval: $[\nabla(s_b(z_l)), \triangle(s_b(z_u))]^1$. In practice, some hardware implementations may reduce the precision with which (2) and (3) are calculated, and force over-approximation or under-approximation, as required to guarantee a correct interval result. The straightforward implementation using double precision indicates the maximum improvement when switching from IFP to ILNS.

To complete the interval support, it is necessary to define how a value is assigned to an interval variable so that it can be used with the interval functions. The assignment converts the value to the number representation being used (FP or LNS). If it has an exact representation, then the interval has zero-width, as in $B = [1, 1]$, which has an exact representation in both FP and LNS. If the value cannot be represented exactly, the nearest upper and lower exactly-representable values are used as the endpoints for the generated interval.

5. Algorithm Comparison

The algorithms chosen to compare the width of result intervals for IFP and ILNS illustrate the viability of using ILNS as an alternative for IFP in certain real-world applications. Due to the nature of the representation, ILNS has the greatest advantage for algorithms that make extensive use of multiplications, divisions, squares and square roots. These include the proposed target applications, such as gravity-force computation in astronomical simulations or normalized-distance calculation for graphics applications. It is also shown that ILNS produces interval widths that are comparable to IFP for other important algorithms, such as the Fast Fourier Transform, where the number of multiplications and divisions is significantly less than the number of additions and subtractions.

5.1. Fast Fourier Transform

The Fast Fourier Transform (FFT), which has been implemented successfully using LNS [28], is a very important algorithm in digital signal processing (DSP), due both its usefulness in changing from the time-domain to

---
[1]Based on data presented in [25], this seems likely to be true, but we have not yet had a chance to verify it.

the frequency-domain and to its optimized implementation. The FFT requires significantly fewer multiplications than additions, which would seem to favor IFP over ILNS.

The selected implementation is a 32-point, real-input, decimation-in-frequency FFT. The FFT algorithm has five radix-2 stages, after which bit-reversing is applied to the outputs. The complex values involved in the computation of the FFT are represented with a real and an imaginary component. Each of these components is an interval defined by its two endpoints.

Two experiments involving different inputs are presented. The first experiment uses a sinusoidal wave of amplitude 10, plus white noise of amplitude 0.5. The second uses a square wave, also with an amplitude of 10, and the same noise. Both are real-valued inputs, in which the imaginary part of each complex input interval is set to zero. The frequency of both waves is 1.0. Although there is a random component involved, the randomly generated values are stored, and identical inputs are supplied to the IFP and ILNS FFT algorithms. Both the input values and the twiddle factors are generated as double-precision FP numbers. These values are then transformed into intervals with 32-bit endpoints. The width of these intervals depends on the corresponding number system, IFP or ILNS, as is explained in Section 4.

The simulation is executed 10,000 times on noisy input waves. Thus, for each input sequence, there are 10,000 FFT computations, producing 320,000 points, each consisting of two intervals (Re and Im) that are defined by two endpoint values. The outputs vary between simulation runs due to the spectral components introduced by the noise signal.

To establish a fair comparison, we take into account that roundoff errors introduced in the interval calculations are relative to the number being represented, as shown in Section 4. The *relative width* of each interval is defined as

$$\text{width}_{rel}(A) = \frac{\text{width}(A)}{\text{midpoint}(A)}, \quad (18)$$

where width(A) is computed using (6) and midpoint(A) is computed using (7). This is adequate for algorithms in which catastrophic cancellation (i.e., subtraction of similar values, yielding an inaccurate result close to zero) does not occur. In the FFT, however, this problem exists, since intervals whose endpoints surround zero may produce extremely large relative errors, and mask the behavior of the other outputs.

Therefore, the selected measure is the ratio of the IFP and ILNS output interval widths for the same input. The ratio obtained is a measure of how much better ILNS performed for a particular output (i.e., how much wider the IFP output is than the ILNS output). Although the ratios for every interval are comparable, they can be slightly misleading, since a ratio of 0.125 has the same relevance but in the opposite direction as a ratio of 8. Therefore, the $\log_2()$ of each

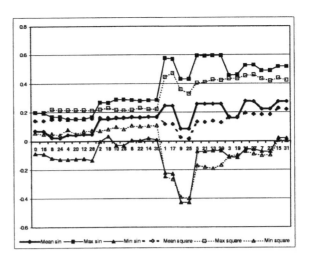

Figure 1. Logarithmic ratios (IFP/ILNS) of real FFT components for sine and square waves.

ratio is calculated. These measures are called the *logarithmic ratios*. This produces an intuitive result, where positive values represent narrower intervals for ILNS, and negative values indicate narrower intervals for IFP.

The results for the real part of the outputs are presented in Figure 1. This graph combines the outputs of both the sine-wave and square-wave experiments. The data presented corresponds to the maximum, minimum and mean of the logarithmic ratios, at each output index for the 32-point FFT. The reason for analyzing the outputs in this way is that the 32-point FFT algorithm represents 32 different sub-algorithms; one for each output. These outputs go through a different number of additions and multiplications, in which different twiddle factors are used. Outputs with the same index, but corresponding to different periods of the input waves, show a very similar behavior, because they are generated by the same sub-algorithm.

The outputs are ordered according to their index before being bit-reversed, since this is helpful in analyzing the results. It is obvious that outputs with an even index (the leftmost sixteen outputs in the plot) present a lower, less variable logarithmic-ratio mean. This is because these outputs are not multiplied in the first FFT stage. The first stage is especially important, because intervals generated in it expand in every consecutive stage. Since even-indexed FFT outputs do not have multiplication in the first stage, ILNS and IFP tend to produce similar intervals. Outputs in the right half of the plot (odd indices) experience a multiplication in the first stage and show, in general, more advantage for ILNS. Multiplication in early stages increase this advantage. Differences in the midpoint values of the intervals being added cause a variability in these results that cannot be predicted just by taking multiplications into account.

The significant observation from Figure 1 is that for the sine- and square-wave inputs given, the mean of the logarithmic ratios are positive for all FFT outputs, indicating similar or slightly better ILNS results. There is variation in the logarithmic ratios between the different points produced by the FFT. For example, some outputs (like 1, 3 and 31) are noticeably better in ILNS, while others (like 8, 24, 9 and 25) are not. We attribute much of this variation to the presence or absence of multiplication (which favors ILNS) in the first stage of the FFT.

5.2. Gravitational-force computation

N-body simulation is a technique that consists of simulating the evolution of a system with N bodies, where the force exerted on each body depends on its interaction with every other body in the system. Typical applications for this technique are studies of astrophysical and molecular systems, where the size of the bodies (stars and molecules, respectively) is negligible compared to the distances between them, which allows bodies to be represented by simple points in a 3-D space.

In astrophysical N-body simulation, the trajectories of stars are calculated by integrating the force (acceleration) due to gravitational interaction. One straightforward method is to calculate, for each star, the component due to each other star present in the simulation. This is accomplished by evaluating the expression

$$\vec{a}_i = \sum_{j=1}^{N} \frac{\vec{x}_j - \vec{x}_i}{(r_{ij}^2 + \epsilon^2)^{3/2}}, \quad (19)$$

where $\vec{a}_i$ is the gravitational acceleration at the position of particle i, $\vec{x}_i$ is the position of particle i, r_{ij} is the distance between particles i and j, and ϵ is the artificial-potential softening used to suppress the divergence of the force as $r_{ij} \to 0$ [19]. The x-coordinate terms in $\vec{a}_i$ are calculated as

$$\frac{x}{(x^2 + y^2 + z^2 + \epsilon^2)^{3/2}}, \quad (20)$$

where $\vec{x}_j - \vec{x}_i = (x, y, z)$ is computed in fixed point and then x, y and z are converted to LNS. This approach is used in the GRAPE-3 processor designed by Makino [19]. Equation (20) is used in this paper to compare the interval widths obtained when using IFP or ILNS. In a complete system, the final addition of every gravity vector exerted on a given body should be taken into account. The whole N-body algorithm has not been implemented for this paper due to its complexity; however, note that the result of operations on intervals of non-minimum width is much more dependent on the widths of the inputs than on the associated rounding. Therefore evaluating equation (20) is considered significant.

The abundance of squares in (20) suggests ILNS will show a clear improvement over IFP. To test this, (20) is executed 100,000 times using both IFP and ILNS. The set of inputs, x, y, z and ϵ, are generated randomly for each computation. The variables x, y, and z take values from $[0, 1)$ and the error variable ϵ takes values from $[0, 0.001)$ (to reflect the smaller magnitude it usually has). This produces two sets of 100,0000 output intervals, one for each number representation. These outputs are similar to the calculations in one step of an N-body astrophysical simulation, where 100,000 stars are uniformly distributed in a normalized tridimensional cube.

To compare results when the inputs are not as narrow as minimum-width intervals, a similar experiment was conducted which increases the width of the input intervals. A new value x_W was generated for input x as

$$x_W = x + x \cdot 2^{-23+W} = x(1 + 2^{-23+W}), \quad (21)$$

and the endpoints of the interval were obtained by rounding x and x_W to the selected 32-bit number representation. This created input intervals with an approximately constant relative width of 2^{-23+W} (it is not perfectly constant because of rounding).

Input Intervals	Output Logarithmic Ratio			
	mean	σ^2	Max	Min
W = 0	0.676	0.043	1.508	-0.273
W = 1	0.395	0.032	2.036	-0.322
W = 2	0.234	0.010	0.909	-0.410
W = 3	0.130	0.003	0.469	-0.118
W = 4	0.069	0.001	0.246	-0.048
W = 5	0.035	0.000	0.126	-0.036

Table 1. Logarithmic-ratio statistics for the gravity equation with 100,000 sets of random inputs.

The comparison between IFP and ILNS output data sets was performed using the logarithmic ratio ($\log_2$(IFP/ILNS)). Statistical analysis was performed on the logarithmic ratios, and the results are shown in Table 1. As explained above, the experiment was repeated for minimum-width input intervals, with $W = 0$, and wider input intervals, with W ranging from 1 to 5. Since this algorithm only adds positive numbers, catastrophic cancellation never occurs. Statistics obtained using relative widths as defined in equation 18 agree with these results.

Table 1 shows that ILNS produces narrower intervals. Using the logarithmic ratios, the mean for minimum-width inputs is 1.598 ($\approx 2^{0.676}$) times wider for IFP output intervals than it is for ILNS output intervals. Calculating the ratio with the mean of the relative widths, the value obtained is 1.606. This is nearly a 60% wider FP interval on average, showing a very important advantage to using ILNS. Also

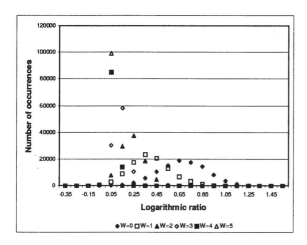

Figure 2. Absolute frequency of the logarithmic ratios of the gravity equation for various input interval widths.

widths. The data show a behavior similar to Table 1, except the mean of the logarithmic ratios increases from 0.676 to 0.903 for minimum-width intervals. (22) favors ILNS over IFP more than (20), since (20) has three additions, rather than two.

Input Intervals	Output Logarithmic Ratio			
	Mean	σ^2	Max	Min
W = 0	0.903	0.083	1.736	-0.269
W = 1	0.499	0.041	1.494	-0.281
W = 2	0.297	0.016	0.878	-0.212
W = 3	0.164	0.006	0.459	-0.130
W = 4	0.086	0.001	0.222	-0.095
W = 5	0.044	0.000	0.119	-0.043

Table 2. Logarithmic-ratio statistics for the normalized distance equation, for 100,000 sets of random inputs.

relevant is that the difference between ILNS and IFP decreases with an increase in the width of the input intervals. This is expected, as the width of the input interval, which is equal for both number representations, becomes the most important factor for the width of the output interval. The interval widening due to roundoff errors, where ILNS gets its advantage, become less and less relevant. This effect is summarized in Figure 2, which shows the graph for each input interval width. For minimum-width inputs ($W = 0$), the logarithmic ratios present a Gaussian-bell-like distribution around 0.676. As the widths of the input intervals increase, the mean moves toward zero and measures are more concentrated around zero, as IFP and ILNS become almost equivalent.

5.3. Normalized distance computation

Calculating normalized distances is a common operation in graphic applications. It consists of normalizing each component of a position vector, $\vec{r}$, by dividing it by $|\vec{r}|$. This computation is similar to the one considered in Section 5.2, and again the normalization of only one component is simulated, since components are random and it does not matter which component is calculated. The equation that is evaluated is

$$x_{norm} = \frac{x}{(x^2 + y^2 + z^2)^{1/2}}, \quad (22)$$

where $\vec{r} = (x, y, z)$. Values for x, y and z are randomly selected from the interval $[0, 1)$.

Equation (22) is similar to (20), yet there are important differences in the results. We performed the same analysis as in Section 5.2, and the results are presented in Table 2, with the simulation repeated for several input interval

Relative widths for each set of input interval widths, both for ILNS and IFP, again agree with the logarithmic ratios. These data show a similar behavior as (20) in Section 5.2, only with a larger difference in favor of ILNS. The output interval width increases by a factor of approximately 2 when W is incremented, showing how the input width directly impacts the output width.

6. Implementation

As described earlier, ILNS multiplication, division, squaring, and square root are trivial to implement using simple fixed-point circuits. For very limited-precision ILNS systems, direct lookup of $\triangledown(s_b(z_l))$ and $\triangle(s_b(z_u))$ from precomputed ROMs is possible.

Daumas and Matula [9] have studied rounding of elementary functions in floating point. Their method involves the introduction of extra flags in the floating-point ALU for correct directed rounding and faithful rounding. Although such an approach could be used to obtain $\triangledown(s_b(z_l))$ and $\triangle(s_b(z_u))$, the cost may be excessive.

Many approximations have been proposed in the conventional LNS literature, some of which support high precision

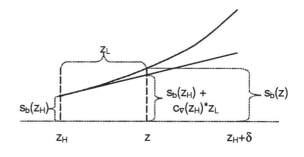

Figure 3. Linear-Taylor Interpolation.

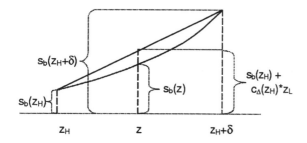

Figure 4. Linear-Lagrange Interpolation.

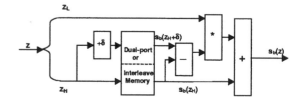

Figure 5. Lagrange Interpolation with Dual-port memory.

[7]. One way to approximate $s_b(z)$ that supports moderate precision is unpartitioned linear interpolation [6]:

$$s_b(z) \approx s_b(z_H) + c(z_H) \cdot z_L \qquad (23)$$

where z_L is the low portion of z such that $0 \le z_L < \delta$, z_H is the high portion of z such that $z_H \le z < z_H + \delta$, and $c(z_H)$ is the slope for the line used in that linear-interpolation region. The pre-computed values of $c(z_H)$ and $s_b(z_H)$ can be stored in ROMs that require $N + K$ address bits each, where K is the number of bits for the integer part of the z_H bus, and N is the number of bits for the fractional part of the z_H bus. The term *unpartitioned* here means $\delta = 2^{-N}$ is the constant distance between tabulated points, in contrast to partitioned methods [16] in which δ varies in an effort to minimize memory.

As is apparent in Figure 3, linear-Taylor interpolation forms a lower bound:

$$\nabla(s_b(z_H) + c_\nabla(z_H) \cdot z_L) \le s_b(z) \qquad (24)$$

where $c_\nabla(z_H) = s'(z) = 1/(b^{-z} + 1)$. Similarly, as shown in Figure 4, linear-Lagrange interpolation forms an upper bound on s_b:

$$s_b(z) \le \Delta(s_b(z_H) + c_\Delta(z_H) \cdot z_L) \qquad (25)$$

where $c_\Delta(z_H) = (s_b(z_H) - s_b(z_H + \delta))/\delta$. The linear-Lagrange method has the further advantage that the c_Δ memory can be eliminated when a dual-port [6] or interleaved [17] memory is used to hold $s_b(z_H)$, as shown in Figure 5. Of course, this technique only works with Lagrange interpolation.

Although two separate prior art techniques like these represented in Figures 3 and 4 could implement ILNS, this paper proposes a novel technique that achieves the same effect using less memory. Rather than requiring separate memories to compute $\nabla(s_b(z_l))$ and $\Delta(s_b(z_u))$, the proposed technique (illustrated in Figure 5) is able to generate both correct upper and lower bounds from a memory that only contains $s_b(z_H)$. These tabulated $s_b(z_H)$ points are stored with an additional G guard bits, making the memory size $2^{K+N} \times (N + G)$ bits. Computing $\nabla(s_b(z_l))$ and $\Delta(s_b(z_u))$ from the same dual-ported memory eliminates about $2^{K+N} \times (2N + 2)$ additional bits that would otherwise be required to store $c_\nabla(z_H)$ and $c_\Delta(z_H)$ with the prior approach.

In the proposed method, the values of $c_\nabla(z_H)$ and $c_\Delta(z_H)$ are reasonably close to each other, and only need to be computed with about $N + 1$ bits of accuracy [4] in order for (23) to produce a reasonable approximation. Therefore, the proposed design computes $c_\nabla(z_H)$ from $c_\Delta(z_H)$, which, in turn, is computed using the two values obtained from the dual-ported memory. The key to this novel interpolation approach is the following relationship between the slopes of the upper and lower bounds:

$$c_\Delta(z_H) - c_\nabla(z_H) \approx s_b''(z_H) \cdot \frac{\delta}{2} = b^{\log_b(s_b''(z_H)) + \log_b(\frac{\delta}{2})}. \qquad (26)$$

In order to capitalize on this relationship, we use a special property of the s_b function [5]:

$$\log_b(s_b''(z_H)) = z_H - 2 \cdot s_b(z_H) + \log_b(\ln b). \qquad (27)$$

Substituting (27) into (26) we have

$$c_\Delta(z_H) - c_\nabla(z_H) \approx b^{z_H - 2 \cdot s_b(z_H) + \log_b(\ln b) + \log_b(\frac{\delta}{2})}. \qquad (28)$$

For $b = 2$, $c_\Delta(z_H) - c_\nabla(z_H) \approx 2^x$, where $x \approx z_H - 2 \cdot s_b(z_H) - (N + 1.5)$ which can be implemented to sufficient accuracy using two $(K + G)$-bit adders. Furthermore, Mitchell's method [20] can approximate the base-two antilogarithm to sufficient accuracy (4 bits) as

$$c_\Delta(z_H) - c_\nabla(z_H) \approx 2^{\text{int}(x)}(1 + \text{frac}(x)) \qquad (29)$$

using only a $(G + 1)$-bit shifter. Thus, a small amount of extra logic allows the same dual-ported memory to be used for computing the upper and lower bounds.

7. Conclusions

Sections 5.1 to 5.3 show the comparison between the interval widths obtained with IFP and ILNS for three algorithms. For gravity-force calculations in N-body simulation, ILNS improves the interval widths over IFP. IFP produces intervals an average of 60% wider, and the worst-case relative interval width is roughly 2.5 times wider in IFP in our experiments. These facts support the decision to implement ILNS for applications where the proportion of multiplications and squares is high. The advantage of ILNS is even more pronounced in the normalized distance calculation, primarily due to fewer additions in the algorithm.

The FFT algorithm is not favorable toward ILNS, because of the low number of multiplications compared to additions, and the poorer representation that ILNS provides for twiddle factors. Furthermore, it appears the truncation required by ILNS is inherently inferior to the truncation required by IFP [18]. Despite these limitations, it has been shown that, on average for typical inputs, the FFT interval results are narrower with ILNS, although a particular interval output may be wider.

Implementation of multiplication, division and square root are significantly cheaper and faster in ILNS than in IFP. Addition and subtraction are more dificult. All but low-precision LNS implementations require table lookups and interpolation to implement addition. This is also the case for ILNS, but a naïve implementation would require separate memories for the upper and lower bounds of the s_b function. This paper has proposed a novel method to implement these upper and lower bounds using no more memory than required for a non-interval LNS.

References

[1] G. Alefeld and J. Herzberger, *Introduction to Interval Computations*, New York Academic Press, New York, 1983.

[2] M. G. Arnold, T. A. Bailey, J. R. Cowles and M. D. Winkel, "Applying Features of IEEE 754 to Sign/Logarithm Arithmetic," *IEEE Trans. Comput.*, vol. 41, pp. 1040–1050, Aug. 1992.

[3] M. G. Arnold, T. A. Bailey, J. R. Cowles and M. D. Winkel, "Arithmetic Cotransformations in the Real and Complex Logarithmic Number Systems," *IEEE Trans. Comput.*, vol. 47, no. 7, pp. 777-786, July 1998.

[4] M. G. Arnold and C. D. Walter, "Unrestricted Faithful Rounding is Good Enough for Some LNS Applications," *15th International Symposium on Computer Arithmetic*, Vail, Colorado, pp. 237-245, 11-13 June 2001.

[5] M. G. Arnold and M. D. Winkel, "A Single-Multiplier Quadratic Interpolator for LNS Arithmetic," *Proceedings of the 2001 International Conference on Computer Design: ICCD 2001*, Austin, TX, pp. 178-183, 23-26 September, 2001.

[6] M. Arnold, "Slide Rules for the 21st Century: Logarithmic Arithmetic as a High-speed, Low-cost, Low-power Alternative to Fixed Point Arithmetic," *Second Online Symposium for Electronics Engineers*, www.techonline.com/community/20140.

[7] C. Chen and C. H. Yang, "Pipelined Computation of Very Large Word-Length LNS Addition/Subtraction with Polynomial Hardware Cost," *IEEE Trans. Comput.*, vol. 49, no. 7, pp. 716-726, July 2000.

[8] E. I. Chester and J. N. Coleman, "Matrix Engine for Signal Processing Applications Using the Logarithmic Number System," *Proceedings of the IEEE International Conference on Application-Specific Systems, Architectures and Processors*, San Jose, CA, pp. 315-324, 17-19 July 2002.

[9] M. Daumas and D. W. Matula, *Rounding of Foating Point Intervals*, Research Report no. 93-06, Laboratoire de l'Informatique du Parallélisme, Ecole Normale Superiéure de Lyon, Lyon, France, March 1993.

[10] E. Hansen, *Global Optimization Using Interval Analysis*, Marcel Dekker, New York, 1992.

[11] T. Hickey, Q. Ju, and M. H. Van Emden, "Interval Arithmetic: From Principles to Implementation," *Journal of the ACM*, vol. 48, no. 5, pp 1038–1068, September, 2001.

[12] O. Holzmann, B. Lang, and H. Schütt, "Newton's Constant of Gravitation and Verified Numeric Quadrature," *Reliable Computing*, no. 3, 1996.

[13] IEEE Standard for Binary Floating-Point Arithmetic, ANSI/IEEE Std 754-1985, IEEE, 1985.

[14] N. G. Kingsbury and P. J. W. Rayner, "Digital Filtering Using Logarithmic Arithmetic," *Electron. Lett.*, vol. 7, no. 2, pp. 56-58, Jan., 1971.

[15] V. Kreinovich and A. Bernat, "Is Solar System Stable? A Remark,' *Reliable Computing*, vol. 3, no. 2, pp. 149-154, 1997.

[16] D. M. Lewis, "An Architecture for Addition and Subtraction of Long Word Length Numbers in the Logarithmic Number System," *IEEE Trans. Comput.*, vol. 39, pp. 1325-1336, Nov. 1990.

[17] D. M. Lewis, "Interleaved Memory Function Interpolators With Application to an Accurate LNS Arithmetic Unit," *IEEE Trans. Comput.*, vol. 43, no. 8, pp. 974-982, Aug. 1994.

[18] J. D. Marasa and D. W. Matula, "A Simulative Study of Correlated Error in Various Finite-Precision Arithmetics," *IEEE Transactions on Computers*, vol. C-22, pp. 587-597, June 1973.

[19] J. Makino and M. Taiji, *Scientific Simulations with Special-Purpose Computers—the GRAPE Systems*, John Wiley and Sons, Chichester, 1998.

[20] J. N. Mitchell, "Computer Multiplication and Division using Binary Logarithms," *IEEE Transactions on Electronic Computers*, vol. EC-11, pp. 512-517, August 1962.

[21] R. E. Moore, *Interval Analysis*, Prentice Hall, Englewood Cliffs, N.J., 1966.

[22] R. E. Moore, "Computing to Arbitrary Accuracy," *Computational and Applied Mathematics I: Algorithms and Theory*, pp. 327-336, North-Holland, Amsterdam, 1992.

[23] V. Paliouras and T. Stouraitis, "Low Power Properties of the Logarithmic Number System," *Proceedings of the 15th IEEE Symposium on Computer Arithmetic*, Vail, Colorado, pp. 229-236, 11–13 June 2001.

[24] N. Pollard and D. May, "Using Interval Arithmetic to Calculate Data Sizes for Compilation to Multimedia Instruction Sets," *Proceedings of the Sixth ACM International Conference on Multimedia*, pp. 279-284, 1998.

[25] M. Schulte and E. Swartzlander, "Exact Rounding of Certain Elementary Functions," 11th IEEE Symposium on Computer Arithmetic, Windsor, Ontario, pp. 138-145, June, 1993.

[26] M. Schulte and E. Swartzlander, "A Family of Variable-Precision Interval Arithmetic Processors," *IEEE Trans. Comput.*, vol. 49, no. 5, pp. 387-397, May 2000.

[27] G. L. Steele, Jr., *Common LISP, 2nd ed.*, Digital Press, 1990.

[28] E. E. Swartzlander, et al., "Sign/logarithm Arithmetic for FFT Implementation," *IEEE Trans. Comput.*, vol. C-32, pp. 526-534, 1983.

[29] A. Wrigley, *Real-Time Ray Tracing on a Novel HDTV Framestore*, Ph. D. University of Cambridge, England, 1993.

Scaling an RNS number using the core function

Neil Burgess,
Cardiff School of Engineering,
Cardiff University,
Queen's Buildings,
The Parade,
CARDIFF CF24 3TF
U.K.

Abstract

This paper introduces a method for extracting the core of a Residue Number System (RNS) number within the RNS, this affording a new method for scaling RNS numbers. Suppose an RNS comprises a set of co-prime moduli, m_i, with $\prod m_i = M$. This paper describes a method for approximately scaling such an RNS number by a subset of the moduli, $\prod m_j = M_J \approx \sqrt{M}$, with the characteristic that all computations are performed using the original moduli and one other non-maintained short wordlength modulus.

1. Background and Motivation

The Residue Number System (RNS) has great potential for accelerating arithmetic operations, achieved by breaking operands into several smaller residues and operating on the residues independently and in parallel. RNS implementations were studied extensively in the 1970's, particularly for DSP applications [1], and led to Inmos' production of an RNS 2-D convolver chip in 1989 [2]. However, wider take-up of RNS for DSP was limited because of a number of fundamental difficulties:

- Conversion to binary representation from RNS is difficult (the inverse operation is simple)
- Direct magnitude comparison and sign determination of RNS numbers is impossible
- Square root operations are not available, and division operations, although available [3], are not practical due to their complexity

These difficulties place major constraints on the possible applications of RNS arithmetic.

Recently, however, DSP chips using RNS have enjoyed something of a renaissance for a variety of reasons:

- They offer high-performance implementations of arithmetic-intensive applications at reduced power supply voltages, important for mobile and wearable computer and communication systems [4]
- They avoid lengthy on-chip interconnects, which now represent the major constraint on the realisation of high-performance digital VLSI circuits [5]
- They afford hardware-efficient complex multipliers ("QRNS multiplication") comprising two independent multiplications instead of four multiplications and two additions [1]
- The component arithmetic operations in an RNS implementation can, without exception, be reduced to short adders and small look-up tables [1]

All the items in the above list are applicable to custom VLSI implementations, and the last two also apply advantageously to FPGA implementations [6,7]. Recent industrial interest in RNS confirms the existence and scale of problems faced in implementing DSP algorithms in digital microelectronic fabrics at high clock rates but with low power consumption. For example, reference [8] describes an FIR filter in RNS designed by Texas Instruments because of its low-power capability, and reference [9] discusses a general-purpose DSP engine developed by Siemens that incorporates an RNS vector processor with a considerably higher data processing bandwidth than its binary counterpart.

The fundamental difficulties with RNS arithmetic listed earlier have been overcome to some extent by recent innovations in RNS theory. For example, the core function has been shown to be advantageous in converting an RNS number to binary [10], and for adding extra moduli to an RNS in order to increase its dynamic range ("base extension") [11]. The outstanding problem with RNS processing that prevents its wider take-up is reducing an RNS number's wordlength through scaling − that is, dividing − by a constant with low latency and minimal hardware cost. In binary arithmetic, the scaling constant is invariably set to a power of two so that wordlength reduction is achieved simply by truncating (or rounding) a number. There is no equivalent operation in an RNS with the consequence that the wordlength growth of an accumulated result through a sequence of multiplications, such as is encountered in a multiple-point FFT or in an IIR filter, is very difficult to manage.

A number of algorithms for scaling RNS numbers have been reported, but as yet none operates entirely within an RNS. Early attempts at scaling fell into two categories: scaling by one modulus, whereby the RNS number was adjusted to be divisible by one of the moduli, dividing by that modulus in all the other moduli in a single step, and finally base extending the scaled number back into the "scaling modulus" (e.g. [12]); or performing a truncated conversion to binary − that is, scaling by a power of two − followed by conversion back into RNS representation (e.g. [13]). However, these methods are generally slow and require processing of longer word-length numbers outside the RNS.

A major advance was made by Shenoy and Kumaresan [14], who devised a novel decomposition of the Chinese Remainder Theorem that enabled scaling by the product of several moduli. However, their scheme was not optimal in that an extra modulus with a similar wordlength to the existing moduli outside the RNS was employed, requiring extra hardware (typically >10%) for its maintenance, and two redundant channels of residue computation were necessary in the scaling algorithm itself. The total hardware count for Shenoy and Kumaresan's RNS scaler operating on k moduli was $k \cdot (k+4)$ modulo arithmetic multiply-accumulates (MACs). Recent work has concentrated on removing the extra modulus in Shenoy and Kumaresan's scheme at the expense of increasing the logical depth of the scaler [15], or of reducing the accuracy of the scaler by limited use of binary arithmetic outside the RNS channels [16].

This paper introduces a novel technique for scaling an RNS number, based on the core function. The method consists simply of extracting the core of the RNS number within the RNS. All computations reduce to inner products within the moduli of the RNS, with one extra inner product using a modulus outside the RNS but not requiring maintenance of the corresponding residue. The paper is structured as follows: first some preliminaries regarding the core function are dealt with; then, the proposed scaling algorithm is introduced along with an example; next, difficulties with the proposed algorithm are identified and a workaround described; finally, the paper concludes with a brief discussion of possible further avenues of research.

2 Scaling an RNS number using the Core Function

2.1 The Core Function

The core function is defined for an integer, n, as:

$$C(n) = \sum_i w_i \cdot \left\lfloor \frac{n}{m_i} \right\rfloor = \sum_i (n - n_i) \cdot \frac{w_i}{m_i} \quad — (1)$$

where n_i denotes $n \bmod m_i$ and w_i denotes the ith weight. Setting $n = M$ in (1) gives:

$$C(M) = \sum_i M \cdot \frac{w_i}{m_i} \quad — (2)$$

so that:

$$\frac{C(M)}{M} = \sum_i \frac{w_i}{m_i} \quad — (3)$$

This implies some values of w_i must be negative to obtain small values of $C(M)$. Substituting (3) into (1) gives:

$$C(n) = n \cdot \frac{C(M)}{M} - \sum_i n_i \cdot \frac{w_i}{m_i} \quad — (4)$$

Equation (4) indicates that a plot of $C(n)$ against n should reveal a straight line with slope $C(M)/M$ with some "furriness" due to the superimposed summation term. The magnitude of the furriness is set by the magnitude of the weights, in turn related to the particular value of $C(M)$ for a given RNS modulus set.

The weights, w_i, are determined by re-arranging (3) and reducing both sides modulo m_j:

$$\langle C(M) \rangle_{m_j} = \left\langle \sum_i \frac{w_i \cdot M}{m_i} \right\rangle_{m_j} = \langle w_j \cdot M_j^* \rangle_{m_j} \quad — (5)$$

($M_i^* \bmod m_j = 0$ for all m_i except m_j.) Re-arranging (5):

$$w_j = \langle C(M) \cdot M_j^{*-1} \rangle_{m_j} \quad — (6)$$

so the weights may be derived once $C(M)$ has been chosen, but with the proviso that (2) is satisfied, implying that some of the weights must be negative to ensure $C(M) \ll M$. Note that if $C(M)$ is a multiple of m_j, the corresponding weight, $w_j = 0$. Finally, from [10], the range of a core function, $G(M)$ is given as:

$$G(M) = C(M) + \sum_i |w_i| \quad — (7)$$

Now, the Chinese Remainder Theorem for converting RNS numbers back to positional (i.e. decimal or binary) representation can be expressed as:

$$n = \sum_i n_i \cdot B_i - R(n) \cdot M \quad — (8)$$

where $R(n)$ is known as the rank function, and B_i denotes the ith base of the RNS:

$$B_i = \frac{M}{m_i} \cdot \left\langle \left(\frac{M}{m_i} \right)^{-1} \right\rangle_{m_i} = M_i^* \langle M_i^{*-1} \rangle_{m_i} \quad — (9)$$

Substituting (8) into (4) gives:

$$C(n) = \frac{C(M)}{M} \cdot \left\{ \sum_i n_i \cdot B_i - R(n) \cdot M \right\} - \sum_i n_i \cdot \frac{w_i}{m_i} \quad — (10)$$

Simplifying and re-arranging (10) yields:

$$C(n) = \sum_i n_i \cdot \left(\frac{B_i \cdot C(M)}{M} - \frac{w_i}{m_i} \right) - R(n) \cdot C(M) \quad — (11)$$

Setting $n = B_i$ in (4):

$$C(B_i) = B_i \cdot \frac{C(M)}{M} - \frac{w_i}{m_i} = \frac{C(M) \cdot \langle M_i^{*-1} \rangle_{m_i} - w_i}{m_i}$$
— (12)

Whence:

$$C(n) = \sum_i n_i \cdot C(B_i) - R(n) \cdot C(M)$$ — (13)

which is known as the Chinese Remainder Theorem of Core Functions. However, owing to the unfeasibility of computing $R(n)$ independently, the preferred form of the Chinese Remainder Theorem for Core Functions is:

$$\langle C(n) \rangle_{C(M)} = \left\langle \sum_i n_i \cdot C(B_i) \right\rangle_{C(M)}$$ — (14)

An example should help make things clearer. Consider an RNS with the modulus set, $m_i = \{2, 3, 5, 7, 11, 13\}$, giving $M = 30{,}030$, and $M_i^{*-1} = \{1, 2, 1, 6, 6, 3\}$. Next, choose $C(M) = 165$. Then, from (5), the weights are found as:

$w_1 = \langle 165 \times 1 \rangle_2 = +1$ or -1
$w_2 = \langle 165 \times 2 \rangle_3 = 0$
$w_3 = \langle 165 \times 1 \rangle_5 = 0$
$w_4 = \langle 165 \times 6 \rangle_7 = +3$ or -4
$w_5 = \langle 165 \times 6 \rangle_{11} = 0$
$w_6 = \langle 165 \times 3 \rangle_{13} = +1$ or -12

In order to minimise the "furriness" in the core function, weights with small magnitudes should be chosen. In this example, the weight set $w_i = \{-1, 0, 0, 3, 0, 1\}$ is chosen, and its legitimacy can be checked against equation (2):

$C(M) = -1 \times 15015 + 3 \times 4290 + 1 \times 2310 = 165$

This core function is plotted in Figure 1. An alternative weight set $w_i = \{1, 0, 0, -4, 0, 1\}$ is available that has the useful property $C(n) \geq 0$ if $n \geq 0$. The legitimacy of this weight set can also be checked as follows:

$C(M) = 1 \times 15015 + -4 \times 4290 + 1 \times 2310 = 165$

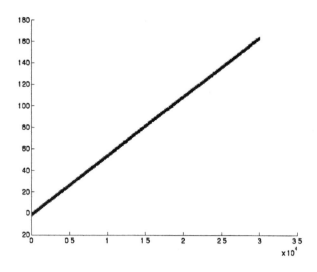

Figure 1 Plot of typical core function: $M = 30{,}030$; $C(M) = 165$

2.2 RNS Scaling Method

This paper proposes an RNS scaling method that consists of extracting the core of a number within the RNS. From equation (4):

$$C(n) = n \cdot \frac{C(M)}{M} - \sum_i n_i \cdot \frac{w_i}{m_i}$$ — (4)

Thus, if $C(n)$ could be computed within the RNS, an approximate scaled version of n is obtained. This can be achieved by splitting a modulus set into two sub-sets, M_J and M_K, such that $M_J M_K = M$ and $M_J/M_K \approx 1$. Then, it is possible to perform scaling by either M_J or M_K (in other words, extract $C(n)$ with $C(M) = M_J$ or $C(M) = M_K$) as follows.

First, set $C_J(M) = M_J = \prod m_j$. Then, from (14):

$$\langle C_J(n) \rangle_{C_J(M)} = \left\langle \sum_i n_i \cdot C_J(B_i) \right\rangle_{C_J(M)}$$ — (15)

for the sub-set of moduli that make up M_J. But, m_j is a factor of $C_J(M)$, so that:

$$\left\langle \langle C_J(n) \rangle_{C_J(M)} \right\rangle_{m_j} = \langle C_J(n) \rangle_{m_j} = \left\langle \sum_i n_i \cdot C_J(B_i) \right\rangle_{m_j}$$
— (16)

However, for the for the sub-set of moduli that make up M_K - namely the moduli, m_k, that do not divide M_J - the same simplification is not possible:

$$\left\langle \langle C_J(n) \rangle_{C_J(M)} \right\rangle_{m_k} = \left\langle \left\langle \sum_i n_i \cdot C_J(B_i) \right\rangle_{C_J(M)} \right\rangle_{m_k}$$ — (17)

Thus, equation (15) may be computed within the subset of moduli, m_j, but (16) may not. Similarly, if $C_K(M) = M_K = \prod m_k$:

$$\langle C_K(n) \rangle_{m_k} = \left\langle \sum_i n_i \cdot C_K(B_i) \right\rangle_{m_k} \quad \text{---(18)}$$

which is computable within the subset of moduli, m_k. Hence, $C_J(n) \approx n/M_K$ may be calculated within the moduli sub-set, M_J, but not the sub-set, M_K; also, $C_K(n) \approx n/M_J$ may be calculated within the moduli sub-set, M_K, but not the sub-set, M_J. However, if the difference between the cores $\Delta C(n) = C_J(n) - C_K(n)$ can be calculated, $C_K(n)$ modulo the subset M_K can be extended into $C_J(n)$ modulo the subset M_K. In other words, by adding $\Delta C(n)$ to (or subtracting it from) the values of one sub-set of scaled residues either $C_J(n)$ or $C_K(n)$ is available across all the residues, and a scaled value of n is obtained within the RNS. A simple expression for $\Delta C(n)$ is obtained from (13) as:

$$\Delta C(n) = \sum_i n_i \cdot C_J(B_i) - R(n) \cdot C_J(M) - \left\{ \sum_i n_i \cdot C_K(B_i) - R(n) \cdot C_K(M) \right\} \quad \text{---(19)}$$

which may be simplified to read:

$$\Delta C(n) = \sum_i n_i \cdot \Delta C(B_i) - R(n) \cdot \Delta C(M) \quad \text{---(20)}$$

where $\Delta C(B_i) = C_J(B_i) - C_K(B_i)$, and $\Delta C(M) = C_J(M) - C_K(M)$. However, given the difficulty of determining the value of $R(n)$ from the residues, a more useful form of equation (20) is:

$$\langle \Delta C(n) \rangle_{\Delta C(M)} = \left\langle \sum_i n_i \cdot \Delta C(B_i) \right\rangle_{\Delta C(M)} \quad \text{---(21)}$$

This expression will be most conveniently evaluated if $\Delta C(M)$ is of similar wordlength to the moduli, m_i (or a not-so-small power of two).

2.3 Worked example of proposed RNS scaling algorithm

Suppose an RNS has the moduli set, {7, 11, 13, 17, 19, 23}: then $M = 7,436,429$. The moduli set is split into two groups, $M_J = 7 \times 17 \times 23 = 2737$ and $M_K = 11 \times 13 \times 19 = 2717$, to give $\Delta C(M) = 20$. Scaling a residue number by either 2737 or 2717 is practical because $\Delta C(M)$ has a similar wordlength to the moduli.

The values of M_i^* and M_i^{*-1} are {1,062,347, 676,039, 572,033, 437,437, 391,391, 323,323} and {6, 1, 2, 12, 2, 2} respectively. The weight set for $C_J(M) = 2737$ is $w_i = $ {0, -2, 1, 0, 2, 0) and for $C_K(M) = 2717$, $w_i = $ {-1, 0, 0, -2, 0, 6}. The two sets of $C(B_i)$ then follow from (12): $C_J(B_i) = $ {2346, 249, 421, 1932, 288, 238}; $C_K(B_i) = $ {2329, 247, 418, 1918, 286, 236}; finally, $\Delta C(B_i) = $ {17, 2, 3, 14, 2, 2}.

The number $n = 1,859,107$ is to be approximately scaled by 2717 to yield ≈ 684. That is, we will compute $C_J(n)$ wholly within the RNS. n is represented as (5, 8, 3, 4, 14, 17) by this set of moduli. First, compute $C_J(n)$ moduli 7, 17, and 23, and $C_K(n)$ moduli 11, 13, and 19 using (15) and (17):

$C_J(n)$ mod 7 = (5×2346 + 8×249 + 3×421 + 4×1932 + 14×288 + 17×238) mod 7 = (5×1 + 8×4 + 3×1 + 4×0 + 14×1 + 17×0) mod 7 = 5

$C_J(n)$ mod 17 = (5×2346 + 8×249 + 3×421 + 4×1932 + 14×288 + 17×238) mod 17 = (5×0 + 8×11 + 3×13 + 4×11 + 14×16 + 17×0) mod 17 = 4

$C_J(n)$ mod 23 = (5×2346 + 8×249 + 3×421 + 4×1932 + 14×288 + 17×238) mod 23 = (5×0 + 8×19 + 3×7 + 4×0 + 14×12 + 17×8) mod 23 = 17

$C_K(n)$ mod 11 = (5×2329 + 8×247 + 3×418 + 4×1918 + 14×286 + 17×236) mod 11 = (5×8 + 8×5 + 3×0 + 4×4 + 14×0 + 17×5) mod 11 = 5

$C_K(n)$ mod 13 = (5×2329 + 8×247 + 3×418 + 4×1918 + 14×286 + 17×236) mod 13 = (5×2 + 8×0 + 3×2 + 4×7 + 14×0 + 17×2) mod 13 = 0

$C_K(n)$ mod 19 = (5×2329 + 8×247 + 3×418 + 4×1918 + 14×286 + 17×236) mod 19 = (5×11 + 8×0 + 3×0 + 4×18 + 14×1 + 17×8) mod 19 = 11

In parallel, calculate $\langle \Delta C(n) \rangle_{\Delta C(M)}$ using (21):
$\Delta C(n)$ mod 20 = (5×17 + 8×2 + 3×3 + 4×14 + 14×2 + 17×2) mod 20 = 8

Finally, add $\Delta C(n)$ to the $C_K(n)$ values to obtain the remaining scaled moduli:
$C_J(n)$ mod 11 = $C_K(n) + \Delta C(n)$ mod 11 = 5+8 mod 11 = 2
$C_J(n)$ mod 13 = $C_K(n) + \Delta C(n)$ mod 13 = 0+8 mod 13 = 8
$C_J(n)$ mod 19 = $C_K(n) + \Delta C(n)$ mod 19 =11+8 mod 19 = 0

Hence, the RNS value of $n = 1,859,107$ (or (5, 8, 3, 4, 14, 17) in RNS format) after being approximately scaled by 2717 is $C_J(n) = $ (5, 2, 8, 4, 0, 17).

We can check the result by converting $C_J(n)$ back to decimal using the Chinese Remainder Theorem:

$C_J(n) = \Sigma n_i \times B_i$ mod M = 5×1062347×6 + 2×676039×1 + 8×572033×2 + 4×437437×12 + 0×391391×2 + 17×323323×2 mod 7436429 = 31870410 + 1352078 + 9152528 + 20996976 + 0 + 10992982 = 74364974 mod 7436429 = 684.

A block diagram of this calculation method, emphasising the consistent use of short wordlength arithmetic and making explicit the degree of available parallelism, is presented in Figure 2.

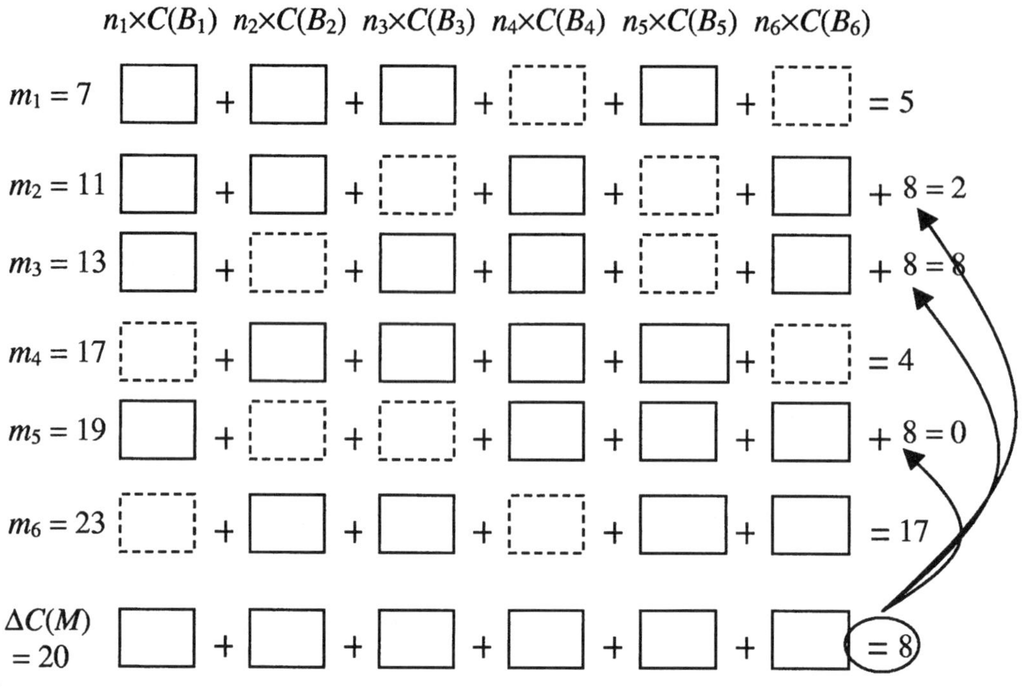

$$n_1 \times C(B_1) \quad n_2 \times C(B_2) \quad n_3 \times C(B_3) \quad n_4 \times C(B_4) \quad n_5 \times C(B_5) \quad n_6 \times C(B_6)$$

$m_1 = 7$ ☐ + ☐ + ☐ + ┆ ┆ + ☐ + ┆ ┆ = 5

$m_2 = 11$ ☐ + ☐ + ┆ ┆ + ☐ + ┆ ┆ + ☐ + 8 = 2

$m_3 = 13$ ☐ + ┆ ┆ + ☐ + ☐ + ┆ ┆ + ☐ + 8 = 8

$m_4 = 17$ ┆ ┆ + ☐ + ☐ + ☐ + ☐ + ┆ ┆ = 4

$m_5 = 19$ ☐ + ┆ ┆ + ┆ ┆ + ☐ + ☐ + ☐ + 8 = 0

$m_6 = 23$ ┆ ┆ + ☐ + ☐ + ┆ ┆ + ☐ + ☐ = 17

$\Delta C(M) = 20$ ☐ + ☐ + ☐ + ☐ + ☐ + ☐ = 8

Figure 2 Example RNS scaling calculation

Each box in Figure 2 represents a ROM look-up table storing the possible multiples of $n_p \times C(B_p) \bmod m_q$, where p denotes the column number, and q the row number. Note that some of the ROM's in Figure 2 could be removed because the corresponding $C(B_i)$ coefficients are 0 - these ROM's are indicated by dotted lines. The total number of MAC operations is $k \cdot (k/2 + 1) + k + k/2 = k \cdot (k+5)/2$, almost half that of Shenoy and Kumaresan's method.

2.4 Error analysis of the scaling algorithm

The error, ε, that is incurred by employing the core function to represent a scaled number rather than performing a rounded division can be estimated as follows:

$$\varepsilon_J = rnd(n/C_K(M)) - C_J(n) \quad \text{---} \quad (22)$$

where ε_J is some integer. Substituting in equation (4) yields:

$$\varepsilon_J = rnd(n/C_K(M)) - (n/C_K(M) - \Sigma n_i \cdot w_i/m_i) \quad \text{---} \quad (23)$$

The remainder of the division $n/C_K(M)$ is given by:

$$rem_K = n/C_K(M) - rnd(n/C_K(M)) \quad \text{---} \quad (24)$$

where $-1 \leq rem_K < 1$. Then,

$$\varepsilon_J = \Sigma n_i \cdot w_i/m_i - rem_K \quad \text{---} \quad (25)$$

Now, the largest positive error in ε_J will occur when all the residues of moduli with positive weights are at their largest:

$$\varepsilon_{J(max+)} = \Sigma w_i^+ \cdot (m_i-1)/m_i - rem_K \approx \Sigma w_i^+ \quad \text{---} \quad (26)$$

where w_i^+ denotes the positive weights. Similarly, the largest negative error in ε_J will occur when all the residues of moduli with negative weights are at their largest:

$$\varepsilon_{J(max-)} = \Sigma w_i^- \cdot (m_i-1)/m_i - rem_K \approx \Sigma w_i^- \quad \text{---} \quad (27)$$

where w_i^- denotes the negative weights. Thus, to minimise the error in the scaled number, the core function with the smaller total magnitude weight set (i.e. smaller value of $\Sigma |w_i|$) should be chosen.. In the above example, where the weight set for $C_J(M) = 2737$ is $\{0, -2, 1, 0, 2, 0\}$ and for $C_K(M) = 2717$ is $\{-1, 0, 0, -2, 0, 6\}$, that would be $C_J(M) = 2737$.

3 Ambiguity in core function extraction

A number of difficulties related to the "furriness" of the core function exist that if left unresolved may restrict the technique:

- $C_J(n)$, $C_K(n)$ or $\Delta C(n)$ may be negative for some values of n.

- $C_J(n)$ or $C_K(n)$ may exceed $C_J(M)$ or $C_K(M)$ respectively
- $\Delta C(n)$ may exceed $\Delta C(M)$.

These occurrences can give rise to difficulties because equations (14) and (21) are both computed over finite fields, so that out-of-range results alias onto in-range but erroneous results. For example, if $C_J(n)$, which is implicitly calculated mod $C_J(M)$, is negative, $C_J(n) + C_J(M)$ is incorrectly returned. Similarly, if $C_J(n) > C_J(M)$, $C_J(n) - C_J(M)$ is incorrectly returned.

3.1 Examples of ambiguity

By way of illustration of these issues, suppose an RNS has the moduli set comprised of the six smallest prime numbers: $m_i = \{2, 3, 5, 7, 11, 13\}$, giving $M = 30{,}030$. The moduli set is split into two groups, $M_J = 3 \times 5 \times 11 = 165$ and $M_K = 2 \times 7 \times 13 = 182$, to give $\Delta C(M) = 17$. Scaling a residue number by either 165 or 182 is practical because $\Delta C(M)$ has a similar wordlength to the moduli.

The values of M_i^* and M_i^{*-1} are $\{15015, 10010, 6006, 4290, 2730, 2310\}$ and $\{1, 2, 1, 6, 6, 3\}$ respectively. The weight set for $C_J(M) = 165$ is $\{-1, 0, 0, 3, 0, 1\}$ and for $C_K(M) = 182$ is $\{0, -2, 2, 0, 3, 0\}$. The two sets of $C(B_i)$ then follow from (11): $C_J(B_i) = \{83, 110, 33, 141, 90, 38\}$; $C_K(B_i) = \{91, 122, 36, 156, 99, 42\}$; finally, $\Delta C(B_i) = \{8, 12, 3, 15, 9, 4\}$. Figure 1 (presented earlier in Section 2) is a plot of $C_J(n)$ against n calculated using (1): that is, outside the RNS, where ambiguity cannot occur.

Figure 3 is a plot of the same function but now computed modulo $C_J(M)$, in accordance with equation 14. Note the two small regions of ambiguity at either end of the x-axis. In many RNS applications, these areas could be avoided by selecting a modulus set with a greater dynamic range than that of the application.

However, in this use of the core function, ambiguity is avoided for values of $n \approx M$ (i.e. as $C(n) \approx C(M)$) because the core is being extracted effectively over modulo M, not modulo $C(M)$. Consequently, there is no ambiguity due to aliassing arising from equation (4). However, aliassing can occur for values of $n \approx 0$ (i.e. as $C(n) \approx 0$), because $C(n)$ could be negative. Two possible solutions to this are (i) select a weight set that prevents $C(n)$ from being negative; (ii) add a small bias after the scaling technique equivalent to the most negative value that $C(n)$ could take.

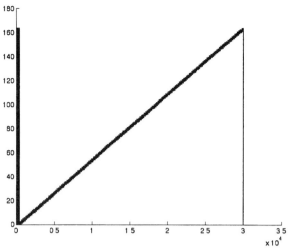

Figure 3 Plot of the core function $\langle C_J(n) \rangle_{165}$

Figure 4 is a plot of $\Delta C(n)$ for the previous example.

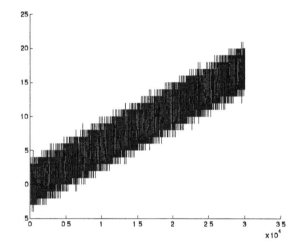

Figure 4 Plot of $\Delta C(n) = C_K(n) - C_J(n)$

Figure 5 is also a plot of $\Delta C(n)$ but computed modulo $\Delta C(M)$, as in the scaling algorithm. Note how the ambiguity region is much greater than in the core function plot. This is because the "furriness" of the plot, which is approximately given by the sums of the magnitudes of the weights [10], is a much greater proportion of the modulus. That is, $\Delta C(M) \ll C_J(M), C_K(M)$. The proposed scaling algorithm aims to reproduce the core function of Figure 1 within the RNS. However, the ambiguity illustrated in Figure 5 prevents this from occurring over much of the range of n. Thus, the major obstacle to RNS scaling using the core function lies in the ambiguity in computing $\Delta C(n)$.

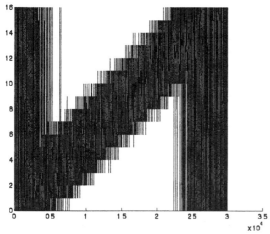

Figure 5 Plot of $\Delta C(n)$ mod $\Delta C(M)$

3.2 Removing ambiguity from core function difference calculations

In [10], two methods for overcoming ambiguity in core function computations were proposed both of which were based on retaining a parity bit at all stages of an RNS processing system. This paper proposes a similar idea for overcoming the ambiguity in computing $\Delta C(n)$, by employing the parity bit to afford calculation of $\Delta C(n)$ over the finite field $2 \cdot \Delta C(M)$.

Equation (21), which is used to calculate $\Delta C(n)$ was given earlier as:

$$\Delta C(n) = \left\langle \sum_i n_i \cdot \Delta C(B_i) \right\rangle_{\Delta C(M)} \quad\text{— (21)}$$

Now, ambiguity occurs in this equation because the range of $\Delta C(n)$ exceeds $\Delta C(M)$. Earlier, a simple expression for the range of a core function, $G(M)$, was shown to be:

$$G(M) = C(M) + \sum_i |w_i| \quad\text{— (7)}$$

Hence, an expression for the maximum range of $\Delta C(n)$ is:

$$\Delta G(M) = C_K(M) + \sum_k |w_k| - \left(C_J(M) + \sum_j |w_j| \right) \quad\text{— (28)}$$
$$\approx \Delta C(M) + \sum |w_k| + \sum |w_j|$$

This follows because the two weight sets are orthogonal: for each modulus in the RNS, one of the weights in the two core functions must be zero. This implies that $\Delta C(n)$ should be calculated to a number greater than modulus $\Delta G(M)$ to avoid ambiguity.

A simple way to achieve this is to calculate $\Delta C(n)$ modulus $2 \cdot \Delta C(M)$:

$$\Delta C(n) = \left\langle \sum_i n_i \cdot \Delta C(B_i) - R(n) \cdot \Delta C(M) \right\rangle_{2\Delta C(M)} \quad\text{— (29)}$$

This method avoids ambiguity provided $2 \cdot \Delta C(M) > \Delta G(M)$, or equivalently provided $\Delta C(M) > \Sigma |w_i|$, which is readily achievable in practice. Now, equation (29) apparently requires $R(n)$ to be calculated; however, the expression can be rewritten such that only the parity of $R(n)$ is needed:

$$\Delta C(n) = \left\langle \sum_i n_i \cdot \Delta C(B_i) - \langle R(n) \rangle_2 \cdot \Delta C(M) \right\rangle_{2\Delta C(M)} \quad\text{— (30)}$$

The rank function, $R(n)$, is defined by the Chinese Remainder Function from (8) as:

$$R(n) = \frac{\sum_i n_i \cdot B_i - n}{M} = \sum_i \frac{n_i \cdot \langle M_i^{*-1} \rangle_{m_i}}{m_i} - \frac{n}{M} \quad\text{— (31)}$$

Hence, the parity of the rank function is given by:

$$\langle R(n) \rangle_2 = \left\langle \sum_i \frac{n_i \cdot \langle M_i^{*-1} \rangle_{m_i}}{m_i} - \frac{n}{M} \right\rangle_2$$
$$= \left\langle \sum_i n_i \cdot \langle M_i^{*-1} \rangle_{m_i} - p \right\rangle_2 \quad\text{— (32)}$$

where p denotes the parity of n, and M and hence all the moduli are assumed to be odd. (If M is even, equation (32) is undefined.) This calculation of the rank function parity can occur in parallel with the proposed scaling algorithm, so that $\Delta C(M)$ (now computed over the finite field $2 \cdot \Delta C(M)$) appears at the same juncture in the algorithm as before. An example should help make things clearer.

3.3 Worked example of unambiguous scaling method

The number $n = 6{,}432{,}750$ is to be scaled by 2717 to yield ≈ 2368, again using an RNS with the moduli set, $\{7, 11, 13, 17, 19, 23\}$, but this time also using $p = n \bmod 2 = 0$. The values of M_i^{*-1} are $\{6, 1, 2, 12, 2, 2\}$, as before. In the RNS, n is represented by $(2, 5, 12, 1, 15, 18)$. First, compute $C_J(n)$ moduli 7, 17, and 23, and $C_K(n)$ moduli 11, 13, and 19 using (15) and (17):

$C_J(n) \bmod 7 = (2 \times 1 + 5 \times 4 + 12 \times 1 + 1 \times 0 + 15 \times 1 + 18 \times 0) \bmod 7 = 0$

$C_J(n) \bmod 17 = (2\times0 + 5\times11 + 12\times13 + 1\times11 + 15\times16 + 18\times0) \bmod 17 = 3$

$C_J(n) \bmod 23 = (2\times0 + 5\times19 + 12\times7 + 1\times0 + 15\times12 + 18\times8) \bmod 23 = 20$

$C_K(n) \bmod 11 = (2\times8 + 5\times5 + 12\times0 + 1\times4 + 15\times0 + 18\times5) \bmod 11 = 3$

$C_K(n) \bmod 13 = (2\times2 + 5\times0 + 12\times2 + 1\times7 + 15\times0 + 18\times2) \bmod 13 = 6$

$C_K(n) \bmod 19 = (2\times11 + 5\times0 + 12\times0 + 1\times18 + 15\times1 + 18\times8) \bmod 19 = 9$

In parallel, calculate $\langle \Delta C(n) \rangle_{\Delta 2C(M)}$ using equations (30) and (32):

$\langle R(n) \rangle_2 = (2\times6 + 5\times1 + 12\times2 + 1\times12 + 15\times2 + 18\times2 - 0) \bmod 2 = (0\times0 + 1\times1 + 0\times0 + 1\times0 + 1\times0 + 0\times0 - 0) \bmod 2 = 1$.

$\Delta C(n) \bmod 40 = (2\times17 + 5\times2 + 12\times3 + 1\times14 + 15\times2 + 18\times2 + 1\times20) \bmod 40 = 20$

Finally, add $\Delta C(n)$ to the $C_K(n)$ values to obtain the remaining scaled moduli:

$C_J(n) \bmod 11 = C_K(n) + \Delta C(n) \bmod 11 = 3+20 \bmod 11 = 1$
$C_J(n) \bmod 13 = C_K(n) + \Delta C(n) \bmod 13 = 6+20 \bmod 13 = 0$
$C_J(n) \bmod 19 = C_K(n) + \Delta C(n) \bmod 19 = 9+20 \bmod 19 = 10$

Hence, the RNS value of $n = 6{,}432{,}750$ (or (2, 5, 12, 1, 15, 18) in RNS format) after being scaled by 2717 is $C_J(n) = (0, 1, 0, 3, 10, 20)$. We can check the result by converting $C_J(n)$ back to decimal using the Chinese Remainder Theorem:

$C_J(n) = \Sigma n_i \times B_i \bmod M = 0\times1062347\times6 + 1\times676039\times1 + 0\times572033\times2 + 3\times437437\times12 + 10\times391391\times2 + 20\times323323\times2 \bmod 7436429 = 37184511 \bmod 7436429 = 2366$.

In the previous worked example, $\Delta C(n)$ would have been calculated as 0 because it equalled $\Delta C(M)$, thus leading to an error in the final scaled result. In order to maintain $p = \langle n \rangle_2$, only one extra XOR gate or one extra AND gate is needed for each addition or multiplication respectively in the rest of the RNS hardware. This obviously represents a tiny overhead.

4 Summary and future work

This paper has introduced a new method for scaling RNS numbers by extracting the core of an RNS number within an RNS. In fact two core functions are extracted, and the difference between them used to help provide the scaled result. A simple technique requiring the maintenance of a parity bit was shown to be effective in removing errors due to ambiguity in computing the difference between the pair of core functions.

Future work could be undertaken to try and make the scaling method more flexible: it may prove possible to scale by M_K / m_i for example. Alternatively, scaling algorithms using more than two core functions might be tried.

Work exploring other uses of computing a core function within an RNS would be of interest: for example, base extension is much simpler if $C(n)$ is known. Finally, it would be of interest to use this method in a typical application (DSP, or long wordlength cryptography) to assess performance benefits.

5 References

[1] "*RNS arithmetic: Modern applications in DSP*", ed. M. Soderstrom *et al*, IEEE Press, 1986

[2] S.R. Barraclough *et al*: "The Design and Implementation of the IMS A110 Image and Signal Processor", Proc. IEEE CICC, San Diego, May 1989, pp. 24.5/1-4

[3] A.A. Hiasat and H.S. Zohdy, "Design and implementation of an RNS division algorithm", Proc. 13th IEEE Symposium on Computer Arithmetic, Asilomar CA, June 1997, pp. 240-249

[4] A.P. Preethy and D. Radhakrishnan, "A 36-bit balanced moduli MAC architecture", Proc. 42nd IEEE Midwest Symp. Circuits & Systems, Las Cruces NM, August 1999, pp. 380-383

[5] H.A. Al-Twaijry and M. J. Flynn, "Technology scaling effects in multipliers", *IEEE Trans.*, vol. 47, pp. 1201-1215 (Nov. 1998)

[6] L. Maltar *et al*, "Implementation of RNS addition and RNS multiplication into FPGAs", Proc. IEEE FCCM, Napa Valley CA, April 1998, pp. 331-332

[7] M. Re, A. Nannarelli, G.C. Cardarilli, R. Lojacono, "FPGA realization of RNS to binary signed conversion architecture", Proc. IEEE International Symposium on Circuits and Systems (ISCAS), Sydney, May 2001, pp. 350-353

[8] M.N Mahesh and M. Mehendale, "Low power realization of residue number system based FIR filters", 13th Int. Conf. on VLSI Design, Bangalore, India, January 2000, pp. 30-33

[9] M. Bhardwaj and B. Ljusanin, "The Renaissance – a residue number system based vector co-processor", Proc. 32nd Asilomar Conference on Signals, Systems & Computers, Asilomar CA, November 1998, pp. 202-207

[10] N. Burgess, "Scaled and unscaled residue number system to binary conversion techniques using the core function", Proc. 13th IEEE Symposium on Computer Arithmetic, Asilomar CA, June 1997, pp. 250-257

[11] D.D. Miller *et al*: "Analysis of the residue class core function of Akushskii, Burcev and Pak", in "*RNS arithmetic: Modern applications in DSP*", *op. cit.*

[12] F.J. Taylor and C. Huang, "An autoscale residue multiplier", *IEEE Trans. Comp.*, vol. C-31, pp. 321-325 (April 1982)

[13] G. Jullien, "Residue number scaling and other operations using ROM arrays", *IEEE Trans.*, vol. C-27, pp. 325-336 (April 1978)

[14] A. Shenoy and R. Kumaseran, "A fast and accurate RNS scaling technique for high speed signal processing", *IEEE Trans.*, vol. ASSP-37, pp. 929-937 (June 1989)

[15] F. Barsi and M.C. Pinotti, "Fast base extension and precise scaling in RNS for look-up table implementations", *IEEE Trans.*, vol. SP-43, pp.2427-2430 (Oct. 1995)

[16] Z.D. Ulman and M. Czyzak, "Highly parallel, fast scaling of numbers in nonredundant residue arithmetic", *IEEE Trans. Sig. Proc.*, vol. SP-46, pp.487-496 (Feb. 1998)

Session 11:
Modeling and Design of Arithmetic Components

Chair: Peter-Michael Seidel

Energy-Delay Estimation Technique for High-Performance Microprocessor VLSI Adders

Vojin G. Oklobdzija[1], Bart R. Zeydel[1], Hoang Dao[1], Sanu Mathew[2], Ram Krishnamurthy[2]

[1]ACSEL
University of California
Davis, CA 95616
www.ece.ucdavis.edu/acsel

[2]Intel Corporation
Circuit Research Labs
Hillsboro, OR 97124
Sanu.k.Mathew@intel.com

Abstract

In this paper, we motivate the concept of comparing VLSI adders based on their energy-delay trade-offs and present a technique for estimating the energy-delay space of various high-performance VLSI adder topologies. Further, we show that our estimates accurately represent tradeoffs in the energy-delay space for high-performance 32-bit and 64-bit processor adders in $0.13\mu m$ and $0.10\mu m$ CMOS technologies, with an accuracy of 8% in delay estimates and 20% in energy estimates, compared with simulated data.

1. Introduction

In the course of VLSI processor design it is very important to choose the adder topology that would yield the desired performance. However, the performance of a chosen topology will be known only after the design is finished. Therefore a lingering question remains: could we have achieved a higher performance, or could we have had a better VLSI adder topology? The answers to those questions are generally not known. There is no consistent and realistic speed estimation method employed today by the computer arithmetic community. Most of the algorithms are based on out-dated methods of counting the number of logic gates in the critical path producing inaccurate and misleading results. The importance of loading and wire delay is not taken into account by most. Knowles has shown how different topologies may influence fan-out and wiring density thus influencing design decisions and yielding better area/power than known cases [1]. This work has further emphasized a disconnect existing between algorithms that are used to derive VLSI adder topologies and the final result. In previous work we have shown the importance of accounting for fan-in and fan-out on the critical path, not merely the number of logic levels [2]. This has led to the

This work has been supported by SRC Research Grant No. 931.001 and California MICRO 01-063

method of Logical Effort (LE) [3], which has been popularized by Harris [4]. Recently, we used Logical Effort to estimate the speed of various VLSI adders and we compared those results with those obtained using a more complex circuit simulation tool H-SPICE [5]. This comparison showed a good match and pointed to the right direction. However, the process of analysis was now time consuming and did not provide a comparison for various circuit sizing that could have been applied. This paper is organized as follows: the second section discusses speed estimation using more realistic measures such as logical effort, the third section introduces the energy effects and discusses the performance in the energy-delay space, the fourth section describes the estimation tool that was developed, the fifth section shows results applied to several well known adder topologies and compares them with simulated results in $0.13\mu m$ technology.

2. Speed Estimation

The speed of a VLSI adder depends on many factors: the technology of implementation (and its own internal rules), circuit family used for the implementation, sizing of transistors, chosen topology of the VLSI adder, and many other second order effect parameters. There were no simple rules that could be applied when estimating VLSI adder speed. Skilled engineers are capable of fine-tuning the design by carefully selecting transistor sizes, obtaining the best performance and energy trade-off. Therefore it is very difficult, if not impossible, to predict which of the topologies developed by the computer arithmetic community is best, even if it is really useful.

2.1 Logical Effort

Logical Effort methodology takes into account the fact that the speed of a digital circuit block is dependent on its output load (fan-out) and its topology (fan-in). Further, LE introduces technology independence by normalizing the speed to that of a minimal size inverter which makes the comparisons of different topologies, implemented in different technologies, possible. For proper understanding and further reading of this paper the reader should be familiar with the LE methodology [3,4]. We will briefly describe some of the main features of LE in this sub-section. The delay expression of a logic block in LE is given as:

$$d = f + p \quad (1)$$

where p = parasitic delay, f = effort or stage delay. Further f = gh where g is defined as logical effort and h as electrical effort. Thus:

$$d = gh + p \quad (2)$$

This dependency is illustrated in Fig. 1.

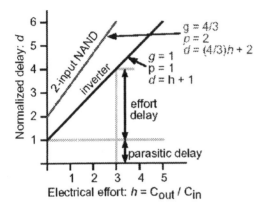

Fig. 1. Delay expressed in terms of a minimal size inverter [3,4]

An important result of LE is that it provides a way of determining appropriate transistor sizing of the critical path to minimize delay. LE also provides an estimate of the critical path delay. Logical Effort results are summarized in Table 1.

Table 1: Logical Effort Equations

Path Logical Effort	$G = \prod g_i$
Path Electrical Effort:	$H = \prod p_i = \dfrac{C_{out}}{C_{in}}$
Branching Effort	$b = \dfrac{C_{off-path} + C_{on-path}}{C_{on-path}}$
Path Branching Effort:	$B = \prod b_i$
Path parasitic delay	$P = \sum p_i$
Path Effort:	$F = GBH$

Logical Effort tells us that the delay will be minimal when each stage bears equal effort given as:

$$\hat{f} = g_i h_i = F^{\frac{1}{N}} \quad (3)$$

In such a case, delay of the path will be equal to:

$$D = N\hat{f} + P \quad (4)$$

In order to calculate optimal transistor sizes to achieve minimal delay, we start from the output and calculate C_{in} for each stage, which determines the sizing of each stage.

2.2. 64-bit Adder Speed Comparison using Logical Effort

We used several representative topologies and performed critical path analysis using Logical Effort technique to compare performance. The adders that were examined were: (a) Static: Kogge-Stone (KS) radix-2 [6], Mux-based carry-select [7], and Han-Carlson (HC) radix-2 [8,9] (b) Dynamic: KS radix-2, Ling Adder [10], HC radix-2 and CLA adder with 4-bit grouping.

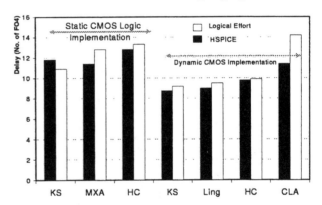

Fig. 2. Speed estimation of various VLSI adders using Logical Effort vs. H-SPICE results

The results obtained using Logical Effort were compared with the results obtained using H-SPICE simulation. The comparison results are shown in Fig. 2. Wire delays were accounted for by estimating the length of the wire and assigning appropriate delay to it, however, the portion of the wire delay was not significant (less than 10% of the total delay) due to the proximity of the cells. The first obvious observation is that there is a huge difference between Static CMOS and Dynamic CMOS implementations. This demonstrates the dependency on logic design style, which obscures any differences between different VLSI adders. This fact has been known by practitioners and rarely would we see a Static CMOS adder in places where high-speed is required. The prediction error is under 10% in most cases.

We are still in the process of refining the LE calculation in order to gain better accuracy. However, our objective is to have a simple "back of the envelope" method for quick estimation and evaluation of different VLSI adder topologies without venturing into CAD tool complexity. Therefore, we compromised by using MS-Excel as a tool for comparison, because of its simplicity and ability to perform complex calculations. An example of the Excel tool used is shown in Table 3 (CM – represents Carry-Merge cell, Dk1ND2 – represents a "footed" dynamic NAND). Before the analysis, it is necessary to characterize the technology used. This step needs to be done only once, but it improves the accuracy of the LE since the characteristics of the technology are taken into account.

Table 2: Normalized LE parameters
0.10μm technology, FO4=19pS

Gate type	LE (g)	Parasitics (p inv)
Inverter	1	1
Dyn. Nand	0.6	1.34
Dyn. CM	0.6	1.62
Dyn. CM-4N	1	3.71
Static CM	1.48	2.53
Mux	1.68	2.93
XOR	1.69	2.97

Table 3: Delay Comparison of Static and Dynamic implementation of Kogge-Stone Prefix-2 Adder

Prefix-2 Kogge-Stone (Static)

Stages	Bit Span	Branch Effort (b_i)	LE (g_i)	Parasitic (p_i)	Total Branch (B)	Total LE (G)	Path Effort (F)	Fopt (f)	Effort Delay (ps)	Parasitic Delay (ps)	Wire Delay (ps)	Total Delay (ps)	Total Delay (FO4)
g0 (NAND2)	0	2.0	1.11	1.84									
C0 (OAI)	2	2.2	1.55	2.26									
C2 (AOI)	4	2.4	1.52	2.76									
C6 (OAI)	8	2.8	1.55	2.26									
C14 (AOI)	16	3.6	1.52	2.76	1.66E+03	2.26E+01	3.76E+04	3.22	106	88	14	209	11.0
C30 (OAI)	32	5.2	1.55	2.26									
C62 (AOI)	0	1.0	1.52	2.76									
S63 (TGXORs)	0	1.0	1.56	2.59									
INV (INV)	0	3.0	1.00	1.00									

Prefix-2 Kogge-Stone (Dynamic)

Stages	Bit Span	Branch Effort (b_i)	LE (g_i)	Parasitic (p_i)	Total Branch (B)	Total LE (G)	Path Effort (F)	Fopt (f)	Effort Delay (ps)	Parasitic Delay (ps)	Wire Delay (ps)	Total Delay (ps)	Total Delay (FO4)
g0 (Dk1ND2)	0	2.0	1.02	1.34									
C0 (OAI)	2	2.2	1.36	1.69									
C2 (DAOI)	4	2.4	0.68	1.33									
C6 (OAI)	8	2.8	1.36	1.69									
C14 (DAOI)	16	3.6	0.68	1.33	1.66E+03	1.26E+00	2.09E+03	2.34	77	60	14	151	8.0
C30 (OAI)	32	5.2	1.36	1.69									
C62 (DAOI)	0	1.0	0.68	1.33									
S63 (TGXORs)	0	1.0	1.56	2.59									
INV (INV)	0	3.0	1.00	1.00									

Characterization is performed using SPICE simulation of the gate delay for various output loads driving a copy of itself, according to the LE rules. This is repeated for each cell used in the logic library. Characterization of dynamic gates requires special attention due to the fact that only one transition is of interest. Obtained results are compared to that of an inverter and parameters such as parasitic delay (p) and effort (g) were normalized with respect to that of an inverter. Select results are shown in Table 2. This step preserves LE features, allowing delay results to be presented in terms of fan-out of 4 (FO4) delay, relatively independent of the technology of implementation. The LE-based delay estimation tool works on the logic stages in the critical path, assigning branch effort (b_i), logical effort (g_i) and parasitic effort (p_i) to each gate (Table 3). In computing the branch effort, we take into consideration the worst-case interconnect at each stage. In a 64-bit Kogge-Stone adder, the worst-case interconnect in stage 6 (CM C30) spans 32 bit-slices. We make an assumption that the adder bit-pitch is 10um, which would result in a 320μm wire. To account for the propagation delay through a wire we incorporate an Elmore delay model in Table 3, which corresponds to the critical-path interconnect delay in the adder. A comparison of representative VLSI adders implemented in static and dynamic CMOS design style is presented in Table 4. It is interesting to see that there are indeed very small speed differences between the three fastest dynamic adders: KS, HC and Quarternary (QT). The advantage of KS is achieved by reduced parasitic delays resulting from fewer stages. It is also very difficult to determine the fastest VLSI adder from the results presented in Table 4.

Table 4. Comparison of representative VLSI adders using Logical Effort (wire delay estimate included)

Adders	Stages	Total Branch (B)	Total LE (G)	Path Effort (F)	f_{opt}	Effort Delay (pS)	Parasitic Delay (pS)	Wire Delay (pS)	Total Delay (pS)	Total Delay (FO4)
Static MXA	15	11600	0.369	4280	1.75	96	93	14	203	10.7
Static KS	9	1660	22.6	37600	3.22	106	88	14	209	11
Static HC	10	1660	22.6	37600	2.87	105	92	14	212	11.1
Dynamic KS	9	1660	1.26	2090	2.34	77	60	14	151	8.0
Dynamic HC	10	1660	1.26	2090	2.15	79	64	14	157	8.26
Dynamic QT	10	1540	2.08	3220	2.24	82	68	8	158	8.3
Dynamic LNG	10	1430	0.973	1400	2.06	76	70	15	161	8.47
Dynamic CLA	14	20600	0.627	12900	1.97	101	81	12	195	10.26

The differences between presented topologies are small and fall within the margin of error introduced by inaccuracy of the estimation method. This further emphasizes difficulties in comparing VLSI adder topologies and determining the best one.

3. Energy-Delay Tradeoffs

Comparing VLSI adders becomes more difficult when the notion of power or energy used for computation is introduced. Suppose that an adder A is compared with an adder B and that adder A is faster than adder B. Based on speed only, our inclination would be to use adder A in our design. However, this is not the complete picture. If the energy consumed is considered, and if adder B turns out to be using less energy, we may chose adder B, depending on the power requirements imposed on our design. However, power (energy) can be traded for speed and vice versa. Fig. 3 illustrates the hypothetical energy-delay dependencies of adders A and B.

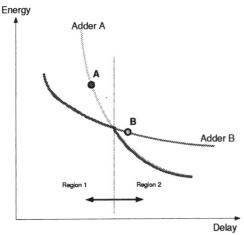

Fig. 3. Energy-Delay dependency

In this example, adder A would be chosen as a better adder topology if we were just to compare two design points A and B with respect to speed and disregard the energy aspect. As the curves show, adder B has more room for improvement and with further energy-delay optimization this adder would move to the point where better performance is obtained by using the adder topology B than topology A (Region 1). However, if low-energy of operation is our objective, we see that adder topology A is better because it can achieve lower energy for the same delay (Region 2). Thus, Fig. 3 illustrates the importance of taking the energy into account, not just merely the speed of the adder.

3.1. Energy Estimation

Logical Effort method does include an estimation of energy. It only provides one point on the Energy-Delay curve corresponding to a sizing optimized for speed. Where this point lies on the Energy-Delay curve remains an unknown. In order to generate Energy-Delay estimates it is first necessary to include some way of estimating energy into the Logical Effort method. We start by characterizing each cell in terms of energy. The energy depends on at least two parameters: output load and cell size. Energy of a two-input NAND gate as a function of its size and fan-out load is shown in Fig. 4

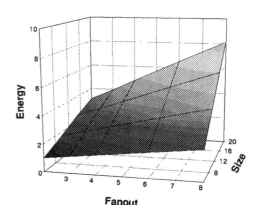

Fig. 4. Energy dependency of a 2-input NAND cell

Thus, the energy estimate will depend on the sizing determined by logical effort as well as the fan-out load on the output of the cell. Each cell used in the design is characterized and parameters determining the energy dependency on the cell size and fan-out load were stored in the table from which dependency parameters were determined. The method of logical effort simplifies the energy calculations as it roughly equalizes the input and output slopes of each gate. This implies that for a given output load and gate size, which defines the output slope, there is only one input slope that is possible. Although it is true that parasitics and unequal stage effort due to wiring will result in some variations in the slopes, it provides enough accuracy for the analysis being performed.

Optimal sizing for speed is determined for each adder by using a modified logical effort methodology. From the information about the size and topology of an adder, the energy consumption is determined. We report worst-case energy for the estimates, defined as the energy consumed when every internal node is switching.

3.2. Sizing

In general, producing various points on the Energy-Delay curve poses a sizing problem. Assuming that we start from some given size - e.g. minimal, Logical Effort should give us an answer to how the circuit blocks should be sized to achieve minimal delay. From the assigned sizes, we can calculate the energy that will be consumed by the adder for a given input activity. However, this is just one point on the Energy-Delay curve. From this point we can move in both directions; toward smaller and toward larger sizes. Such an Energy-Delay curve, produced for two 64-bit implementations of KS and HC adders is shown in Fig. 5.

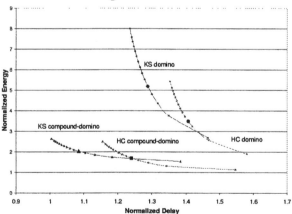

Fig. 5. Estimated Energy-Delay dependency of 64-bit Prefix-2 KS and HC Adders for Domino and Compound-Domino implementations

In order to stay on the same curve we vary the size of the inputs by assigning different values to the input capacitance C_{in}, or by assigning different values to H (electrical effort). This results in different sizes (all optimal in terms of speed) determined by LE methodology using different energies. The adders were implemented as regular domino and compound-domino. The single points obtained when using the same input size for each adder implementation are shown in Fig. 5. The difference between regular domino and compound-domino is in the Carry-Merge stage. Given that dynamic CMOS Domino logic is used for implementation of Carry-Merge blocks as in Fig. 6.

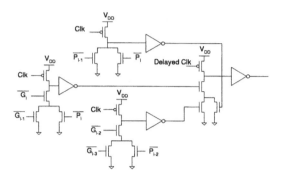

Fig. 6. Carry-Merge: Regular Domino Implementation

It has been realized that the inverter, which is necessary in the CMOS Domino logic block, can be replaced with a static AND-NOR gate, referred to as compound-domino. Thus, two domino blocks are merged into one with the advantage that an additional function is achieved by replacing the inverter. This is shown in Fig. 7.

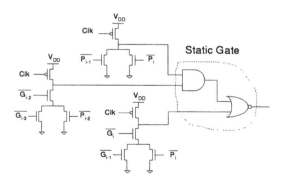

Fig. 7. Carry-Merge: Compound-Domino Implementation

When using compound-domino circuits for adders, it is important to note that the dynamic and the static outputs are potentially attached to long wires, while in domino only the inverter outputs are attached to long wires. Note that "footless domino", i.e. a circuit where the bottom transistor is eliminated, is used. A 64-bit critical path contains 6 carry-merge stages; none of which contain a stack of more than two n or p transistors. A critical path in the HC adder is shown in Fig. 8 [9]. It contains one more stage in the critical path, but it eliminates approximately one half of the blocks in the carry-merge tree, thus bringing some potential energy advantages over a KS implementation.

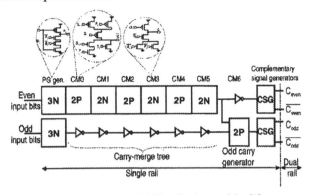

Fig. 8. Critical Path in the 64-bit Han-Carlson adder [9]

The energy-delay chart in Fig. 5 shows domino HC having an advantage over domino KS in terms of lower energy, but domino KS can stretch further in terms of lower delay at the expense of increased energy. When comparing compound-domino implementations, both HC and KS result in lower energy than the regular domino implementations. If one is concerned about energy, HC is better. However, as the speed becomes increasingly important the advantage moves in favor of KS. Both schemes suffer from high gate counts, large transistor sizes, and excessive wiring, resulting in large layout areas and high-energy consumption. In an attempt to arrive at a denser design, a Ladner-Fischer adder described in [1,11] trades off wiring complexity by exponentially growing the fan-outs of successive carry-merge gates to 1,2,4,8 and 16 respectively. However, this does not address the problem of high gate counts and large-transistor sizes. The idea of reducing energy by breaking the scheme into 4-bit blocks, which are conditionally added, was further developed by Intel [12,13]. It was realized that pruning down the main tree to obtain a sparse-tree that propagates P and G signals for 4-bit sections of the adder and combining this with a modest increase in hardware complexity due to the conditional sum technique, might be a good energy-delay trade-off. A 32-bit Quaternary-Tree (QT) adder core as described in [12], consists of a sparse-tree that generates 1 in 4 carries (Fig. 9) and a parallel side-path of 4-bit conditional sum blocks, resulting in an 8-stage design (same number of stages as KS).

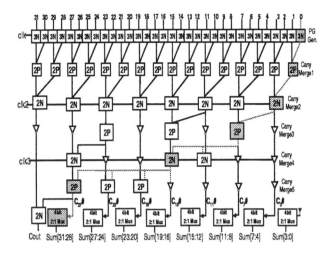

Fig. 9. 32-bit Quaternary-tree Adder Core [12]

Carries generated by the sparse-tree select the final sum using a 2:1 multiplexer. QT adder achieves energy reduction due to two factors: (a) reduced fan-outs and reduced wiring as compared to KS and HC structure (resulting in smaller transistor sizes) and (b) use of non-critical ripple-carry conditional sum blocks (with much smaller transistor sizes than main tree). In contrast to a Ladner-Fischer design, the QT adder tree has unit fan-outs on all generate gates, except the 3 highlighted gates in the Carry-Merge3, Carry-Merge4 and Carry-Merge5 stages. These 3 gates have fan-outs of 2,3 and 4 respectively (In both cases, we ignore the presence of the buffering inverters at the LSBs). Parallel-prefix logic removed from the main-tree is performed in the conditional sum blocks using an energy-efficient ripple-carry scheme with smaller transistor sizes. The structure of conditional sum blocks is shown in Fig. 10.

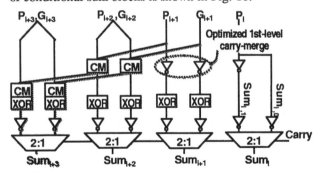

Fig. 10. Quaternary-tree Adder: Conditional Sum [12]

A 64-bit QT adder (Fig. 11) has 2 levels of conditional carry generation with the main tree generating 1 in 16 carries, which select betwwen the 1 in 4 conditional carries generated by the intermediate generator. The 1 in 4 carries then choose the appropriate 4-bit conditonal sum. Such a design has 10 stages (same as HC).

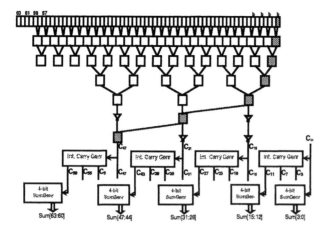

Fig. 11. 64-bit Quaternary-tree Adder Core

4. Estimation Method

The goal of the estimation method is to provide a simple, yet accurate, method for comparing designs in the energy-delay space. Logical effort has been shown to yield reasonably accurate results for the parallel prefix adders that were tested, thus finding it suitable for delay estimation and sizing. One of the issues with Logical Effort is branching, specifically with regards to long internal wires and different number of stages in the off path. These effects are difficult to account for and accurate accounting makes LE complex, thus simplifications are required to make the analysis linear. A more detailed account of these issues would result in greater delay accuracy when applying LE to adders.

4.1 Energy-Delay Curve Estimation

As previously described, the different points on the energy-delay curve are obtained by varying the size of the input gate for each adder. This provides a delay estimate and the sizing of the critical path, which in turn gives an energy estimate for the gates on the critical path. However, this estimation does not provide an energy estimate for the entire adder. The estimated energy for the critical paths is shown in Fig.12. From these estimates, it can be seen that the critical path of KS uses less energy for a given delay then HC. This is explained by the one extra stage that HC uses. However as mentioned previously, HC uses approximately half of the gates in the CM-section than KS. This must be accounted for in the total energy estimate for the adders. The critical path energy-delay estimate provides an idea of the minimal delay that can be achieved, which is shown by the vertical asymptote of the curves in Fig.12. Since we are interested in comparing adders not only by delay, but also in the energy-delay space, we need to determine a method for estimating the energy of an adder. By determining the number of gates per stage and assuming the same energy for those gates as the corresponding energy of the critical path gate in the same stage, we were able to obtain reasonably accurate results (within 20%). This provided a simple method to apply to most parallel prefix designs, however for a design like QT where the number of gates on each path is different, more care must be taken in determining what energy the off-path circuits consume.

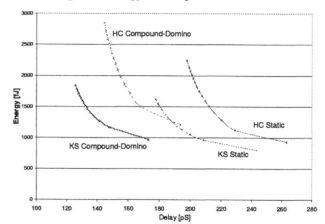

Fig. 12. Energy Delay comparison of 64-bit KS and HC Compound-Domino and Static adders

5. Results

The accuracy of our energy-delay estimation was tested on 32-bit KS and QT compound-domino adder implementations, with comparison to simulation results in 0.13μm technology [12]. Each of the energy-delay simulation points was obtained using a circuit optimizer. The energy reported was obtained by running the worst-case energy vector for each topology. Since the simulated data points were in 0.13μm technology, we needed to extrapolate the results and normalize them to the same technology as our estimates (0.10μm). We used a rule of thumb of 30% performance improvement per generation as well as a 50% energy improvement. The energy-delay estimates were obtained by varying the size of the input gate to the adder, as previously described. The estimated energy assumes an estimated worst-case switching activity associated with the topology of the adder and the logic design style. A more accurate method would require detailed modeling of the switching activity within each adder core, which is not possible prior to implementation. The comparison of the estimated energy-delay versus simulated is shown in Fig. 13. At iso-delay the simulations show 55% difference in energy at the knee of the curves, while estimation shows 35% difference. At iso-energy, the simulations show 21% delay improvement at the knee of the curves, while estimation shows 13%.

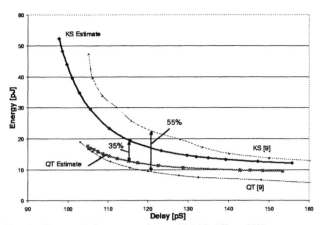

Fig. 13. Energy-Delay comparison of 32-bit QT and KS adders: estimated vs. simulation in 0.10μm technology

As the intent of this method is to compare the energy-delay characteristics of a given design, this comparison shows the estimation accurately represents tradeoffs in the energy-delay space for different architectures. To further explore the energy-delay design space we analyzed 64-bit KS, HC and QT compound-domino and static topologies, shown in Fig. 14.

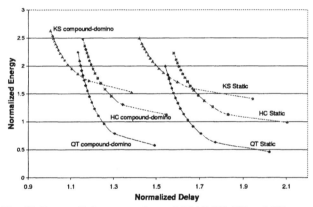

Fig. 14. Energy-Delay comparison of 64-bit KS, HC and QT adders

The 64-bit QT implementation used has one extra stage than the KS, which accounts for the higher performance that can be achieved by KS for both compound-domino and static topologies. The difference in energy for iso-delay is more significant in the 64-bit design space. This is due to the increased number of stages over which the delay is evenly distributed, resulting in increased gate sizes, and increased number of gates, e.g. a 32-bit KS has 417 gates, while a 64-bit KS has 1025 gates. Thus in the 64-bit design space, energy efficient designs display greater savings, which explains why QT uses substantially less energy than either KS or HC. The QT compound-domino design achieves an 86% reduction in energy vs. HC compound-domino at the knee of the curve.

6. Conclusion

We have shown that an LE based analysis of logical circuits is an effective tool for a quick exploration of the energy-delay space for comparing the performance of high-performance adders. Further, a tool was developed based on this technique to quickly estimate the energy-delay space of 32/64-bit Kogge-Stone, Han-Carlson, and Quaternary-tree adders implemented in 0.13μm and 0.10μm CMOS technologies using static and dynamic circuit styles, thereby, accurately representing tradeoffs in the energy-delay space with an accuracy of 8% in delay estimates and 20% in energy estimates, compared with simulated data.

References

[1] S. Knowles, "A Family of Adders", Proceedings of the 14th Symposium on Computer Arithmetic, Australia. April 1999.
[2] V. G. Oklobdzija and E. R. Barnes, "On Implementing Addition in VLSI Technology," IEEE Journal of Parallel and Distributed Computing, No. 5, 1988 pp. 716-728.
[3] R. F. Sproull, and I. E. Sutherland, "Logical Effort: Designing for Speed on the Back of an Envelop," IEEE Adv. Research in VLSI, C. Sequin (editor), MIT Press, 1991.
[4] D. Harris, R.F. Sproull, and I.E. Sutherland, "Logical Effort Designing Fast CMOS Circuits," Morgan Kaufmann Pub., 1999.
[5] H.Q. Dao, V. G. Oklobdzija, "Application of Logical Effort Techniques for Speed Optimization and Analysis of Representative Adders," 35th Annual Asilomar Conference on Signals, Systems and Computers, 2001.
[6] P.M. Kogge and H.S. Stone, "A parallel algorithm for the efficient solution of a general class of recurrence equations", IEEE Trans. on Comp. Vol. C-22, No.8, Aug. 1973, pp.786-793.
[7] A. Farooqui, V. G. Oklobdzija, F. Chehrazi, "Multiplexer Based Adder for Media Signal Processing", International Symp on VLSI Technology, Systems, and Applications, 1999
[8] T. Han, D. A. Carlson, and S. P. Levitan, "VLSI Design of High-Speed Low-Area Addition Circuitry," Proceedings of the IEEE International Conference on Computer Design: VLSI in Computers and Processors,1987, pp.418-422.
[9] S.K. Mathew et al, "Sub-500-ps 64-b ALUs in 0.18μm SOI/bulk CMOS: design and scaling trends," IEEE Journal of Solid-State Circuits, vol.36, Nov. 2001, pp.1636-46.
[10] H. Ling, "High Speed Binary Adder", IBM Journal of Research and Development, Vol. 25, No 3, 1981, p.156-166.
[11] R.E. Laddner and M.J. Fischer, "Parallel prefix computation", Journal of ACM, Vol.27, No.4, 1980, pp.831-38.
[12] S.K. Mathew et al, "A 4GHz 130nm Address Generation Unit with 32-bit Sparse-tree Adder Core," 2002 Symposium on VLSI Circuits Digest of Technical Papers, pp.126-127.
[13] J. Sklansky, "Conditional-sum addition logic." IRE Transactions on Electronic Computers, vol.9, 1960, pp. 226-231.

Tutorial:

Design of Power Efficient VLSI Arithmetic: Speed and Power Trade-offs

Given by:

Prof. Vojin G. Oklobdzija, *University of California*
Dr. Ram Krishnamurthy, *Intel Advanced Microprocessor Research Laboratories*

This tutorial will talk about issues related to performance of arithmetic algorithms when implemented in silicon. Most of the algorithms in use today are based on old and antiquated methods of counting the number of logic gates in the critical path. This produces inaccurate and misleading results. The importance of loading and wire delay is not taken into account by most. As the technology scales further into the sub-micron range, wire starts to dominate the delay. We will show how differed topologies of VLSI adders may influence fan-out and wiring density thus influencing design decisions and yielding to better area/power then known cases. This tutorial will further emphasize a disconnect that exists between algorithms that are used and the final result. The importance of accounting for Fan-In and Fan-Out on the critical path has been demonstrated in Logical Effort (LE) method which can be used for quick speed estimation. However, Logical Effort is not complete if the energy is treated separately. The power is starting to limit the speed of VLSI processors. But, even if we are able to manage the total power, the power density remains a problem.

In the course of VLSI processor design it is very important to choose the circuit topology that would yield desired performance for a given power budget. However, the performance and power of a chosen topology will be known only after the fact, i.e. after the design is finished. Therefore a question remains, could a higher performance have been achieved had a different topology been chosen? The answer is generally not known because there is no consistent nor realistic speed estimation method employed today.

We motivate the concept of comparing VLSI structures based on their energy-delay trade-offs and we discuss techniques used for estimating the energy-delay space of various high-performance VLSI topologies.

Author Index

Aharoni, M. 158
Arnold, M. 253
Asaf, S. 158
Bajard, J.-C. 181
Boldo, S. 79, 129
Boullis, N. 20
Bruguera, J. 204
Burgess, N. 262
Cotofana, C. 245
Cowlishaw, M. 104
Daneshbeh, A. 174
Danysh, A. 12
Dao, H. 272
Daumas, M. 79, 129
Dimitrov, V. 238
Ercegovac, M. 4, 204
Erdem, S. 28
Even, G. 165
Fahmy, H. 95
Ferguson, W. 165
Fit-Florea, A. 63
Flynn, M. 95
Frougny, C. 212
Garcia, J. 253
Gerwig, G. 87
Haess, J. 87
Harrison, J. 148
Hasan, A. 174, 188
Huang, Z. 4
Hughey, R. 54
Imbert, L. 181
Iordache, C. 122
Jullien, G. 238
Kaihara, M. 220
Koç, Ç. 28
Koren, I. 228
Koren, Y. 228
Kornerup, P. 38
Krishnamurthy, R. 272, 280
Kwon, S. 196
Lageweg, C. 245
Lefèvre, V. 142
Li, R.-C. 129
Liebelt, M. 12
Markstein, P. 137
Mathew, S. 272
Matula, D. 2, 63
McCann, M. 46
Muller, J.-M. 114
Nègre, C. 181
Oklobdzija, V. 272, 280
Oomman, B. 228
Piñeiro, J.-A. 204
Pippenger, N. 46
Plantard, T. 181
Reyhani-Masoleh, A. 188
Rice, E. 54
Schmookler, M. 70
Schulte, M. 253
Schwarz, E. 70, 87, 112
Seidel, P.-M. 165
Stehlé, D. 142
Surarerks, A. 212
Takagi, N. 220
Tan, D. 12
Tang, P. 122
Tisserand, A. 20
Trong, S. 70
Vassiliadis, S. 245
Wahid, K. 238
Wetter, H. 87
Zeydel, B. 272
Zimmermann, P. 142
Ziv, A. 158

Press Operating Committee

Chair
Mark J. Christensen
Independent Consultant

Editor-in-Chief
Mike Williams
Department of Computer Science, University of Calgary

Board Members

Roger U. Fujii, *Vice President, Logicon Technology Solutions*
Richard Thayer, *Professor Emeritus, California State University, Sacramento*
Sallie Sheppard, *Professor Emeritus, Texas A&M University*
Deborah Plummer, *Group Managing Editor, Press*

IEEE Computer Society Executive Staff
David Hennage, *Executive Director*
Angela Burgess, *Publisher*

IEEE Computer Society Publications

The world-renowned IEEE Computer Society publishes, promotes, and distributes a wide variety of authoritative computer science and engineering texts. These books are available from most retail outlets. Visit the CS Store at *http://computer.org* for a list of products.

IEEE Computer Society Proceedings

The IEEE Computer Society also produces and actively promotes the proceedings of more than 160 acclaimed international conferences each year in multimedia formats that include hard and softcover books, CD-ROMs, videos, and on-line publications.

For information on the IEEE Computer Society proceedings, please e-mail to csbooks@computer.org or write to Proceedings, IEEE Computer Society, P.O. Box 3014, 10662 Los Vaqueros Circle, Los Alamitos, CA 90720-1314. Telephone +1-714-821-8380. Fax +1-714-761-1784.

Additional information regarding the Computer Society, conferences and proceedings, CD-ROMs, videos, and books can also be accessed from our web site at *http://computer.org/cspress*

Revised October 29, 2001